Exercise Manual in Probability Theory

Mathematics and Its Applications (*East European Series*)

J. Stoyanov, I. Mirazchiiski, Z. Ignatov, and M. Tanushev
Institute of Mathematics, Sofia, Bulgaria

Exercise Manual in Probability Theory

Edited by K. Kocherlakota

Kluwer Academic Publishers
Dordrecht / Boston / London

Library of Congress Cataloging in Publication Data

Exercise manual in probability theory.

(Mathematics and its applications. East European series)
Bibliography: p.
Includes index.
1. Probabilities--Problems, exercises, etc.
I. Stoĭanov, Ĭordan. II. Kocherlakota, K.
III. Series.
QA273.25.E94 1988 519.2'076 87-38125
ISBN 90-277-2687-6

Published by Kluwer Academic Publishers,
P.O. Box 17, 3300 AA Dordrecht, The Netherlands.

Kluwer Academic Publishers incorporates
the publishing programmes of
D. Reidel, Martinus Nijhoff, Dr W. Junk and MTP Press.

Sold and distributed in the U.S.A. and Canada
by Kluwer Academic Publishers,
101 Philip Drive, Norwell, MA 02061, U.S.A.

In all other countries, sold and distributed
by Kluwer Academic Publishers Group,
P.O. Box 322, 3300 AH Dordrecht, The Netherlands.

This English edition is a revised and updated version of 'Rukovodstvo za uprazhnenia po teoria na veroyatnostite' (1985)

Translated from the Bulgarian by L. Boneva, O. Enchev, V. Kaishev, and L. Markov

Table of Contents

CONTENTS

SERIES EDITOR'S PREFACE

Approach your problems from the right end and begin with the answers. Then one day, perhaps you will find the final question.

'The Hermit Clad in Crane Feathers' in R. van Gulik's *The Chinese Maze Murders.*

It isn't that they can't see the solution. It is that they can't see the problem.

G.K. Chesterton. *The Scandal of Father Brown* 'The point of a Pin'.

Growing specialization and diversification have brought a host of monographs and textbooks on increasingly specialized topics. However, the "tree" of knowledge of mathematics and related fields does not grow only by putting forth new branches. It also happens, quite often in fact, that branches which were thought to be completely disparate are suddenly seen to be related.

Further, the kind and level of sophistication of mathematics applied in various sciences has changed drastically in recent years: measure theory is used (non-trivially) in regional and theoretical economics; algebraic geometry interacts with physics; the Minkowsky lemma, coding theory and the structure of water meet one another in packing and covering theory; quantum fields, crystal defects and mathematical programming profit from homotopy theory; Lie algebras are relevant to filtering; and prediction and electrical engineering can use Stein spaces. And in addition to this there are such new emerging subdisciplines as "experimental mathematics", "CFD", "completely integrable systems", "chaos, synergetics and large-scale order", which are almost impossible to fit into the existing classification schemes. They draw upon widely different sections of mathematics. This programme, Mathematics and Its Applications, is devoted to new emerging (sub)disciplines and to such (new) interrelations as exempla gratia:

- a central concept which plays an important role in several different mathematical and/or scientific specialized areas;
- new applications of the results and ideas from one area of scientific endeavour into another;
- influences which the results, problems and concepts of one field of enquiry have and have had on the development of another.

The Mathematics and Its Applications programme tries to make available a careful selection of books which fit the philosophy outlined above. With such books, which are stimulating rather than definitive, intriguing rather than encyclopaedic, we hope to contribute something towards better communication among the practitioners in diversified fields.

Because of the wealth of scholarly research being undertaken in the Soviet Union, Eastern Europe, and Japan, it was decided to devote special attention to the work emanating from these particular regions. Thus it was decided to start three regional series under the umbrella of the main MIA programme.

Possibly the best way to learn any topic at all is to work in it, to do it. According to some this is the only way. With that I do not agree, especially in the earlier stages, the first years of study; in that period the method is to slow and laborious.

However for those who have had basic mathematical training and especially for those who are already professionals in another branch of mathematics there are few better tools to acquire skills and knowledge than a book full of problems which also contains just enough basic definitions and theory to make it usuable as a standard-alone.

This is precisely such a book for probability and it will surely help many specialists from other fields to acquire active knowledge of this large and immensely useful subject.

The unreasonable effectiveness of mathematics in science ...

Eugene Wigner

Well, if you know of a better 'ole, go to it.

Bruce Bairnsfather

What is now proved was once only imagined.

William Blake

As long as algebra and geometry proceeded along separate paths, their advance was slow and their applications limited.

But when these sciences joined company they drew from each other fresh vitality and thenceforward marched on at a rapid pace towards perfection.

Joseph Louis Lagrange.

Antibes, June 1988 Michiel Hazewinkel

PREFACE

This book is a collection of 777 exercises covering basic topics of Probability theory. The book is intended mainly to those studying and teaching mathematics at university level, but can be used also by every one who teaches or studies elements of Probability theory. The material is organized and presented in a way which makes the book suitable for self-education as well. Many of the widely known examples and exercises which form the so-called 'probability classics' are included, as well as several new and original ones. Some of the exercises are elementary but other ones require serious probabilistic reasoning. The exercises of higher degree of difficulty are asterisked.

The material is distributed into 24 sections which are grouped in the following four chapters: 1. Elementary probabilities. 2. Probability spaces. Random variables. 3. Characteristics of random variables. 4. Limit theorems.

Each section comprises introductory notes (basic definitions and statements), illustrative examples and formulation of the exercises followed by complete solutions, hints or answers. In order to facilitate the solution of some of the problems we have included two tables (Normal and Poisson distributions). The list of references consists of titles which have been used and could be recommended for further work.

The present book is a result of our nearly twenty years long experience in teaching Probability theory at the University of Sofia. The first Bulgarian edition appeared in 1976 and in 1982 it was translated into Polish. The present English translation is based on the second improved Bulgarian edition (1985), but we have enlarged the introductory notes to the sections and added new illustrative examples.

The book as a whole is a result of the joint work of the authors; however, each of them was responsible for the following sections: I. Mirazchiiski: 1, 2, 3, 4, 5, 8; M. Tanushev: 6, 7, 11, 12, 13, 14; Z. Ignatov: 15, 16, 17, 18, 20; J. Stoyanov: 9, 10, 19, 21, 22, 23, 24. The coordination when preparing both Bulgarian editions as well as the English translation of the book was carried out by J. Stoyanov.

We are very pleased that the present English edition is planned to appear in 1988 when the University of Sofia is going to celebrate its 100 anniversary.

Our special thanks are addressed to Prof. K. Kocherlakota, the editor of the translation. His very useful comments and suggestions helped us to avoid several omissions of different kind thus improving both the content and the style of the presentation.

Finally, it is our pleasant duty to thank Kluwer Academic Publishers for their collaboration.

Sofia, March 1987 The authors

BASIC NOTATIONS AND ABBREVIATIONS

$(\Omega, \mathfrak{A}, P)$	probability space
A, B	random events (events)
$\bar{A}$	event complementary to A
ω	simple event (elementary outcome)
Ω	set of elementary outcomes (sample space)
$\mathfrak{A}, \mathcal{F}$	algebras or σ-algebras of events
$\mathfrak{C}$	class of events
$\sigma(\mathfrak{C})$	σ-algebra generated by $\mathfrak{C}$
$P(A)$	probability of the event A
$P(A\|B)$	conditional probability of A given B
$\mathbb{R}_n$	n-dimensional Euclidean space
$\mathfrak{B}_n$	Borel σ-algebra in $\mathbb{R}_n$
ξ, η, ζ	random variables
$E\xi$	expectation (mean value) of ξ
$V\xi$	variance of ξ
$E\{\xi\|\mathcal{F}\}$	conditional expectation of ξ given $\mathcal{F}$
F	distribution function: $F(x) = P\{\omega : \xi(\omega) < x\}$, $x \in \mathbb{R}_1$
f	density: $F(x) = \int_{-\infty}^{x} f(u)\, du$, $x \in \mathbb{R}_1$
$F_1 * F_2$	convolution of F_1 and F_2: $(F_1*F_2)(x) = \int_{-\infty}^{\infty} F_1(x - u)\, dF_2(u)$
$f_1 * f_2$	convolution of f_1 and f_2: $(f_1*f_2)(x) = \int_{-\infty}^{\infty} f_1(x - u) f_2(u)\, du$
ϕ	characteristic function: $\phi(t) = E\{\exp(it\xi)\}$, $t \in \mathbb{R}_1$
Φ	standard normal distribution function: $\Phi(x) = \frac{1}{\sqrt{2\pi}} \int_{-\infty}^{x} e^{-u^2/2}\, du$
ϕ_0	standard normal density: $\phi_0(x) = \frac{1}{\sqrt{2\pi}} e^{-x^2/2}$
μ_n	number of successes in Bernoulli scheme (n, p)
$P_n(k)$	binomial probability: $P_n(k) = P\{\mu_n = k\} = \binom{n}{k} p^k (1 - p)^{n-k}$
$\mathcal{B}(n, p)$	set of binomially distributed random variables with parameters n and p
$\mathcal{P}(\lambda)$	set of Poisson distributed random variables with parameter λ
$\mathcal{N}(a, \sigma^2)$	set of normally distributed random variables with parameters a and σ^2

$E(\lambda)$	set of exponentially distributed random variables with parameter λ
L_r	space of random variables with integrable r-th power
r.v.	random variable
d.f.	distribution function
ch.f.	characteristic function
g.f.	generating function
a.s.	almost sure
(P-a.s.)	almost sure with respect to the probability P
$\xrightarrow{d}$	convergence in distribution
$\xrightarrow{P}$	convergence in probability
$\xrightarrow{a.s.}$	convergence almost surely
$\xrightarrow{L_r}$	convergence in L_r-sense
$F(x_0 + 0)$	limit of $F(x)$ as $x \downarrow x_0$
$F(x_0 - 0)$	limit of $F(x)$ as $x \uparrow x_0$
$\Gamma(\alpha)$	Euler's gamma function: $\Gamma(\alpha) = \int_0^\infty x^{\alpha-1} e^{-x}\, dx$
$B(\alpha, \beta)$	Euler's beta function: $B(\alpha, \beta) = \int_0^1 x^{\alpha-1} (1 - x)^{\beta-1}\, dx$
I_A, 1_A	indicator of the set A: $1_A(\omega) = 1$ if $\omega \in A$ and $1_A(\omega) = 0$ if $\omega \bar{\in} A$
[x]	the integer part of the real number x
Re z	the real part of the complex number z
Res $g(z_0)$	residue of the function $g(z)$ at $z = z_0$
□	end

Chapter 1

ELEMENTARY PROBABILITY

1. Combinatorics

Introductory Notes

In this section we consider only *finite sets*; i.e., sets with finite number of elements. We denote the set M, consisting of the elements a_1, a_2, ..., a_n, by the notation $M = \{a_1, a_2, ..., a_n\}$. When the context is clear, we shall denote the elements a_i of M only by their indices; i.e., $M = \{1, 2, ..., n\}$. The *number of elements* of the set M will be denoted by $\nu(M)$; hence, in the example above we have $\nu(M) = n$. If $\nu(M) = 0$, we say that M is an empty set and denote it by $\emptyset$. With each of two sets A and B we can associate two other sets, $A \cup B$ and $A \cap B$ called, respectively, the *union* (sum) and the *intersection* (product). The set $A \cup B$ consists of the elements belonging to at least one of the sets A and B. The set $A \cap B$ consists of the elements belonging both to A and B. For simplicity we shall write AB instead of $A \cap B$. If A and B do not have any common elements ($AB = \emptyset$), we call them *mutually exclusive* or *disjoint*. Only in such a case we shall denote their union by $A + B$ instead of $A \cup B$. The symbol $A + B$ should remind us that the sets A and B are disjoint.

Let $M = \{1, 2, ..., n\}$. The subset $\{i_1, i_2, ..., i_k\}$, consisting of any k elements of M, we shall call a k-*tuple* or a *sample of size* k. We can form the following four different sets of k-tuples depending on whether ordering of the elements of the k-tuples is of importance and also whether repetition of one or more elements is allowed:

$\mathbf{K}_k^n$ = {unordered k-tuples without repetition}, $k = 0, 1, ..., n$;

$\tilde{\mathbf{K}}_k^n$ = {unordered k-tuples with repetition}, k is arbitrary;

$\mathbf{B}_k^n$ = {ordered k-tuples without repetition}, $k = 0, 1, ..., n$;

$\tilde{\mathbf{B}}_k^n$ = {ordered k-tuples with repetition}, k is arbitrary.

The elements of the sets $\mathbf{K}_k^n$ and $\tilde{\mathbf{K}}_k^n$ are called *combinations of n different elements taken k at a time*, without and with repetition, respectively. The elements of the sets $\mathbf{B}_k^n$ and $\tilde{\mathbf{B}}_k^n$ are called *permutations of n elements taken k at a time* (or k-permutations of n distinct elements), without and with repetition, respectively. In the particular case $k = n$ we denote the set $\mathbf{B}_k^n$ by Π_n and call it a set of *permutations of n elements*. The following notations will be used for the number of elements of these four sets

$$\nu(\mathbf{K}_k^n) = C(n, k), \quad \nu(\tilde{\mathbf{K}}_k^n) = \tilde{C}(n, k), \quad \nu(\mathbf{B}_k^n) = P(n, k),$$

$$\nu(\tilde{\mathbf{B}}_k^n) = \tilde{P}(n, k), \quad \nu(\Pi_n) = P_n.$$

It is well-known that:

$$C(n, k) = \binom{n}{k} = \frac{n(n-1)\ \dots\ (n-k+1)}{1 \cdot 2 \cdot \dots \cdot k} = \frac{n!}{k!(n-k)!}; \tag{1.1}$$

$$\tilde{C}(n, k) = \binom{n+k-1}{k}; \tag{1.2}$$

$$P(n, k) = n(n-1) \cdot \dots \cdot (n-k+1) = \frac{n!}{(n-k)!}; \tag{1.3}$$

$$\tilde{P}(n, k) = n^k; \tag{1.4}$$

$$P_n = n!. \tag{1.5}$$

Illustrative Examples

Note. In some examples and exercises below we shall often prove identities and propositions using a *combinatorial proof*. Such a proof consists of establishing a one-to-one correspondence between elements in appropriately chosen sets; hence, it follows that these sets have an equal number of elements.

Example 1.1. Let $M = \{a_1, a_2, \dots, a_n\}$. We wish to calculate the number of combinations, with repetition, formed from the elements of M taken k at a time.

Solution. Adding $k - 1$ new elements to the set M we get the set

$$M^* = \{a_1, \dots, a_n, a_{n+1}, \dots, a_{n+k-1}\}.$$

Assign the following k-tuple: $\{a_{i_1+0}, a_{i_2+1}, \dots, a_{i_k+k-1}\}$ of elements of M^* to an arbitrary k-tuple $\{a_{i_1}, a_{i_2}, \dots, a_{i_k}\}$ with repetition, of the elements of M, where $i_1 \leqslant i_2 \leqslant \dots \leqslant i_k$. It is easily seen that we have defined a one-to-one correspondence between the set of all k-tuples with repetition, constructed from the elements of M, and the set of all k-tuples without repetition, constructed from the elements of M^*. Thus we conclude that $\nu(\tilde{\mathbf{K}}_k^n) = \nu(\mathbf{K}_k^{n+k-1})$; hence

$$\tilde{C}(n, k) = C(n + k - 1, k) = \binom{n+k-1}{k}.$$

Example 1.2. Let the sets $A_i = \{a_1^i, a_2^i, \ldots, a_{n_i}^i\}$, $i = 1, \ldots, k$, be given. The set of ordered k-tuples of the type $\{a_{i_1}^1, a_{i_2}^2, \ldots, a_{i_k}^k\}$, where $a_{i_s}^s \in A_s$, $s = 1, \ldots, k$, is called a *Cartesian product* of the sets $A_1, A_2, \ldots, A_k$ and is denoted by $A_1 \times A_2 \times \ldots \times A_k$. (a) Calculate $\nu(A_1 \times A_2 \times \ldots \times A_k)$. (b) If $A_1 = A_2 = \ldots = A_k = A$, then the set $A \times A \times \ldots \times A = A^{\times k}$ is said to be the kth Cartesian degree of the set A. Calculate $(A^{\times k})$.

Solution. (a) Let us begin with the particular case $k = 2$. In this case all elements of the product $A_1 \times A_2$ can be disposed in the following table:

$A_1 \backslash A_2$	a_1^2	a_2^2	a_3^2	...	$a_{n_2}^2$
a_1^1	(a_1^1, a_1^2)	(a_1^1, a_2^2)	(a_1^1, a_3^2)	...	$(a_1^1, a_{n_2}^2)$
a_2^1	(a_2^1, a_1^2)	(a_2^1, a_2^2)	(a_2^1, a_3^2)	...	$(a_2^1, a_{n_2}^2)$
⋮	⋮	⋮	⋮	⋮	⋮
$a_{n_1}^1$	$(a_{n_1}^1, a_1^2)$	$(a_{n_1}^1, a_2^2)$	$(a_{n_1}^1, a_3^2)$	...	$(a_{n_1}^1, a_{n_2}^2)$

Hence, $\nu(A_1 \times A_2) = \nu(A_1)\nu(A_2) = n_1 n_2$. Consider further the case $k = 3$. It is easy to see, that the set $(A_1 \times A_2) \times A_3$, which is a set of ordered pairs with first elements also ordered pairs, and the set $A_1 \times A_2 \times A_3$, which is a set of ordered triplets, contain the same number of elements. Therefore, $\nu(A_1 \times A_2 \times A_3) = n_1 n_2 n_3$. Proceeding in the same way we get finally that $\nu(A_1 \times A_2 \times \ldots \times A_n) = n_1 n_2 \ldots n_k$.
(b) Suppose $\nu(A) = n$. The above reasoning shows that we have $\nu(A^{\times k}) = n^k$. Note further that the set $A^{\times k}$ coincides with the set of permutations of n elements taken k at a time, with repetition allowed. Thus formula (1.4.) is proved.

Example 1.3. Let the set A^* containing n elements be

$$\{a_1, a_2, \ldots, a_{n_1}, b_1, b_2, \ldots, b_{n_2}, \ldots, c_1, c_2, \ldots, c_{n_r}\},$$

where $n_1 + n_2 + \ldots + n_r = n$. Form all permutations of the elements of A^* when the elements $a_1, \ldots, a_{n_1}$ are indistinguishable as are the elements b, ..., and the elements c. The obtained arrangements we call *n-permutations with repetitions* formed from the elements of the set A = {a, b, ..., c}, in which the element a is repeated n_1 times, b is repeated n_2 times, ..., c is repeated n_r times. Denote the set of these permutations by $\tilde{\Pi}_n$. Prove that

$$\nu(\tilde{\Pi}_n) = \frac{n!}{n_1!n_2! \dots n_r!}, \quad n_1 + n_2 + \dots + n_r = n. \tag{1.6}$$

Solution. *Method 1*: The element a can be placed in n_1 positions of the total number of n positions in $\binom{n}{n_1}$ ways. The element b can be placed in n_2 of the remaining $n - n_1$ positions in $\binom{n-n_1}{n_2}$ ways. The procedure is continued until c can be placed in the remaining $n - n_1 - n_2 - \dots - n_{r-1}$ positions in $\binom{n - n_1 - n_2 - \dots - n_{r-1}}{n_r} = 1$ way. Therefore,

$$\nu(\tilde{\Pi}_n) = \binom{n}{n_1}\binom{n - n_1}{n_2} \dots \binom{n - n_1 - \dots - n_{r-1}}{n_r} =$$

$$= \frac{n!}{n_1!n_2! \dots n_r!}.$$

Method 2: If the elements of A^* were distinguishable, then the total number of all permutations of the elements of A^* would be equal to n!. If the elements $a_1, \dots, a_{n_1}$ are indistinguishable, then $n_1!$ of all permutations coincide; i.e., there remain $\frac{n!}{n_1!}$ different permutations. If $b_1, \dots, b_{n_2}$ are also indistinguishable, then there remain $\frac{n!}{n_1!n_2!}$ different permutations, ..., if the elements $c_1, \dots, c_{n_r}$ are indistinguishable, then there remain $\frac{n!}{n_1!n_2! \dots n_r!}$ different permutations.

Exercises

1.1. Two finite sets of A and B are given, and AB = ∅. Prove that

$$\nu(A + B) = \nu(A) + \nu(B).$$

1.2. Two finite sets A and B are given, and AB ≠ ∅. Prove that

$$\nu(A \cup B) = \nu(A) + \nu(B) - \nu(AB).$$

1.3. Given the finite sets $A_1, A_2, \dots, A_n$, prove that

$$\nu\left(\bigcup_{i=1}^{n} A_i\right) = \sum_{i=1}^{n} \nu(A_i) - \sum_{i<j} \nu(A_iA_j) + \sum_{i<j<k} \nu(A_iA_jA_k) - \dots + (-1)^{n-1}\nu(A_1A_2 \dots A_n),$$

where as usual $\sum_{i<j} \nu(A_iA_j)$ denotes the double sum $\sum_{i=1}^{n} \sum_{j=i+1}^{n} \nu(A_1A_j)$ and $\sum_{i<j<k} \nu(A_1A_jA_k)$, ..., have an analogous meaning.

1.4. Let $A \cup B \cup C = \Omega$ and $\nu(\Omega) = 1000$. Show that the following data are inconsistent: $\nu(A) = 510$, $\nu(B) = 490$, $\nu(C) = 427$, $\nu(AB) = 189$, $\nu(AC) = 140$ and $\nu(BC) = 85$.

1.5. A collection consists of 100 cubes. The six faces of each cube have been painted either red, or blue, or green. It is known that among the given cubes 80 have at least one red face, 85 have at least one blue face and 75 have at least one green face. What is the smallest number of cubes having faces of all three colours?

1.6. Let M_n be a set with $\nu(M_n) = n$. Denote by $\mathbf{M}_n$ the set of all subsets formed from the elements of M_n. Prove that $\nu(\mathbf{M}_n) = 2^n$: (a) analytically; (b) by induction; (c) using the notion of a Cartesian product.

1.7. Consider the set $M_{m+n} = \{a_1, a_2, \dots, a_m, b_1, b_2, \dots, b_n\}$. How many subsets of M_{m+n} contain at least one element "a" and at least one element "b"?

1.8. An urn contains n balls numbered from 1 up to n. Draw k times in succession a ball and each time note its number. The results obtained (i.e., the k recorded numbers) represent a sample of size k from the n elements. If after each drawing and recording of its number, we return the ball into the urn (before the next selection) the sample is with replacement. If, however, we remove the chosen ball after noting its number, the sample is without replacement. If the order of recording the k drawn numbers is essential, we say that the sample is ordered; otherwise it is called unordered.

Calculate the number of all samples if: (a) the samples are unordered and the selection is without replacement; (b) the samples are unordered and the selection is with replacement; (c) the samples are ordered and the selection is without replacement; (d) the samples are ordered and the selection is with replacement.

1.9. (*Fermi-Dirac statistics*) Let k indistinguishable particles be distributed in n different cells. Find the number of all possible ways of distributing the particles if every cell can contain at most one particle.

1.10. (*Bose-Einstein statistics*) Let k indistinguishable particles be distributed in n different cells. Find the number of all possible ways of distributing the particles if there are no restrictions about the number of particles which can be placed into one cell.

1.11. (*Maxwell-Boltzmann statistics*) Let k distinguishable particles be distributed in n different cells. Find the number of all possible ways of placing them if there are no restrictions on the number of particles which can be placed in one cell.

1.12. Let k distinguishable particles be distributed in n different cells. Find the number of all possible ways of placing them if one cell can contain at most one particle.

1.13. Let Ω be a set of ordered n-tuples in which repetitions of the digits 1, 2 and 3 are permitted. Now consider the following restrictions and find $\nu(\Omega)$ if the elements of Ω: (a) begin with 1; (b) begin and end with 1 and contain the digit 1 exactly k + 2 times ($k + 2 \leqslant n$); (c) contain the digit 2 exactly k times ($k \leqslant n$); (d) contain the digit 1, k_1 times, the digit 2, k_2 times and the digit 3, k_3 times, where $k_1 + k_2 + k_3 = n$.

1.14. Give a combinatorial proof of the identities:

$$\binom{n}{k} = \binom{n}{n-k} \quad (a); \qquad \binom{n}{k} = \binom{n-1}{k-1} + \binom{n-1}{k} \quad (b).$$

*1.15†. Give a combinatorial proof of the identities:

$$P(n, k) = P(n-1, k) + kP(n-1, k-1); \tag{a}$$

$$\tilde{P}(n, k) = \binom{k}{0}\tilde{P}(n-1, k) + \binom{k}{1}\tilde{P}(n-1, k-1) + \binom{k}{2}\tilde{P}(n-1, k-2) + \ldots + \binom{k}{k}\tilde{P}(n-1, 0); \tag{b}$$

$$\tilde{C}(n, k) = \tilde{C}(n-1, k) + \tilde{C}(n-1, k-1) + \ldots + \tilde{C}(n-1, 0); \tag{c}$$

$$\tilde{P}(n, k) = n\tilde{P}(n, k-1); \tag{d}$$

$$\binom{n}{k} = \sum_{i=r}^{n-k+r} \binom{n-i}{k-r}\binom{i-1}{r-1}, \text{ where } r \text{ is a fixed integer, } 1 \leqslant r \leqslant k. \tag{e}$$

1.16. Consider a set of m + n different elements, say $a_1, a_2, \ldots, a_m$ (type a) and $b_1, b_2, \ldots, b_n$ (type b) elements. From this form an unordered sample without replacement of size k, where $k \leqslant \min[m, n]$. Let A_i be the set of the samples, which contain exactly i elements of type a, i = 0, 1, ..., k. Analogously, denote by B_i the set with respect to the elements of type b. (a) Calculate $\nu(A_i)$ and $\nu(B_i)$. (b) Prove that $\nu(A_i) = \nu(B_{k-i})$. (c) Give a combinatorial proof of $\sum_{i=0}^{k} \binom{m}{i}\binom{n}{k-i} = \binom{m+n}{k}$. (d) Denote by A the set of the samples, which contain at least one element of type a. Calculate $\nu(A)$. (e) Denote by D the set of the

† The exercises of higher degree of difficulty are asterisked.

samples which contain at least one element of type a and at least one element of type b. Calculate $\nu(D)$.

1.17. Calculate the number of possible ways of placing k balls into n different cells $a_1, a_2, \ldots, a_n$, where for i = 1, 2, ..., n it is given that cell a_i contains exactly k_i balls with $k_1 + k_2 + \ldots + k_n = k$, if: (a) the balls are distinguishable; (b) the balls are indistinguishable.

1.18. Let k distinguishable balls $b_1, b_2, \ldots, b_k$ be distributed in n different cells provided that each cell may contain no more than 1 ball with $k \leqslant n$. Calculate the number of ways of distributing all balls such that: (a) ball b_1 is located in cell a_1; (b) balls b_1 and b_2 are located in cells a_1 and a_2, respectively; (c) balls b_1 and b_2 are located in cells a_1 and a_2.

1.19. A telephone number may begin with any one of the digits 0, 1, 2, ..., 9. Calculate the number of the six-figure telephone numbers of which: (a) all figures are different; (b) all figures are odd.

1.20. Ten people are arranged in a row. How many arrangements there are in which three particular people stand side by side?

1.21. How many planes are equidistant from four given points not lying on one and the same plane?

1.22. How many different outcomes there are when two dice are thrown if: (a) the dice are distinguishable; (b) the dice are indistinguishable; (c) the outcomes are distinguished according to the sum of the dots on the two faces?

1.23. How many different ways can forms be filled when playing sport-lotto if one has to guess: (a) the possible outcomes in a competition between 13 pairs of football teams with three possibilities (win, lose and draw) for each team; (b) the possible outcomes in crossing out any six sports from a list of 49?

1.24. Suppose that an individual's initials consist of three letters. Prove that in Sofia there are at least two persons with the same initials.

1.25. Consider the function $f(x, y, z) = x^3 e^{y+z\cos x} + (xy \sin z)^{10}$. Find the number of all different partial derivatives of sixth order.

1.26. How many integer non-negative solutions are there for the equation

$$x_1 + x_2 + \ldots + x_m = n?$$

1.27. Let f be an infinitely differentiable function of m arguments. How many different partial derivatives of f of nth order are there?

1.28. In how many ways can nine people be distributed into three groups, each one containing three people, if the distributions differ only in the structure of the groups? The order of the groups and the order of the people in the groups are not of any importance.

1.29. In how many ways can a group of three boys and six girls be distributed into three groups of three people each, so as each triplet to have one boy? The order of the groups and the arrangement in each of them are of no importance.

1.30. There are m girls and n boys at a dancing party. In how many ways can k dancing pairs be composed, $k \leqslant \min[m, n]$?

1.31. Let n red and r green balls be distributed in m different cells, each cell containing at most one ball, $n + r \leqslant m$. What is the number of the different distributions if: (a) the balls are indistinguishable; (b) the balls are distinguishable?

1.32. There are n chess-players using k chess-boards with $n \geqslant 2k$. In how many different ways can k chess-pairs be formed for the first round of a chess-tournament?

*1.33. Let $N_a(n, m)$ be the number of different ways in which n indistinguishable balls may be placed in m different cells, each cell containing at most a balls, $m \leqslant n \leqslant am$. Prove the following recurrence relation:

$$N_a(n, m) = \sum_{j=1}^{m} \binom{m}{j} N_{a-1}(n - j, j).$$

*1.34. Evaluate the number of ways of placing k distinguishable balls into n different cells if cell a_1 contains an arbitrary number of balls and the remaining cells contain no more than one ball, when: (a) in cell a_1 there are exactly s balls; (b) in cell a_1 there are no more than s balls; (c) in cell a_1 there are at least s balls.

*1.35. The numbers $i_1, \ldots, i_k$, where $k \leqslant N$, are chosen from the set $\{1, 2, \ldots, N\}$. Denote by $\min[i_1, \ldots, i_k]$ and $\max[i_1, \ldots, i_k]$ the smallest and the largest of the k chosen numbers, respectively. Using combinatorial reasoning prove the following relations:

$$\sum^{(1)} (\min[i_1, \ldots, i_k] + \max[i_1, \ldots, i_k]) = N^k(N + 1), \qquad (a)$$

where the sum $\Sigma^{(1)}$ is taken over all integers $i_1, \ldots, i_k$ such that $1 \leqslant i_1 \leqslant N, \ldots, 1 \leqslant i_k \leqslant N$;

$$\sum^{(2)} (\min[i_1, \ldots, i_k] + \max[i_1, \ldots, i_k]) = \frac{(N + 1)!}{(N - k)!}, \qquad (b)$$

where the sum $\Sigma^{(2)}$ is taken over all integers $i_1, \ldots, i_k$ such that $1 \leqslant i_1 \leqslant N, \ldots, 1 \leqslant i_k \leqslant N$ and $i_m \neq i_s$ for $m \neq s$. (*D. Vladeva*)

*1.36. Use combinatorial reasoning to prove the following identity known as *Euler's formula*:

$$\sum_{k=0}^{n} (-1)^{n-k} \binom{n}{k} k^m = \begin{cases} 0, & \text{if } 0 \leqslant m < n \\ n!, & \text{if } m = n. \end{cases}$$

2. Events and Relations among Them

Introductory Notes

Consider a random experiment whose outcomes will be denoted by the simple (elementary) event or *sample point* ω. We shall refer to the set of these sample points as the *sample space* and denote it by Ω. Now let A be a subset of Ω (it may also be the empty set $\emptyset$ as well as space Ω itself). In probability theory these subsets A are called *events*. We shall denote them by A, B, C, ... The expression "an event A has occurred" means that the outcome of the experiment is one of the points ω of the set A. It is quite natural to call $\emptyset$ an *impossible event* and Ω a *certain event*. Let $\mathfrak{P}(\Omega)$ be the *set of all events* of the space Ω. For any two events A and B of $\mathfrak{P}(\Omega)$ the operations union and intersection are defined (see Section 1). The events $A \cup B$ and AB have the following interpretation in probabilistic language: $A \cup B$ implies the *occurrence of at least one of the events* A and B; AB implies the *joint* (simultaneous) *realization* of A and B.

If $AB = \emptyset$, the events A and B are called *mutually exclusive*. If $A \subset B$ (A implies B) and $B \subset A$ we say that the events A and B are *equivalent* and denote that relationship by $A = B$. If $A \in \mathfrak{P}(\Omega)$, then we call the event $\bar{A}$ a *complementary event* to A whenever it satisfies the conditions: $A\bar{A} = \emptyset$ and $A \cup \bar{A} = \Omega$. Some combinations of the operations union, intersection and complementation bear special names. For example, the event $A\bar{B}$ is called a *difference* of the events A and B, and is denoted by $A \setminus B$.

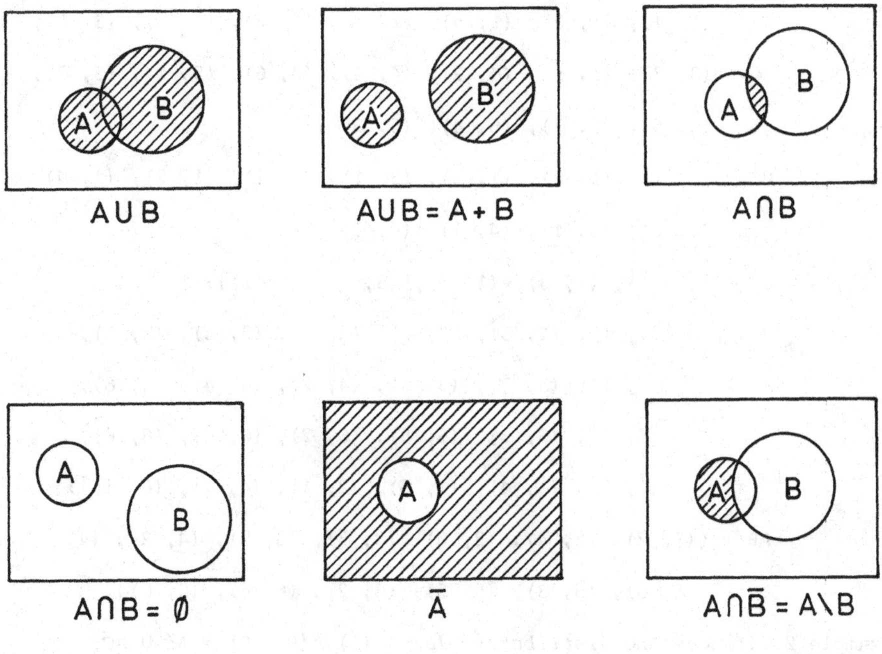

Figure 2.1.

The set $\mathcal{P}(\Omega)$ is closed with respect to the operations of union and intersection (taken for an arbitrary number of its elements) and with respect to the operation complementation.

Figure 2.1 shows the results of applying the operations union, intersection and complementation on the elements of $\mathcal{P}(\Omega)$. The space Ω is represented here, provisionally, as a rectangle, the events as circles, and the result of the corresponding operation is denoted by C and is shaded.

If the events A_i, $i = 1, 2, \ldots$, satisfy the conditions $\bigcup_i A_i = \Omega$ and $A_iA_j = \emptyset$ for $i \neq j$, $i, j = 1, 2, \ldots$, we say that $A_1, A_2, \ldots$ form a *partition* of Ω. If the events A and B are mutually exclusive, sometimes we shall denote their sum $A \cup B$ by the symbol $A + B$.

In all exercises of this section, whenever several events A, B, ... are mentioned, it should always be understood that these events belong to the same set $\mathcal{P}(\Omega)$.

Illustrative Examples

Example 2.1. Two distinguishable dice are thrown. Consider the events A = {the sum of dots on the two faces is an even number} and B = {ace appears at least on one of the dice}. Describe the events: AB, $A \cup B$, $A\bar{B}$.

Solution. The possible outcomes in this experiment are the ordered pairs (i, k), where $i = 1, 2, \ldots, 6$ is the number appearing on the first die and $k = 1, 2, \ldots, 6$ is the number appearing on the second die. The events are:

$$A = \{(1, 1), (1, 3), (1, 5), (2, 2), (2, 4), (2, 6), (3, 1), (3, 3), (3, 5), (4, 2), (4, 4), (4, 6), (5, 1), (5, 3), (5, 5), (6, 2), (6, 4), (6, 6)\};$$

$$B = \{(1, 1), (1, 3), (1, 5), (3, 1), (5, 1), (1, 2), (1, 4), (1, 6), (2, 1), (4, 1), (6, 1)\};$$

$$AB = \{(1, 1), (1, 3), (1, 5), (3, 1), (5, 1)\};$$

$$A \cup B = \{(1, 1), (1, 3), (1, 5), (2, 2), (2, 4), (2, 6), (3, 1), (3, 3), (3, 5), (4, 2), (4, 4), (4, 6), (5, 1), (5, 3), (5, 5), (6, 2), (6, 4), (6, 6), (1, 2), (1, 4), (1, 6), (2, 1), (4, 1), (6, 1)\};$$

$$A\bar{B} = \{(2, 2), (2, 4), (2, 6), (3, 3), (3, 5), (4, 2), (4, 4), (4, 6), (5, 3), (5, 5), (6, 2), (6, 4), (6, 6)\}.$$

Example 2.2. Prove the *distributive laws*: (i) $A(B \cup C) = AB \cup AC$; (ii) $A \cup BC = (A \cup B)(A \cup C)$.

Solution. We shall prove only the first relationship. The second one can be proved analogously.

Method 1: We shall show that $A(B \cup C) \subset AB \cup AC$, and conversely that $AB \cup AC \subset A(B \cup C)$; hence the statement follows.

(a) Let $x \in A(B \cup C)$. Then $x \in A$ and $x \in B \cup C$. The following cases are possible: (a_1) $x \in A$, $x \in B$, $x \overline{\in} C$; (a_2) $x \in A$, $x \overline{\in} B$, $x \in C$; (a_3) $x \in A$, $x \in B$, $x \in C$. In case (a_1), $x \in AB$; therefore $x \in AB \cup AC$. In case (a_2), $x \in AC$; therefore $x \in AB \cup AC$. Finally, in case (a_3), $x \in AB$ and $x \in AC$; hence it follows again that $x \in AB \cup AC$. Thus it is proved that $A(B \cup C) \subset AB \cup AC$.

(b) Let $y \in AB \cup AC$. The following cases are possible: (b_1) $y \in AB$, $y \overline{\in} AC$; (b_2) $y \overline{\in} AB$, $y \in AC$; (b_3) $y \in AB$, $y \in AC$. In case (b_1) we have $y \in A$ and $y \in B$. Then $y \in B \cup C$, which implies that $y \in A(B \cup C)$. In case (b_2) we have $y \in A$, $y \in C$ and hence $y \in B \cup C$; i.e., $y \in A(B \cup C)$ again. Finally, in case (b_3) $y \in A$, $y \in B$, $y \in C$; hence $y \in A(B \cup C)$. Thus it is proved that $AB \cup AC \subset A(B \cup C)$.

The conclusions in case (a) and (b) imply that $A(B \cup C) = AB \cup AC$.

Method 2. Both sides of the relations $A(B \cup C) = AB \cup AC$ represent the same event. Indeed, the event $A(B \cup C)$ occurs only when A occurs together with the occurrence of at least one of the events B and C; i.e., when AB or AC occurs. This, however, is equivalent to the occurrence of the event $AB \cup AC$.

Example 2.3. Let A, B and C be arbitrary events. Simplify the expression for the event D defined by $D = (A \cup B)(\overline{A} \cup \overline{B}) \cup (A \cup \overline{B})(\overline{A} \cup B)$.

Solution. For the simplification of analytical expressions which are, in fact events we apply the formulas proved in Example 2.2. and in Exercise 2.7. For the event D we obtain the following chain of equalities:

$$\begin{aligned} D &= [A(\overline{A} \cup \overline{B}) \cup B(\overline{A} \cup \overline{B})] \cup [A(\overline{A} \cup B) \cup \overline{B}(\overline{A} \cup B)] = \\ &= A\overline{A} \cup A\overline{B} \cup \overline{A}B \cup B\overline{B} \cup A\overline{A} \cup AB \cup \overline{A}\overline{B} \cup B\overline{B} = \\ &= \emptyset \cup A\overline{B} \cup \overline{A}B \cup \emptyset \cup \emptyset \cup AB \cup \overline{A}\overline{B} \cup \emptyset = \\ &= [A\overline{B} \cup \overline{A}\overline{B}] \cup [AB \cup \overline{A}B] = [\overline{B}(A \cup \overline{A})] \cup [B(A \cup \overline{A})] = \\ &= \overline{B}\Omega \cup B\Omega = \overline{B} \cup B = \Omega. \end{aligned}$$

In the first two equalities we apply one of the distributive laws, namely $X(X \cup Z) = XY \cup XZ$, the commutative law for multiplication: $XY = YX$, and the associative law for addition: $X \cup (Y \cup Z) = (X \cup Y) \cup Z = X \cup Y \cup Z$; in the third equality we use the fact that $X\overline{X} = \emptyset$; in the fourth that $X \cup \emptyset = X$ and the associative and commutative law for addition; in the fifth we use the same distributive law used at the beginning; in the sixth and in the seventh equalities we use respectively the relations $X \cup \overline{X} = \Omega$ and $X\Omega = X$; the last equality is a consequence of the relation $X \cup \overline{X} = \Omega$.

Exercises

2.1. A target consists of ten concentric circles with radii r_k, $k = 1, 2, \ldots, 10$, where $r_1 < r_2 < \ldots < r_{10}$. The event A_k indicates a hit in the circle with radius r_k, $k = 1, 2, \ldots, 10$. Describe in words the events $B = \bigcup_{k=1}^{6} A_k$ and $C = \bigcap_{k=5}^{10} A_k$.

2.2. Let A be the event "at least one of three checked items is defective" and B the event "all three items are good". Describe in words the events: (a) $A \cup B$; (b) AB; (c) $\overline{A}$; (d) $\overline{B}$.

2.3. There are three different editions each containing at least three volumes. The events A, B and C respectively indicate that at least one book is taken from the first, the second and the third edition. Let A_s = {s volumes are taken from the first edition} and B_k = {k volumes are taken from the second edition}. What is the meaning of the events: (a) $A \cup B \cup C$; (b) ABC; (c) $A \cup B_3$; (d) A_2B_2; (e) $A_1B_3 \cup A_3B_1$?

2.4. One number is chosen from the set of natural numbers. Let A = {the chosen number is divisible by 5} and B = {the chosen number ends with 0}. What is the meaning of the events $A \setminus B$ and $A\overline{B}$?

2.5. For a game of bridge let N_k, $k = 1, 2, 3, 4$, be the event that the first player has received at least k aces. Let E_k, S_k and W_k be the analogous events for the second, third and fourth players, respectively. How many aces does the fourth player have if the following events have occurred: (a) $\overline{W}_1$; (b) N_2S_2; (c) $\overline{N}_1\overline{S}_1\overline{E}_1$; (d) $W_2 \setminus W_3$; (e) $N_1S_1E_1W_1$; (f) N_3W_1; (g) $(N_2 \cup S_2)E_2$.

Note. *Bridge* is a game played with a deck of 52 cards in which each of four players receive 13 cards. Each card of the deck has two characteristics: suit and value. There are four suits: spades, hearts, diamonds and clubs. These suits can be further classified by colour: spades and clubs are black while diamonds and hearts are red. Each suit consists of 13 cards: 2, 3, ..., 10, 11 (knave), 12 (queen), 13 (king) and 1 (ace). Usually a complete deck of 52 cards is used; however, it is possible to construct incomplete decks. For example, a deck of 32 cards usually consists of 7, 8, 9, 10, knave, queen, king and ace.

2.6. Let A, B and C be three arbitrary events. Find the analytical expressions for the following events. From A, B anc C: (a) only A occurs; (b) both A and B, but not C, occur; (c) all three events occur; (d) at least one of the events occurs; (e) at least two of the events occur; (f) one and only one event occurs; (g) two and only two events occur; (h) none of the events occurs; (i) no more than two events occur.

2.7. The following properties illustrate the relationship between events and sets. Prove that: (a) $A\Omega = A$, $A \cup \emptyset = A$; (b) $A \cup \Omega = \Omega$, $A\emptyset = \emptyset$; (c) $\overline{\overline{A}} = A$; (d) $\overline{\Omega} = \emptyset$, $\overline{\emptyset} = \Omega$; (e) $A \cup A = A$, $AA = A$ (*idempotency*); (f) $A \cup (B \cup C) = (A \cup B) \cup C$, $A(BC) = (AB)C$ (*associative laws*); (g) $\overline{A \cup B} = \overline{A} \cap \overline{B}$, $\overline{A \cap B} = \overline{A} \cup \overline{B}$ (*De Morgan's laws*); (h) $\overline{\bigcup_{k\in I} A_k} = \bigcap_{k\in I} \overline{A}_k$, $\overline{\bigcap_{k\in I} A_k} = \bigcup_{k\in I} \overline{A}_k$ (*De Morgan's laws*) (Here I denotes a set of finite or infinite, but countable, number of indices); (i) $A \cup AB = A$, $A(A \cup B) = A$ (*absorbtion law*).

Note. In the distributive laws (Example 2.2) and in the associative laws (Exercise 2.7 (f)) the following *duality principle* (holding for events) appears: If everywhere in an equality of events we exchange the symbols $\cup$, $\cap$, Ω and $\emptyset$ with $\cap$, $\cup$, $\emptyset$ and Ω, respectively, the obtained equality (dual to the first one) is also valid.

2.8. Let A, B and C be arbitrary events. Prove that: (a) if $A \subset B$ and $B \subset C$, then $A \subset C$; (b) $AB \subset A$; (c) $A \subset A \cup B$.

2.9. Let A, B and C be events. Discuss the meaning of the equalities: (a) $ABC = A$; (b) $A \cup B \cup C = A$.

2.10. Let A, B and C be events. Simplify the expressions:
(a) $\overline{(A \cup B)}(B \cup C)$; (b) $(A \cup B)(A \cup \bar{B})$; (c) $(A \cup B)(A \cup \bar{B})(\bar{A} \cup B)$;
(d) $(A \cup B)(A \cup B) \cup (\bar{A} \cup B)(\bar{A} \cup \bar{B})$; (e) $(A \cup B)(A \cup \bar{B})(\bar{A} \cup B)(\bar{A} \cup \bar{B})$.

2.11. Let $A \subset B$. Find simple expressions for: (a) AB; (b) $A \cup B$; (c) ABC; (d) $A \cup B \cup C$.

2.12. Show that the following events form a partition of the sample space Ω: A, $\bar{A}B$ and $\overline{A \cup B}$.

2.13. Find a condition under which the events $A \cup B$, $\bar{A} \cup B$ and $A \cup \bar{B}$ are mutually exclusive.

2.14. Let A, B and C be arbitrary events belonging to $\mathcal{P}(\Omega)$. Does it follow that A and B are equivalent if: (a) $\bar{A} = \bar{B}$; (b) $A \cup C = B \cup C$; (c) $AC = BC$?

2.15. Prove that the events A and B are equivalent if: (a) $A \cup C = B \cup C$ and $AC = BC$; (b) $A \cup C = B \cup C$ and $A \cup \bar{C} = B \cup \bar{C}$.

2.16. Determine the event X in the equality $\overline{(X \cup A)} \cup \overline{(X \cup \bar{A})} = B$.

2.17. Let A, B, C and D be events. Assume that: $X = AB \cup \bar{A}D$, $Y = \bar{A}B \cup AD$ and $Z = \bar{A}BC \cup ABD$. Show that, no matter how the event D is chosen, the equations: (a) $AX = AB$; (b) $A \cup Y = A \cup B$; (c) $AB \cup Z = (A \cup C)(B \cup C)$, are satisfied.

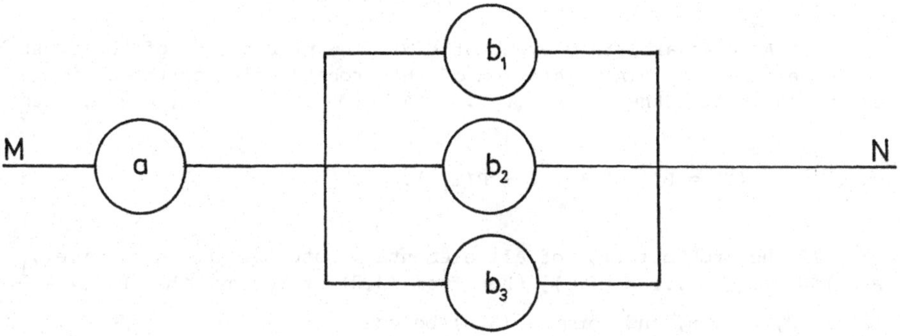

Figure 2.2.

2.18. Figure 2.2 represents an electric circuit between the points M and N. Consider the events A = {a failure of the element a} and B_k = {a failure of the element b_k} for k = 1, 2, 3. If the event C implies a cut-off of the chain, then express C and $\bar{C}$ in terms of A and B_k.

2.19. A steamer has one steering-gear, four steam-boilers and two turbines. Let the event A denote that the steering-gear is in a good condition, B_k that the kth steam-boiler is in a good condition, k = 1, 2,

3, 4, and C_j that the jth turbine is in a good condition, j = 1, 2. The event D means that the steamer is steerable, which occurs only when the steering-gear, at least one of the steam-boilers and at least one of the turbines are simultaneously in a good condition. Express the events D and $\overline{D}$ in terms of A, B_k and C_j.

2.20. A device consists of two blocks of type I and of three blocks of type II. Let A_k = {the kth block of type I is in order}, k = 1, 2, and B_j = {the jth block of type II is in order}, j = 1, 2, 3. The device works if at least one block of type I and no less than two blocks of type II are in order. Express the event C = {the device works} in terms of A_k and B_j.

3. Classical Definition of Probability

Introductory Notes

Let $\Omega = \{\omega_1, \omega_2, \ldots, \omega_n\}$ be a space with a finite number of *elementary outcomes* (*simple events*) in a random *experiment*. To each simple event ω_i a weight $P(\omega_i)$ is assigned. The weights $P(\omega_i)$, i = 1, ..., n, are called *probabilities* of the outcomes $\omega_1, \omega_2, \ldots, \omega_n$ if they satisfy the following conditions:

$$P(\omega_i) > 0 \quad \text{for } i = 1, 2, \ldots, n; \tag{3.1}$$

$$P(\omega_1) + P(\omega_2) + \ldots + P(\omega_n) = 1. \tag{3.2}$$

Let A be an arbitrary event of $\mathcal{P}(\Omega)$. The probability of the event A, which we denote by P(A), is a sum of the probabilities of those simple events which form the set A; i.e., if $A = \{\omega_{i_1}, \ldots, \omega_{i_m}\}$, $m \leqslant n$, then

$$P(A) = P(\omega_{i_1}) + \ldots + P(\omega_{i_m}). \tag{3.3}$$

If the probabilities of all elementary outcomes are equal, i.e., $P(\omega_1) = P(\omega_2) = \ldots = P(\omega_n)$, then from (3.2) it follows that $P(\omega_i) = \frac{1}{n}$, i = 1, 2, ..., n, and formula (3.3) becomes

$$P(A) = \frac{m}{n}, \tag{3.4}$$

which is known as a *classical definition of probability*. Here number m is the number of the *favourable outcomes* for the event A and n is the number of *all possible outcomes* of the experiment.

We shall notice now the following more important properties of the events in the classical model of probability:

$$0 \leqslant P(A) \leqslant 1 \quad \text{for every} \quad A \in \mathcal{P}(\Omega); \tag{3.5}$$

$$P(\Omega) = 1; \tag{3.6}$$

$$P(A) + P(\bar{A}) = 1 \quad \text{for every} \quad A \in \mathcal{P}(\Omega). \tag{3.7}$$

Random choice or *selection at random* from $E_1, \ldots, E_n$ involves an experiment in which the possible outcomes are $\omega_1, \ldots, \omega_n$, where ω_j = {the object E_j is selected}, $j = 1, \ldots, n$, and each one occurs with the same probability $1/n$.

Illustrative Examples

Example 3.1. Each face of a ten-inch cube is coloured. The cube is then cut into 1000 one-inch cubes. The 1000 cubes are thoroughly mixed and one cube is drawn at random. Find the probability that the chosen cube will have: (i) exactly one coloured face; (ii) exactly two coloured faces; (iii) exactly three coloured faces.

Solution. For the 1000 cubes, there are four types of cubes distinguishable from the number of coloured faces: none, one, two, three. There will be 8^3 with no coloured face, 6×8^2 with one, 12×8 with two and 8 with three. Thus the probabilities p_1, p_2 and p_3 in cases (i) - (iii), respectively, are as follows:

$$p_1 = \frac{6 \times 8^2}{1000} = 0.384, \quad p_2 = \frac{12 \times 8}{1000} = 0.096, \quad p_3 = \frac{8}{1000} = 0.008.$$

Example 3.2. An urn contains M black and N - M white balls, $M < N$. An unordered random sample of size n is selected without replacement, $n < N$, $n < N - M$ (see Exercise 1.8). Find the probability that the sample contains exactly k black balls where $k \leq n$ and $k \leq M$.

Solution. There are $\binom{N}{n}$ possible samples of which we are interested in those containing k black and $n - k$ white balls. The number of the samples with k black balls is equal to $\binom{M}{k}$. The remaining $n - k$ white balls could be chosen in $\binom{N-M}{n-k}$ ways from the set of $N - M$ white balls. Hence the number of favourable outcomes is equal to $\binom{M}{k}\binom{N-M}{n-k}$. The desired probability is

$$p = \binom{M}{k}\binom{N-M}{n-k}\Big/\binom{N}{n}.$$

Example 3.3. Two fair dice are rolled. Find the probability that at least one "six" will appear if: (a) the dice are distinguishable and we assume that the different outcomes are equally probable; (b) the dice are indistinguishable and we assume that the different outcomes are equally probably. (c) Can both assumptions (a) and (b) describe the state of nature?

Solution. The result of rolling k dice corresponds to a placement of k balls (dice) into six different cells (faces): 1, 2, 3, 4, 5, 6. We calculate the desired probability $P(A)$ by the formula $P(A) = 1 - P(\bar{A})$ which follows from (3.7). Here $\bar{A}$ = {both faces are different from "six"}.

(a) In this case the number of possible outcomes n is equal to the number of the different ways of distributing two distinguishable balls into six different cells; i.e., $n = 6^2$. The event $\overline{A}$ occurs when no ball is placed into cell "number 6", which is equivalent to the number of ways of placing two distinguishable balls in the remaining five cells; i.e., $m = 5^2$. Then

$$P(A) = 1 - 5^2/6^2 \approx 0.306.$$

(b) Analogously to case (a), if the balls are indistinguishable we have

$$n = \binom{6 + 2 - 1}{2} = \binom{7}{2}, \quad m = \binom{5 + 2 - 1}{2} = \binom{6}{2},$$

and hence

$$P(A) = 1 - P(\overline{A}) = 1 - \binom{6}{2}/\binom{7}{2} \approx 0.285.$$

(c) Suppose that both assumptions (a) and (b) adequately describe the state of nature. Then the probality of at least one "six" is 0.306 if the dice are distinguishable and 0.285 if they are indistinguishable. Such a difference should not depend on a subjective factor - the ability of the player to distinguish between the two dice. Actually, the outcome (1, 6) may occur in two different ways which are described by the ordered pairs (1, 6) and (6, 1), while the outcome (6, 6) can occur in only one way. This implies that the probability of the occurrence of (1, 6) ought to be twice as big as the one of (6, 6). Hence, assumption (b) is not adequate.

If in case (b) we no longer assume that the outcomes are equally likely, then the model can be made equivalent to that in case (a). Now P(A) can be evaluated by formula (3.3). The space Ω of the outcomes in this experiment is: $\Omega = \{11, 12, 13, 14, 15, 16, 22, 23, 24, 25, 26, 33, 34, 35, 36, 44, 45, 46, 55, 56, 66\}$, where the outcome ω_{ij} is denoted only by ij for short. Let $P(\omega_{ii}) = x$ for $i = 1, \ldots, 6$. As we saw above, $P(\omega_{ij}) = 2x$ if $i \neq j$ for $i, j = 1, \ldots, 6$. From (3.2) it follows that $6 \times x + 15 \times 2x = 1$ and $x = 1/36$. Since $A = \{16, 26, 36, 46, 56, 66\}$, then

$$P(A) = 5 \times 2x + x = 11/36,$$

which coincides with the result in case (a).

Example 3.3 illustrates that in exercises involving dice, it is more convenient to use a model in which the dice are assumed to be distinguishable. Then in the corresponding sample space Ω, the elementary outcomes are equally probable and the classical definition of probability is applicable.

In connection with this example we shall discuss de Méré's paradox which was explained by the great French mathematician B. Pascal, one of the founders of the theory of probability.

De Méré's paradox. In games of chance involving dice, de Méré noticed that, when three dice were rolled simultaneously, a sum of 11 occurred more frequently than a sum of 12. De Méré reasoned that a total of 11 could occur in six different ways: 6-4-1, 6-3-2, 5-5-1, 5-4-2, 5-3-3 and 4-4-3. Likewise a total of 12 could occur in six ways: 6-5-1, 6-4-2, 6-3-3, 5-5-2, 5-4-3 and 4-4-4. Thus the events A_{11} and A_{12} should have equal their probabilities. Where was the flaw in de Méré's reasoning? Pascal explained de Méré's mistake was in thinking that in the latter case the dice were indistinguishable. In such a case the elementary outcomes are not equally probable and the classical definition of probability is not applicable. In fact the classical definition of probability could be applied if the dice are considered as distinguishable; i.e., the outcomes are ordered triplets and have equal probabilities. In that case there are 27 outcomes favourable to the event A_{11}, while for the event A_{12} there are only 25. Verify!

De Méré's paradox provides an experimental confirmation for the validity of the Maxwell-Boltzmann model (see Exercises 1.11 and 3.49).

In physics (more precisely in statistical mechanics) the probabilistic models of Bose-Einstein (see Exercises 1.10 and 3.47) and of Fermi-Dirac (see Exercises 1.9 and 3.48) are accepted and with their help the behaviour of different classes of particles is satisfactorily described. The first one is valid for photons, atomic nuclei and atoms containing an even number of elementary particles. The model of Fermi-Dirac applies to electrons, protons and neutrons. These examples are very instructive because very often, for choosing a probabilistic model for describing a natural phenomenon, our everyday notions about the macrocosmos are not sufficient.

Exercises

3.1. Two distinguishable dice are thrown. What is the probability of getting two "threes" if the total of the faces is a multiple of 3?

3.2. Three distinguishable dice are thrown. What is the probability that "three" will appear on at least one of them if the total of the three faces is equal to 10?

3.3. There are five segments with lengths 1, 3, 5, 7 and 9 units, respectively. What is the probability that three of them chosen at random can be sides of a triangle?

3.4. Find the probability that in throwing n dice the total of the faces will not be less than $6n - 1$.

3.5. From the numbers 1, 2, ..., n, two numbers are chosen at random. What is the probability that one of them will be strictly smaller, and the other will be greater than a given number k, where $1 < k < n$?

3.6. A domino piece chosen at random contains a different number on each half. Find the probability that a piece, taken at random from the remaining ones, will contain at least one of the numbers of the first piece.

Note. The numbers marked on each half of a domino piece form a combination (repetition allowed) of the numbers 0, 1, ..., 6. The set of dominos contains every one of these combinations only once.

3.7. Winning in an ancient dice game consists of getting a total greater than ten when three dice are thrown. Find the probability of: (a) getting a total of 11; (b) getting a total of 12; (c) winning in this game.

3.8. The numbers 1, 2, 3, 4, 5 are written on five cards. Three cards are chosen at random one after another and the selected digits are noted successively in the order of their selection. Find the probability that the obtained three-digit number will be even.

3.9. Two dice are thrown. Find the probability that the product of the faces will be even.

3.10. What is the probability that in a random permutation of n elements two given elements (a and b) will not be adjacent?

3.11. From an urn containing n white and n black balls, a sample containing an even number of balls is taken at random. Notice that all distinguishable samples containing an even number of balls should be regarded as equally probable (including samples with size zero). Find the probability that among the drawn balls there will be an equal number of black and white ones.

3.12. Suppose m zeros and n ones are arranged at random in a sequence. What is the probability that the sequence starts with exactly k zeros and ends with exactly s ones ($k \leqslant m$, $s \leqslant n$)?

3.13. From a lot with n good and m defective items, s items are taken at random for examination. Suppose that the first k of the s examined items are good, $k < s$. Find the probability that the $(k + 1)$st item will turn out to be good.

3.14. A safety lock contains five disks fixed to a common axis. Each one of the disks is divided into six sectors with different letters marked on them. The lock can be opened only when each disk comes to a fixed position towards the body of the lock. Find the probability that the lock will be opened by putting an arbitrary combination of letters.

3.15. The letters L, I, T, E, R, A are written on six pieces of cardboard. Select four of them and put them in a row. What is the probability that the four letters spell the word TIRE?

3.16. From an urn containing N balls with numbers 1, 2, ..., N select n balls taking one at a time. Find the probability that the numbers of the selected balls, written in their order of appearance will form an increasing sequence if after each drawing the chosen ball: (a) is replaced into the urn before the next drawing; (b) is not replaced.

3.17. A lot contains m items of first quality and n items of second quality. An examination of the first b items, $b < n$, taken at random from the lot, has shown that all of them are of second quality. Find the probability that at least one of the next two items, taken at random out of the still unchecked ones, will turn out to be of second quality.

3.18. Ten books are placed at random on a shelf. Find the probability that: (a) three particular books will be adjacent; (b) k particular books will be adjacent, $2 \leqslant k \leqslant 10$.

3.19. Find the probability that the number of a randomly chosen bond will not contain equal digits, if this number can be any five-digit number starting from 00001.

3.20. There are nine passengers and three carriages, which are labeled A, B and C. Each passenger is choosing a carriage at random. Find the probability that: (a) each of the carriages will have three passengers; (b) in the carriages A, B and C, respectively four, three and two passengers will get on.

3.21. Suppose that each of n sticks is broken into one long piece and one short piece. The 2n piece are then joined at random into pairs. Find the probability that: (a) the pieces will be joined in their original order; (b) each one of the long pieces will be paired with a short one.

3.22. Suppose that n dice are thrown. Find the probability of getting n_1 "ones", n_2 "twos", ..., n_6 "sixes", where $n_1 + \ldots + n_6 = n$.

3.23. Suppose that n dice are thrown. Find the probability that the total of the faces is equal to: (a) n; (b) n + 1; (c) a given number s where $n \leqslant s \leqslant 6n$.

3.24. Find the probability that in a simultaneous throwing of m dice and n coins only "sixes" and "heads" will appear.

3.25. What is the probability that: (a) the birthdays of 12 persons will fall in 12 different calendar months (assuming equal probabilities for the 12 months); (b) the birthdays of six persons will fall in exactly two calendar months?

3.26. If we select people at random, how many people will have to be sampled in order that the probability that the birthdays of at least two of them coincide, will not be less than 1/2? (The years of birth can be different. It is assumed that 29th of February cannot be a birthday and the remaining 365 days are considered as equally probable birthdays.)

3.27. Suppose that n persons are standing at random in a row. Find the probability that between two particular persons, say A and B, there will be exactly r people, where $r \leqslant n - 2$.

3.28. A group of n people is sitting at a round table. Find the probability that two people specified in advance will: (a) be neighbours; (b) not be neighbours.

3.29. On each of n benches m persons are seated at random. What is the probability that two particular persons will be sitting next to each other?

3.30. In a hall with n + k seats n people are choosing seats at random. Find the probability that m specified seats will be occupied, $m \leqslant n$.

3.31. Assume that the car registration numbers consist of four digits, from 0000 up to 9999, with repetition. Find the probability that the registration number of the first car coming toward us: (a) will not contain any repetition of the digits; (b) will have one pair of repetitions; (c) will have three repetitions; (d) will have two pairs of repetitions; (e) will have the sum of the first two digits equal to the sum of the last two digits; (f) will consist of four repetitions.

3.32. A box contains 4N marbles labeled 1, 2, ..., 4N. The marbles are divided at random into two groups of 2N marbles. Find the probability that: (a) in each group there will be an equal number of marbles labeled with even numbers and odd ones; (b) all labels divisible by N will fall into one group; (c) the two groups will contain an equal number of marbles labeled with numbers divisible by N.

3.33. In bridge which is more probable: your partner's hand and yours will together contain (i) all 13 diamonds or (ii) no diamond?

3.34. From an urn containing n different balls a sample is selected at random without replacement. Compute the probability that the same will contain (i) an even number of balls? (ii) an odd number of balls? (Note that a sample of size zero is not possible.) Compare the probabilities obtained in (i) and (ii).

3.35. A deck of ten cards consists of five numbered with a "1", three numbered with a "3" and two numbered with a "5". Three cards are chosen at random without replacement. Find the probability that: (a) at least two of them will have the same number; (b) the sum of the three numbers will be seven.

3.36. From n pairs of shoes, 2r shoes are chosen at random with $2r < n$. Find the probability that among the selected shoes there will be: (a) no complete pair; (b) exactly one complete pair; (c) exactly two complete pairs.

3.37. A fair coin is tossed n times. Determine the probability that a head appears an odd number of times.

3.38. Consider a regular $(2n + 1)$ polygon where n is a natural number. Three of its vertices are chosen in a random way such that all choices are equally likely. What is the probability that the centre of the given polygon lies inside a triangle formed from the three chosen points?

3.39. An $n \times n$ square is cut into n^2 equal squares. Of these n^2 squares, $n + k$ with $k < n(n - 1)$ are painted red and the remaining are white. The n^2 squares are arranged at random in the form of an $n \times n$ square. What is the probability that n of the red squares will fall into a row, column or diagonal of the constructed square?

3.40. An integer N is chosen at random. Find the probability that the last two digits of the cube of the chosen number will both be one.

3.41. An integer N is chosen at random. Find the probability that one can obtain a number terminating with 1, when: (a) N is raised to the second power; (b) N is raised to the fourth power; (c) N is multiplied by another arbitrary integer which has been chosen at random.

*3.42. From the set M^{x2} where $M = \{1, 2, \ldots, n\}$ and $n > 3$, an ordered pair (x, y) is chosen at random. (a) Find the probabilities of the events $A_2 = \{x^2 - y^2$ is divisible by two$\}$ and $A_3 = \{x^2 - y^2$ is divisible by three$\}$. (b) Which one of the two events is more probable?

3.43. From the set $\{1, 2, \ldots, N\}$, a sample of n numbers $x_1 < x_2 < \ldots < x_n$ is taken at random. Find the probability of the event $\{x_m = M\}$ with m and M fixed integers, $1 < m < n$, $1 < M < N$.

3.44. Ten distinguishable dice are thrown. What is the probability to appear an equal number of ones and sixes?

3.45. A group of people consists of five men and ten women. Five groups of three people each are formed at random. Find the probability that there will be exactly one man in each group.

3.46. The number of participants in a tournament is 2^n. It is known that among them there is just one, say A, who is the best and there is another one, say B, who is the best among the others. At the beginning the position of each player in the list is chosen at random. The scheme of the tournament is as usual; i.e., similar to that shown in Figure 3.1 in the case $n = 3$. What is the probability that B will play with A at the final match?

3.47. Let n indistinguishable particles be distributed at random into n different cells. Each cell may contain an arbitrary number of particles. Suppose that all distinguishable distributions are equally probable (Bose-Einstein statistics, see Exercise 1.10). Determine the probability that: (a) a fixed cell will contain exactly k particles, $k \leqslant n$; (b) exactly m cells will be empty, $m < n$; (c) in each cell there will be at least two particles.

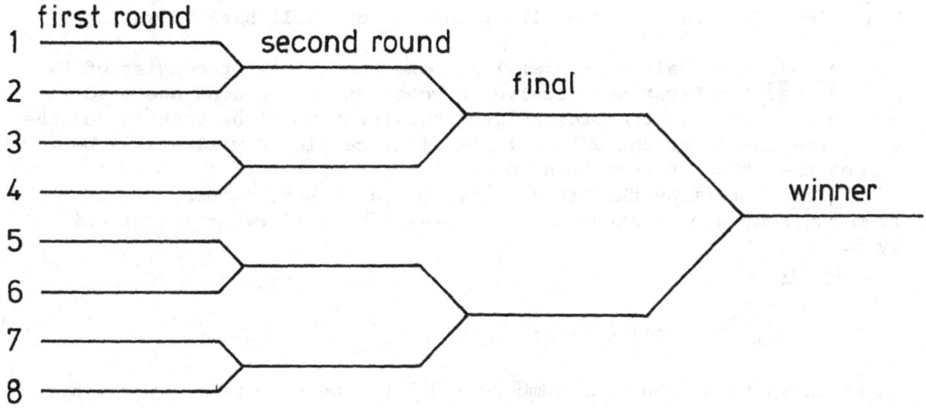

Fig. 3.1

3.48. Let r indistinguishable particles be placed at random into n different cells. Each cell may contain at most one particle. Assume that all distinguishable distributions are equally probable (Fermi-Dirac statistics, see Exercise 1.9). Determine the probability that a fixed cell will be empty, $r < n$.

3.49. Let r distinguishable particles be distributed at random into n different cells. Each cell may contain an unlimited number of particles. Assume that all distinguishable distributions are equally probable (Maxwell-Boltzmann statistics, see Exercise 1.11). Determine the probability that: (a) the first cell will contain k_1 particles, the second k_2 particles, etc., where $k_1 + k_2 + \dots + k_n = r$; (b) for $n = r$ no one cell will remain empty; (c) for $n = r$ exactly one cell will remain empty; (d) exactly m cells will be empty, $m < n$, $r \geqslant n - m$.

3.50. Three fair distinguishable dice are thrown. What is the probability of the event $A = \{$the sum and the product of the faces are equal$\}$?

*3.51. From an urn containing N balls numbered 1, 2, ..., N, n balls are drawn in succession with the selected ball being replaced before the next drawing. The numbers of the selected balls are rearranged in a non-decreasing sequence. Compute the probability of the event: the number occupying the mth position in this sequence is equal to M, where $1 \leqslant m \leqslant n$, $1 \leqslant M \leqslant N$.

*3.52. Let a, b, c, d, e and f be integers selected at random such that their absolute values do not exceed the natural number n. Let p_n be the probability that the system $\left| \begin{array}{l} ax + by = e \\ cx + dy = f \end{array} \right.$ has exactly one solution. Find $\lim_{n\to\infty} p_n$.

*3.53. There are n + m people waiting in a line to buy tickets which cost one lev[1)] each. Of these, n have only one-lev notes and m have only two-lev notes with $m \leqslant n + 1$. Each person in the queue buys

1) One lev is the Bulgarian monetary unit. (Translator's note.)

only one ticket and pays in cash. The ticket office has no cash at the beginning. What is the probability that no one will have to wait for change?

*3.54. (*The ballot problem.*) Suppose that a ballot consist of two candidates. The first one receives n votes and the second one m votes with $n > m$. What is the probability that throughout the counting of the votes the number of the votes of the first candidate will always be larger than that of the second one?

3.55. Determine the probability that a number, chosen at random from the sequence of natural numbers, will be: (a) even; (b) divisible by 5.

Note. Let

$$a_1, a_2, \dots, a_n, \dots \tag{3.8}$$

be an infinite sequence of numbers and let α be some property which is either valid or not for each of these numbers. We are interested in the probability of the event $A = \{$a number chosen at random from the sequence (3.8) possesses the property $\alpha\}$. In this case the classical definition of probability can be reformulated: Consider the first N members of the sequence (3.8):

$$a_1, a_2, \dots, a_N \tag{3.9}$$

and denote by $q(N)$ the number of those quantities in (3.9) which possess the property α. Define

$$P(A) = \lim_{n\to\infty} \frac{q(N)}{N} \tag{3.10}$$

if the limit exists.

3.56. Let k be a fixed natural $(n - 1)$-digit number. Denote by S_m the set of all m-digit numbers, where $m \geq n$. Select at random one number from S_m. Determine the probability $P_m(k)$ that the selected number is divisible by k and find the limit

$$\lim_{n\to\infty} P_m(k).$$

3.57. Each of two persons independently chooses one number from the set $\{1, 2, \dots, N\}$. Let ξ and η be the chosen numbers and $P\{\xi^2 + \eta^2 \leq N^2\} = p_N$. Find $\lim_{n\to\infty} p_N$.

4. Conditional Probability. Independence of Events

Introductory Notes

Let B be an event with $P(B) > 0$. The *conditional probability of the event A given B* is denoted by $P(A|B)$ and defined by the relation

$$P(A|B) = \frac{P(AB)}{P(B)} . \tag{4.1}$$

If A and B are events with $P(A) > 0$ and $P(B) > 0$, the probability of their simultaneous occurrence is calculated by the rule for multiplying probabilities (or the theorem for compound probabilities):

$$P(AB) = P(B)P(A|B) = P(A)P(B|A). \tag{4.2}$$

This result can be extended to n events: if the events $A_1, \ldots, A_n$ are such that $P(A_1A_2 \ldots A_{n-1}) > 0$, then

$$P\left(\bigcap_{k=1}^{n} A_k\right) = P(A_1)P(A_2|A_1)P(A_3|A_1A_2) \ldots$$

$$\ldots P(A_n|A_1A_2 \ldots A_{n-1}). \tag{4.3}$$

The events $A_1, \ldots, A_n$ are called *mutually independent* (or just *independent*) if for any k of them ($2 \leqslant k \leqslant n$) the following relation holds:

$$P(A_{i_1}A_{i_2} \ldots A_{i_k}) = P(A_{i_1})P(A_{i_2}) \ldots P(A_{i_k}). \tag{4.4}$$

In the particular case when $k = 2$, the events A_1 and A_2 are independent if

$$P(A_1A_2) = P(A_1)P(A_2). \tag{4.5}$$

If (4.4) is fulfilled only for $k = 2$, the given n events are called *pairwise independent*.

If the events A and B are independent, the events in the following pairs: $\bar{A}$ and B, A and $\bar{B}$, $\bar{A}$ and $\bar{B}$ are also independent.

Illustrative Examples

Example 4.1. Show that if two events are independent and each of them has a positive probability, then these events are not mutually exclusive.

Solution. We have $P(AB) = P(A)P(B)$. As $P(A) > 0$ and $P(B) > 0$, then $P(AB) > 0$, and hence $AB \neq \emptyset$; i.e., A and B are not mutually exclusive.

Example 4.2. It is known that 4% of the items of a lot are defective, and 75% of the good items are first-grade. Calculate the probability that a chosen at random item will be first-grade.

Solution. Let A = {the chosen item is not defective} and B = {the chosen item is a first-grade one}. We are looking for $P(AB)$. By assumption $P(A) = 1 - P(\bar{A}) = 1 - 0.4 = 0.96$; $P(B|A) = 0.75$. Hence $P(AB) = P(A)P(B|A) = 0.96 \times 0.75 = 0.72$.

Example 4.3. An urn contains five white and four black balls. Two balls are drawn in succession without replacement. Find the probability that: (a) both balls will be white; (b) the first one will be black and the second white.

Solution. Let A_i = {a white ball is drawn at the ith trial} and B_i = {a black ball is drawn at the ith trial}, where i = 1, 2. Then:

$$P(A_1A_2) = P(A_1)P(A_2|A_1) = \frac{5}{9} \cdot \frac{4}{8} = \frac{5}{18}\;; \quad \text{(a)}$$

$$P(B_1A_2) = P(B_1)P(A_2|B_1) = \frac{4}{9} \cdot \frac{5}{8} = \frac{5}{18}\,. \quad \text{(b)}$$

Example 4.4. How many times must two fair coins be tossed in order that one can claim with probability no less than 0.99 that two heads will appear at least once?

Solution. In a single toss of the two coins, let the event H_1 = {head appears on the first coin} and the event H_2 = {head appears on the second coin}. The events H_1 and H_2 are independent and $P(H_1H_2) = P(H_1)P(H_2) = 0.5 \times 0.5 = 0.25$. Let now A_k = {the occurrence of H_1H_2 at the kth toss}, where k = 1, 2, ..., n and n is the desired number of trials. Denote by $A = A_1 \cup A_2 \cup \ldots \cup A_n$. Since the events A_k are independent and $P(A_k) = 0.25$, then $P(A) = 1 - P(\bar{A}) = 1 - P(\overline{A_1 \cup A_2 \cup \ldots \cup A_n}) = 1 - (0.75)^n$. We can now determine n from the condition $1 - (0.75)^n \geqslant 0.99$. Thus $(0.75)^n \leqslant 0.01$ and $n\ln 0.75 \leqslant \ln 0.01$, hence $n \geqslant \frac{\ln 0.01}{\ln 0.75} \approx 16.01$ or $n \geqslant 17$.

Exercises

4.1. The probability of producing a first-grade item on a turning-lathe is 0.7 and the probability of producing the same type of item on another lathe is 0.8. On the first lathe two items are produced, and on the second lathe three are produced. What is the probability that all five items will be first-grade?

4.2. The probability of a marksman hitting a target is 2/3. If he manages to hit the target with his first shot then he earns the right to fire a second time at a second target. The probability of hitting both targets with two shots is 0.5. Find the probability of hitting the second target if the marksman has earned the right for a second shot.

4.3. The six letters of the word CARTER are written on six pieces of cardboard. The pieces of cardboard are shuffled and then are selected at random one after another. What is the probability that the letters drawn will form the word TRACER?

4.4. A box contains three poker chips: the first has both faces painted white, the second has both black and the third has one white and one black. A chip is selected at random and tossed on a table. If the face that appears is white, what is the probability that the other face, which is not visible, is also white?

4.5. An urn has one white and one black ball. A ball is selected at random with replacement until a black ball appears. Each time a white ball is chosen it is replaced and two more white balls are added. Find the probality that in the first 50 trials no black ball will be drawn.

4.6. An urn contains n balls numbered from 1 to n. The balls are drawn at random without replacement. What is the probability that in the first k drawings the numbers on the balls will coincide with the numbers of the drawings?

4.7. In a group of 2n people the number of the men equals that of the women. They take seats at random around a table. Find the probability that no two people of the same sex will be neighbours.

4.8. The names of n boys are written on n cards and the names of m girls on another m cards ($m \geq n$). The cards are placed in a box and after a thorough mixing two cards are drawn n times in succession without replacement. What is the probability that each time a pair of cards "boy-girl" will be drawn?

4.9. An urn contains m white and n black balls. From them k are drawn at random without replacement and all are found to be unicoloured. What is the probability that all the selected balls are black?

4.10. Somebody has forgotten the last digit of a telephone number and is dialing the digit at random. (a) Find the probability that he will have to dial to no more than three subscribers. (b) How does this probability change if it is known that the last digit is odd?

4.11. The probability of a break-down of the kth block of a given machine during an interval of time T is equal to p_k, where $k = 1, 2, \ldots, n$. Find the probability that during the indicated interval of time at least one of the n blocks of the machine will break down if all the blocks work independently.

4.12. The probability of at least one occurrence of the event A during the performance of four trials is 1/2. Find the probability of the occurrence of A during the performance of one trial if this probability is the same in all trials.

4.13. In each trial, an event occurs with probability p. The trials are performed successively until the event occurs. Find the probability that the event will occur exactly at the kth trial, where $k = 1, 2, \ldots$

4.14. A cut-off in an electrical circuit may occur due to a break-down either of the element K or of the two elements K_1 and K_2. The break-downs of the three elements are mutually independent, and their probabilities are 0.3, 0.2 and 0.2, respectively. Find the probability of a cut-off in the circuit.

4.15. Which is more probable: A_1 = {at least one ace in rolling four dice} or A_2 = {in 24 rolls of two dice to get two aces at least once}?

4.16. What is the minimal number of people who must be selected at random so that the probability of the birthday of at least one of them coinciding with yours is greater than 1/2? (As in Exercise 3.26, assume that the year of birth is of no importance, the 29th of February is not a birthday and all remaining 365 days are equally probably as birthdays.)

4.17. Let the event A be the occurrence of at least one of the events $A_1, \ldots, A_n$ whose probabilities are $p_1, \ldots, p_n$, respectively. Prove that $P(A) \geq 1 - \exp(-p_1 - p_2 - \ldots - p_n)$.

4.18. How many times should a die be rolled so as the probability of at least one "six" is greater than: (a) 0.5; (b) 0.8; (c) 0.9?

4.19. How many numbers have to be taken from the table of random numbers so that, with probability no less than 0.9, at least one of them will be even?

Note. Let $\mathbf{K}$ be the set of all k-digit integers excluding the one with zeros in all k positions. There will be $10^k - 1$ such numbers.

A *table of random numbers* should be constructed in such a way that with the help of an appropriate selecting rule to be possible to obtain a sequence $a_1, a_2, \ldots, a_n, \ldots$ of elements of $\mathbf{K}$ having the following properties: (1) $P\{a_n = \kappa\} = 10^{-k}$ for each $\kappa \in \mathbf{K}$ and for every n; (2) $P\{a_n = \kappa | a_m = \kappa_1\} = P\{a_n = \kappa\}$ for all values of $\kappa \in \mathbf{K}$, $\kappa_1 \in \mathbf{K}$ and with $n = m$.

4.20. How many times should two dice be thrown in order, with probability no less than r, one can claim that the event A will occur at least once equal to 12? (*Ch. de Méré*)

4.21. How many times should an experiment be repeated so that, with probability no less than r, one can claim that the event A will occur at least once? The probability of A is assumed to be p at each independent repetition of the experiment?

4.22. Prove that if A and B are events for which $P(A) > 0$, $P(B) > 0$ and $P(A|B) > P(A)$, then $P(B|A) > P(B)$.

4.23. The events A and B are mutually exclusive with $P(A) \neq 0$ and $P(B) \neq 0$. Are these events independent?

4.24. Prove that if the events A and B are independent and $A \subset B$, then either $P(A) = 0$ or $P(B) = 1$.

4.25. The faces of a regular balanced tetrahedron are coloured as follows: the first white, the second green, the third red and the fourth simultaneously with white, green and red. Let W be the event that the tetrahedron falls on a face which is white. The analogous events for the green and red faces are G and R, respectively. Are the events W, G and R pairwise independent? Are these events mutually independent? (*S. N. Bernstein*)

4.26. Let every two of the events $A_1, A_2, \ldots, A_n$ be independent $(n > 2)$. Does it follow that the events $A_1, A_2, \ldots, A_n$ are mutually independent?

4.27. The sample space of some experiment is $\Omega = \{\omega_1, \omega_2, \omega_3, \omega_4, \omega_5\}$, where ω_1 has a probability 1/8, ω_2, ω_3 and ω_4 have probabilities 3/16 each and ω_5 has a probability of 5/16. Consider the events $A_1 = \{\omega_1, \omega_2, \omega_3\}$, $A_2 = \{\omega_1, \omega_2, \omega_4\}$ and $A_3 = \{\omega_1, \omega_3, \omega_4\}$. Show that A_1, A_2 and A_3 are mutually independent but not pairwise independent. (*G. Roussas*)

4.28. Two dice are thrown. Let A_1 = {odd face appears on the first die}, A_2 = {odd face appears on the second die} and A_3 = {the total of the faces is odd}. Are these events pairwise independent? Are A_1, A_2 and A_3 mutually independent?

4.29. From a deck of 36 cards, made up of {6, 7, ..., ace} of each suit, one card is drawn at random. Consider the events A_1 = {the drawn card is a spade}, A_2 = {the drawn card is a queen}. Are these two events independent? What is the answer if the deck contains 52 cards?

4.30. On n cards, n different real numbers are written. The cards are put in a box, mixed well, and then drawn one by one without replacement. Let A_k = {the kth drawn number is greater than all previously drawn ones}. (1) Show that $P(A_k) = 1/k$, $k = 1, \ldots, n$. (2) Prove that the events $A_1, \ldots, A_n$ are independent.

4.31. The sample space Ω consists of N elements. What is the greatest n for which one can define on the subsets of Ω a probability and such n events $A_1, \ldots, A_n$ which are mutually independent and $0 < P(A_i) < 1$, $i = 1, \ldots, n$.

4.32. A fair coin is tossed successively three times. Let A = {head turns up at the first toss}, B = {at least two heads appear on the three tosses} and C = {the same result turns up on all three tosses}. Consider the pairs A and B, A and C, B and C. Are these three events pairwise independent?

4.33. A fair coin is tossed either until a head appears or until three tosses are carried out. What is the probability that the coin must be tossed three times if it is known that the first toss was a tail?

4.34. A fair coin is tossed either until a head appears or until k tosses are performed. Find the probability of tossing the coin k times if it is known that a tail has appeared at the first two tosses ($k \geqslant 3$).

4.35. Consider a game two persons A and B are playing. Player A has the first move and he wins with probability 0.3. If he wins, the game is terminated. Otherwise B makes a move and he wins with probability 0.5. Again this winning causes the game to terminate. If B does not succeed in winning, then A has the final move of the game. In fact, A can win at this second move and hence wins the whole game with probability 0.4. Calculate the probabilities of the events: A = {A wins the game}; B = {B wins the game} and O = {no one wins the game}.

*4.36. Consider the fraction formed by drawing the numerator and the denominator at random and independently from one another from the sequence of the natural numbers. What is the probability that the fraction will be irreducible? (*P. L. Chebyshev*)

4.37. Let $x_1, \ldots, x_n$ be arbitrary numbers in the interval (0, 1). On the base of probabilistic reasoning, prove the inequalities:

$$1 - \sum_{k=1}^{n} x_k < \prod_{k=1}^{n} (1 - x_k) < \left(1 + \sum_{k=1}^{n} x_k\right)^{-1}; \qquad \text{(a)}$$

$$1 + \sum_{k=1}^{n} x_k < \prod_{k=1}^{n} (1 + x_k) <$$

$$< \left(1 - \sum_{k=1}^{n} x_k\right)^{-1}, \quad \text{if } \sum_{k=1}^{n} x_k < 1. \qquad \text{(b)}$$

5. Probability of a Sum of Events. Formula for Total Probability. Bayes' Formula

Introductory Notes

Let A and B be arbitrary events belonging to $\mathcal{P}(\Omega)$. Then by the probability of the occurrence of at least one of them is determined by the following formula:

$$P(A \cup B) = P(A) + P(B) - P(AB). \tag{5.1}$$

In general, the probability of the occurrence of at least one of the events $A_1, A_2, \ldots, A_n$, with $A_i \in \mathcal{P}(\Omega)$, $i = 1, \ldots, n$, is:

$$P\left(\bigcup_{i=1}^{n} A_i\right) = \sum_{i=1}^{n} P(A_i) - \sum_{i<j} P(A_i A_j) + \sum_{i<j<k} P(A_i A_j A_k) + \ldots + (-1)^{n-1} P(A_1, \ldots, A_n). \tag{5.2}$$

(*formula of A. Poincaré*)

If the events $A_1, \ldots, A_n$ are mutually exclusive, then

$$P\left(\bigcup_{i=1}^{n} A_i\right) = \sum_{i=1}^{n} P(A_i) \quad (\textit{additive property}). \tag{5.3}$$

Let the events $H_1, \ldots, H_m$ form a partition of the sample space Ω; i.e., $\bigcup_{i=1}^{m} H_i = \Omega$ and $H_i H_j = \emptyset$ for $i \neq j$, $i, j = 1, \ldots, m$. Then an arbitrary event $A \in \mathcal{P}(\Omega)$ is represented in the form $A = AH_1 + \ldots + AH_m$ and its probability is calculated by the so-called *formula for total probability*:

$$P(A) = \sum_{k=1}^{m} P(H_k) P(A|H_k). \tag{5.4}$$

Assume the probabilities $P(H_1), \ldots, P(H_m)$ are known. These probabilities are usually referred as *a priori probabilities* of the hypotheses $H_1, \ldots, H_m$, respectively. If we are informed that some event A has occurred, then we are interested in the calculation of the conditional probabilities $P(H_1|A), \ldots, P(H_m|A)$ called *a posteriori probabilities* of $H_1, \ldots, H_m$. For this calculation the following formula, known as the *Bayes formula*, can be used:

$$P(H_k|A) = \frac{P(H_k)P(A|H_k)}{\sum_{i=1}^{m} P(H_i)P(A|H_i)}, \quad k = 1, \ldots, m. \tag{5.5}$$

The sample space Ω is said to be *discrete* if it consists of finitely or countably many elementary events (sample points); i.e., if $\Omega = \{\omega_1, \ldots, \omega_N\}$ for some natural number N, or if $\Omega = \{\omega_1, \omega_2, \ldots\}$. Associated with each sample point ω_i is a weight $p_i = \vec{P}(\omega_i)$. Then the sequence $\{p_i\}$, finite or infinite, has the properties:

$$0 < p_i \leqslant 1 \quad \text{and} \quad \sum_i p_i = 1.$$

If $A \in \mathcal{P}(\Omega)$ is an arbitrary event, then its probability P(A) is given as follows (see also (3.3)):

$$P(A) = \sum_{k:\omega_k \in A} p_k.$$

Illustrative Examples

Example 5.1. From a deck of 52 cards one card is drawn at random. What is the probability that this card is spade or face card of an arbitrary suit? (A face card is a knave, a queen or a king.)

Solution. Let A_1 = {the drawn card is spade} and A_2 = {the drawn card is a face card}. We are looking for the probability of the event $A = A_1 \cup A_2$, and we have

$$P(A) = P(A_1) + P(A_2) - P(A_1A_2) = \frac{13}{52} + \frac{12}{52} - \frac{3}{52} = \frac{11}{26}.$$

Example 5.2. We have two urns. In the first urn m_1 white and n_1 black balls are contained. The second urn contains m_2 white and n_2 black balls. From each urn one ball is drawn at random, and then one of these two balls is chosen also at random. What is the probability that this ball will be white?

Solution. *Method 1*: The possible outcomes from the two urns may be symbolized by H_{ww}, H_{wb}, H_{bw} and H_{bb}, where, for example, H_{wb} = {a white ball is drawn from the first urn and a black ball from the second one}. Hence

$$P(H_{ww}) = \frac{m_1m_2}{(m_1 + n_1)(m_2 + n_2)}, \quad P(H_{bw}) = \frac{n_1m_2}{(m_1 + n_1)(m_2 + n_2)},$$

$$P(H_{wb}) = \frac{m_1n_2}{(m_1 + n_1)(m_2 + n_2)}, \quad P(H_{bb}) = \frac{n_1n_2}{(m_1 + n_1)(m_2 + n_2)}.$$

Let A = {the ball chosen from the two drawn is white}. Then

$$P(A|H_{ww}) = 1, \quad P(A|H_{bw}) = P(A|H_{wb}) = \frac{1}{2}, \quad P(A|H_{bb}) = 0,$$

and thus

$$P(A) = \frac{2m_1m_2 + n_1m_2 + m_1n_2}{2(m_1 + n_1)(m_2 + n_2)}.$$

Method 2: The event A = {the ball chosen from the two drawn balls is white} may occur together with one of the following two hypotheses concerning the choice: H_i = {the chosen ball is the one drawn from the ith urn}, where i = 1, 2. We have

$$P(H_1) = P(H_2) = \frac{1}{2}, \quad P(A|H_1) = \frac{m_1}{m_1 + n_1},$$

$$P(A|H_2) = \frac{m_2}{m_2 + n_2}.$$

Thus

$$P(A) = \frac{1}{2}\left(\frac{m_1}{m_1 + n_1} + \frac{m_2}{m_2 + n_2}\right),$$

which coincides with the result obtained by Method 1.

Example 5.3. There is a group of k_1 urns containing m_1 white and n_1 black balls each. There is also another group of k_2 urns each containing m_2 white and n_2 black balls. A ball is drawn at random from a randomly chosen urn among the urns from the two groups. If the ball drawn is white, then what is the probability that an urn from the first group was chosen?

Solution. Let A = {the drawn ball is white} and let H_1 = {an urn from the first group is chosen} and H_2 = {an urn from the second group is chosen}. Then we have

$$P(H_i) = \frac{k_i}{k_1 + k_2} \quad \text{and } P(A|H_i) = \frac{m_i}{m_i + n_i} \quad \text{for } i = 1, 2.$$

Now we calculate the desired probability $P(H_1|A)$ using Bayes' formula:

$$P(H_1|A) = \frac{\frac{k_1}{k_1 + k_2}\frac{m_1}{m_1 + n_1}}{\frac{k_1}{k_1 + k_2}\frac{m_1}{m_1 + n_1} + \frac{k_2}{k_1 + k_2}\frac{m_2}{m_2 + n_2}} =$$

$$= \left(1 + \frac{k_2 m_2 (m_1 + n_1)}{k_1 m_1 (m_2 + n_2)}\right)^{-1} .$$

Example 5.4. Three players A, B and C take turns at the following game: A and B play with each other and then the winner plays with C; the new winner plays with the loser of the first round, etc. All three players are equally skilled and the possibility of ties for the individual games is disregarded, i.e., in each round one of the players wins. The tournament ends when one of the players wins two successive rounds. (a) Describe the space of the possible outcomes of the tournament. (b) Prove that the probability that the tournament will never end is zero while the probability that it will terminate after a finite number of moves is 1. (c) Find the probabilities of each of the players for winning the tournament. (d) If the first round is won by A, what are now the probabilities that B and C will win? (e) Calculate the probability that the tournament will not be finished prior to the nth round.

Solution. (a) If we denote, respectively, by a, b and c the events "a win by player A, B or C in the corresponding round", then the space of the possible outcomes of the tournament is to be described as follows:

$$\begin{cases} \text{aa, acc, acbb, acbaa, acbacc, acbacbb, acbacbaa, ...} \\ \text{bb, bcc, bcaa, bcabb, bcabcc, bcabcaa, bcabcabb, ...} \end{cases} \tag{5.6}$$

$$\text{acbacbacb, ...,} \quad \text{bcabcabca, ...,} \tag{5.7}$$

where in (5.6) the outcomes with which the tournament terminates after a finite number of rounds are given and in (5.7) the two possibilities at which the tournament never ends are described. The space of the elementary outcomes is infinite but denumerable because we can arrange its points on a line. (b) Since the events a, b and c are independent in the corresponding rounds, and each of them has a probability 1/2, the probability of a simple event in (5.6), which may be described, say, by k symbols a, b or c, is $1/2^k$. The probability that the tournament will not terminate is P(acbacb... ∪ bcabca...) = P(acbacb...) + P(bcabca...) = $2 \cdot \frac{1}{2} \cdot \frac{1}{2} \ldots = \lim_{k\to\infty}(1/2)^k = 0$. The probability of the tournament to terminate after a finite number of moves is equal to the sum of the probabilities of the points of (5.6); i.e., it is equal to $2\left(\frac{1}{2^2} + \frac{1}{2^3} + \ldots\right) = 2 \cdot \frac{1}{2^2} \cdot \frac{1}{1 - \frac{1}{2}} = 1$. (c) *Method 1*: The probability that the player C will win the tournament equals the sum of the probabilities of those points of (5.6), which end in two symbols c. From (5.6) it is easily seen that $P(C) = 2\left(\frac{1}{2^3} + \frac{1}{2^6} + \frac{1}{2^9} + \ldots\right) = \frac{2}{7}$. Thus $P(A) = P(B) = \frac{1}{2}(1 - P(C)) = \frac{5}{14}$.

Method 2: Denote A = {A wins the tournament} and A_i = {A wins the ith round}. The symbols B, C, B_j and C_k have an analogous meaning. By the formula for the total probability we find

$$P(A) = P(A_1)P(A|A_1) + P(B_1)P(A|B_1) =$$

$$= \frac{1}{2}(P(A|A_1) + P(A|B_1)) = P(B),$$

$$P(C) = P(A_1C_2)P(C|A_1C_2) + P(B_1C_2)P(C|B_1C_2) =$$

$$= \frac{1}{4}(P(C|A_1C_2) + P(C|B_1C_2)).$$

But $P(C|A_1C_2) = P(C|B_1C_2) = P(A|A_1)$. Therefore, $P(C) = \frac{1}{2}P(A|A_1)$. Further: $P(A|B_1) = P(C_2)P(A|B_1C_2) + P(B_2)P(A|B_1B_2) = P(C_2)P(A|B_1C_2) = P(C_2)(P(A_3)P(A|B_1C_2A_3) + P(C_3)P(A|B_1C_2C_3)) = P(C_2)P(A_3)P(A|B_1C_2A_3) = \frac{1}{4}P(A|A_1)$. Here we used the fact that $P(A|B_1B_2) = P(A|B_1C_2C_3) = 0$ and $P(A|B_1C_2A_3) = P(A|A_1)$. We obtain

$$P(A) = P(B) = \frac{1}{2}(P(A|A_1) + P(A|B_1)) = \frac{5}{8}P(A|A_1),$$

$$P(C) = \frac{1}{2} \cdot P(A|A_1) = \frac{1}{2} \cdot \frac{8}{5}P(A) = \frac{4}{5}P(A).$$

Taking into account that $P(A) + P(B) + P(C) = 1$ we finally get $P(A) = P(B) = \frac{5}{14}$, $P(C) = \frac{2}{7}$. (d) Applying some of the intermediate results of Method 2 in (c), we obtain

$$P(B|A_1) = P(A|B_1) = \frac{1}{4}P(A|A_1) = \frac{1}{4} \cdot \frac{8}{5}P(A) = \frac{2}{5} \cdot \frac{5}{14} = \frac{1}{7},$$

$$P(C|A_1) = P(C) = \frac{2}{7}.$$

(e) Let K_i = {the tournament ends at the ith round}, $i = 2, 3, \ldots, n$ and K = {the tournament has not ended up to the nth round}. Since $K_2 + \ldots + K_n + K = \Omega$ and $P(K_i) = 2/2^i = 1/2^{i-1}$, then

$$P(K) = 1 - P(K_2 + \ldots + K_n) = 1 - P(K_2) - \ldots - P(K_n) =$$

$$= 1 - \frac{1}{2} \cdot \frac{1 - 2^{-(n-1)}}{1 - 2^{-1}} = \frac{1}{2^{n-1}}.$$

Exercises

5.1. A lot contains N items from which n items are chosen for examination. The lot is accepted if, among these n items, there are less than m defective items. Find the probability of accepting the lot if it contains M defective items.

5.2. One of two urns contains five white, eleven black and eight green balls, and the other contains ten white, eight black and six green balls. From each of the two urns one ball is drawn at random. What is the probability the two drawn balls to be of the same colour?

5.3. In a single attempt a sportsman improves his own performance with probability p. What is the probability of improving the performance if two attempts are permitted? (The results of the attempts are assumed to be independent.)

5.4. A participant in a lottery has to fix six different numbers from 1 to 49. Then six balls are drawn at random from an urn containing 49 balls labeled 1, ..., 49. If there are at least three coincidences among the drawn numbers and the numbers fixed by the participant, then he wins some amount a money, which depends on the number of coincidences. What is the probability for a fixed participant to win?

5.5. There are n persons. Independently of the others, each person chooses a number from the set {1, 2, 3, 4, 5, 6, 7, 8, 9}. Let S_n and Π_n be respectively the sum and the product of the numbers chosen. Find the probability of the events $A = \{S_n \geqslant n + 4\}$ and $B = \{\Pi_n$ is divisible by $70\}$.

5.6. Two persons are engaged in a match. For winning the match the first and the second player are to win m and n games, respectively. The first player wins each individual game with probability p, and the second one wins with probability $q = 1 - p$. Find the probability that the first player wins the match.

*5.7. Two players A and B have each placed a levs in a stake. The total of 2a levs is won by the person who first manages to win a predetermined number of individual games. The game is interrupted when A needs to win m games and B needs to win n games. How should they divide the stake of 2a levs if the probability of any of them to win an individual game is 0.5? (*Ch. de Méré*)

5.8. Each of N urns contains m white and n black balls, respectively. One ball is drawn at random from the first urn and is transfered to the second one. After that one ball is drawn from the second urn and is transfered to the third one, etc. What is the probability of drawing a white ball from the last urn?

*5.9. Two balanced dice are thrown. Let us denote by r_i, $i = 2, 3, \dots, 12$, the probability that the total of the faces will be equal to i:

i	2	3	4	5	6	7	8	9	10	11	12
r_i	$\frac{1}{36}$	$\frac{2}{36}$	$\frac{3}{36}$	$\frac{4}{36}$	$\frac{5}{36}$	$\frac{6}{36}$	$\frac{5}{36}$	$\frac{4}{36}$	$\frac{3}{36}$	$\frac{2}{36}$	$\frac{1}{36}$

Prove that there do not exist two unbalanced dice such that $r_2 = r_3 = \dots = r_{12} = r$. (*E. Parzen*)

5.10. Three distinguishable n-faced dice are numbered with the natural numbers 1, 2, ..., n. The three dice are thrown once and S_3 denotes the total of the faces on which the dice have fallen. Prove that, with probability no less than 1/4, the total S_3 is a multiple of 3.

5.11. A number is chosen at random from the sequence of natural numbers. Find the probability that this number will not be divisible: (a) by 2 and by 3; (b) by 2 or by 3.

5.12. A point O is fixed in the interior of a triangle and a line l is drawn at random through it. We attach each vertex to only one side of the triangle, so that if l is passing through a vertex, it crosses only one of the two sides intersecting into that vertex. Let the probability that l will have a common point with the first, second and third side be p_1, p_2 and p_3, respectively. Prove that $p_1 + p_2 + p_3 = 2$.

5.13. An urn contains n balls numbered from 1 up to n. The balls are drawn one by one without replacement. Calculate the probability that at least in one draw the number of the ball drawn will coincide with the number of the draw.

*5.14. Let $A_1, \ldots, A_n$ be arbitrary events with $p_i = P(A_i)$, $p_{ij} = P(A_iA_j)$, $p_{ijk} = P(A_iA_jA_k)$, etc. Put $S_0 = 1$, $S_1 = \sum_{i=1}^{n} p_i$, $S_2 = \sum_{i<j} p_{ij}$, $S_3 = \sum_{i<j<k} p_{ijk}$, ..., $S_n = p_{12\ldots n}$. Let P_r be the probability that exactly r of the events $A_1, \ldots, A_n$ will occur, and Q_r be the probability that at least r of them will occur. Express P_r and Q_r as functions of $S_0, S_1, \ldots, S_n$.

*5.15. (The problem of the absent minded secretary) A secretary has written n letters which she had to send to n persons $M_1, \ldots, M_n$. She inserted the letters into envelopes and sealed them before writing the addresses, so that the letters got mixed up. Nevertheless, she has written an address at random on each envelope and has sent the letters. Find the probability that: (a) every one of the persons $M_1, \ldots, M_n$ will receive his own letter; (b) exactly n - 1 among the persons will receive the letters intended for them; (c) no one will receive his own letter; (d) exactly k from the n persons $M_1, \ldots, M_n$ will receive their own letters.

*5.16. A secretary wrote n letters which she had to send to n persons. Each letter was written in m copies, so she took mn envelopes and wrote m times each of the n addresses. After that she put these mn letters at random into the envelopes and sent them. Let $p_n(m)$ be the probability that no one of the n people will receive the letter intended for him. Find the limit of $p_n(m)$ as $n \to \infty$. (*O. Enchev*)

5.17. In a train, made up of n carriages, k passengers ($k \geqslant n$) choose a carriage at random. Calculate the probability that at least one passenger will get into each carriage.

*5.18. Using probabilistic reasoning prove the following inequality: $(1 - x^m)^n + (1 - (1 - x)^n)^m \geqslant 1$, where $x \in [0, 1]$, and both m and n are natural numbers.

5.19. Prove that

$$P\left(\bigcap_{k=1}^{n} A_k\right) = \sum_{k=1}^{n} P(A_k) - \sum_{k=1}^{n-1} \sum_{j=k+1}^{n} P(A_k \cup A_j) +$$

$$+ \sum_{k=1}^{n-2} \sum_{j=k+1}^{n-1} \sum_{i=j+1}^{n} P(A_k \cup A_j \cup A_i) - \ldots + (-1)^{n-1} P\left(\bigcup_{k=1}^{n} A_k\right).$$

5.20. Let A and B be arbitrary events. Prove that $P(AB) \geqslant 1 - P(\bar{A}) - P(\bar{B})$. *(G. Boole)*

5.21. Consider the events A, B and C. If A and B are independent, A and C are independent and B and C are mutually exclusive, show that the events A and B + C are independent.

5.22. The probabilities of the events A and AB are known. Find the probability of the event $A\bar{B}$.

5.23. Show that for two arbitrary events A and B

$$P(\bar{A}\bar{B}) = 1 - P(A) - P(B) + P(AB).$$

5.24. Prove that if $P(B|A) = P(B|\bar{A})$, then the events A and B are independent.

5.25. Suppose the events A and B are independent. Prove that the following pairs of events are also independent: (a) A and $\bar{B}$; (b) $\bar{A}$ and B; (c) $\bar{A}$ and $\bar{B}$.

5.26. Let A and B be mutually exclusive events. Prove that $P(A) \leqslant P(\bar{B})$.

5.27. If $P(A) = a$ and $P(B) = b$, then $P(A|B) \geqslant (a + b - 1)/b$.

5.28. Find a lower bound for the conditional probability $P(A|B)$ if $P(A) = 0.99$ and $P(B) = 0.5$.

5.29. Consider the events A_1, A_2 and A_3 with $A_1 \subset A_2 \subset A_3$ and $P(A_1) = \frac{1}{4}$, $P(A_2) = \frac{5}{12}$, $P(A_3) = \frac{7}{12}$. Calculate the probabilities of the events $\bar{A}_1A_2$, $\bar{A}_1A_3$, $\bar{A}_2A_3$, $A_1\bar{A}_2\bar{A}_3$, $\bar{A}_1\bar{A}_2\bar{A}_3$.

5.30. Let A_1, A_2, A_3, A_4 be independent events with probabilities p_1, p_2, p_3, p_4, respectively. Find the probabilities of the events: (a) $A_1\bar{A}_3A_4$; (b) $A_1 \cup A_2$; (c) $(A_1 \cup A_2) \cap (\bar{A}_3 \cup \bar{A}_4)$.

5.31. The simultaneous occurrence of the events A_1 and A_2 implies necessarily the occurrence of the event A. Prove that: (a) $P(A) \geqslant P(A_1) + P(A_2) - 1$, where the sign of this inequality remains the same as well in the particular case $A = A_1A_2$; (b) $P(A_2|A_1) \geqslant 1 - P(\bar{A}_2)/P(A_1)$.

5.32. Prove that if $ABC \subset D$, then $P(D) \geqslant P(A) + P(B) + P(C) - 2$.

5.33. Two urns have the following contents: the first has five white and nine black balls and the second has three white and four black balls. An urn is chosen at random and two balls are drawn in succession without replacement. Consider the events A = {at least one of the two balls is white}, B = {at least one of the two balls is black} and C = {the two balls are of different colours}. (a) Do the events A, B and C form a partition of the sample space? (b) Are the events A and C independent? (c) Are the events C and A ∪ B independent? (d) Calculate the conditional probability $P(A|C)$. (e) If we know that two white balls are drawn, find the probability that these balls are drawn from the first urn.

5.34. The events A, B and C have positive probabilities and are such that A, BC and B ∪ C are independent; B and AC are independent; C and AB are independent. Show that the three given events are independent.

5.35. Let $A_1, \ldots, A_n$ be independent events with $P(A_i) > 0$, for $i = 1, \ldots, n$ and $P\left(\bigcup_{i=1}^{n} A_i\right) = 1$. Show that the probability of their simultaneous occurrence does not exceed the number n^{-n}.

*5.36. The events A, B and C are pairwise independent but all three cannot occur at the same time. Furthermore, let $P(A) = P(B) = P(C) = x$. Find the largest possible value of x.

*5.37. (*The problem of the liars*) Player A gets information expressed as "yes" or "no" and lets B know it. In the same manner B delivers it to C and C to D who announces the received information. Every one of the four players tells the truth in one of three cases. What is the probability that A has told the truth if it is known that D has announced a true result?

5.38. In the first of three urns there are two white and four black balls, in the second three white and five black balls, and in the third four white and six black balls. From the first urn, two balls are drawn at random whitout replacement and transferred into the second one. Similarly, two balls are drawn from the second urn and transferred into the third urn. Finally, two balls from the third urn are transferred into the first one. What is the probability that: (a) in all three urns the content of the balls will not be changed in colour; (b) the content in all three urns will be changed?

5.39. Each of $N + 1$ urns contains N balls. The urn numbered k contains k white and $N - k$ black balls, for $k = 0, 1, \ldots, N$. An urn is chosen at random and n balls are selected at random with replacement. What is the probability that all of the selected balls will be white?

5.40. There are n students in a college, and n_k $(k = 1, 2, 3)$ are in the kth year. Two students are selected at random and one turns out to be senior to the other. What is the probability that the senior student will be a third year one?

5.41. All items in one of two lots are non-defective, whiel 1/4 of the items in the other one are defective. A lot is chosen at random and an item selected at random from it is non-defective. Calculate the probability that a second item, taken at random from the same lot, will turn out to be defective, if after being checked, the first item has been replaced into its lot.

*5.42. A box contains n tickets, m of which are winning tickets. Each of n players then successively chooses at random a ticket. What are the chances for winning for each player? When is it more advantageous to chose a ticket?

*5.43. A player knows that one marked ball may be found with probability α in the first of two urns and with probability $1 - \alpha$ in the second one. The probability of drawing the marked ball from the urn in which it is located equals p with $p < 1$. At each move the player chooses one of the two urns and draws a ball from it. If the drawn ball is not a marked one, it is replaced back into the same urn. A player wins if he manages to draw the marked ball no later than the nth move. How many times should the player choose the first urn and how many times the second one so that the probability of his winning the game will be maximal?

5.44. Let $A_1, A_2, \ldots, A_5$ be a partition of the space Ω, A be an event, and let $P(A_j) = j/15$, $P(A|A_j) = (5 - j)/15$, $j = 1, \ldots, 5$. Calculate $P(A_j|A)$, $j = 1, \ldots, 5$. For which j is the probability $P(A_j|A)$ the greatest?

5.45. Consider 13 urns $Y_1, \ldots, Y_{13}$, where Y_i contains i white and 13 - i black balls, $i = 1, \ldots, 13$. An urn is chosen with probability proportional to the number of white balls it contains. From this urn two balls are drawn at random. If they are of different colours, from which one of the urns it is more likely that the two balls were drawn?

5.46. An urn contains n balls, each of which is equally likely to be white or black. From them, k balls are taken at random with replacement. What is the probability that the urn contains only white balls if no black ball appears in the k draws?

5.47. Two players toss a fair coin in succession. Player A starts the game. The player who first succeeds in getting a head is the winner. (a) Describe the sample space Ω and compute the probabilities of the sample points (elementary outcomes). (b) Show that with probability 1 the game terminates after a finite number of moves. (c) What is the probability of each of the players A and B winning?

5.48. Three people are tossing a coin in succession. The first who succeeds in getting a head is the winner. Find the probability that each of the players wins.

5.49. An urn contains n white, m black and 1 green balls which are drawn at random one by one: (a) without replacement; (b) with replacement. Find, in both cases, the probability of the event A = {a white ball is drawn before a black one}.

5.50. (*The gambler's ruin problem*) A coin is tossed in succession. If a tail appears the player wins 1 lev, while if a head appears he loses 1 lev. At the beginning of the game the player has x levs. The game finishes when either the player manages to accumulate a predetermined amount of a levs, or when he has gambled away all his money. What is the probability that the player will be ruined?

6. Urn Models. Polya Urn Model

Introductory Notes

The basic urn model can be described as follows: a known number of identical balls are placed in an urn. They differ only in colour or some other mark which can be determined only after the ball is drawn. When drawn, all the balls that are in the urn have equal chances of getting into the sample. When balls are consecutively drawn from the urn there are two possibilities:

(a) the drawn ball is returned to the urn (sampling with replacement);

(b) the drawn ball is removed (sampling without replacement).

In case (a) the probability for drawing a ball with a definite colour is equal for all the trials. In case (b) this probability depends upon the colour of the balls that were drawn by the previous trials. This basic urn model is an appropriate model for the binomial and the multinomial distribution and of some kind of relationship among the events (see also Exercises 4.5, 4.6, 5.2, 5.3).

For modelling of more sophisticated phenomena the so-called *Polya urn model* is considered. An urn contains w white and b black balls. After each trial the drawn ball is returned into the urn and c balls of this same colour are added, $c > 0$. The Polya urn model is a particular case of the more general *Friedman model* in which balls of both colours are added to the urn: c balls of the same colour as the drawn one and d balls of the other colour. When $c = -1$ and $d = 0$, the model for sampling without replacement (hypergeometrical distribution) is obtained. When $c = -1$ and $d = 1$, the *Ehrenfest model* of heat exchange between two isolated bodies is obtained. In the general case it is difficult to obtain brief expressions for the probabilities of events associated with consecutive trials; however, in the Polya urn model this problem has a simple solution, as can be seen from the examples below.

Illustrative Examples

Example 6.1. In the Polya urn model find the probability that in a sequence of n trials k white and n - k black balls will appear.

Solution. Let us consider the events A_i = {by the ith trial a white ball is drawn}, $i = 1, \ldots, n$. The probability of drawing a white ball in the first k trials and a black one in the remaining ones is

$$P(A_1 \ldots A_k\bar{A}_{k+1} \ldots \bar{A}_n) = P(A_1)P(A_2|A_1)P(A_3|A_1A_2) \ldots$$

$$\ldots P(A_k|A_1A_2 \ldots A_{k-1})P(\bar{A}_{k+1}|A_1A_2 \ldots A_k) \ldots$$

$$\ldots P(\bar{A}_n|A_1 \ldots A_k\bar{A}_{k+1} \ldots \bar{A}_{n-1}) =$$

$$= \frac{w}{w+b} \cdot \frac{w+c}{w+b+c} \cdots \frac{w+(k-1)c}{w+b+(k-1)c} \cdot$$

$$\cdot \frac{b}{w+b+kc} \cdots \frac{b+(n-k-1)c}{w+b+(n-1)c} .$$

Obviously, the result will be the same for probability of drawing a white ball in any k fixed trials (not necessarily the first k trials) and a black ball in the remaining n - k trials. The factors in the numerator and in the denominator will be the same but in a different order. Since k of the n places can be chosen by $\binom{n}{k}$ different ways, the desired probability is

$$p = \binom{n}{k} \frac{w(w+c) \ldots (w+(k-1)c)b(b+c) \ldots (b+(n-k-1)c)}{(w+b)(w+b+c) \ldots (w+b+(n-1)c)} .$$

Example 6.2. An urn contains w white and r red balls. The balls are drawn with replacement until two balls of the same colour appear consecutively. Find the probability that the first two consecutively drawn balls of the same colour are white.

Solution. Let us denote W_i = {the ith ball drawn is white}, $i = 1, 2, \ldots$ Since the drawings are mutually independent, then for any i we have $P(W_i) = \frac{w}{w+r} = p$. Let W be the event the probability of which we seek. It is easy to see that

$$W = W_1W_2 + \overline{W}_1W_2W_3 + W_1\overline{W}_2W_3W_4 + \overline{W}_1W_2W_3W_4W_5 + \dots,$$

and therefore

$$P(W) = p^2 + qp^2 + qp^3 + q^2p^3 + q^2p^4 + \dots = p^2(1 + q)/(1 - pq).$$

The reader should try to find the probability for more than two consecutive appearances of white balls before the same number of red balls appears.

Exercises

6.1. From an urn, containing ten white, seven black and six red balls, one ball is drawn. What is the probability that the ball will be: (a) white; (b) black; (c) red?

6.2. An urn contains eight white and four red balls. Two balls are drawn simultaneously. Which event is more probable: that the two balls are white or of different colour?

6.3. From an urn, containing twelve white and eight red balls, two balls are drawn simultaneously. Find the probability that: (a) both balls are white; (b) both balls are red; (c) the two balls are of different colour.

6.4. In an urn there are w white and b black balls. A ball is drawn with replacement two times in succession. Let us denote by A_1 the event "the first drawn ball is white" and by B_2 the event "the second drawn ball is black". Find the probabilities of the events: (a) $A_1\overline{B}_2$; (b) $\overline{A}_1B_2$; (c) $\overline{A}_1\overline{B}_2$.

6.5. From an urn containing 2k white and 2l black balls, k + l balls are drawn simultaneously. What is the probability that k white and l black balls have remained in the urn?

6.6. In an urn there are w_1 white and r_1 red balls, and in another one w_2 white and b_2 black balls. One ball is drawn from the first urn and is then placed into the second urn, without looking at its colour; after this a ball is drawn from the second urn. Find the probability that the ball drawn from the second urn is white.

6.7. In an urn there are w white and b black balls. Two players draw consecutively one ball each, and the drawn ball is eliminated. The winner is the one to draw a white ball first. What is the probability of winning for the player who starts the game first?

6.8. By choosing an appropriate urn model, prove the identity

$$1 + \frac{a - b}{a - 1} + \frac{(a - b)(a - b - 1)}{(a - 1)(a - 2)} + \dots + \frac{(a - b)(a - b - 1)\dots 2 \cdot 1}{(a - 1)(a - 2)\dots(b + 1)b} = \frac{a}{b},$$

where a and b are arbitrary natural numbers such that $a > b$.

6.9. In the Polya urn model find the probability that: (a) in two consecutive trials balls of different colours will be drawn; (b) in three consecutive trials all the three balls will be of the same colour.

6.10. In the Polya urn model find the probability that the first ball was black conditional on the second drawn ball being black.

*6.11. In the Polya urn model find the probability of a white ball appearing on the nth trial. (*K. Jordan*)

*6.12. In the Polya urn model prove that for m and n arbitrary, $m < n$, the probability that: (a) a white ball appears on the mth trial and a black one on the nth is $wb/((w + b)(w + b + c))$; (b) white balls appear on the mth and on the nth trial is $w(w + c)/((w + b)(w + b + c))$.

*6.13. In the Polya urn model let A_m and A_n $(m < n)$ denote the events "drawing of a white ball" respectively on the mth and nth trial. Prove the following equalities: (a) $P(A_n|A_m) = P(A_m|A_n)$; (b) $P(\bar{A}_n|\bar{A}_m) = P(\bar{A}_m|\bar{A}_n)$; (c) $P(\bar{A}_n|A_m) = P(\bar{A}_m|A_n)$; (d) $P(A_n|\bar{A}_m) = P(A_m|\bar{A}_n)$.

6.14. Consider two urns and 20 balls of which ten are white and ten are black. How should these balls be distributed in the two urns by number and by colour so that if one of the urns is chosen at random and one ball is drawn from it, the probability that this ball is white is maximized?

6.15. In an urn there are n white balls, numbered with the integers from one to n, and n black balls, also numbered with the integers from one to n. From the urn 2m balls are drawn simultaneously. Find the probability that among the drawn balls there are exactly k pairs numbered with one and the same integer. On the basis of the result obtained calculate the sum

$$S = \sum_{k=0}^{m} \binom{n-k}{2m-2k} \binom{n}{k} 2^{2m-2k}.$$

*6.16. An urn contains w white and b black balls. The balls are drawn consecutively without replacemend. Find the probability that the number of the white balls drawn will be equal to the number of the black ones drawn at least once in the sequence of draws.

*6.17. An urn contains n balls. All possible hypotheses concerning the number of the white balls are equally likely. One white ball is dropped into the urn and a ball is then selected at random. What is the probability that this ball will be white?

7. Geometric Probability

Introductory Notes

Consider the sample space Ω of random outcomes which is a set of geometrical objects, e.g. points, straight lines, planes. The determination of probabilities associated with such sample spaces are referred to as problems in geometric probability.

Let X be a set of geometrical objects. Then for any subset $\Omega \subset X$ define a function $\mu(\Omega)$ satisfying the conditions $\mu(\Omega) \geqslant 0$ and $\mu(\Omega_1 + \Omega_2) = \mu(\Omega_1) + \mu(\Omega_2)$ if $\Omega_1\Omega_2 = \emptyset$. Any such function is called an *additive measure*.

Let Ω have a finite measure; i.e., $\mu(\Omega) < \infty$. For an arbitrary subset $A \subset \Omega$ a function is defined by means of the formula

$$P(A) = \frac{\mu(A)}{\mu(\Omega)}, \tag{7.1}$$

which possesses the properties (3.5)-(3.7) that are valid for the classical probability. The function $P(A)$, $A \subset \Omega$, thus defined, is called *geometric probability* in the set Ω. In the particular case, if X is the real line $\mathbb{R}_1$, Ω is some finite interval and μ is the Lebesgue measure on $\mathbb{R}_1$ (in this case the value of μ for an interval $[a, b]$, $a < b$, is equal to the length $b - a$ of this interval), then with the formula (7.1) geometric probability is defined over the interval $[a, b]$. This probability is known as the uniform distribution on $[a, b]$. If Ω is a domain from the plane $\mathbb{R}_2$ and $\mu(A)$ for $A \subset \Omega$ is defined as a surface of A with $\mu(\Omega) \leqslant \infty$, then by means of the formula (7.1) geometric probability is defined in the domain $\Omega \subset \mathbb{R}_2$. This probability is known as the uniform distribution of a point in Ω. Analogously by means of the formula (7.1) geometric probability is defined in domains from an arbitrary Euclidean space $\mathbb{R}_n$, $n = 3, 4, \ldots$

In some geometric-probability problems it is possible that the elementary events are not points but are other geometrical objects such as segments or planes (see Exercise 7.1). Such problems can be reduced to other problems, associated with the random choice of a point from an arbitrary domain in Euclidean space of corresponding dimension. When several possible parametric representations of the elements of Ω exist, the initial problem may usually reduce to different "random choice of a point" and then different answers are obtained (see for instance Exercise 7.1). This means that the condition of the problem is not completely specified.

When the geometrical objects are points, straight lines or planes in an Euclidean space, unique to within a constant, the so-called kinematic measure is invariant with respect to rotation and translation of the sets under consideration. The problems, in which the probability is defined by this measure, and the statistical applications associated with them, form a part of stochastic geometry known as stereorology. For instance, if Ω is a set of points in the plane, the kinematic measure coincides with the Lebesque measure of the set of the polar coordinates (ρ, ϕ) of the lines, intersecting Ω with respect to an arbitrary point from the plane. In this case the probability $P(A)$ or $A \subset \Omega$, defined by (7.1), does not depend upon the location and orientation of A (see Exercises 7.31, 7.32, 7.34).

Geometric-probability problems, in which $\mu(\Omega) = \infty$, are solved by means of passage to the limit. First, some subset Ω_N is considered, where $\mu(\Omega_N) < \infty$; the desired probabilities are calculated and after that their limit when $\Omega_N \uparrow \Omega$ is obtained.

Some geometric-probability problems have the following form: n points in some domain Ω of an Euclidean space are chosen randomly and independently one from the other. The probability P is sought such that the aggregate of these n points will possess a given property, depending only upon the mutual position of the points in Ω. One such a property might be that the largest of the distances among the points should not exceed a fixed value.

Suppose the domain Ω can vary and is a function of some parameter α. In this case the probability P, considered as a function of α, satisfies the relation

$$dP = n(P_1 - P)\frac{d\mu(\Omega)}{\mu(\Omega)} \tag{7.2}$$

(*Crofton's formula*).

In (7.2) $\mu(\Omega)$ is the Lebesgue measure of Ω (length, surface, volume, etc.); P_1 is the probability for the n points to possess the given property, when one of the points is chosen on the contour of Ω and the remaining $n - 1$ points are chosen at random in Ω.

When solving problems by means of Crofton's formula, the probability P_1 is found first and then the required probability is found as a solution of the differential equation (7.2) (see Example 7.2 and Exercise 7.30).

Illustrative Examples

Example 7.1. From the inside of a regular hexagon with side a a point is chosen at random. What is the probability that the distance from this point to the centre of the hexagon will not exceed x, where $0 < x < a\sqrt{3}/2$?

Solution. The surface of the hexagon is $3a^2\sqrt{3}/2$. The distance from the points to the centre of the hexagon will not exceed x, if the point lies in the circle with radius x and the same centre. The surface of this circle is πx^2 and therefore the desired probability is $2\pi x^2\sqrt{3}/(9a^2)$.

Example 7.2. On a segment of length s two points are chosen at random. What is the probability P that the distance between them will not exceed x, where $0 \leqslant x \leqslant s$?

Solution. Apply Crofton's formula (7.2). Let one end of the segment be fixed and the other be variable. Take the length s as a parameter. Here $n = 2$, and P_1 is the probability that the distance between the points will not exceed x, when one of them is at the end of the segment. It is evident that $P_1 = x/s$. According to (7.2), the probability P satisfies the equation

$$dP = 2\left(\frac{x}{s} - P\right)\frac{ds}{s} .$$

The solution of this equation is $s^2P = 2sx + c$, c = constant. From the boundary condition for $s = x$, $P = 1$, we find for the constant $c = -x^2$. Hence $P = (2sx - x^2)/s^2$, $0 \leqslant x \leqslant s$.

Exercises

7.1. A chord is drawn at random in a circle with radius R. Find the probability that its length will exceed the side of the inscribed regular triangle. (*Bertrand's paradox*)

7.2. Let t_x and t_y defined on the interval $[0, T]$ be the times of occurrence of two independent events X and Y. The events are said to coincide if $-a \leqslant t_x - t_y \leqslant b$ $(a > 0, b > 0)$. Find the probability that: (a) X will occur before Y; (b) X and Y will coincide; (c) Y will occur before X conditional on X and Y are coinciding.

7.3. A point is chosen at random from the unit square $0 \leqslant x \leqslant 1$, $0 \leqslant y \leqslant 1$. Find the probability that: (a) $x \geqslant 1/2$ when $x + y \geqslant 1/3$; (b) $x \geqslant y$ when $xy \geqslant 1/4$; (c) $y \leqslant 1/2$ when $x = 1/2$; (d) $x^2 + y^2 \leqslant 1/4$ when $xy \leqslant 1/16$.

7.4. On the line segment AB three points are chosen randomly and independently one from the other. Find the probability that a triangle can be constructed from the three line segments equal to the distances from A to these points.

7.5. What is the probability that the roots of the quadratic equation $x^2 + 2ax + b = 0$ will be real, if the values of the coefficients are uniform over the rectangle $-k \leqslant a \leqslant k$, $-s \leqslant b \leqslant s$?

7.6. From a line segment of length one two points are chosen at random. What is the probability that none of the three parts of the segment will be less than a, where $0 \leqslant a \leqslant 1/3$?

7.7. On a line segment two points are chosen randomly to divide it into three parts. What is the probability that a triangle can be constructed from these three parts?

7.8. In a horizontal plane two sheaves of parallel lines are drawn which divide it into rectangles with sides a and b $(a \leqslant b)$. A coin with diameter $2r < a$ is thrown at random on the plane. Find the probability that the coin will intersect none of the lines.

7.9. On a parquet, formed from equilateral triangles with a side a, a coin with radius r $(r < a\sqrt{3}/6)$ is thrown at random. Find the probability that the coin will intersect the contour of none of the triangles.

7.10. From a line segment of length s a point is chosen at random that divides it into two parts. What is the probability that from the two parts obtained and from a line segment of length s/2 a triangle can be constructed?

7.11. An infinite number of parallel lines are drawn in a plane with the distances between the lines being 3 and 16 cm alternatively. What is the probability that a circle with radius 5 cm, thrown at random on the plane, will intersect none of the lines?

7.12. A circle revolves with a constant velocity. A line segment of length 2h is placed in front of the circle in its plane in such a way that the end points of the line segment are equidistant from the centre of the circle. In an arbitrary moment a particle flies from the circumference tangent to it. Find the probability that the particle will hit the line segment if the radius of the circle is R and the middle of the line segment is at a distance s from the centre of the circle.

7.13. Two points are chosen at random from a line segment of length one. What is the probability that the distance between them will be less than x for $0 < x < 1$?

7.14. Two points L and M are chosen randomly from the line segment AB of length one. Find the probability that L will be closer to M than to A.

7.15. Three points A, B and C are chosen at random on a circumference. What is the probability that the triangle ABC will be acute-angled?

7.16. Parallel lines at a distance 2L from one another are drawn on a horizontal plane. A needle of length 2s, $s < L$, is thrown at random onto the plane. Find the probability of the needle intersecting a line. (*Buffon's needle problem*)

7.17. The plane is covered with rectangles with sides a and b. A needle of length s, $s < \min[a, b]$, is thrown at random on the plane. Find the probability that the needle does not intersect any side of a rectangle. (*P.-S. Laplace*)

7.18. Find the probability that the three roots of the cubic equation $x^3 - 3ax + 2b = 0$ will be real if all values of the coefficients are uniform in the rectangle: $|a| \leqslant k$, $|b| \leqslant s$.

7.19. The points A, B and C are arranged in this order on a straight line. A point M is chosen at random from the segment AB of length a; a point N is chosen at random from the segment BC of length b, where $a < b$. Find the probability that a triangle can be constructed from the segments AM, MN and NC.

7.20. Let a and b, where $a < b$, be the lengths of two of the sides of a triangle. The length of the third side is determined by choosing a point at random from the interval $(a - b, a + b)$. What is the probability that the triangle obtained will be acute-angled?

7.21. Suppose N points are distributed randomly and independently one from the other in a ball with radius R. Find the probability that the distance from the centre of the ball to the nearest point will be no less than a, $0 < a < R$. What is the limit of this probability when $R \to \infty$ and $N/R^3 \to 4\pi\lambda/3$? (*A problem from the astronomy*)

7.22. The event A is equally likely to occur at any time during the interval $[0, T]$. The probability that A occurs at all during this interval is p. It is known that up to the moment t, $0 < t < T$, the event A has not occurred. Calculate the probability that A will occur during the interval $[t, T]$.

7.23. A metal wire of length 20 cm is bent at a randomly chosen point. After this a rectangular frame is made by bending the larger part at two more points. Find the probability that the area of the frame obtained does not exceed 21 cm^2.

7.24. A square is inscribed into a circle. Find the probability that: (a) a point chosen at random in the circle will lie in the square; (b) if five points are chosen at random in the circle, one will lie in the circle and one in each of the four segments outside of the square.

7.25. A geographical coordinate network is plotted over a ball. The ball is thrown on a plane. Find the probability that the point of the first contact of the ball with the plane will be: (a) between 0° and 90° eastern longitude; (b) between 45° and 90° northern latitude; (c) in the common part of the regions in (a) and (b).

*7.26. The points A and B are chosen randomly and independently from one another from the inside of a circle. Find the probability that the circumference of the circle with centre A and radius AB will lie inside of the circle.

*7.27. The points A and B are chosen randomly from the inside of a ball. Find the probability that the sphere with centre A and radius AB will lie inside of the ball.

*7.28. Three points are chosen at random on a circumference. Find the probability that an arc of size α radians exists on which all the three points lie.

*7.29. Countable points are chosen at random in the interval (0, 1). Prove that the event A = {each subinterval of (0, 1) contains at least one of the random points} has a probability P(A) = 1. (In this case we say that the randomly chosen points are placed with probability one in the interval (0, 1) dense everywhere.)

*7.30. Two points A and B are chosen randomly and independently one from the other on the surface of a hemisphere with radius R. Let us denote by s_{AB} the length of the arc of the large circumference which passes through these two points. Prove that $P\{s_{AB} < 2R \arccos(1/\sqrt{2})\} = 1/2$. (*A. Obretenov*)

7.31. Let $K(O, R)$ and $K_1(O, R_1)$ be two concentric circumferences in the plane, and let $R_1 < R$. A random straight line from the plane intersects K. Find the probability that this line intersects K_1 as well.

7.32. Consider a square Q in the plane; the midpoints of the sides of Q are the vertices of the square Q_1. A random straight line from the plane intersects the square Q. Find the probability that this line intersects Q_1 as well.

7.33. A circumference K(O, R) and one of the diameters CD of a circle are given in the plane. A random straight line from the plane intersects the circumference K. What is the probability that this line will intersect the diameter CD?

7.34. A circumference and one of the radii of a circle are given in the plane. Find the probability that a random straight line which intersects the circumference will intersect the radius as well.

*7.35. Consider n points chosen randomly and independently from one another on a segment of length s. For the event A = {the distance between each two of the n points is no less than d} prove that

$$P(A) = (1 - (n - 1)\, d/s)^n, \qquad 0 \leqslant d \leqslant s/(n - 1).$$

8. Bernoulli Trials: Binomial and Multinomial Distributions

Introductory Notes

A sequence of trials are called the *Bernoulli trials* if the following conditions are satisfied: (1) there are two possible outcomes, say A and $\bar{A}$, for each trial; (2) the trials are independent; i.e., the result of any trial does not influence the result of the remaining trials; (3) the probability P(A) remains the same throughout all trials.

It is convenient to call the events A and $\bar{A}$ a *success* and a *failure*, respectively, and denote their probabilities by $p = P(A)$ and $q = P(\bar{A})$ where $p + q = 1$.

The Bernoulli scheme (n, p) is defined as a sequence of n Bernoulli trials with probability of success equal to p. Let us denote by μ_n the *number of successes* in the Bernoulli scheme (n, p). Then the probability $P_n(k)$ that exactly k successes will occur, is calculated by the formula

$$P_n(k) = P\{\mu_n = k\} = \binom{n}{k} p^k q^{n-k} \quad \text{for } k = 0, 1, \ldots, n. \tag{8.1}$$

The set of numbers $\{P_n(k),\ k = 0, 1, \ldots, n\}$ is called a *binomial distribution*. According to Newton's formula, the probability $P_n(k)$ is equal to the coefficient s^k in the expansion of the function $(ps + q)^n$. The function $g(s) = (ps + q)^n$ is said to be the *probability generating function* (p.g.f.) of the binomial distribution.

We shall consider also such sequences of trials which satisfy conditions (1) and (2) for the Bernoulli scheme, but for which the probability of success changes from trial to trial. For such a sequence of trials, let us denote by p_i the probability of success at the ith trial. Then the probability $\tilde{P}_n(k)$ of k successes in n trials is equal to the coefficient of s^k in the expansion of the corresponding p.g.f. g(s) which is given by

$$g(s) = \prod_{i=1}^{n} (p_i s + q_i) = \sum_{k=0}^{n} \tilde{P}_n(k) s^k \tag{8.2}$$

where $p_i + q_i = 1$ for $i = 1, \ldots, n$.

The probabilities $\tilde{P}_n(k)$ can be calculated either directly from (8.2) or by the relation

$$\tilde{P}_n(k) = \frac{1}{k!} \left. \frac{d^k g(s)}{ds^k} \right|_{s = 0} . \tag{8.3}$$

If the numbers of successes k is varying from 0 to n, then the binomial probability $P_n(k)$ first increases and then decreases, reaching its greatest value $P_n(m)$, where

$$m = [(n + 1)p]. \tag{8.4}$$

Here [x] denotes the function "the integer part of x". The integer m from (8.4) is called the *most probable number of successes*. It satisfies the inequalities

$$(n + 1)p - 1 \leqslant m \leqslant (n + 1)p. \tag{8.5}$$

If (n + 1)p is an integer, then there are two values for the most probable number of successes: $m_1 = (n + 1)p$ and $m_2 = (n + 1)p - 1$. In such cases we have $P_n(m_1) = P_n(m_2)$.

Now suppose each of the n independent trials can result in one of r outcomes, say $A_1, A_2, \ldots, A_r$, with probabilities $p_1, p_2, \ldots, p_r$, respectively, where $\sum_{i=1}^{r} p_i = 1$. Then the probability that the event A_1 will occur exactly k_1 times, the events A_2 will occur exactly k_2 times, ..., the event A_r will occur exactly k_r times, $k_1 + k_2 + \ldots + k_r = n$, is

$$P_n(k_1, k_2, \ldots, k_r) = \frac{n!}{k_1!k_2! \ldots k_r!} p_1^{k_1} p_2^{k_2} \ldots p_r^{k_r}. \qquad (8.6)$$

The *multinomial distribution* is defined as the set of the numbers (8.6) where $k_1, k_2, \ldots, k_r$ are arbitrary non-negative integers with $k_1 + k_2 + \ldots + k_r = n$. The probability (8.6) is equal to the coefficient $s_1^{k_1} s_2^{k_2} \ldots s_r^{k_r}$ in the expansion of the corresponding p.g.f. which in that case has the form

$$g(s_1, s_2, \ldots, s_r) = (p_1 s_1 + p_2 s_2 + \ldots + p_r s_r)^n. \qquad (8.7)$$

The use of the p.g.f.'s (8.2) and (8.7) facilitates the calculation of the required probability. For example, for m + n independent trials the corresponding p.g.f.'s satisfy the relation $g_{m+n}(s) = g_m(s)g_n(s)$.

The solution of some rather complicated combinatorial problems can be easily obtained if in (8.7) an appropriate change of the variables is made. The following two cases serve as a good illustration: (i) Consider n independent trials with r possible outcomes at each trial: A_1, $A_2, \ldots, A_r$, and with corresponding probabilities $p_1, p_2, \ldots, p_r$. Suppose we are interested in the probability that the number of times A_1 occurs is exceeding the number of times A_2 occurs by exactly k. In (8.7) we put $s_1 = s$, $s_2 = s^{-1}$ and $s_3 = s_4 = \ldots = s_r = 1$. Then the desired probability is equal to the coefficient of s^k in the expansion of the p.g.f.

$$g(s) = \left(p_1 s + \frac{p_2}{s} + \sum_{i=3}^{r} p_i\right)^n. \qquad (8.8)$$

(ii) Here n independent trials are carried out, where the possible outcomes at each trial are $A_1, A_2, \ldots, A_r$ and their probabilities are $p_1 = p_2 = \ldots = p_r = 1/r$. Suppose we wish to find the probability that the sum of the numbers of the occurred events in these n trials will be equal to a given number k. If in (8.7) we put $s_i = s^i$ for $i = 1, 2, \ldots, r$, then the desired probability is the coefficient of s^k in the expansion of the p.g.f.

$$g(s) = \frac{s^n}{r^n}\left(\frac{1 - s^r}{1 - s}\right)^n, \quad |s| < 1. \tag{8.9}$$

Using the Newton's binomial formula for $(1 - s^r)^n$ and for $(1 - s)^{-n}$ we can find that the desired probability, say p_k, is given by:

$$p_k = \frac{1}{r^n} \sum_{j=0}^{[(k-n)/r]} (-1)^j \binom{n}{j} \binom{k - 1 - jr}{k - n - jr}.$$

Illustrative Examples

Example 8.1. Assume that the probabilities for a birth of a boy and of a girl are both equal to 0.5. Calculate the probability that in a family with six children there will be: (a) three boys; (b) no less than one and no more than five boys.

Solution. (a) Let A_k = {exactly k successes occur in a Bernoulli scheme $(6, \frac{1}{2})$}, $k = 0, 1, \ldots, n$. Then

$$P(A_3) = P_6(3) = \binom{6}{3}\left(\frac{1}{2}\right)^3\left(\frac{1}{2}\right)^3 = \frac{5}{16} = 0.3125.$$

(b) Since the events A_k, $k = 0, 1, \ldots, n$, are mutually exclusive, then the event B we are interested in, has probability

$$P(B) = 1 - P(\bar{B}) = 1 - P(A_0 + A_6) = 1 - P(A_0) - P(A_6) =$$

$$= 1 - \binom{6}{0}\frac{1}{2^6} - \binom{6}{6}\frac{1}{2^6} = \frac{31}{32} = 0.96875.$$

Example 8.2. How many numbers must be chosen from a table of random numbers so that the probability that exactly three of them will end with the digit seven is maximized? (See the note after Exercise 4.19.)

Solution. The most probable number in a Bernoulli scheme (n, 0.1) is $m = [(n + 1) \cdot 0.1]$. We want to find the smallest n such that $[(n + 1) \cdot 0.1] = 3$. From inequalities (8.5) we obtain $(n + 1) \cdot 0.1 - 1 \leqslant m \leqslant (n + 1) \cdot 0.1$; i.e., $n + 1 - 10 \leqslant 30$ and $n + 1 \geqslant 30$. The solutions of these inequalities are the integers $29 \leqslant n \leqslant 39$. Therefore, the smallest number with the required property is $n = 29$.

Example 8.3. An urn contains three balls which are identical but different in colour: white, green and red. Five times we draw a ball with replacement. What is the probability that the white and the red ball will be drawn at least twice each?

Solution. If we denote the probabilities of drawing a white, green and red ball by p_1, p_2 and p_3, respectively, then $p_1 = p_2 = p_3 = \frac{1}{3}$. Let A_{ijk} = {i white, j green and k red balls are drawn}, where $i, j, k = 0, 1, \ldots, 5$. By formula (8.6) we obtain

$$P(A_{212} + A_{302} + A_{203}) = P_5(2, 1, 2) + P_5(3, 0, 2) +$$

$$+ P_5(2, 0, 3) = \frac{5!}{2!1!2!} \cdot \frac{1}{3^5} + \frac{5!}{3!0!2!} \cdot \frac{1}{3^5} +$$

$$+ \frac{5!}{2!0!3!} \cdot \frac{1}{3^5} = \frac{50}{243} .$$

Example 8.4. The probabilities that the first, second and third bulbs will burn out are 0.1, 0.2 and 0.3, respectively. The probabilities that a device fails when one, two and three bulbs have, respectively, burnt out are 0.25, 0.6 and 0.9. Calculate the probability that the device fails.

Solution. Let A = {the device fails}, A_i = {bulb number i burns out}, i = 1, 2, 3 and H_k = {exactly k bulbs burnt out}, k = 0, 1, 2, 3. Then

$$H_0 = \bar{A}_1\bar{A}_2\bar{A}_3, \quad H_1 = A_1\bar{A}_2\bar{A}_3 + \bar{A}_1A_2\bar{A}_3 + \bar{A}_1\bar{A}_2A_3,$$

$$H_2 = A_1A_2\bar{A}_3 + A_1\bar{A}_2A_3 + \bar{A}_1A_2A_3;$$

hence we can calculate the probabilities of each of the hypotheses H_k. We can obtain these probabilities more easily using the p.g.f. given by (8.2). To use this formulation we shall interprete the failing of the first, second and third bulbs as three independent trials with varying probability of success at each trial: $p_1 = 0.1$, $p_2 = 0.2$ and $p_3 = 0.3$. According to formula (8.2) we obtain $g(s) = (0.1s + 0.9)(0.2s + 0.8)(0.3s + 0.7) = 10^{-3}(2s^2 + 26s + 72)(3s + 7) = 10^{-3}(6s^3 + 92s^2 + 398s + 504)$. Hence $P(H_3) = 0.006$, $P(H_2) = 0.092$, $P(H_1) = 0.398$ and $P(H_0) = 0.504$. By assumption, $P(A|H_1) = 0.25$, $P(A|H_2) = 0.6$ and $P(A|H_3) = 0.9$. Hence we find $P(A) = 0.398 \times 0.25 + 0.092 \times 0.6 + 0.006 \times 0.9 = 0.1601$.

Example 8.5. Each of two players tosses a fair coin n times. Calculate the probability that both will get the same number of heads.

Solution. *Method 1*: Let A_{ik} = {the ith player gets k heads}, where i = 1, 2, k = 0, 1, ..., n. The events A_{1i} and A_{2i} are independent. The aim is to find the probability of the event $A = \bigcup_{k=0}^{n} A_{1k}A_{2k}$. Thus we get

$$P(A) = \sum_{k=0}^{n} P(A_{1k})P(A_{2k}) = \sum_{k=0}^{n} \left[\binom{n}{k} \frac{1}{2^k} \frac{1}{2^{n-k}}\right]^2 =$$

$$= (1/2^{2n}) \sum_{k=0}^{n} \binom{n}{k}^2 = \binom{2n}{n}/4^n .$$

Method 2: The p.g.f. of the probabilities for the numbers of the appearances of a head for the first and second players, respectively, are $g(s_1) = (0.5s_1 + 0.5)^n = 2^{-n}(s_1 + 1)^n$ and $g_2(s_2) = 2^{-n}(s_2 + 1)^n$. The p.g.f. of the compound experiment consisting of two independent sequences of n trials each is $g(s_1, s_2) = 2^{-2n}(s_1 + 1)^n(s_2 + 1)^n = 4^{-n}(s_1s_2 + s_1 + s_2 + 1)^n$. Since we are interested in the probability that both players will get an equal number of heads, we have to substitute $s_1 = s$ and $s_2 = 1/s$ in $g(s_1, s_2)$. Thus, we obtain $g(s) = 4^{-n}(1 + s + s^{-1} + 1)^n = 4^{-n}s^{-n}(1 + s)^{2n}$. The desired probability is equal to the coefficient of s^0 (the constant term) in the expansion of g(s) in powers of s; i.e., $\binom{2n}{n}/4^n$.

Exercises

8.1. In a bridge game using 52 cards, one of the four players did not receive a single ace in three consecutive hands. Does he have a reason to complain of ill luck?

8.2. From a deck of 52 cards 13 cards are drawn at random with replacement. (a) What is the probability that two of them will be red? (b) Find the same probability when the cards are drawn without replacement.

8.3. In a library there are only technical and mathematical books. Each reader chooses a mathematical book with probability 0.7 and a technical one with probability 0.3. He is allowed to choose only one book at a time. Find the probability that five readers in succession will choose either only mathematical books or only technical ones.

8.4. When playing with an adversary of equal skill (if ties are not possible), which is more probable: (a) winning three out of four games or five out of eight; (b) winning no less than three out of four or no less than five out of eight games?

8.5. For a Bernoulli scheme (n, p) with $p = \frac{1}{2}$, prove that

$$\frac{1}{2\sqrt{n}} < P_{2n}(n) < \frac{1}{\sqrt{2n + 1}} .$$

8.6. Four radio signals are sent in succession. The probabilities of receiving each separate signal do not depend on whether the remaining signals are received, and are 0.1, 0.2, 0.3 and 0.4, respectively. (a) Calculate the probability of receiving k signals for k = 0, 1, 2, 3, 4. (b) Find the probability of establishing a two-way radio communication if the probabilities for the occurrence of such an event, given that one, two, three or four signals are received, are 0.2, 0.6, 1 and 1.

*8.7. There is a sequence of n independent trials with probability p_k of a success at the kth trial, where k = 1, 2, ..., n. Consider the functions

$$g_1(s) = \left[1 - \prod_{k=1}^{n} (p_k s + q_k)\right](1 - s)^{-1},$$

$$g_2(s) = \left[\prod_{k=1}^{n} (p_k s + q_k) - s^{n+1}\right](1 - s)^{-1}.$$

Prove that: (a) the coefficient of s^m in $g_1(s)$ is equal to the probability of the occurrence of more than m successes for $m = 0, 1, \ldots, n - 1$; (b) the coefficient of s^m in $g_2(s)$ is equal to the probability of the occurrence of no more than m successes for $m = 0, 1, \ldots, n$.

*8.8. Consider a Bernoulli scheme (n, p). Prove that the coefficient of s^k in the function $g(s) = [1 - (ps + q)^n](1 - s)^{-1}$ is equal to the probability of the occurrence of no more than k successes, where $k = 0, 1, \ldots, n - 1$.

8.9. In testing the reliability of certain devices, the probability of a failure of each device is 0.2. How many devices should be checked so as the probability of the occurrence of at least three failures to be less than 0.9?

8.10. A telephone exchange A has to be connected with ten subscribers of a telephone exchange B. Every subscriber occupies the trunk-line 12 minutes per hour. The calls of any two subscribers are independent. In order that all the demands will be met with probability at least 0.99, what is the minimal number of lines that are necessary?

8.11. Measurements are made with either positive or negative errors. The probabilities associated with these measurements are 2/3 and 1/3, respectively. If four measurements are made, find the most probable number of positive and negative errors and then calculate the corresponding probabilities.

8.12. How many Bernoulli trials are necessary so that the most probable number of successes will be 51 if the probability of success is $p = 0.64$?

8.13. The most probable number of the good items in a lot of 90 items is equal to 82. What is the probability that one item of the lot is good?

*8.14. Let P be the maximal probability in a Bernoulli scheme (n, p) and P' be the maximal probability in a Bernoulli scheme (n + 1, p). Prove that $P' \leqslant P$. When will the equality be attained?

8.15. Two basketball players have to shoot three shots each from the free-throw line. The probability of making a basket at each shot is 0.6 for one of them and 0.7 for the other. Compute the probability that: (a) both players will make equal number of baskets; (b) the first will have more baskets than the second one.

8.16. In a sequence of Bernoulli trials with the probability of success, given by p, find the probability that the rth success will occur exactly at the (k + r)th trial, $k = 0, 1, 2, \ldots$

*8.17. A certain smoker always carries two match-boxes in the pocket. When he wants to light a cigarette, he takes a match from one of the two boxes chosen at random. After some time, he again chooses one of the two boxes and he finds that it is empty. What is the probability that the other box will contain at that moment k matches if at the beginning both boxes did contain n matches each? (*S. Banach*)

8.18. In a sequence of Bernoulli trials with probability p of success, find the probability that a successes will occur before b failures.

8.19. Consider a sequence of Bernoulli trials with probability of success p. Suppose that it is known that the first two trials were failures. What is the conditional probability that the first success will occur after the fifth trial but before the tenth one?

8.20. Consider an unbalanced coin such that "head" and "tail" occur with probabilities p and 1 - p, respectively. The coin is tossed n times. Let A = {at the first toss a head occurs}, A_k = {in n trials exactly k heads appear}. For what relation between n, p and k are the events A and A_k independent?

8.21. The probability of putting a certain aggregate under repairs, after m damages, should be computed by the formula $G(m) = 1 - (1 - 1/c)^m$, where c is the average number of the damages which have happened before the repairs. Find the probability of putting the aggregate under repairs after n production cycles if during one cycle there occurs at most one damage with probability p for occurrence.

8.22. Before a certain trial there are two equally probable and exhaustive hypotheses concerning the probability of success at one trial: $H_1 : p = 1/2$ and $H_2 : p = 2/3$. Which one of the two hypotheses has a greater a posteriori probability if 116 successes have occurred in 200 trials?

8.23. An urn contains k white, m green and n red balls. Let the $k_1 + m_1 + n_1$ balls be drawn one by one. Calculate the probability of drawing: (a) first l_1 white, then m_1 green and finally n_1 red balls; (b) k_1 white, m_1 green and n_1 red balls in such a way that the balls of the same colour appear consecutively, the succession of the colours being unspecified; (c) k_1 white, m_1 green and n_1 red balls in any order.

8.24. A factory worker produces: a non-defective item with probability 0.9, an item with a reparable defect with probability 0.09 and an item with an irreparable defect with probability 0.01. Three items are produced. Find the probability that at least one of them will be non-defective and at least one with reparable defect.

8.25. A lighting device has four blocks with electronic tubes. If it is known that one tube is damaged, the probabilities that this tube belongs to a given block are $p_1 = 0.6111$, $p_2 = p_3 = 0.0664$ and $p_4 = 0.2561$, and they do not depend on how many tubes have been damaged up to that time. Compute the probability that the device will fail when four tubes are damaged, if it is known that the device fails when either at least one tube of the first block is damaged or at least one tube of each one of the second and third block is damaged.

8.26. Let $P_n(\tilde{k}_1, \tilde{k}_2, \ldots, \tilde{k}_r)$ be the maximal probability of the multinomial distribution (8.6). Prove that for i = 1, 2, ..., r the inequalities $np_i - 1 < \tilde{k}_i \leqslant (n + r - 1)p_i$ are fulfilled.

8.27. What is the probability that if we select a six-digit integer at random from the set of numbers from 000 000 up to 999 999, the sum of the first three digits of this integer will be equal to the sum of the last three digits?

8.28. Two players play 20 games. The probability of winning any of them is 0.2 for each of the players. What is the probability that the whole game will terminate with result 12 : 8?

8.29. An urn contains 16 balls, two of them have two dots, eight have one dot and six have no dots. Of the two with two dots, one has white dots and one has black dots; of the eight with one dot, four have a white dot and four have a black dot. A ball is drawn at random n times and each time it is replaced in the urn before the next draw after both its colour and the number of dots on it are noted. Calculate the probability that the number of the white dots will not exceed by more than m the number of the black dots, $m \leqslant 2n$.

8.30. Each of n variables $x_1, x_2, \ldots, x_n$ takes values 1, 2, ..., m with equal probabilities. Calculate the probability that the sum $x_1 + x_2 + \ldots + x_n$ will be: (a) equal to a given number k, where $n \leqslant k \leqslant mn$; (b) no less than a given number k.

8.31. What is the probability that the number of successes in a Bernoulli scheme (n, p) will be even?

8.32. An urn contains one white and one black ball. A ball is drawn successively with replacement. This procedure is continued until the absolute value of the difference between the number of the white balls drawn and the black balls drawn equals m, where m is fixed. What is the probability p_n that exactly n drawings should be performed, if:

(a) $m = 2$; (b) $m = 3$?

8.33. (*The gambler's ruin problem*) A player A has m levs and wins each game with probability p, and a player B has n levs and wins with probability q, where $p + q = 1$. At the end of each game the defeated person pays one lev to the winner. Calculate the probability of each one of the players being ruined.

8.34. Let $P_n(k)$ be the probability of k successes in a Bernoulli scheme (n, p). Prove the following identity: (a) analytically; (b) by probabilistic reasoning:

$$\sum_{k=0}^{r} P_{n_1}(k) P_{n_2}(r - k) = P_{n_1 + n_2}(r).$$

*8.35. Let $Q_n(k)$ be the probability of no more than r successes in a Bernoulli scheme (n, p). On the base of probabilistic reasoning prove the equalities:

$$\text{(a)} \quad Q_{n+1}(r) = Q_n(r) - pP_n(r); \qquad \text{(b)} \quad Q_{n+1}(r + 1) = Q_n(r) + qP_n(r).$$

*8.36. Let $0 < p < 1/2$ and the integer n be such that $np = m$ is also an integer. At one trial the event A occurs with probability p. Prove that in n independent trials the probability that A occurs no more than m times is greater than the probability that A occurs more than m times. (*A. Simons*)

9. Discrete Random Variables and Their Characteristics

Introductory Notes

Let Ω be the sample space of all elementary outcomes ω of a random experiment and let $\xi(\omega)$ be a function whose domain is Ω and which assumes its values on the real line $\mathbb{R}_1$ or in the m-dimensional Euclidean space $\mathbb{R}_m$. Suppose the range of ξ is the finite or infinite but countable set $\{x_1, x_2, \ldots\}$ which is a subset of $\mathbb{R}_1$ or $\mathbb{R}_m$. Let ξ assume the value x_i with probability $p_i = P\{\omega : \xi(\omega) = x_i\}$, $i = 1, 2, \ldots$, where $p_i > 0$ and $\sum_i p_i = 1$. Then we say that ξ is a *discrete random variable* (discrete r.v.). The set $\{p_1, p_2, \ldots\}$ forms the *distribution* of ξ. Obviously, the discrete r.v. ξ is completely defined by the table:

ξ	x_1	x_2	$\ldots$	x_k	$\ldots$
p_ξ	p_1	p_2	$\ldots$	p_k	$\ldots$

(9.1)

Note that if ξ assumes only a finite number of values, then ξ is called an elementary (or simple) r.v.

Next we shall describe some of the frequently used discrete distributions.

(a) *The hypergeometric distribution*. The range of a hypergeometric r.v. ξ is the set $\{0, 1, 2, \ldots, \min[M, n]\}$ and ξ assumes these values respectively with probability $p_k = P\{\xi = k\} = \binom{M}{k}\binom{N-M}{n-k}/\binom{N}{n}$ for $k = 0, 1, \ldots, \min[M, n]$. Here M, n and N are positive integers with $M < N$ and $n < N$ (see Example 3.2).

(b) *The binomial distribution*. The range of a binomial r.v. ξ is the set $\{0, 1, \ldots, n\}$ and $P_n(k) = P\{\xi = k\} = \binom{n}{k}p^k(1-p)^{n-k}$ for $k = 0, 1, \ldots, n$. (See Section 8.) Alternatively, ξ might be interpreted as the number of successes in n Bernoulli trials with probability of success at each trial given by p.

The collection of all r.v.'s whose distribution law is binomial with parameters n and p is denoted by $\mathcal{B}(n, p)$.

(c) *The multinomial distribution*. A multinomial random variable ξ assumes the values $(k_1, k_2, \ldots, k_r)$ with probability given respectively by $P_n(k_1, k_2, \ldots, k_r) = \frac{n!}{k_1!k_2! \ldots k_r!} p_1^{k_1} p_2^{k_2} \ldots p_r^{k_r}$, where k_i are integers, $0 \leqslant k_i \leqslant n$, $i = 1, 2, \ldots, r$ (see Section 8).

(d) *The geometric distribution*. The range of a geometrically distributed random variable ξ is the set of all non-negative integers $\{0, 1, \ldots\}$ and these values are assumed with probabilities $p_k = q^k p$, $0 < p < 1$, $q = 1 - p$, $k = 0, 1, 2, \ldots$ Alternatively, ξ might be interpreted as the number of failures preceding the occurrence of the first success in a sequence of Bernoulli trials (see Exercise 4.5).

(e) *The negative binomial distribution (Pascal's distribution)*. The range of a random variable ξ with a negative binomial distribution is the set of all non-negative integers 0, 1, 2, ... and these values are assumed with probabilities $p_k = \binom{-r}{k}(-q)^k p^r$, $k = 0, 1, 2, \ldots$, $0 < p < 1$, $q = 1 - p$, where r is an arbitrary positive number. If r is an integer, then ξ might be interpreted as the number of failures preceding the occurrence of the rth success in a sequence of Bernoulli trials (see Exercise 8.16).

(f) *The Poisson distribution.* A random variable ξ having such a distribution assumes the values 0, 1, 2, ... respectively with probability $p_k = \lambda^k e^{-\lambda}/k!$, $k = 0, 1, 2, \ldots$, $\lambda > 0$.

By $\mathcal{P}(\lambda)$ we shall denote the collection of all random variables ξ, which are Poisson distributed with parameter λ.

In all definitions which follow we have in mind only real-valued discrete random variables.

If the series $\Sigma_i x_i p_i$ is absolutely convergent, then the quantity

$$E(\xi) = \Sigma_i x_i p_i \tag{9.2}$$

is referred to as the *expectation* (the *mean value* or the *first moment*) of the r.v. ξ.

Let $g(x) \in \mathbb{R}_1$ for $x \in \mathbb{R}_1$, be any function for which the series $\Sigma_i g(x_i)p_i$ is absolutely convergent. Then the quantity

$$E\{g(\xi)\} = \Sigma_i g(x_i)p_i \tag{9.3}$$

is referred to as the expectation of the r.v. $g(\xi)$.

The numbers $E\{\xi^k\}$ and $E\{(\xi - E(\xi))^k\}$, $k \geqslant 0$, assuming that the corresponding series are absolutely convergent, are referred to as the *kth moment* about the origin and the *kth central moment*, respectively.

The second central moment

$$V(\xi) = E\{(\xi - E(\xi))^2\} \tag{9.4}$$

is called *variance* and sometimes it is denoted by σ^2 instead of $V(\xi)$. Equivalently to (9.4), $V(\xi)$ might also be written as

$$V(\xi) = E(\xi^2) - \{E(\xi)\}^2, \tag{9.5}$$

which is sometimes more convenient to use.

The number $\sigma = \sqrt{V(\xi)}$ is called the *standard deviation*.

The joint or *bivariate distribution* of the r.v.'s ξ and η, which are assumed to be defined on one and the same space Ω, is given by the table

$\xi \backslash \eta$	y_1	y_2	y_3	
x_1	$p(x_1, y_1)$	$p(x_1, y_2)$	$p(x_1, y_3)$	...
x_2	$p(x_2, y_1)$	$p(x_2, y_2)$	$p(x_2, y_3)$	...
$\vdots$	...	...	...	...

where $p(x_i, y_i)$ is the probability of the event $\{\omega : \xi(\omega) = x_i;\ \eta(\omega) = y_j\}$. It is assumed that $p(x_i, y_j) > 0$, $\Sigma_{i,j} p(x_i, y_j) = 1$. Also $\Sigma_i p(x_i, y_j) = P\{\eta = y_j\}$, $\Sigma_j p(x_i, y_j) = P\{\xi = x_i\}$.

The *marginal distribution* of the r.v. ξ is given by the numbers $P\{\xi = x_i\}$, $i = 1, 2, \ldots$ and the marginal distribution of the r.v. η is given by $P\{\eta = y_j\}$, $j = 1, 2, \ldots$

Note that multivariate extensions of the notions considered in this section are given in Section 16.

The random variables ξ and η defined above are said to be *independent* if

$$P\{\xi = x_i, \eta = y_j\} = P\{\xi = x_i\}P\{\eta = y_j\}$$

for all x_i and y_j.

The mean value has the following properties:

$$E\{c_1\xi + c_2\} = c_1E(\xi) + c_2 \tag{9.6}$$

for arbitrary constants c_1 and c_2;

$$E\{\xi + \eta\} = E(\xi) + E(\eta) \tag{9.7}$$

for arbitrary r.v.'s ξ and η;

$$E\{\xi\eta\} = (E\{\xi\})(E\{\eta\}) \tag{9.8}$$

for arbitrary independent r.v.'s ξ and η.

The *covariance* of two discrete r.v.'s ξ and η is defined by

$$\text{cov}(\xi, \eta) = E\{(\xi - E[\xi])(\eta - E[\eta])\} = E\{\xi\eta\} - (E[\xi])(E[\eta]) =$$

$$= \Sigma_{i,j}(x_i - E[\xi])(y_j - E[\eta])p(x_i, y_j). \tag{9.9}$$

The *correlation coefficient* of ξ and η is defined by

$$\rho(\xi, \eta) = \frac{\text{cov}(\xi, \eta)}{\sqrt{V\xi V\eta}}. \tag{9.10}$$

For any two r.v.'s ξ and η it holds that

$$V\{a\xi \pm b\eta\} = a^2V(\xi) + b^2V(\eta) \pm 2ab\ \mathrm{cov}(\xi, \eta). \tag{9.11}$$

If ξ and η are independent r.v.'s, then

$$V\{\xi \pm \eta\} = V(\xi) + V(\eta). \tag{9.12}$$

Let ξ be a discrete r.v. Then ξ is said to be an *integer-valued r.v.* if it takes only positive (or non-negative) integer values. For any integer-valued r.v. ξ, its *probability generating function* (p.g.f.) is defined by

$$g(s) = E\{s^{\xi}\} = \Sigma_i p_i s^i, \tag{9.13}$$

where $p_i = P\{\xi = i\}$ and $|s| \leqslant 1$.

If $g'(s)$ and $g''(s)$ exist for $s = 1$, then $E(\xi)$ and $V(\xi)$ can be calculated as follows:

$$E(\xi) = g'(1); \tag{9.14}$$

$$V(\xi) = g''(1) + g'(1) - [g'(1)]^2. \tag{9.15}$$

Illustrative Examples

Example 9.1. Calculate directly $E(\xi)$ and $V(\xi)$, where ξ is a hypergeometric r.v.

Solution. According to (9.2), for $m = \min[M, n]$ one has

$$E(\xi) = \sum_{k=0}^{m} k\binom{M}{k}\binom{N-M}{n-k}/\binom{N}{n} =$$

$$= \binom{N}{n}^{-1} \sum_{k=1}^{m} \frac{M!}{(k-1)!(M-k)!}\binom{N-M}{n-k} =$$

$$= M\binom{N}{n}^{-1} \sum_{k=1}^{m} \binom{M-1}{k-1}\binom{N-M}{n-k} =$$

$$= M\binom{N}{n}^{-1} \sum_{k=1}^{m-1} \binom{M-1}{k}\binom{N-M}{n-k-1}.$$

Now we compare coefficients at x^{n-1} in both sides of the equality

$$(1 + x)^{M-1}(1 + x)^{N-M} = (1 + x)^{N-1}.$$

This yields

$$\sum_{k=0}^{m-1} \binom{M-1}{k}\binom{N-M}{n-k-1} = \binom{N-1}{n-1}$$

(see the solution of Exercise 9.1(a)). Thus, we have:

$$E(\xi) = M\binom{N-1}{n-1}/\binom{N}{n} = Mn/N;$$

$$E\{\xi^2\} = \sum_{k=0}^{m} k^2\binom{M}{k}\binom{N-M}{n-k}/\binom{N}{n} =$$

$$= \binom{N}{n}^{-1} \sum_{k=1}^{m} k^2 \frac{M!}{k!(M-k)} \binom{N-M}{n-k} =$$

$$= \binom{N}{n}^{-1} \sum_{k=1}^{m} (k - 1 + 1) \frac{M!}{(k-1)!(M-k)!} \binom{N-M}{n-k} =$$

$$= \binom{N}{n}^{-1}[M(M-1) \sum_{k=2}^{m} \frac{(M-2)!}{(k-2)!(M-k)!} \binom{N-M}{n-k} +$$

$$+ M \sum_{k=1}^{m} \frac{(M-1)!}{(k-1)!(M-k)!} \binom{N-M}{n-k}] =$$

$$= \binom{N}{n}^{-1}[M(M-1) \sum_{k=0}^{m-2} \binom{M-2}{k}\binom{M-2}{n-k-2} +$$

$$+ M \sum_{k=0}^{m-1} \binom{M-1}{k}\binom{N-M}{n-k-1}] = [M(M-1)\binom{N-2}{n-2} +$$

$$+ M\binom{N-1}{n-1}]/\binom{N}{n} = \frac{Mn[Mn + N - M - n]}{N(N-1)} .$$

Finally we get

$$V(\xi) = E\{\xi^2\} - [E\{\xi\}]^2 = n \frac{M}{N} \frac{N-M}{N} \frac{N-n}{N-1} .$$

Example 9.2. Find the p.g.f. of a binomial r.v. ξ with parameters (n, p) and then find the mean $E(\xi)$ and the variance $V(\xi)$.

Solution. We have

$$g(s) = E\{s^{\xi}\} = \sum_{k=0}^{n} s^k \binom{n}{k} p^k q^{n-k} =$$

$$= \sum_{k=0}^{n} \binom{n}{k} (ps)^k q^{n-k} = (ps + q)^n.$$

Thus $g'(s) = np(ps + q)^{n-1}$ and $g''(s) = n(n - 1)p^2(ps + q)^{n-2}$. Then from (9.14) and (9.15) we obtain:

$$E(\xi) = g'(1) = np;$$

$$V(\xi) = g''(1) + g'(1) - [g'(1)]^2 = n(n - 1)p^2 + np - n^2p^2 =$$

$$= np(1 - p) = npq.$$

Remark. Substituting $p = M/N$ and $q = 1 - p$ in Example 9.1 one can easily obtain that expectations of the binomial and the hyperyeometric distribution coincide, while their variances differ only in the multiple $(N - n)/(N - 1)$. We shall again discuss the difference between these two distributions in Exercise 9.20 below.

Example 9.3. Suppose n independent trials are carried out. The probability for success at the kth trial is given by p_k, $k = 1, \ldots, n$. Find $E(\xi)$ and $V(\xi)$, where the r.v. ξ is defined as the number of successes in the n trials.

Solution. Let ξ_k be the number of successes at the kth trial. Then ξ_k assumes the values 1 and 0 with probabilities p_k and $q_k = 1 - p_k$, respectively. Moreover, $\xi_1, \ldots, \xi_n$ are independent r.v.'s with $E(\xi_k) = p_k$, $V(\xi_k) = p_k - p_k^2 = p_k q_k$. Obviously $\xi = \xi_1 + \xi_2 + \ldots + \xi_n$ and generalizing (9.6) and (9.12) for n random variables we obtain:

$$E(\xi) = p_1 + p_2 + \ldots + p_n;$$

$$V(\xi) = p_1q_1 + p_2q_2 + \ldots + p_nq_n.$$

Remark. If we compare the above n trials with the standard Bernoulli scheme (n, p) we come to a conclusion which is contrary to our intuition. Indeed, if we denote by $p = \frac{1}{n}(p_1 + p_2 + \ldots + p_n)$ the average probability of success, then the variance of ξ can be written as

$$V(\xi) = \sum_{k=1}^{n} p_k(1 - p_k) = n \cdot \frac{1}{n} \sum_{k=1}^{n} p_k - \sum_{k=1}^{n} p_k^2 = np - \sum_{k=1}^{n} p_k^2 .$$

It can be verified quite easily (inductively or by elementary calculations) that $\sum_{k=1}^{n} p_k^2$ is a minimum when all the p_k are equal, and hence $V(\xi)$ is a maximum. Any deviation of p_k from p, the average probability, decreases the variance! For example, if for n machines the average percentage of quality products is known to be p, then the production of the whole complex will be most variable when each of the machines produces the same percentage of quality products.

Example 9.4. A target has a circle with a concentric ring around it. If a marksman hits the circle, he gets ten marks and if he hits the ring, he gets five marks. A hit outside results in a loss of one mark. For each shot the probabilities of hitting the circle or ring are 0.5 and 0.3 respectively. Let the r.v. ξ be the sum of marks for three independent shots. Find the probability distribution of ξ.

Solution. Let A_k = {the sum of the marks, obtained in three independent shots equal k}. If ξ_1, ξ_2 and ξ_3 are the marks at these three shots, then $P(A_k) = P\{\xi_1 + \xi_2 + \xi_3 = k\}$. Let g(s) be the p.g.f. of the integer-valued r.v. $\xi_1 + \xi_2 + \xi_3$; i.e., $g(s) = E\left\{s^{\xi_1} + s^{\xi_2} + s^{\xi_3}\right\}$. The independence of ξ_1, ξ_2 and ξ_3 and property (9.8) imply that $g(s) = \left(E\left\{s^{\xi_1}\right\}\right)^3$. On the other hand, $E\left\{s^{\xi_1+\xi_2+\xi_3}\right\} = \sum_k s^k P\{\xi_1 + \xi_2 + \xi_3 = k\}$, where the summation is taken over all possible values of $\xi_1 + \xi_2 + \xi_3$. Hence the desired probability $p_k = P(A_k)$ is exactly the coefficient of s^k in the expansion of the following function:

$$g(s) = (\frac{5}{10} s^{10} + \frac{3}{10} s^5 + \frac{2}{10} \frac{1}{s})^3 = 10^{-3} s^{-3} (5s^{11} + 3s^6 + 2)^3 =$$

$$= 10^{-3} s^{-3} (25s^{22} + 30s^{17} + 9s^{12} + 20s^{11} + 12s^6 + 4)(5s^{11} +$$

$$+ 3s^6 + 2) = 10^{-6} (125s^{30} + 225s^{25} + 135s^{20} + 150s^{19} +$$

$$+ 27s^{15} + 180s^{14} + 54s^9 + 60s^8 + 36s^3 + 8s^{-3}).$$

The the r.v. ξ has the following distribution:

ξ	-3	3	8	9	14	15	19	20	25	30
p_ξ	0.008	0.036	0.060	0.054	0.180	0.027	0.150	0.135	0.225	0.125

Exercises

9.1. Check by direct calculations that condition $\Sigma_i p_i = 1$ is fulfilled for the following distributions: (a) the hypergeometric distribution; (b) the binomial distribution; (c) the multinomial distribution; (d) the geometric distribution; (e) the negative binomial distribution; (f) the Poisson distribution.

Prove the same result using probabilistic reasoning only.

9.2. Let ξ be a r.v. which assumes only non-negative integer values and for which $P\{\xi = j\} = c/3^j$, j = 0, 1, ... Find c and calculate: (a) $P\{\xi \geqslant 10\}$; (b) $P\{\xi \in A\}$, where A = {j : j = 2k + 1, k = 0, 1, ...}; (c) $P\{\xi \in B\}$, where B = {j : j = 3k + 1, k = 0, 1, ...}.

9.3. A fair die is rolled. Let ξ and η denote the number of spots obtained on the upper face and lower face, respectively. Find the distribution of the r.v. $\zeta = \xi\eta$. (Recall that the sum of the spots on opposite faces of a die is equal to 7.)

9.4. The chess-man 'horse' is placed at random on a 8×8 chess board. Denote by ξ the number of positions the horse is allowed to attack. Describe the r.v. ξ.

9.5. An urn contains N balls, labeled 1, 2, ..., N. A sample of size n with $n \leqslant N$, is drawn without replacement. Denote by η the maximum ball-number obtained in the sample. Find the distribution of η.

9.6. Let ξ and η be independent r.v.'s whose distributions $\{p_k\}$ and $\{r_k\}$ are given by: $p_k = 1/n$, $k = 1, 2, \ldots, n$, $n \geqslant 1$; $r_k = \frac{1}{2}$, $k = 0$ or $k = n$. Show that $\xi + \eta$ is uniformly distributed over the set $\{1, 2, \ldots, 2n\}$.

9.7. Two distinghuishable dice are rolled. Let ξ_1 be the outcome of the first die and let ξ_2 be the greater outcome of the two dice. Find the joint distribution of ξ_1 and ξ_2.

9.8. A random variable ξ, which assumes finitely many values x_1, $x_2, \ldots, x_n$, $n \geqslant 1$, with equal probabilities, is said to be *uniformly distributed* over the set $\{x_1, x_2, \ldots, x_n\}$. For such a r.v. ξ calculate: (a) $P\{\xi = x_i\}$ and (b) $E(\xi)$, $V(\xi)$ when $x_i = (i - 1)/(n - 1)$ for $i = 1, \ldots, n$; (c) $E(\xi)$ and $V(\xi)$ when $x_i = a + (i - 1)(b - a)/(n - 1)$, $b > a$, for $i = 1, 2, \ldots, n$; (d) $\lim_{n\to\infty} 3\sigma$, where σ is the standard deviation in (b) and (c).

9.9. Using the method of Example 9.3, calculate $E(\xi)$ and $V(\xi)$, where the r.v. ξ is distributed: (a) binomially; (b) hypergeometrically.

9.10. Find the p.g.f. of the r.v. ξ and then using the p.g.f. calculate $E(\xi)$ and $V(\xi)$ if the distribution of ξ is: (a) geometric; (b) negative binomial; (c) Poisson.

9.11. Let ξ be an integer-valued r.v. whose p.g.f. is $g(s)$. Calculate the p.g.f. of the r.v. η, where: (a) $\eta = \xi + 1$; (b) $\eta = \xi + k$ with k an integer; (c) $\eta = 2\xi$.

9.12. Let ξ be an integer-valued r.v. with p.g.f. $g(s)$. Express the quantity $h(s) = \sum_{n=0}^{\infty} a_n s^n$ in terms of $g(s)$ if: (a) $a_n = P\{\xi \leqslant n\}$; (b) $a_n = P\{\xi < n\}$; (c) $a_n = P\{\xi \geqslant n\}$; (d) $a_n = P\{\xi > n + 1\}$; (e) $a_n = P\{\xi = 2n\}$.

9.13. Find the distribution of the sum of the digits of a randomly chosen integer number between 0 and 999.

9.14. For producing an article one can choose among n different materials. Each piece of material is sufficient for making exactly one article. The article is of first quality with some fixed probability p, which is the same for all materials. Assume a first quality article is obtained and denote by ξ and η respectively the number of materials still unused and the number of materials already used. Find the distribution of ξ and η. Find also $E(\xi)$, $E(\eta)$, $V(\xi)$, $V(\eta)$, $\lim_{n\to\infty} E(\xi)$, $\lim_{n\to\infty} E(\eta)$, $\lim_{n\to\infty} V(\xi)$,

$\lim_{n\to\infty} V(\eta)$ and the p.g.f.'s $g_\xi(s)$, $g_\eta(s)$.

9.15. Find the distribution of an integer-valued r.v. ξ whose p.g.f. $g(s)$ is given by: (a) $(1 - 4\alpha^2)/(1 - 4\alpha^2)s^2 - 4s + 4)$, $|\alpha| < \frac{1}{2}$; (b) $2s/(18 - 27s + 13s^2 - 2s^3)$; (c) $\mathrm{ch}\lambda\sqrt{s}/\mathrm{ch}\lambda$.

9.16. Prove that if $\xi \in P(\lambda)$ and $E(\xi)$ is an integer, then ξ assumes only two values each with probability $\frac{1}{2}$.

9.17. Let $\xi \in P(\lambda)$. Which value does ξ assume with greatest probability?

9.18. Consider successive independent trials with probability of occurrence of success in each trial given by p. Find: (a) the distribution of the number of trials preceding the occurrence of the first success; (b) the most probable number of trials preceding the first success.

9.19. Let the r.v. ξ have a negative binomial distribution with parameters (r, p); i.e., $P\{\xi = k\} = p_r(k) = \binom{-r}{k}p^r(-q)^k$, $k = 0, 1, \ldots,$ where r is positive integer, $0 < p < 1$ and $q = 1 - 0$. (a) Prove that

$$\text{(i)} \qquad \frac{p_r(k)}{p_r(k+1)} = 1 + \frac{p}{k}[(r-1)\frac{q}{p} - k] \qquad \text{for } k = 1, 2, \ldots$$

(ii) as k increases, then $p_r(k)$ first increases, reaches its maximum value and then decreases. (b) Find the value m of ξ which is most likely to occur.

9.20. An urn contains N balls, M of which are black and the remaining N - M are white, $M < N$. One by one n balls are drawn without replacement for $n < N$. From Example 3.1 the probability that a sample of n balls contains exactly k black balls for $k = 0, 1, \ldots, \min[M, n]$, is $p_k = \binom{M}{k}\binom{N-M}{n-k}/\binom{N}{n}$.

Let the number N of all balls and the number M of the black balls be both increasing in such a manner that $\lim_{N\to\infty} (M/N) = p$, $0 < p < 1$. Then prove that $\lim_{N\to\infty} p_k = \binom{n}{k}p^k(1-p)^{n-k}$.

Remark. Consider the occurrence of a black ball in each drawing as a "success". Then if the drawings are carried out without replacement, one can interpret p_k as the probability of occurrence of exactly k successes in n *dependent* trials. When the sampling is carried with replacement the trials become independent and then the probability for exactly k successes in n trials is the binomial probability $P_n(k)$. Exercise 9.20 shows that when the sample is taken from a set with large number of elements, then it is practically unimportant whether the sample is with or without replacement assuming that the draws are random. This observation plays an important role in statistical quality control.

9.21. Assume that $n \to \infty$ and $p_n \to 0$ in such a way that $np_n = \lambda$ where λ is constant. Prove that

$$\lim_{n\to\infty} \binom{n}{k} p_n^k (1 - p_n)^{n-k} = \lambda^k e^{-\lambda}/k! \quad k = 0, 1, \ldots$$

9.22. Two basketball players A and B consecutively throw the ball in the basket until one of them scores. For each attempt the probability that A scores is p_1 and that B scores is p_2. Player A throws first. Let ξ be number of trials necessary for A to score and let η be the same number for B. We assume that, when one of the two players scores, the other one leaves the game with no successes. Find the distributions of the r.v.'s ξ and η.

9.23. Let $\xi_1, \ldots, \xi_m$ be independent r.v.'s. Prove that: (a) if $\xi_i \in \mathcal{B}(n_i, p)$ for $i = 1, \ldots, m$, then $\xi_1 + \ldots + \xi_m \in \mathcal{B}(n_1 + \ldots + n_m, p)$; (b) if $\xi_i \in \mathcal{P}(\lambda_i)$ for $i = 1, \ldots, m$, then $\xi_1 + \ldots + \xi_m \in \mathcal{P}(\lambda_1 + \ldots + \lambda_m)$.

9.24. Let ξ and η be independent r.v.'s. Find: (a) the conditional distribution of ξ under the condition $\xi + \eta = n$ when $\xi \in \mathcal{P}(\lambda)$ and $\eta \in \mathcal{P}(\mu)$; (b) the distribution of the r.v. $\xi - \eta$ when $\xi, \eta \in \mathcal{P}(\lambda)$.

9.25. The number of trials ξ is random and can vary from 0 to ∞. It assumes the value k with probability $P\{\xi = k\} = \lambda^k e^{-\lambda}/k!$. In each trial success and failure occur with probability p and $q = 1 - p$, respectively. Find the distribution of the number of successes.

9.26. Let $\xi \in \mathcal{B}(N, p)$ for some given $N > 0$ and p, $0 < p < 1$. Now let N also be a r.v. with $N \in \mathcal{B}(M, q)$, where $0 < q < 1$ and $M > 0$ is a given integer number. Prove that $\xi \in \mathcal{B}(M, pq)$.

9.27. Suppose ξ and η are independent r.v.'s with the following distributions:

ξ	-1	1
P_ξ	0.5	0.5

η	1	3	5
P_η	0.5	0.25	0.25

Consider the r.v.'s $\xi = 2\xi + \eta + 1$ and $\Theta = \xi\eta$. Calculate: (a) $E(\Theta)$; (b) $E(\xi)$; (c) $V(\xi)$.

9.28. The bivariate r.v. (ξ, η) has the following distribution:

η \ ξ	0.01	0.02	0.03	0.04
0.002	0.01	0.02	0.04	0.04
0.004	0.03	0.24	0.15	0.06
0.006	0.04	0.10	0.08	0.08
0.008	0.02	0.04	0.03	0.02

Calculate: (a) the marginal distributions of ξ and η; (b) $E(\xi)$, $E(\eta)$, $V(\xi)$, $V(\eta)$; (c) $\mathrm{cov}(\xi, \eta)$ and the correlation coefficient ρ.

9.29. Let ξ_1, ξ_2, ξ_3, ξ_4 and ξ_5 be the independent r.v.'s with equal variances. Find the correlation coefficient between the r.v.'s: (a) $\eta_1 = \xi_1 + \xi_2$ and $\eta_2 = \xi_3 + \xi_4 + \xi_5$; (b) $\zeta_1 = \xi_1 + \xi_2 + \xi_3$ and $\zeta_2 = \xi_3 + \xi_4 + \xi_5$.

9.30. Let $(\xi_1, \ldots, \xi_r)$ be a multinomial r.v. with parameters $p_1, \ldots, p_r$ and let n be the corresponding number of independent trials. Show that the correlation coefficient $\rho(\xi_i, \xi_j)$ is given by

$$\rho(\xi_i, \xi_j) = -\sqrt{p_i p_j / [(1 - p_i)(1 - p_j)]}, \quad i \neq j;$$

i.e., each pair of r.v.'s is negatively correlated.

9.31. Let ξ and η be independent r.v.'s with $V(\xi) > 0$ and $V(\eta) > 0$. What condition for the moments of ξ and η ensures that the equality $V\{\xi\eta\} = V(\xi)V(\eta)$ will be attained?

9.32. The first player is tossing three coins and the second player is tossing two coins. The player who obtains more heads wins the game and receives all the five coins. If both players obtain the same number of heads, then the tossing continues until one of them wins. What is the average expected gain of each player?

9.33. In a lottery there are m_1 prizes of value k_1, m_2 prizes of value $k_2, \ldots, m_n$ prizes of value k_n. The total number of tickets is N and each ticket is blank or brings one prize only. What should the selling price of the tickets be if the expected gain of each ticket is to be exactly one half of its price?

9.34. Let $p_k = 0$ for $k \leqslant 0$ and let $p_k = ba^k/k$ for $k = 1, 2, \ldots$, where $0 < a < 1$. For what value of b does the set $\{p_k\}$ represent a probability distribution? This is called *Fisher's logarithmic distribution*. For this distribution find the p.g.f., the expectation and the variance.

9.35. Sequentially k numbers are drawn without replacement from the set $\{1, 2, \ldots, n\}$. The numbers in the sample are rearranged in an increasing order. Let ξ_r be the rth number in the rearranged sample. Find the distribution, the expectation and the variance of ξ_r.

*9.36. Let $\xi_1, \ldots, \xi_n$ be independent r.v.'s which are uniformly distributed over the set $\{1, 2, \ldots, N\}$. Prove that

$$E\{\max[\xi_1, \ldots, \xi_n]\} + E\{\min[\xi_1, \ldots, \xi_n]\} = N + 1.$$

*9.37. Let k be a fixed natural number. Denote by τ_k the minimal number of tossing of a fair coin for which each outcome "head" and "tail" occurs at least k times. Find the distribution of the r.v. τ_k and calculate $E(\tau_k)$.

*9.38. Prove that for every n the function $g_n(s) = (n!)^{-1}s(s + 1) \ldots (s + n - 1)$ is a p.g.f. of a probability distribution. If $a_1(n)$ denotes the expectation of this distribution, then show that $\lim_{n\to\infty} (a_1(n)/\ln n) = 1$.

10. Normal and Poisson Approximations for the Binomial Distribution

Introductory Notes

Let μ_n be the number of successes in a Bernoulli scheme (n, p) (see Section 8). The probability for exactly k successes is given by

$$P\{\xi_n = k\} = P_n(k) = \binom{n}{k} p^k q^{n-k}, \quad q = 1 - p, \; k = 0, 1, \ldots, n.$$

The calculation of the above quantity for fixed n, p and k is technically difficult even for small values of n and k. The approximations given below facilitate this calculation. With these approximative methods and the tables given at the end of the Manual, many practical problems can be solved easily.

The following two functions:

$$\phi_0(x) = \frac{1}{\sqrt{2\pi}} e^{-x^2/2} \quad \text{and} \quad \Phi(x) = \frac{1}{\sqrt{2\pi}} \int_{-\infty}^{x} e^{-u^2/2}\, du, \quad x \in \mathbb{R}_1,$$

are playing an important role in probability theory and statistics (ϕ_0 is called *standard normal density* and Φ is called *standard normal distribution function*). They obey the following properties:

$$\phi_0(x) \geqslant 0, \quad \int_{-\infty}^{\infty} \phi_0(u)\, du = 1, \quad \phi_0(-x) = \phi_0(x),$$

$$0 \leqslant \Phi(x) \leqslant 1, \quad \Phi(x) = \int_{-\infty}^{x} \phi_0(u)\, du, \quad \Phi(-x) = 1 - \Phi(x), \; x \in \mathbb{R}_1.$$

The values of $\phi_0(x)$ and $\Phi(x)$ for some positive values of x are given in Table 1 (p. 346).

Theorem (*De Moivre-Laplace*). Let $0 < p < 1$ and let $n \to \infty$. Then for every fixed k, k = 0, 1, ..., the following relation holds:

$$P_n(k) \approx \frac{1}{\sqrt{npq}} \phi_0(x_k), \quad x_k = \frac{k - np}{\sqrt{npq}}. \tag{10.1}$$

Also for every $x \in \mathbb{R}_1$,

$$P\left\{\frac{\mu_n - np}{\sqrt{npq}} < x\right\} \to \Phi(x). \square \tag{10.2}$$

Relations (10.1) and (10.2) are called respectively the *local* and *integral theorem of de Moivre-Laplace*.

Theorem. (*Poisson*) Consider independent trials with probability p_n of success at the nth trial. If $\tilde{P}_n(k)$ is the probability to get k successes in the first n trials and $a_n = np_n$, then

$$\left| \tilde{P}_n(k) - \frac{a_n^k e^{-a_n}}{k!} \right| \to 0 \quad \text{as } n \to \infty, \tag{10.3}$$

for every fixed k, k = 0, 1, ...□

If $p_n \to 0$ as $n \to \infty$ in such a way that $a_n = np_n \to \lambda$, $\lambda \geqslant 0$, then one can write (10.3) in the form

$$\tilde{P}_n(k) \to \frac{\lambda^k e^{-\lambda}}{k!}, \quad k = 0, 1, \ldots \tag{10.4}$$

(see also Exercise 9.3).

The numbers $\lambda^k e^{-\lambda}/k!$, k = 0, 1, ..., correspond to the *Poisson distribution* (see Section 9). The values of $\lambda^k e^{-\lambda}/k!$ for some values of λ and k are given in Table 2 (p. 347).

When n is large and k_1 and k_2 are fixed, the most commonly used versions of (10.2) and (10.4) respectively are the following two relations:

$$P\{k_1 \leqslant \mu_n \leqslant k_2\} \approx \Phi\left(\frac{k_2 - np}{\sqrt{npq}}\right) - \Phi\left(\frac{k_1 - np}{\sqrt{npq}}\right); \tag{10.5}$$

$$P\{k_1 \leqslant \mu_n \leqslant k_2\} \approx \sum_{k=k_1}^{k_2} \frac{\lambda^k e^{\lambda}}{k!}. \tag{10.6}$$

Illustrative Examples

Example 10.1. The probability that a defective article will be produced is 0.005. What is the probability that out of 10,000 randomly chosen articles there will be: (a) exactly 40 defective articles; (b) not more than 70 defective articles?

Solution. Consider a Bernoulli scheme (n, p) with n = 10,000 and p = 0.005. (a) It is necessary to find the probability $P_n(k)$ for k = 40. Since np = 50, $\sqrt{npq} = \sqrt{49.75} \approx 7.05$ and $x_{40} = \dfrac{40 - np}{\sqrt{npq}} \approx -1.4$, then according to (10.1),

$$P_n(40) = \binom{10000}{40} 0.005^{40} 0.995^{9960} \approx \frac{1}{7.05} \phi_0(x_{40}).$$

From Table 1 we get $\phi_0(x_{40}) = \phi_0(-1.4) = \phi_0(1.4) \approx 0.1497$. Hence $P_n(40) \approx 0.0197$. (b) Now the probability $P\{\mu_n \leqslant 70\}$ has to be found. Using (10.2) and (10.5) we have

$$P\{\mu_n \leqslant 70\} = P\left\{-\frac{50}{\sqrt{49.75}} \leqslant \frac{\mu_n - np}{\sqrt{npq}} \leqslant \frac{20}{\sqrt{49.75}}\right\} =$$

$$= P\left\{-7.09 \leqslant \frac{\mu_n - np}{\sqrt{npq}} \leqslant 2.84\right\} \approx \Phi(2.84) - \Phi(-7.09) =$$

$$= \Phi(2.84) + \Phi(7.09) - 1.$$

However, Table 1 does not include the values of $\Phi(x)$ for $x = 2.84$ and $x = 7.09$. We can use the fact that to within 10^{-3} we have $\Phi(2.84) \approx 0.997$ and $\Phi(7.09) \approx 1$. Then $P\{\mu_n \leqslant 70\} \approx 0.997$.

Example 10.2. The probability of hitting a target with a single shot is 0.001. To destroy the target at least two hits are necessary. What is the probability of destroying the target with 5,000 shots?

Solution. Let μ be the number of hits with 5,000 shots. Obviously, the probability $P\{\mu \geqslant 2\}$ is to be determined. We have

$$P\{\mu \geqslant 2\} = 1 - P\{\mu < 2\} = 1 - P_n(0) - P_n(1),$$

where $n = 5{,}000$ and $p = 0.001$. In our case $a_n = np_n = 5$ and according to Poisson's theorem and using Table 2 we find

$$P\{\mu \geqslant 2\} \approx 1 - e^{-5} - 5e^{-5} \approx 0.9596.$$

Exercises

10.1. Consider an unfair coin for which a head and tail occur with probability $\frac{3}{4}$ and $\frac{1}{4}$, respectively. The coin is tossed 2,000 times. What is the probability that the total number of heads obtained lies between 1,465 and 1,535?

10.2. Prove that when n is large, then with probability close to 0.5: (a) the number of heads obtained in n tossings of a fair coin is contained in the interval $(\frac{1}{2}n - 0.335\sqrt{n}, \frac{1}{2}n + 0.335\sqrt{n})$; (b) the number of times a "six" will appear when a fair die is rolled n times is contained in the interval $(\frac{1}{6}n - 0.251\sqrt{n}, \frac{1}{6}n + 0.251\sqrt{n})$.

10.3. The Bernoulli scheme (n, p) is considered. The number $\tilde{p}_n = \mu_n/n$ is called the *relative frequency of success*. (a) For $p = 0.4$ and $n = 1{,}500$ what is the probability that $\tilde{p}_n$ is contained in the interval $(0.40, 0.44)$? (b) If $p = 0.375$, how many independent trials are needed to ensure the probability of the event $\{|\tilde{p}_n - p| \leqslant 0.001\}$ will be at least 0.995? (c) Let $p = \frac{2}{3}$ and $n = 800$. Determine ε so that $|\tilde{p}_n - p| < \varepsilon$ with a probability of at least 0.985. (d) If $n = 144$ find p so that the probability of the event $\{|\tilde{p}_n - p| \leqslant 0.005\}$ is not less than 0.88.

10.4. A fair coin was tossed 14,400 times and a head occurred 7,428 times. What is the probability of the deviation of the observed number of heads from np? What is the probability of a deviation greater in absolute value than that obtained?

10.5. An urn contains white and black balls in the ratio 3 : 1. Suppose n balls are chosen at random. After each drawing the colour of the ball drawn is noted and the ball is returned to the urn. How many drawings are necessary so that with a probability of at least 0.92 the deviation of the relative frequency of the occurrence of a white ball from its expectation will not exceed 0.05?

10.6. A unit is constructed from 10,000 components of three different types: $n_1 = 1{,}000$, $n_2 = 2{,}000$ and $n_3 = 7{,}000$. Each component fails independently of the others and the probability of failure depends only on its type: $p_1 = 7/10^4$, $p_2 = 3/10^4$ and $p_3 = 1/10^4$. The unit is out of order if at least two components fail. Find the probability of failure.

10.7. In some amount of coffee beans 5,000 beans are not roasted. All the coffee is distributed in 10,000 different packs. Each coffee bean has equal chances to fall into the different packs. What is the probability that a fixed pack will contain at least one unroasted coffee bean?

10.8. There are 50 misprints in a book of 500 pages. The probability of a misprint is the same for each page. Find the probability that a page chosen at random has at least three misprints.

10.9. A fair coin is tossed n times. How large must n be for the number of heads to lie between 0.48n and 0.52n with a probability of 0.9?

10.10. Suppose that the probability of a boy at birth is approximately 0.512. What is the probability that among 10,000 newly born babies: (a) the number of girls is not less than the number of boys; (b) the number of boys is at least 200 more than the number of girls?

10.11. Let $\tilde{p}_n = \mu_n/n$ be the relative frequency of success in a Bernoulli scheme (n,p). Prove that $\tilde{p}_n \xrightarrow{P} p$ as $n \to \infty$; i.e., for every $\varepsilon > 0$ we have $P\{|p_n - p| < \varepsilon\} \to 1$ as $n \to \infty$. (The *Bernoulli law of large numbers*; see also Section 22.)

10.12. For what m will the number of heads obtained in 10,000 tosses of a fair coin lie between 440 and m with probability approximately equal to 0.5?

10.13. In a Bernoulli scheme (n, p), where n is large, find the approximate values of $P\{|\mu_n - np| < 1.96\sqrt{npq}\}$, $P\{|\mu_n - np| < 3\sqrt{npq}\}$ and $P\{|\mu_n - np| \geqslant 3\sqrt{npq}\}$.

10.14. In 20,000 tosses of a coin a head occurred 10,800 times. Is there enough reason to consider that the coin may be biased?

10.15. Let μ be the number of heads, obtained in 2n tosses of a fair coin. Prove that: (a) $P\{\mu = n\} \approx \frac{1}{\sqrt{\pi n}}$; (b) $P\{\mu = n + 1\} = P\{\mu = n - 1\} \approx \frac{1}{\sqrt{\pi n}} e^{-1/n}$; (c) $P\{\mu = n + 2\} = P\{\mu = n - 2\} \approx \frac{1}{\sqrt{\pi n}} e^{-4/n}$.

10.16. A Bernoulli scheme (n, p) is considered with $p = 1/2$. Let $\xi_n = 2\mu_n - n$. (a) Find the distribution of the r.v. ξ_n. (b) Prove that as $n \to \infty$, $P\{n^{-1/2}\xi_n < x\} \to \Phi(x)$, $x \in \mathbb{R}_1$ and $P\{n^{-1/2}|\xi_n| < x\} \to 2\Phi(x) - 1$,

$x \geqslant 0$.

10.17. There are N tickets in a lottery, M of which are winning tickets. For a given probability r, $0 < r < 1$, determine how many tickets must be bought in order to obtain at least one winning ticket with probability greater than r.

10.18. Let μ_n denote the number of successes obtained in a Bernoulli scheme $(n, \frac{1}{2})$. Prove that when n is large, then the probability of the event $A_n = \{|\mu_n - \frac{1}{2}n| \leqslant \frac{1}{2}\sqrt{n}\}$ is more than twice the probability of the event $B_n = \{|\mu_n - \frac{1}{2}n| \geqslant \frac{1}{2}\sqrt{n}\}$.

10.19. For an arbitrary p, $0 < p < 1$, and for an arbitrary natural number n define

$$a_n(p) = \sum_{k=0}^{n} \binom{2n+1}{k} p^k (1-p)^{2n+1-k}.$$

Prove the convergence of the sequence $\{a_n(p)\}$ as $n \to \infty$. Prove also that for the limit $a(p) = \lim_{n\to\infty} a_n(p)$, we have

$$a(p) = \begin{cases} 1, & \text{if } 0 < p < \frac{1}{2}, \\ \frac{1}{2}, & \text{if } p = \frac{1}{2}, \\ 0, & \text{if } \frac{1}{2} < p < 1. \end{cases}$$

(*A. Philippou*)

Chapter 2

PROBABILITY SPACES AND RANDOM VARIABLES

11. General Definition of Probability and σ-Algebra of Events

Introductory Notes

Let Ω be an arbitrary non-empty set and $\mathfrak{P}(\Omega)$ be the aggregate of all its subsets. The class $\mathfrak{A} \subset \mathfrak{P}(\Omega)$ is said to be a *Boolean algebra* (an *algebra*), if: (a) $\Omega \in \mathfrak{A}$; (b) $\mathfrak{A}$ is closed with respect to the operations of union, intersection and complementation; i.e., from A, B $\in \mathfrak{A}$ it follows that $A \cup B \in \mathfrak{A}$, $AB \in \mathfrak{A}$ and $\bar{A} \in \mathfrak{A}$.

If the algebra $\mathfrak{A}$ is closed with respect to the operations of countable union and intersection; i.e., if for any sequence $A_1, A_2, \ldots,$ where $A_n \in \mathfrak{A}$, $n = 1, 2, \ldots,$ we have $\bigcup_{n=1}^{\infty} A_n \in \mathfrak{A}$ and $\bigcap_{n=1}^{\infty} A_n \in \mathfrak{A}$, then $\mathfrak{A}$ is said to be a *σ-Boolean algebra* (simply a *σ-algebra*).

According to the generally accepted terminology in the theory of probability, the set Ω is called a *sample space* (a space of the outcomes of some random experiment), the pair $(\Omega, \mathfrak{A})$ a measurable space (assuming that $\mathfrak{A}$ is a σ-algebra), $\mathfrak{A}$ *σ-algebra of the events*, and the elements A, B, ... of $\mathfrak{A}$ *events*.

For an arbitrary sequence of events $\{A_n\}$ we are going to define the following two events:

$$\limsup_n A_n = \bigcap_{n=1}^{\infty} \bigcup_{k=n}^{\infty} A_k, \qquad \liminf_n A_n = \bigcup_{n=1}^{\infty} \bigcap_{k=n}^{\infty} A_k.$$

(The notations A^* and A_* or $\overline{\lim}\, A_n$ and $\underline{\lim}\, A_n$ are also used.)

Let $\{A_n\}$ be a sequence of events, for which $A_n \subset A_{n+1}$ or $A_n \supset A_{n+1}$, $n \geqslant 1$. Such a sequence is said to be monotonically increasing or decreasing, respectively ($A_n\uparrow$ and $A_n\downarrow$). If $A = \limsup_n A_n = \liminf_n A_n$, then the sequence $\{A_n\}$ is said to be convergent, and the event A is called its limit ($A = \lim_n A_n$). Any monotone sequence is convergent and

$$\lim_n A_n = \bigcup_{n=1}^{\infty} A_n, \quad \text{if } A_n\uparrow; \quad \text{and} \quad \lim_n A_n = \bigcap_{n=1}^{\infty} A_n, \quad \text{if } A_n\downarrow.$$

If $\{A_n\}$ is a sequence of events in the σ-algebra $\mathfrak{A}$, then $\mathfrak{A}$ contains the events A^* and A_* as well.

For any class $\mathfrak{C} \subset \mathfrak{P}(\Omega)$ there exists a *minimal σ-algebra*, containing

$\mathfrak{C}$ (generated by $\mathfrak{C}$). The notation $\sigma(\mathfrak{C})$ is used for it.

Any finite or countable set of events $\{A_n\}$ such that $A_i A_j = \emptyset$, $i \neq j$ and $\bigcup_n A_n = \Omega$, is called an *exhaustive set of events* or a *partition* of the sample space Ω. (See also Section 2.)

Any event $A \neq \emptyset$ such that for an arbitrary $B \in \mathfrak{A}$ either $AB = \emptyset$, or $AB = A$, is called an *atom* of the algebra $\mathfrak{A}$ (of the σ-algebra $\mathfrak{A}$).

For instance, if $\{A_n\} = \mathfrak{C}$ is a partition of Ω and $\mathfrak{A} = \sigma(\mathfrak{C})$, then $\mathfrak{C}$ is the set of the atoms of $\mathfrak{A}$.

Any numerical function $P(A)$, defined for each $A \in \mathfrak{A}$ is called *probability* on the measurable space $(\Omega, \mathfrak{A})$, if it satisfies the axioms: (1) $P(A) \geqslant 0$ for each $A \in \mathfrak{A}$ and $P(\Omega) = 1$; (2) $P\left(\sum_{k=1}^{n} A_k\right) = \sum_{k=1}^{n} P(A_k)$ for an arbitrary finite number of pairwise mutually exclusive events A_1, ..., A_n (*finite additivity*); (3) $P(A_n) \to 0$ for $A_n \downarrow \emptyset$ when $n \to \infty$ (*continuity*).

The triple $(\Omega, \mathfrak{A}, P)$ is said to be a *probability space*, and $(\mathfrak{A}, P)$ is said to be a *probability field*.

The set (the event) $N \subset \Omega$ is said to be a *negligible* with respect to the probability P, if there exists $A \in \mathfrak{A}$ with $P(A) = 0$ and $N \subset A$. If $\mathfrak{A}$ contains all P-negligible sets in Ω, then it is said to be a complete σ-algebra with respect to P and $(\Omega, \mathfrak{A}, P)$ is said to be a *complete probability space*.

Note. The event $A \Delta B = A\bar{B} + \bar{A}B = (A \setminus B) + (B \setminus A)$ is said to be a *symmetrical difference* of A and B (see also Section 2).

Illustrative Examples

Example 11.1. Let Ω be the real line $\mathbb{R}_1$ and let $\mathfrak{A} = \mathcal{P}(\mathbb{R}_1)$. Let us consider the events $A_n = [0, 1 + (-1)^{n+1} \cdot \frac{1}{n}]$, $n = 1, 2, \ldots$ Describe the events

$$A^* = \limsup_n A_n \quad \text{and} \quad A_* = \liminf_n A_n .$$

Is the sequence $\{A_n\}_{n=1}^{\infty}$ convergent?

Solution. According to the definition, $A^* = \limsup_n A_n = \bigcap_{n=1}^{\infty} \bigcup_{k=n}^{\infty} A_k$. But

$$\bigcup_{k=n}^{\infty} A_k = \begin{cases} [0, 1 + \frac{1}{n}], & \text{if } n \text{ is odd,} \\ [0, 1 + \frac{1}{n+1}], & \text{if } n \text{ is even,} \end{cases}$$

and therefore $A^* = \bigcap_{n=1}^{\infty} [0, 1 + \frac{1}{n}] = [0, 1]$.

In the same way $A_* = \liminf_n A_n = \bigcup_{n=1}^{\infty} \bigcap_{k=n}^{\infty} A_k$,

$$\bigcap_{k=n}^{\infty} A_k = \begin{cases} [0,\ 1 - \frac{1}{n}], & \text{if n is even} \\ [0,\ 1 - \frac{1}{n+1}], & \text{if n is odd,} \end{cases}$$

and therefore $A_* = \bigcup_{n=1}^{\infty} [0,\ 1 - \frac{1}{n+1}] = [0,\ 1)$.

Since $1 \in A^*$ but $1 \overline{\in} A_*$, then $A_* \neq A^*$, and hence the sequence $\{A_n\}_{n=1}^{\infty}$ is not convergent.

Example 11.2. Let $\Omega = (0,\ 1]$. Let $\mathbb{A} \subset \mathbb{P}(\Omega)$ be the class of such subsets of Ω such that they either are countable sets or their complements are countable sets. Let Q be the set of the rational numbers in Ω. Each $r \in Q$ can be represented in the form $r = m/n$, where m and n are integers and $(m,\ n) = 1$; i.e., the fraction m/n is irreducible. (Here (m, n) denotes the greatest common divisor of m and n.)

Let us define $p_r = 1/(s_n 2^n)$ for any $r \in Q$, where s_n is the total number of all the irreducable fractions from (0, 1] with a denominator n. Prove that: (i) $\mathbb{A}$ is a σ-algebra; (ii) the equality $P(A) = \sum_{r \in A \cap Q} p_r$ defines a probability in $\mathbb{A}$, but the probability space $(\Omega,\ \mathbb{A},\ P)$ is not complete.

Solution. (i) We are going to make use of the fact that a countable union of countable sets is a countable set as well.

Now let $A = \bigcup_{k=1}^{\infty} A_k$, $A_k \in \mathbb{A}$, and let at least one set (for instance A_{k_0}) be the complement of the countable set B (i.e., $A_{k_0} = \overline{B}$). Then $\overline{A} = \bigcap_{k=1}^{\infty} \overline{A}_k \subset \overline{A}_{k_0} = B$; hence, A is a countable set and $A \in \mathbb{A}$. According to our definition, $\mathbb{A}$ is closed with respect to complementation as well, and therefore $\mathbb{A}$ is a σ-algebra.

(ii) The proof is obtained from the following assertions:

(a) Since $p_r > 0$ for each $r \in Q$, then $P(A) \geq 0$ for each $A \in \mathbb{A}$.

(b) $P(\Omega) = \sum_r p_r = \sum_{n=1}^{\infty} s_n/(s_n 2^n) = \sum_{n=1}^{\infty} 2^{-n} = 1$.

(c) To prove the σ- additivity of P, let us put $A = \sum_{i=1}^{\infty} A_i$, $A_i \in \mathbb{A}$. We have

$$\sum_{i=1}^{\infty} P(A_i) = \sum_{i=1}^{\infty} \sum_{r \in A_i} p_r = \sum_{r \in \sum_{i=1}^{\infty} A_i} p_r = \sum_{r \in A} p_r = P(A).$$

(Here the communative rule for convergent series with positive terms is used.)

Let $C = (0,\ \frac{1}{2}] \setminus Q$. Then neither C nor $\overline{C}$ are countable, and hence

$C \overline{\in} \mathcal{A}$. But $C \subset \Omega \setminus Q$, and $P(\Omega \setminus Q) = 0$. This shows that $(\Omega, \mathcal{A}, P)$ is not a complete probability space.

Exercises

11.1. For which events A and B are the following identities true: (a) $A \cup \overline{A} = A$; (b) $AA = A$; (c) $A \cup B = A$; (d) $AB = A$; (e) $A \cup B = \overline{A}$; (f) $AB = \overline{A}$; (g) $A \cup B = AB$?

11.2. Prove that for arbitrary events A, B, C and D the following equalities are true: (a) $A \setminus B = A \setminus (AB) = (A \cup B) \setminus B$; (b) $A \Delta B = (A \cup B) \setminus (AB)$; (c) $(A \setminus B)(C \setminus D) = (AC) \setminus (BD)$; (d) $(AB) \Delta (AC) = A(B \Delta C)$; (e) $(A \Delta B) + (A \Delta \overline{B}) = \Omega$; (f) $A \cup B = A \Delta B \Delta (AB)$.

11.3. Which of the following assertions are true for arbitrary events A, B and C: (a) $(A \cup B) \setminus C = (A \cup (B \setminus C)$; (b) $(A \cup B) \setminus A = B$; (c) $A \cup B = (A \setminus (AB)) + B$; (d) $A \cup BC = C \setminus (C(A \cup B))$; (e) $A \cup B \cup C = A + (B \setminus A) + (C \setminus (A \cup B))$; (f) $ABC \subset AB \cup BC \cup CA \subset A \cup B \cup C$?

11.4. Let $\bigcap_{k=1}^{\infty} A_n \subset B$. Does it follow that $\bigcap_{n=1}^{\infty} \overline{A}_n \subset \overline{B}$?

11.5. Determine the event X from the condition $AX + \overline{BX} = \emptyset$.

11.6. Let $A_1, A_2, \ldots, A_n$ be arbitrary events. Prove that $B_0 = \bigcap_{k=1}^{n} \overline{A}_k$, $B_1 = A_1$, $B_2 = \overline{A}_1A_2$, $B_3 = \overline{A}_1\overline{A}_2A_3$, ..., $B_n = \overline{A}_1 \ldots \overline{A}_{n-1}A_n$ form a partition of the sample space Ω.

11.7. Prove that for arbitrary events $A_1, \ldots, A_n$ the relation

$$\bigcup_{n=1}^{N} A_n = \sum_{n=1}^{N} \left(A_n \bigcup_{j=1}^{n-1} A_j\right)$$

holds when $\bigcup_{j=1}^{0} A_j = \emptyset$.

11.8. Prove that for an arbitrary sequence of events $\{A_n\}$ the following representation is valid:

$$\bigcup_{i=1}^{\infty} A_i = A_1 + \overline{A}_1A_2 + \overline{A}_1\overline{A}_2A_3 + \ldots + \overline{A}_1 \ldots \overline{A}_{n-1}A_n + \ldots$$

11.9. Let $A_1, \ldots, A_n$ be arbitrary events and set $S_k = \bigcap(A_{i_1} \cup A_{i_2} \cup \ldots \cup A_{i_k})$, $U_k = \bigcup(A_{i_1} \cap A_{i_2} \cap \ldots \cap A_{i_k})$, where the union and the intersection are taken over all k-tuples $(i_1, i_2, \ldots, i_k)$. Prove that $U_k = S_{n-k+1}$ for $k = 1, 2, \ldots, n$.

11.10. Prove that: (a) $A \Delta B = B \Delta A$; (b) $(A = B) \Leftrightarrow (A \Delta B = \emptyset)$; (c) $A \Delta (B \Delta C) = (A \Delta B) \Delta C$; (d) $(A \Delta B = M \Delta N) \Leftrightarrow (A \Delta M) = B \Delta N)$; (e) $A \cup B = AB + (A \Delta B)$; (f) $A \Delta A = \emptyset$; (g) $A \Delta \overline{A} = \Omega$; (h) $A \Delta \Omega = \overline{A}$; (i) $A \Delta \emptyset = A$, where A, B, C, M and N are arbitrary events, and $\Leftrightarrow$ is a symbol for equivalence.

11.11. Describe all the events of the algebra $\mathcal{A}$ formed from all the subsets of the space $\Omega = \{\omega_1, \ldots, \omega_n\}$, where n is a natural number; find their total number.

11.12. Let $\{A_n\}$ be an arbitrary sequence of subsets of Ω. Prove that: (a) $\omega \in A^*$ if and only if ω is an element of an infinite number of the sets $\{A_n\}$; (b) $\omega \in A_*$ if and only if ω is an element of all the sets $\{A_n\}$ with the exception of a finite number of them at most; (c) the inclusion $A_* \subset A^*$ is valid.

11.13. Let A and B be events. Let us define the sequence $\{C_n\}$, $n = 1, 2, \ldots$, by means of the equalities $C_{2k-1} = A$ and $C_{2k} = B$, $k = 1, 2, \ldots$ Prove that $C_* = AB$ and $C^* = A \cup B$.

*11.14. Prove that any finite Boolean algebra is isomorphic to the Boolean algebra of all the subsets of some given finite set.

11.15. Prove that the number of the elements of a finite Boolean algebra $\mathbb{A}$ can only be a number of the kind 2^n; i.e., $\nu(\mathbb{A}) = 2^n$ for some natural number n.

11.16. Let $\Omega = \{1, 2, 3, 4, 5, 6\}$, $A = \{1, 2, 3\}$ and $B = \{1, 4\}$. What is the number of the σ-algebras in $\mathcal{P}(\Omega)$, containing A and B? Describe the σ-algebra, generated by A and B.

11.17. Let $\Omega = \mathbb{R}_1^+ = [0, \infty)$, $\mathfrak{C}_1$ be the class of all the intervals of the kind $[a, b)$ or $[a, \infty)$ in Ω and $\mathfrak{C}_2$ be the class of all the finite sums of intervals from $\mathfrak{C}_1$. Prove that: (a) $\mathfrak{C}_1$ is not an algebra; (b) $\mathfrak{C}_2$ is an algebra, but is not a σ-algebra; (c) $\mathfrak{C}_2$ is generated by $\mathfrak{C}_1$; (d) the σ-algebra, generated by $\mathfrak{C}_1$, contains all the intervals in Ω (open, closed and half-open, finite or infinite).

11.18. Let the probability space $(\Omega, \mathbb{A}, P)$ be given. Prove the following properties of the probability P: (a) $P(\emptyset) = 0$; (b) if $A \subset B$, then $P(A) \leqslant P(B)$; (c) $P(A \cup B) = P(A) + P(B) - P(AB)$; (d) if $A \supset B$, then $P(A \setminus B) = P(A) - P(B)$; (e) $P(A \setminus B) = P(A) - P(AB)$; (f) $P(A \Delta B) = P(A) + P(B) - 2P(AB)$.

*11.19. Let $(\Omega, \mathbb{A}, P)$ be an arbitrary probability space. Prove that in Ω there are at most countable elements ω such that $P(\{\omega\}) > 0$.

11.20. Prove that for arbitrary events $A_1, \ldots, A_n$ the following inequalities are valid:

$$P\left(\bigcup_{i=1}^{n} A_i\right) \geqslant 1 - \sum_{i=1}^{n} P(\overline{A}_i); \tag{a}$$

$$P\left(\bigcup_{i=1}^{n} A_i\right) = 1 - P\left(\bigcap_{i=1}^{n} \overline{A}_i\right) \leqslant \sum_{i=1}^{n} P(A_i). \tag{b}$$

11.21. Let $\{A_n\}$ be an infinite sequence of events and $P(A_n) = 1$ for $n = 1, 2, \ldots$ Prove that $P\left(\bigcap_{n=1}^{\infty} A_n\right) = 1$.

11.22. Let $\{A_n\}$ be a monotone sequence of events. Prove that: (a) if $A_n \uparrow A$, then $P(A_n) \uparrow P(A)$; (b) if $A_n \downarrow A$, then $P(A_n) \downarrow P(A)$.

11.23. Prove that the axioms (2) and (3) in the definition of probability are equivalent to the following axiom: (2') $P\left(\sum_{i=1}^{\infty} A_i\right) = \sum_{i=1}^{\infty} P(A_i)$ for an arbitrary sequence $\{A_n\}$ of pairwise mutually exclusive events (σ-*additivity* or *countable additivity*).

11.24. Let $\Omega = [0, \infty)$, and $\mathfrak{C}$ be the class of the intervals of the kind $[0, a)$, $a \geqslant 0$. Let $F(x)$, $0 \leqslant x < \infty$, be a left-continuous, non-decreasing function such that $F(0) = 0$ and $\lim_{x\to\infty} F(x) = 1$. Let us define the function P for the elements of $\mathfrak{C}$ in the following way: $P([0, a)) = F(a)$.

Prove that the σ-algebra $\mathfrak{A} = \sigma(\mathfrak{C})$ contains all the intervals in Ω. How are $P((a, b))$ and $P([a, b])$ to be defined by means of $F(x)$ so that P will be countable additive?

11.25. Prove that for an arbitrary sequence of events $\{A_n\}$ the following inequalities are true:

$$P(A_*) \leqslant \liminf_n P(A_n) \leqslant \limsup_n P(A_n) \leqslant P(A^*).$$

As a corollary prove that if $\lim_{n\to\infty} A_n$ exists, then

$$P(\lim_{n\to\infty} A_n) = \lim_{n\to\infty} P(A_n).$$

11.26. Prove that if $\{A_n\}$ is an arbitrary sequence of events and $\sum_{n=1}^{\infty} P(A_n) < \infty$, then $P(A^*) = 0$. (*Borel-Cantelli lemma*)

11.27. Let $\{A_n\}$, $n = 1, 2, \ldots$, be a sequence of events with the same probability $P(A_n) = p$. Prove that $P(A^*) \geqslant p$.

11.28. Prove that the class of negligible sets in $(\Omega, \mathfrak{A}, P)$ is closed with respect to the operations of countable union and intersection.

*11.29. Let $\mathfrak{N}$ be the class of negligible sets in $(\Omega, \mathfrak{A}, P)$; let $\mathfrak{A}^*$ be the class of sets of the kind $A \cup N$, $A \in \mathfrak{A}$ and $N \in \mathfrak{N}$. Prove that: (a) if $\mathfrak{A}$ is a σ-algebra, then $\mathfrak{A}^*$ is also a σ-algebra; (b) $\mathfrak{A}^*$ is generated by $\mathfrak{A} \cup \mathfrak{N}$; (c) the formula $P^*(A \cup N) = P(A)$ defines probability in $(\Omega, \mathfrak{A}^*)$; (d) P^* is the unique probability in $\mathfrak{A}^*$, which coincides with P in $\mathfrak{A}$.

11.30. Prove that the probability space $(\Omega, \mathfrak{A}^*, P^*)$, defined in Exercise 11.29, is complete.

*11.31. Let $\mathfrak{C} \subset \mathfrak{P}(\Omega)$ be a semialgebra. (The class $\mathfrak{C}$ is called a *semialgebra*, if $\Omega \in \mathfrak{C}$, $\emptyset \in \mathfrak{C}$, $\mathfrak{C}$ is closed with respect to the finite intersections, and the complement of any element from $\mathfrak{C}$ is a finite sum of non-intersecting elements from $\mathfrak{C}$.) Prove that the algebra $\mathfrak{A}$, generated by $\mathfrak{C}$, coincides with the class of all the possible finite sums on non-intersecting sets from $\mathfrak{C}$.

11.32. Prove that for arbitrary events A, B and C the inequality $P(A \Delta B) \leqslant P(A \Delta C) + P(B \Delta C)$ is fulfilled. When is the equality attained?

*11.33. Let A and B be events and $P(A \cup B) > 0$. We put $d(A, B) =$

$P(A \Delta B)/P(A \cup B)$. Prove that for arbitrary events A, B and C such that $P(A \cup B) > 0$, $P(B \cup C) > 0$ and $P(A \cup C) > 0$, the "triangle inequality" holds:

$$d(A, B) \leqslant d(A, C) + d(B, C).$$

*11.34. Let Ω be the set of the rational numbers r from the interval [0, 1]. We denote by $\mathfrak{F}$ the class of the subsets of Ω of the kind $\{r : a \leqslant r \leqslant b\}$, $\{r : a < r \leqslant b\}$, $\{r : a < r < b\}$ or $\{r : a \leqslant r < b\}$, where a and b are rational numbers. The class $\mathfrak{B}$ of the finite sums of non-intersecting sets from $\mathfrak{F}$ is an algebra. Let us define $P(A) = b - a$ for $A \in \mathfrak{F}$ and $P(B) = \sum_{i=1}^{n} P(A_i)$, if $B = \sum_{i=1}^{n} A_i$, $A_i \in \mathfrak{F}$. Show that P is finitely additive, but is not σ-additive in $\mathfrak{B}$.

11.35. Let Ω be an infinite countable set with elements $\omega_1, \omega_2, \ldots, \omega_n, \ldots$ Let us consider the probability space $(\Omega, \mathfrak{P}(\Omega), P)$, where P is defined in the following way: $P(\{\omega_n\}) = p_n$, where $p_n \geqslant p_{n+1} \geqslant 0$, $n = 1, 2, \ldots$, and $\sum_{n=1}^{\infty} p_n = 1$. (a) Prove that the set of the values of the probability P is a perfect set. (b) Prove that the set of the values of the probability P coincides with the interval [0, 1] if and only if $p_n \leqslant \sum_{i=n+1}^{\infty} p_i$ for $n = 1, 2, \ldots$ (c) If $x_1, \ldots, x_m$ are arbitrary non-negative numbers and $\sum_{i=1}^{m} x_i = 1$, the condition $p_n \leqslant (1/m) \sum_{k=n}^{\infty} p_k$, $n = 1, 2, \ldots$ is necessary and sufficient for the existence of events $A_1, \ldots, A_m$ such that $P(A_i) = x_i$, $i = 1, \ldots, m$.

Note. The set C is called *perfect*, if it is closed and does not contain isolated points; i.e., if C contains the limit of each sequence of points belonging to C, and if each point x from C is a limit of a sequence of points, belonging to C, and all terms of which are different from x.

*11.36. Suppose the probability space $(\Omega, \mathfrak{A}, P)$ contains no atoms; i.e., for each event A with $P(A) > 0$ there exists $B \in \mathfrak{A}$ such that $B \subset A$ and $0 < P(B) < P(A)$. Prove that for an arbitrary $\varepsilon > 0$ the space Ω can be partitioned to a finite number of mutually exclusive events $A_1, \ldots, A_n$ such that $P(A_i) < \varepsilon$, $i = 1, \ldots, n$.

*11.37. Let the probability space $(\Omega, \mathfrak{A}, P)$ contain no atoms (see Exercise 11.36). Prove that the set of values of the probability $\vec{P}$ is the interval [0, 1].

*11.38. If $(\Omega, \mathfrak{A}, P)$ is an arbitrary probability space, prove that the set of values of the probability $\vec{P}$ is closed; i.e., if $x_n \to x$ and $x_n = P(A_n)$, $A_n \in \mathfrak{A}$, then there exists $A \in \mathfrak{A}$ with $P(A) = x$.

*11.39. Let $(\Omega, \mathfrak{A}, P)$ and $(\Omega, \mathfrak{A}, Q)$ be probability spaces and let $P(C) = Q(C)$, $C \in \mathfrak{C}$, where $\mathfrak{C} \subset \mathfrak{A}$ is some class of events, closed with respect to the operation of intersection. Prove that $P(A) = Q(A)$ for any $A \in \sigma(\mathfrak{C})$.

*11.40. Let Ω be the set of all ordered samples of different ele-

ments from the set $M = \{1, \ldots, n\}$, the samples containing at least two elements each. Let P be a probability which is defined on Ω. Let $I \subset M$ and $\bar{I}$ be its complement. Let us denote by $A(I, \bar{I})$ the set of those samples from Ω which contain elements both from I and $\bar{I}$ ($A(I, \bar{I}) = \emptyset$, if I or $\bar{I} = \emptyset$). Prove that, for any probability P in Ω, the set $I \subset M$ can be chosen in such a way that the inequality $P(A(I, \bar{I})) \geq \frac{1}{2}$ holds.
(*H. Araki, I. Woods, D. Vandev*)

12. Random Variables and Integration

Introductory Notes

Let Ω_1 and Ω_2 be arbitrary non-empty sets, and $\mathbb{P}(\Omega_1)$ and $\mathbb{P}(\Omega_2)$ be respectively the sets of all their subsets. If the mapping $\xi : \Omega_1 \to \Omega_2$ is given, then the mapping $\xi^{-1} : \mathbb{P}(\Omega_2) \to \mathbb{P}(\Omega_1)$ is defined by means of the equality $\xi^{-1}(A) = \{\omega_1 : \omega_1 \in \Omega_1, \xi(\omega_1) \in A \subset \Omega_2\}$.

Let $(\Omega_1, \mathbb{A}_1)$ and $(\Omega_2, \mathbb{A}_2)$ be two measurable spaces. The mapping $\xi : \Omega_1 \to \Omega_2$ is said to be *measurable*, if the pre-images of the measurable sets in $(\Omega_2, \mathbb{A}_2)$ are measurable in $(\Omega_1, \mathbb{A}_1)$; i.e., if for any $A \in \mathbb{A}_2$, we have $\xi^{-1}(A) \in \mathbb{A}_1$.

Let $(\Omega, \mathbb{A}, P)$ be an arbitrary probability space, $\mathbb{R}_1 = (-\infty, +\infty)$ and $\mathbb{B}_1$ be the Borel σ-algebra in $\mathbb{R}_1$. Any measurable mapping (function) $\xi : (\Omega, \mathbb{A}) \to (\mathbb{R}_1, \mathbb{B}_1)$ is called a *random variable* (r.v.). Obviously, each r.v. ξ induces a probability P_ξ in $(\mathbb{R}_1, \mathbb{B}_1)$ by means of the equality $P_\xi(B) = P(\xi^{-1}(B))$, $B \in \mathbb{B}_1$ (see also Exercise 12.4). The function $F_\xi(x) = P(\xi^{-1}(-\infty, x))$ is called a *distribution function* (d.f.) of the r.v. . (The properties of d.f.'s are considered in Section 15.)

The discrete r.v. ξ is called *simple* or *elementary*, if it takes only a finite number of values. (See also Section 9.) The function

$$1_A(\omega) = \begin{cases} 1, & \text{if } \omega \in A \\ 0, & \text{if } \omega \bar{\in} A \end{cases}$$

is called an *indicator* of the set A; if $A \in \mathbb{A}$, then 1_A is a r.v. The simple r.v. ξ, taking on the values $x_1, \ldots, x_n$, has the representation $\xi(\omega) = \sum_{i=1}^{n} x_i 1_{A_i}(\omega)$, where $A_i = \{\omega : \xi(\omega) = x_i\}$ and $A_1, \ldots, A_n$ form a partition of Ω.

The r.v. ξ is called *non-negative*, if $\xi(\omega) \geq 0$ for any $\omega \in \Omega$. A sequence of r.v.'s, denoted $\{\xi_n(\omega)\}$, is given by

$$\xi_1(\omega), \xi_2(\omega), \ldots, \xi_n(\omega), \ldots, \tag{12.1}$$

where $\xi_n(\omega)$ is a r.v. for any n = 1, 2, ... The sequence (12.1) is called monotonically increasing, if $\xi_n(\omega) \leq \xi_{n+1}(\omega)$ for any n and for any $\omega \in \Omega$. The sequence (12.1) is called *convergent*, if for any $\omega \in \Omega$ the numerical sequence $\xi_1(\omega)$, $\xi_2(\omega)$, ... is convergent. In this case its limit is a function of ω, which we denote by $\xi(\omega)$; i.e., for any $\omega \in \Omega$ we have $\lim_{n\to\infty} \xi_n(\omega) = \xi(\omega)$. The following properties concerning sequences of r.v.'s are important: (i) If $\{\xi_n(\omega)\}$ is a sequence of r.v.'s and if $\lim_{n\to\infty} \xi_n(\omega) = \xi(\omega)$, then $\xi(\omega)$ is a r.v. as well (see Exercise 12.5). (ii) If $\{\xi_n(\omega)\}$ is not convergent for some ω, but the event $A = \{\omega : \{\xi_n(\omega)\}$ is convergent$\}$ has a probability P(A) = 1, then $\{\xi_n\}$ is said to be *convergent with probility 1* or *convergent almost surely* (a.s.). If we denote $\lim_{n\to\infty} \xi_n(\omega) = \xi(\omega)$ for $\omega \in A$, then we can also write $\xi_n(\omega) \xrightarrow{a.s} \xi(\omega)$ as $n \to \infty$ with $\xi(\omega)$ being defined only for $\omega \in A$. If some property is fulfilled for $\omega \in B$ and P(B) = 1, this property is said to be fulfilled with P-probability 1 or P-almost surely (a.s.). If ξ and η are r.v.'s and if $P\{\omega : \xi(\omega) = \eta(\omega)\} = 1$, we write $\xi \overset{a.s.}{=\!=} \eta$. In this case ξ and η are said to be *equivalent* (P-equivalent).

The concept of an integral of a r.v. ξ with respect to the probability P is of fundamental significance in the theory of probability. The following notations are used:

$$\int_\Omega \xi(\omega)\, dP(\omega) = \int_\Omega \xi(\omega) P(d\omega) = \int_\Omega \xi\, dP = E(\xi).$$

This integral is called an *expectation* or the *mean value* of the r.v. (see also Section 17).

Usually the integral is defined in the following way:

(1) If ξ is a simple r.v. and $\xi(\omega) = \sum_{i=1}^{n} x_i 1_{A_i}(\omega)$, then

$$\int_\Omega \xi\, dP = \sum_{i=1}^{n} x_i P(A_i).$$

2. If ξ is a non-negative r.v. and $\{\xi_n\}$ is a monotonically increasing sequence of simple r.v.'s such that $\lim_{n\to\infty} \xi_n = \xi$, then

$$\int_\Omega \xi\, dP = \lim_{n\to\infty} \int_\Omega \xi_n\, dP.$$

Obviously $\int_\Omega \xi\, dP \geq 0$ for $\xi \geq 0$. If this limit is a finite one, then ξ is called an *integrable r.v.* and the integral is said to be convergent.

Otherwise we write $\int_\Omega \xi \, dP = +\infty$ and ξ is called *quasi-integrable*.

3. Let ξ be an arbitrary r.v., $\xi^+ = \max[\xi, 0]$, $\xi^- = \max[-\xi, 0]$. Then ξ^+ and ξ^- are non-negative r.v.'s (see Exercise 12.8) and $\xi = \xi^+ - \xi^-$. The r.v. ξ is said to be *integrable*, if ξ^+ and ξ^- are integrable, and then

$$\int_\Omega \xi \, dP = \int_\Omega \xi^+ \, dP - \int_\Omega \xi^- \, dP.$$

In this case $\int_\Omega |\xi| \, dP < \infty$, since $|\xi| = \xi^+ + \xi^-$. If only one of the variables ξ^+ and ξ^- is integrable, then ξ is called quasi-integrable.

Let $A \in \mathcal{A}$. An integral of a r.v. ξ on the set A is defined by means of the equality

$$\int_A \xi \, dP = \int_\Omega \xi(\omega) 1_A(\omega) \, dP.$$

For instance,

$$\int_A dP = \int_\Omega 1_A \, dP = P(A).$$

If $\{A_i\}$ is a sequence of pairwise mutually exclusive events and $A = \bigcup_{i=1}^{\infty} A_i$, then for any integrable r.v. ξ the equality

$$\int_A \xi \, dP = \sum_{i=1}^{\infty} \int_{A_i} \xi \, dP$$

holds (σ-additivity of the integral).

If for the sequence of the events $\{A_n\}$ we have $A_n \downarrow \emptyset$ as $n \to \infty$ and if ξ is an integrable r.v., then

$$\lim_{n\to\infty} \int_{A_n} \xi \, dP = 0$$

(continuity of the integral).

Let P and Q be probabilities on the measurable space $(\Omega, \mathcal{A})$. Then P is said to be *absolutely continuous* with respect to Q (denoted by $P \ll Q$), if for any $A \in \mathcal{A}$ such that $Q(A) = 0$ we also have $P(A) = 0$.

<u>Radon-Nikodym theorem.</u> If P and Q are probabilities on $(\Omega, \mathcal{A})$ and if $P \ll Q$, then there exists a random variable $\xi = \xi(\omega)$, which is non-negative, finite, integrable with respect to Q and such that for any $A \in \mathcal{A}$,

$$P(A) = \int_A \xi(\omega)\, dQ(\omega).$$

The variable ξ is unique up to Q-equivalence.

We shall give some more properties of the integral.

If ξ and η are integrable r.v.'s and c is a constant, the r.v.'s $\xi + \eta$ and $c\xi$ (see Exercise 12.8) are also integrable and for them we have

$$\int(\xi + \eta)\, dP = \int \xi\, dP + \int \eta\, dP; \tag{a}$$

$$\int(c\xi)\, dP = c\int \xi\, dP; \tag{b}$$

$$\int \xi\, dP \leq \int \eta\, dP, \quad \text{if } \xi \leq \eta \tag{c}$$

Lebesgue theorem. Let $\{\xi_n\}$ be a sequence of r.v.'s and $\xi_n \xrightarrow{a.s.} \xi$. If $|\xi_n| < \eta$ for each n, where η is an integrable r.v., then the limit ξ is integrable, and

$$\int \xi\, dP = \lim_{n\to\infty} \int \xi_n\, dP.$$

The sequence $\{\xi_n\}$ of integrable r.v.'s is said to be *uniformly integrable*, if for any $\varepsilon > 0$ there exists $c > 0$ such that for all n we have

$$\int_{A_c} |\xi_n|\, dP < \varepsilon,$$

where $A_c = \{\omega : |\xi_n(\omega)| > c\}$. This is equivalent to the relation

$$\sup_n \int_{\{|\xi_n|>a\}} |\xi_n|\, dP \to 0 \quad \text{as } a \to \infty.$$

Illustrative Examples

Example 12.1. Let us consider the measurable space $(\mathbb{R}_1, \mathcal{B}_1)$. The mapping $\xi : \mathbb{R}_1 \to \mathbb{R}_1$ is given, where $\xi(x) = \cos x$. Describe the σ-algebra $\xi^{-1}(\mathcal{B}_1)$, generated by ξ.

Solution. Let $B \in \mathcal{B}_1$. If $\cos x \in B$, then $\cos(-x) \in B$, as well as $\cos(x + 2k\pi) \in B$ for $k = \pm 1, \pm 2, \ldots$ Thus for $\xi^{-1}(B)$ we have: $x \in \xi^{-1}(B)$ $\Rightarrow$ $-x \in \xi^{-1}(B)$, as well as $x + 2k\pi \in \xi^{-1}(B)$. But $\cos x$ is a continuous function, and hence a measurable one; therefore, $\xi^{-1}(B) \subset \mathcal{B}_1$. So the pre-images of Borel sets are also Borel sets. They are symmetrical with respect to the origin and periodic with a period 2π. Such a set A has the form

$$A = \pm A^* + 2k\pi, \quad k = \pm 1, \pm 2, \ldots,$$

where

$$A^* \in \mathcal{B}_{[0,\pi]}, \quad -A^* = \{x : -x \in A^*\},$$

$$A^* + 2\pi = \{x + 2\pi, x \in A^*\}.$$

Example 12.2. Let us consider the sequence of r.v.'s $\{\xi_n\}_{n=1}^{\infty}$, where $P\{\xi_n = 2^n\} = 2^{-n}$ and $P\{\xi_n = 0\} = 1 - 2^{-n}$. Prove that the r.v.'s ξ_n are integrable, but the sequence $\{\xi_n\}_{n=1}^{\infty}$ is not uniformly integrable.

Solution. For the expectation of ξ_n we obtain $E(\xi_n) = 2^n \cdot 2^{-n} = 1$; hence, the r.v.'s ξ_n are integrable for each n.

The following relation holds:

$$\int_{\{|\xi_n| \geqslant a\}} |\xi_n| \, dP = \begin{cases} 0, & \text{if } a > 2^n, \\ 1, & \text{if } a \leqslant 2^n. \end{cases}$$

This means that $\int_{\{|\xi_n| \geqslant a\}} |\xi_n| \, dP$ does not tend to zero uniformly in n when $a \to \infty$. Therefore the sequence $\{\xi_n\}_{n=1}^{\infty}$ is integrable, but it is not not uniformly integrable.

Exercises

12.1. Let $(\Omega_1, \mathcal{A}_1)$ and $(\Omega_2, \mathcal{A}_2)$ be measurable spaces and ξ be a mapping of Ω_1 in Ω_2. Prove that: (a) the mapping ξ^{-1} preserves the set-theoretic operations union, intersection and complementation; (b) $\xi^{-1}(\mathcal{A}_2)$ is a σ-algebra; (c) ξ is a measurable mapping if and only if $\xi^{-1}(\mathcal{C}) \subset \mathcal{A}_1$ for some class $\mathcal{C}$, generating $\mathcal{A}_2$.

12.2. Prove that the mapping $\xi : (\Omega, \mathcal{A}) \to (\mathbb{R}_1, \mathcal{B}_1)$ is measurable if the pre-images of the intervals of the kind $(-\infty, x)$, $x \in \mathbb{R}_1$ are measurable.

12.3. Let the mappings $\xi : (\Omega_1, \mathcal{A}_1) \to (\Omega_2, \mathcal{A}_2)$ and $\eta : (\Omega_2, \mathcal{A}_2) \to (\Omega_3, \mathcal{A}_3)$ be measurable. Prove that the mapping $\eta(\xi) : (\Omega_1, \mathcal{A}_1) \to (\Omega_3, \mathcal{A}_3)$ is also measurable.

12.4. An arbitrary probability space $(\Omega, \mathcal{F}, P)$ and an arbitrary measurable space $(\Omega^*, \mathcal{F}^*)$ are given. Let $\xi : (\Omega, \mathcal{F}) \to (\Omega^*, \mathcal{F}^*)$ be a measurable mapping. Prove that the formula $P^*(A^*) = P(\xi^{-1}(A^*))$, $A^* \in \mathcal{F}^*$, defines a probability in $\mathcal{F}^*$. The triple $(\Omega^*, \mathcal{F}^*, P^*)$ is called an *induced probability space.*

12.5. Prove that if the sequence of r.v.'s $\{\xi_n\}$ is convergent, then its limit is also a r.v.

12.6. Let $\{\xi_n\}$ be a sequence of r.v.'s. Represent the sets $A = \{\omega : \lim_n \xi_n(\omega) \text{ exists}\}$ and $B = \{\omega : \lim_n \xi_n(\omega) \text{ does not exist}\}$ in terms of events of the type $\{\omega : |\xi_n(\omega) - \xi_m(\omega)| \geq \varepsilon_k\}$ and the operations of union, intersection and complementation. Show that A and B are events.

12.7. For the indicators of sets prove the properties: (a) $1_A \leq 1_B$ is equivalent to $A \subset B$; $1_A = 1_B$ is equivalent to $A = B$. (b) $1_\emptyset = 0$, $1_\Omega = 1$, $1_{\bar{A}} = 1 - 1_A$; (c) $1_{\bigcup_n A_n} = \sup_n 1_{A_n}$, $1_{\bigcap_n A_n} = \inf_n 1_{A_n}$.

12.8. Let ξ and η be r.v.'s, $c \in \mathbb{R}_1$. Prove that the following are r.v.'s: (a) c; (b) $c\xi$; (c) $\xi + \eta$; (d) $\xi\eta$; (e) $\min[\xi, \eta]$; (f) $\max[\xi, \eta]$; (g) ξ^+ and ξ^-.

12.9. The functions $F_k(x_1, \ldots, x_n)$, $k = 1, \ldots, n$, $(x_1, \ldots, x_n) \in \mathbb{R}_n$, are defined in the following way: $F_k(x_1, \ldots, x_n) = x_{(k)}$, where $x_{(k)}$ is the kth by size number from $x_1, x_2, \ldots, x_n$; i.e., $x_{(1)} \leq x_{(2)} \leq \ldots \leq x_{(n)}$. Prove that the mapping $F_k : \mathbb{R}_n \to \mathbb{R}_1$ is measurable for any k, $k = 1, \ldots, n$.

12.10. Let $\Omega = [a, b]$, $\mathcal{A}$ be the Borel σ-algebra of the subsets of Ω, and $P(A)$, $A \in \mathcal{A}$, be proportional to the Lebesgue measure of A. Let $\xi(\omega) = \omega$, $\omega \in \Omega$. Prove that ξ is a r.v. and find its d.f. (uniform distribution, see Section 15).

12.11. Let the r.v. ξ be integrable. Show that the function $Q(A) = \int_A \xi(\omega)\, dP(\omega)$, $A \in \mathcal{A}$ is σ-additive. Under what conditions would $Q(A)$ be a probability?

12.12. Let the r.v. ξ be integrable on the set $A \in \mathcal{A}$ and let B be a set from $\mathcal{A}$ such that $P(A \Delta B) = 0$. Show that ξ is integrable on B and $\int_B \xi\, dP = \int_A \xi\, dP$.

12.13. Let ξ and η be integrable r.v.'s coinciding almost everywhere; i.e., $P\{\omega : \xi(\omega) \neq \eta(\omega)\} = 0$. Prove that $E(\xi) = E(\eta)$.

12.14. Let $\alpha : (\Omega_1, \mathcal{A}_1) \to (\Omega_2, \mathcal{A}_2)$ be a measurable mapping, P be a probability in $\mathcal{A}_1$, P_α be the induced probability in $\mathcal{A}_2$ (see Exercise 12.4) and ξ be a r.v. in $(\Omega_2, \mathcal{A}_2, P_\alpha)$. Show that

$$\int_{\Omega_2} \xi(\omega_2)\, dP_\alpha(\omega_2) = \int_{\Omega_1} \xi(\alpha(\omega_1))\, dP(\omega_1);$$

i.e., if at least one of the two integrals exists, then the other exists as well and they are equal. (*E. B. Dynkin*)

12.15. Let $\xi = \xi(\omega)$ be a r.v. on the probability space $(\Omega, \mathcal{A}, P)$,

$g(x)$, $x \in \mathbb{R}_1$, be a measurable function, $\eta(\omega) = g(\xi(\omega))$ and P_ξ be the probability in $(\mathbb{R}_1, \mathcal{B}_1)$, induced by ξ. Prove that if $E(\eta)$ exists, then

$$E(\eta) = \int_\Omega \eta(\omega)\, dP(\omega) = \int_{\mathbb{R}_1} g(x)\, dP_\xi(x).$$

12.16. Let ξ be a non-negative r.v. Prove that from $E(\xi) = 0$ it follows that $P\{\xi = 0\} = 1$.

12.17. Let $\mathbb{L}_r = \mathbb{L}_r(\Omega, \mathcal{F}, P)$ be the space of r.v.'s ξ (equivalent r.v.'s are regarded as equal), defined on the probability space $(\Omega, \mathcal{F}, P)$ such that $E\{|\xi|^r\} < \infty$ for $r \in (0, 1]$. Prove that $d_r(\xi, \eta) = E\{|\xi - \eta|^r/(1 + |\xi - \eta|^r)\}$ defines a metric in $\mathbb{L}_r$.

*12.18. Prove that the sequence of r.v.'s $\{\xi_n\}$ is uniformly integrable, if there exists a positive function $g(x)$, $x \geq 0$, such that $\lim_{x\to\infty} (g(x)/x) = +\infty$ and $\sup_n E\{g(|\xi_n|)\} < \infty$. (*C. J. Vallée-Poussin*)

12.19. The sequence $\{\xi_n\}$ with $E\{|\xi_n|\} < \infty$ for $n \geq 1$ is uniformly integrable if and only if the following two conditions are satisfied: (1) $\sup_n E\{|\xi_n|\} < \infty$; (2) for each $\varepsilon > 0$ there exists $\delta > 0$ such that from $P(A) < \delta$ the following inequality $\sup_n \int_A |\xi_n|\, dP \leq \varepsilon$ holds.

13. Conditional Probability, Independence and Martingales

Introductory Notes

Let $(\Omega, \mathcal{A}, P)$ be a probability space. The concept of conditional probability with respect to a given event B with $P(B) > 0$ is considered in Section 4. Let $\mathcal{F}$ be a sub-σ-algebra of $\mathcal{A}$. Let ξ be a random variable. Any $\mathcal{F}$-measurable r.v., defined to within a P-equivalence from the relation

$$\int_B E\{\xi|\mathcal{F}\}\, dP = \int_B \xi\, dP \tag{13.1}$$

for an arbitrary $B \in \mathcal{F}$, is called a *conditional mean* (a *conditional expectation*) of the integrable r.v. ξ with respect to the σ-algebra $\mathcal{F}$. We shall denote it by $E\{\xi|\mathcal{F}\}$.

The existence of $E\{\xi|\mathcal{F}\}$ follows from the Radon-Nikodym theorem (see Section 12).

The *conditional probability* with respect to the σ-algebra $\mathcal{F}$, denoted by $P\{\cdot|\mathcal{F}\}$, is defined by the equality

$$P\{A|\mathcal{F}\} = E\{1_A|\mathcal{F}\}, \quad \text{(P-a.s.)} \tag{13.2}$$

for an arbitrary $A \in \mathcal{A}$.

If ξ is a r.v., it generates the σ-algebra $\mathcal{F}_\xi \subset \mathcal{A}$ (see Exercise

12.1). If $\mathcal{F}_1, \mathcal{F}_2, \ldots$ are σ-algebras, we denote the minimal σ-algebra, containing $\mathcal{F}_1, \mathcal{F}_2, \ldots$, by the symbol $\sigma(\mathcal{F}_1, \mathcal{F}_2, \ldots)$. The σ-algebra $\sigma(\xi_1, \xi_2, \ldots) = \mathcal{F}(\xi_1, \xi_2, \ldots)$, generated by the r.v.'s $\xi_1, \xi_2, \ldots$, is defined analogously.

The conditional mean of the r.v. ξ with respect to the r.v. η is defined in the following way:

$$E\{\xi|\eta\} = E\{\xi|\mathcal{F}_\eta\}, \quad \text{(P-a.s.)}. \tag{13.3}$$

Some properties of the conditional mean are given by:
(1) If $\xi(\omega) \geqslant 0$, then $E\{\xi|\mathcal{F}\} \geqslant 0$, (P-a.s.).
(2) If $\xi(\omega) = c$, a constant, (P-a.s.), then $E\{\xi|\mathcal{F}\} = c$, (P-a.s.).
(3) If ξ_1 and ξ_2 are integrable r.v.'s, then

$$E\{\xi_1 + \xi_2|\mathcal{F}\} = E\{\xi_1|\mathcal{F}\} + E\{\xi_2|\mathcal{F}\}, \quad \text{(P-a.s.)}.$$

(4) If $\mathcal{F}_1 \subset \mathcal{F}_2 \subset \mathcal{A}$, then

$$E\{\xi|\mathcal{F}_1\} = E\{E(\xi|\mathcal{F}_2)|\mathcal{F}_1\}, \quad \text{(P-a.s.)}.$$

For example, $E\{\xi|\eta_1\} = E\{E(\xi|\eta_1, \eta_2)|\eta_1\}$.
(5) $E(\xi) = E\{E(\xi|\mathcal{F})\}$.
(6) If ξ and η are r.v.'s, ξ is $\mathcal{F}$-measurable and ξ and $\xi\eta$ are integrable, then

$$E\{\xi\eta|\mathcal{F}\} = \xi E\{\eta|\mathcal{F}\}, \quad \text{(P-a.s.)}.$$

In particular,

$$E\{\xi g(\eta)|\eta\} = g(\eta)E\{\xi|\eta\}.$$

Analogous properties are true for the conditional probability $P\{\cdot|\mathcal{F}\}$ with respect to a given σ-algebra $\mathcal{F} \subset \mathcal{A}$, as well as for the conditional probability $P\{\cdot|\eta\} = P\{\cdot|\mathcal{F}_\eta\}$ with respect to a given r.v. η.

Recall (see the Introductory Notes of Section 4) that the events $A_1, A_2, \ldots$ (finite or infinite number) are called *independent*, if for every combination of different indices $i_1, i_2, \ldots, i_k$, $k = 2, 3, \ldots$, we have

$$P(A_{i_1} A_{i_2} \ldots A_{i_k}) = P(A_{i_1})P(A_{i_2}) \ldots P(A_{i_k}). \tag{13.4}$$

The independence of classes of events is defined analogously. Let the classes of events $\mathfrak{C}_1, \mathfrak{C}_2, \ldots$ (finite or infinite number) be given. Then they are called *independent*, if the equality (13.4) holds for every combination of different indices $i_1, i_2, \ldots, i_k$, $k = 2, 3, \ldots$ and for every possible choice of the events $A_{i_1}, A_{i_2}, \ldots, A_{i_k}$ such that $A_{i_1} \in$

$\mathfrak{C}_{i_1}$, $A_{i_2} \in \mathfrak{C}_{i_2}$, ..., $A_{i_k} \in \mathfrak{C}_{i_k}$.

The random variables $\xi_1, \xi_2, \ldots$ are said to be *independent*, if the σ-algebras $\mathcal{F}_{\xi_1}, \mathcal{F}_{\xi_2}, \ldots$ generated by them are independent.

For independent and integrable r.v.'s ξ_1 and ξ_2 the relation $E(\xi_1\xi_2) = E(\xi_1)E(\xi_2)$ holds.

The concept of independence can also be defined for uncountable many sets of events, of classes, of r.v.'s and of σ-algebras by means of independence in any of their finite subsets.

Kolmogorov's 0-1 law: Let $\{\mathcal{F}_n\}$ be a sequence of independent σ-algebras and let $\mathcal{F} = \bigcap_{k=1}^{\infty} \sigma(\mathcal{F}_k, \mathcal{F}_{k+1}, \ldots)$. Then for any event $A \in \mathcal{F}$ we have either $P(A) = 0$ or $P(A) = 1$.□

Let $\{\xi_n\}$ be a sequence of r.v.'s defined on the probability space $(\Omega, \mathcal{F}, P)$, and let $\{\mathcal{F}_n\}$ be an increasing sequence of sub-σ-algebras of $\mathcal{F}$. The system $\{\xi_n, \mathcal{F}_n\}$, $n = 1, 2, \ldots$, is called a *martingale*, if for any $n \geqslant 1$ we have: (i) ξ_n is $\mathcal{F}_n$-measurable; (ii) $E\{|\xi_n|\} < \infty$; (iii) $E\{\xi_{n+1}|\mathcal{F}_n\} = \xi_n$, (P-a.s.).

If instead of (iii) the relation $E\{\xi_{n+1}|\mathcal{F}_n\} \leqslant \xi_n$, (P-a.s.) or $E\{\xi_{n+1}|\mathcal{F}_n\} \geqslant \xi_n$, (P-a.s.) is fulfilled, then $\{\xi_n, \mathcal{F}_n\}$ is called a *supermartingale* or a *submartingale*, respectively.

Illustrative Examples

Example 13.1. Let ξ be an integrable r.v. on $(\Omega, \mathfrak{A}, P)$ and $\mathcal{F}$ be a sub-σ-algebra of $\mathfrak{A}$. Prove the following assertions:

(i) $P(A) = \int_\Omega E\{1_A|\mathcal{F}\}\, dP$ for an arbitrary $A \in \mathfrak{A}$.

(ii) $E\{c\xi|\mathcal{F}\} = cE\{\xi|\mathcal{F}\}$, where c = constant.

Solution. (i) Since $\Omega \in \mathcal{F}$, it follows from the definition of conditional expectation that

$$\int_\Omega E\{1_A|\mathcal{F}\}\, dP = \int_\Omega 1_A\, dP = \int_A dP = P(A).$$

(ii) For an arbitrary $B \in \mathcal{F}$ the following equalities hold:

$$\int_B E\{c\xi|\mathcal{F}\}\, dP = \int_B c\xi\, dP, \qquad \int_B E\{\xi|\mathcal{F}\}\, dP = \int_B \xi\, dP.$$

Therefore,

$$\int_B E\{c\xi|\mathcal{F}\}\, dP = \int_B c\xi\, dP = \int_B cE\{\xi|\mathcal{F}\}\, dP.$$

Since the conditional expectation is determined up to P-equivalence, then $E\{c\xi|\mathcal{F}\} = cE\{\xi|\mathcal{F}\}$, (P-a.s.). This property, as well as Property (3) from Introductory Notes, shows that the conditional expectation is a linear operator.

Example 13.2. Let $\{\xi_n\}$ be a sequence of independent identically distributed r.v.'s with $E(\xi_1) = 0$ and $E\{\xi_1^2\} = a^2$. Let $\zeta_n = \left(\sum_{k=1}^{n} \xi_k\right)^2 - na^2$ and $\mathcal{F}_n = \sigma(\zeta_1, \zeta_2, \ldots, \zeta_n)$.

Prove that the sequence $\{\xi_n, \mathcal{F}_n\}_{n=1}^{\infty}$ is a martingale.

Solution. It is easy to see that $\mathcal{F}_i \subset \mathcal{F}_{i+1}$, $i = 1, 2, \ldots$ Let $\mathfrak{X}_n = \sigma(\xi_1, \xi_2, \ldots, \xi_n)$. Then $\mathcal{F}_n \subset \mathfrak{X}_n$ for any n.

We have the following as a consequence of the properties of condtional expectation:

$$E\{\zeta_{n+1}|\mathcal{F}_1\} = E\{E\{\zeta_{n+1}|\mathfrak{X}_n\}|\mathcal{F}_n\};$$

$$E\{\zeta_{n+1}|\mathfrak{X}_n\} = E\left\{\left(\sum_{k=1}^{n+1} \xi_k\right)^2 - (n+1)a^2|\mathfrak{X}_n\right\} =$$

$$= E\left\{\left(\zeta_n + \xi_{n+1}^2 - a^2 + 2\xi_{n+1}\sum_{k=1}^{n} \xi_k\right)|\mathfrak{X}_n\right\} =$$

$$= E\{\zeta_n|\mathfrak{X}_n\} + E\{(\xi_{n+1}^2 - a^2)|\mathfrak{X}_n\} + 2E\left\{\xi_{n+1}\sum_{k=1}^{n} \xi_k|\mathfrak{X}_n\right\}.$$

Since ξ_k, $k = 1, 2, \ldots$, are independent,

$$E\{\xi_{n+1}^2 - a^2|\mathfrak{X}_n\} = E(\xi_{n+1}^2 - a^2) = a^2 - a^2 = 0$$

and

$$E\left\{\xi_{n+1}\sum_{k=1}^{n} \xi_k|\mathfrak{X}_n\right\} = \left(\sum_{k=1}^{n} \xi_k\right)E\{\xi_{n+1}\ \mathfrak{X}_n\} = \left(\sum_{k=1}^{n} \xi_k\right)E\xi_{n+1} = 0.$$

Because ζ_n is $\mathcal{F}_n$-measurable and $\mathfrak{X}_n$-measurable, it follows that

$$E\{\zeta_n|\mathcal{F}_n\} = E\{\zeta_n|\mathfrak{X}_n\} = \zeta_n.$$

Hence $E\{\zeta_{n+1}|\mathcal{F}_n\} = \zeta_n$, which means that the sequence $\{\zeta_n, \mathcal{F}_n\}$ is a martingale.

Recall that all the above relations concerning conditional expectation, are to be understood as P-a.s.

Exercises

13.1. Let $(\Omega, \mathcal{A}, P)$ be an arbitrary probability space and B be a fixed event with $P(B) > 0$. We put $Q_B(A) = P(AB)$, $A \in \mathcal{A}$. Let C be an event with $Q_B(C) > 0$. Prove that for any $A \in \mathcal{A}$ the equality $Q_B(A|C) = P(A|BC)$ is fulfilled.

13.2. Let $A_1, \ldots, A_n$ and $B_1, \ldots, B_n$ be two partitions of the sample space Ω. Prove the following *generalized formula for total probility*: for an arbitrary event C we have

$$P(C) = \sum_{i=1}^{n} \left\{ \sum_{j=1}^{m} P(C|A_iB_j)P(B_j|A_i) \right\} P(A_i).$$

13.3. The events $A_1, \ldots, A_n$ are said to be *symmetrically dependent (mutually exchangeable)*, if the probabilities $P(A_{i_1} A_{i_2} \ldots A_{i_k})$, $1 \leqslant k < n$ and $1 \leqslant i_1 < i_2 < \ldots < i_k \leqslant n$. are functions of k only and do not depend on the concrete choice of the indices. Prove that: (a) if the events $A_1, \ldots, A_n$ are independent and equally likely, they are mutually exchangeable; (b) the independence of the events does not follow from their symmetrical dependence.

13.4. In the Polya urn model (see Section 6), let A_k = {a white ball is drawn at the kth trial}. Show that on n consecutive trials, the events $A_1, \ldots, A_n$ are symmetrically dependent.

13.5. Given m urns U_i, $i = 1, \ldots, m$, each of them containing w_i white and r_i red balls respectively. One urn is chosen at random; U_i being chosen with probability p_i, $p_i \geqslant 0$, $p_1 + \ldots + p_m = 1$. A ball is drawn from the chosen urn n times with replacement. Let A_k = {a white ball is drawn at the kth trial}, $k = 1, \ldots, n$. Prove that: (a) the events $A_1, \ldots, A_n$ are symmetrically dependent; (b) in the general case these events are not even pairwise independent.

13.6. Let $\{\xi_n\}$ be a sequence of independent r.v.'s We put $\mathcal{A}_n = \sigma(\mathcal{F}_{\xi_n}, \mathcal{F}_{\xi_{n+1}}, \ldots)$. Let $f(x_1, \ldots, x_n, \ldots)$ be a function of an infinite number of variables $x_i \in \mathbb{R}_1$, $i = 1, 2, \ldots$, and $f(\xi_1, \ldots, \xi_n, \ldots)$ be $\mathcal{A}_n$-measurable for any n. Prove that the variable $f(\xi_1, \ldots, \xi_n, \ldots)$ is a constant with probability 1.

13.7. Let ξ be an integrable r.v. on the probability space $(\Omega, \mathcal{A}, P)$ and $\mathcal{F}$ be a sub-σ-algebra of $\mathcal{A}$. Let the σ-algebras $\mathcal{F}_\xi$ and $\mathcal{F}$ be independent. Prove that $E(\xi|\mathcal{F}) = E(\xi)$, (P-a.s.).

13.8. Let ξ be a r.v. and $\mathcal{F} = \{\emptyset, \Omega\}$. Prove that $E(\xi|\mathcal{F}) = E(\xi)$, (P-a.s.).

13.9. Let the sample space Ω contain n elements. Prove that no more than $[\ln n/\ln 2]$ independent non-degenerate r.v.'s can be defined on Ω. (The r.v. ξ is called *degenerate*, if for any $A \in \mathcal{F}_\xi$ we have $P(A) = 0$ or

1). (See also Exercise 4.31.)

13.10. Let $(\Omega, \mathbf{A}, P)$ be a probability space and $\mathbf{C}_i$, $i = 1, \ldots, n$, be n independent classes, $\mathbf{C}_i \subset \mathbf{A}$. Prove that the independence of the classes is retained, if to each class are added: (a) the null events and the almost sure events; i.e., the events for which either $P(A) = 0$ or $P(A) = 1$; (B) the proper differences of the events of the same class; i.e., if $B, A \in \mathbf{C}_i$ for some i and $B \subset A$, then $A \setminus B$ is added to $\mathbf{C}_i$. For instance, if $A, \Omega \in \mathbf{C}_i$, then $\bar{A} = \Omega \setminus A$ is added to $\mathbf{C}_i$; (c) the countable sums of mutually exclusive events of the same class; (d) the limits of convergent sequences of events, belonging to the same class.

13.11. The events A, B and C are independent. Check up whether the following pairs of events are independent: (a) AB and C; (b) $A \cup B$ and C; (c) $A \Delta B$ and C; (d) AC and BC.

13.12. The events A_1, A_2, A_3, A_4, A_5, and A_6 are mutually independent. Show that the following triples consist of mutually independent events: (a) A_1A_2, A_3A_4 and A_5A_6; (b) $A_1 \cup A_2$, $A_3 \cup A_4$ and $A_5 \cup A_6$.

13.13. Let A_1, A_2, A_3 be independent events, each of them having a probability 1/2. We put $B_{ij} = \overline{A_i \Delta A_j}$. Show that the events B_{12}, B_{23}, B_{31} are not mutually independent but are pairwise independent.

13.14. Let $\{A_n\}$ be a sequence of independent events such that $P(A_n) = p_n$, $\alpha_n = \min[p_n, 1 - p_n]$, $n = 1, 2, \ldots$, and let the series $\sum_{n=1}^{\infty} \alpha_n$ be divergent. Show that the probability space cannot contain atoms.

13.15. Let $\mathbf{C}_i$, $i = 1, \ldots, n$, be independent classes of events, each class being closed with respect to the finite intersections. Prove that the σ-algebras $\sigma(\mathbf{C}_i)$, $i = 1, \ldots, n$, are also independent; i.e., the σ-algebras, generated by independent algebras, are independent.

13.16. Let the events A_k, $k = 1, \ldots, n$, be independent. Prove that the algebras $\mathcal{F}_k$, where $\mathcal{F}_k = \{\emptyset, A_k, \bar{A}_k, \Omega\}$, are independent.

13.17. Show that: (a) if for all $x, y \in \mathbb{R}_1$ the relation $P(\{\xi < x\} \cap \{\eta < y\} = P\{\xi < x\}P\{\eta < y\}$ is fulfilled, then the r.v.'s ξ and η are independent; (b) for the independence of ξ and η it is sufficient that the above relation holds for rational numbers x and y.

13.18. Let $\{\xi_t\}$, $t \in T$, be an arbitrary set of r.v.'s. Prove that the necessary and sufficient condition for them to be independent is

$$P\{\xi_{t_1} < x_1, \xi_{t_2} < x_2, \ldots, \xi_{t_n} < x_n\} =$$

$$= P\{\xi_{t_1} < x_1\}P\{\xi_{t_2} < x_2\} \ldots P\{\xi_{t_n} < x_n\}$$

for any finite set of indices $t_1, t_2, \ldots, t_n \in T$ and arbitrary $x_1, x_2,$

..., $x_n \in \mathbb{R}_1$. Formulate this result in terms of the corresponding distribution function (see Section 15).

13.19. Let $(\Omega, \mathfrak{D}, P)$ be a probability space and let $\{\mathfrak{A}_n\}$ and $\{\mathcal{F}_n\}$, $n = 1, 2, \ldots$, be monotonically decreasing sequences of sub-σ-algebras of $\mathfrak{D}$. Let the relation

$$\sup_{A_n \in \mathfrak{A}_n, B_n \in \mathcal{F}_n} |P(A_n B_n) - P(A_n)P(B_n)| \to 0$$

be fulfilled as $n \to \infty$. Prove that the σ-algebras $\mathfrak{A} = \lim_{n\to\infty} \mathfrak{A}_n = \bigcap_{n=1}^{\infty} \mathfrak{A}_n$ and $\mathcal{F} = \lim_{n\to\infty} \mathcal{F}_n = \bigcap_{n=1}^{\infty} \mathcal{F}_n$ are independent.

*13.20. Let $\mathfrak{C}$, $\mathfrak{D}$ and $\mathfrak{G}$ be σ-algebras and let $\sigma(\mathfrak{D}, \mathfrak{C})$ (the minimal σ-algebra, containing $\mathfrak{D}$ and $\mathfrak{C}$) be independent from $\mathfrak{G}$. Let ξ be a r.v. with $\mathcal{F}_\xi \subset \mathfrak{C}$ and $E(|\xi|) < \infty$. Prove that $E(\xi|\mathcal{F}) = E(\xi|\mathfrak{D})$, (P-a.s.), where $\mathcal{F} = \sigma(\mathfrak{D}, \mathfrak{G})$.

13.21. Let $\{\xi_n\}$ be a sequence of independent r.v.'s and let $A = \{\omega : \lim_{n\to\infty} \xi_n(\omega)$ exists and is finite$\}$. Prove that either $P(A) = 1$ or $P(A) = 0$.

13.22. Let the r.v. ξ be integrable. Find $E(\xi|\mathcal{F})$, if $\mathcal{F} = \{\emptyset, A, \bar{A}, \Omega\}$, where A is an event with $0 < P(A) < 1$. What would the result be, if $\xi = I_B$, where B is a fixed event?

13.23. The r.v. ξ and the measurable function $f(x)$, $x \in \mathbb{R}_1$, are given. Prove that if the r.v.'s ξ and $\eta = f(\xi)$ are independent, then $f(\xi)$ is a constant with probability 1.

13.24. Let each of the r.v.'s ξ and η take on only two values. Prove that the equality $E(\xi\eta) = E(\xi)E(\eta)$ implies the independence of ξ and η.

13.25. Let ξ be an integrable r.v., defined on the probability space $(\Omega, \mathcal{F}, P)$. Let $\{\mathcal{F}_n\}$ be an increasing sequence of σ-subalgebras of $\mathcal{F}$. We put $\eta_n = E(\xi|\mathcal{F}_n)$. Prove that the system $\{\eta_n, \mathcal{F}_n\}$ forms a martingale.

13.26. Let $\{\xi_n, \mathcal{F}_n\}$ and $\{\eta_n, \mathcal{F}_n\}$ be submartingales. Show that: (a) $\{\max[\xi_n, \eta_n], \mathcal{F}_n\}$ is a submartingale; (b) $\{-\xi_n, \mathcal{F}_n\}$ is a supermartingale.

13.27. Let $\{\xi_n, \mathcal{F}_n\}$ be a martingale and $E\{|\xi_n|^\alpha\} < \infty$ for any n, where $\alpha \geq 1$ is a fixed number. Prove that $\{|\xi_n|^\alpha, \mathcal{F}_n\}$ is a submartingale.

*13.28. Let $\{\xi_n, \mathcal{F}_n\}$ be a submartingale and let $\{\varepsilon_n\}$ be a sequence of r.v.'s any of which takes on a value 0 or 1; ε_n is $\mathcal{F}_n$-measurable. We put $\eta_1 = \xi_1$, $\eta_n = \xi_1 + \varepsilon_1(\xi_2 - \xi_1) + \ldots + \varepsilon_{n-1}(\xi_n - \xi_{n-1})$. Prove that $\{\eta_n, \mathcal{F}_n\}$ is a submartingale and $E(\eta_n) \leq E(\xi_n)$. If $\{\xi_n, \mathcal{F}_n\}$ is a martingale, then $\{\eta_n, \mathcal{F}_n\}$ is a martingale and $E(\eta_n) = E(\xi_n)$.

*13.29. Let $\{\xi_n, \mathcal{F}_n\}$ be a submartingale. Prove that $\xi_n = m_n + a_n$, (P-a.s.), where $\{m_n, \mathcal{F}_n\}$ is a martingale, $0 \leq a_0 \leq a_1 \leq a_2 \leq \ldots$ and

a_{n+1} is $\mathcal{F}_n$-measurable. Prove that this decomposition is unique. (*J. L. Doob*)

<u>*13.30.</u> The sequence $\{\eta_n\}$ of independent r.v.'s with $E(\eta_n) = 0$ and $E(\eta_n^2) = \sigma_n^2 < \infty$ is given. Let $\xi_n = \eta_1 + \ldots + \eta_n$ and $\mathcal{F}_n = \sigma(\xi_1, \ldots, \xi_n) = \sigma(\eta_1, \ldots, \eta_n)$. (a) Show that $\{\xi_n, \mathcal{F}_n\}$ forms a martingale. (b) If $\xi_n^2 = m_n + a_n$ is the decomposition of the submartingale $\{\xi_n^2, \mathcal{F}_n\}$, according to Doob's theorem (see Exercise 13.29), prove that $a_n = \sigma_1^2 + \sigma_2^2 + \ldots + \sigma_n^2$.

<u>13.31.</u> Let $\xi_1, \xi_2, \ldots,$ be identically distributed r.v.'s each with a finite expectation and $S_n = \xi_1 + \ldots + \xi_n$. Prove that

$$E\left\{\frac{S_n}{n} \,\middle|\, S_{n+1}\right\} = \frac{S_{n+1}}{n + 1}$$

with probability 1.

<u>13.32.</u> Let the r.v. ξ be defined on the probability space $(\Omega, \mathcal{F}, P)$ and $\mathfrak{G} \subset \mathcal{F}$ be a σ-algebra with $(\Omega, \mathfrak{G}, P)$ being a complete probability space (i.e., $\mathfrak{G}$ contains all P-null subsets of Ω). Prove that if the r.v. ξ and the r.v. $\eta = E\{\xi|\mathfrak{G}\}$ have one and the same distribution, then ξ is $\mathfrak{G}$-measurable; i.e., $\xi = \eta$ with probability 1. (*D. Hikt and W. Vervaat*)

14. <u>Products of Measurable Spaces and Probabilities on Them</u>

Introductory Notes

Let Ω_1 and Ω_2 be two spaces. The set of the ordered pairs $\omega = (\omega_1, \omega_2)$, $\omega_1 \in \Omega_1$, $\omega_2 \in \Omega_2$, is called a *Cartesian product* $\Omega_1 \times \Omega_2$. The set $A_1 \times A_2 = \{\omega : \omega_1 \in A_1, \omega_2 \in A_2\}$ is called a *rectangle* with sides A_1 and A_2, $A_1 \subset \Omega_1$, $A_2 \subset \Omega_2$. The sets of the kind $A_1 \times \Omega_2$ or $\Omega_1 \times A_2$ are called *cylinders* with bases A_1 and A_2, respectively.

If $(\Omega_1, \mathbb{A}_1)$ and $(\Omega_2, \mathbb{A}_2)$ are measurable spaces, then $\mathbb{A}_1 \times \mathbb{A}_2$ is defined as the σ-algebra, generated by the rectangles in $\Omega_1 \times \Omega_2$, and $(\Omega, \mathbb{A}) = (\Omega_1 \times \Omega_2, \mathbb{A}_1 \times \mathbb{A}_2)$ is called a *product of the measurable spaces* given.

A product of an arbitrary number of measurable spaces is defined analogously.

Let $A \subset \Omega$ and ω_1 be a fixed point from Ω_1. In this case $A_{\omega_1} = \{\omega_2 : (\omega_1, \omega_2) \in A\}$ is called a *section* of the set A in the point ω_1. For any function $f = f(\omega_1, \omega_2)$, $(\omega_1, \omega_2) \in \Omega$, the section f_{ω_1} in the point ω_1 is defined as a function, given in Ω_2 with the equality $f_{\omega_1}(\omega_2) = f(\omega_1, \omega_2)$. The sections A_{ω_2} and f_{ω_2} in the point ω_2 from Ω_2 are defined analogously.

It is known that:

(a) the section of each measurable set is measurable;

(b) the section of each random variable is a random variable as well.

Let $(\Omega, \mathbf{A})$ be the product of the measurable spaces $(\Omega_1, \mathbf{A}_1)$ and $(\Omega_2, \mathbf{A}_2)$. Let $P_1(\cdot, \omega_2)$, $\omega_2 \in \Omega_2$, be a family of probabilities on $(\Omega_1, \mathbf{A}_1)$ such that for any $A_1 \in \mathbf{A}_1$ the function $P_1(A_1, \omega_2)$ is $\mathbf{A}_2$-measurable.

Fubini's theorem. If $P_1(\cdot, \omega_2)$ is the above defined function and if P_2 is a probability on $(\Omega_2, \mathbf{A}_2)$, then: (a) There exists a unique probability P on the product $(\Omega, \mathbf{A})$ such that for arbitrary $A_1 \in \mathbf{A}_1$ and $A_2 \in \mathbf{A}_2$

$$P(A_1 \times A_2) = \int_{A_2} P_1(A_1, \omega_2)\, dP_2(\omega_2).$$

(b) If $\xi = \xi(\omega_1, \omega_2)$ is a non-negative r.v. (i.e., $\xi(\omega_1, \omega_2)$ is a $\mathbf{A}_1 \times \mathbf{A}_2$-measurable function), then the function

$$\eta(\omega_2) = \int_{\Omega_1} \xi(\omega_1, \omega_2) P_1(d\,\omega_1, \omega_2)$$

is a r.v. (i.e., $\eta(\omega_2)$ is $\mathbf{A}_2$-measurable), and

$$\int_{\Omega_1 \times \Omega_2} \xi(\omega_1, \omega_2)\, dP(\omega_1, \omega_2) =$$

$$= \int_{\Omega_2} \left(\int_{\Omega_1} \xi(\omega_1, \omega_2) P_1(d\,\omega_1, \omega_2) \right) dP_2(\omega_2).$$

In the particular case when the function P_1 does not depend on ω_2; i.e., $P_1(\cdot, \omega_2) = P_1(\cdot)$, Fubini's theorem is formulated in the following way:

(a) For arbitrary $A_1 \in \mathbf{A}_1$ and $A_2 \in \mathbf{A}_2$ we have the equality

$$P(A_1 \times A_2) = P_1(A_1) P_2(A_2).$$

In this case the probability P is called a *product of the probabilities* P_1 and P_2. The equalities $P(\Omega_1 \times A_2) = P_2(A_2)$ and $P(A_1 \times \Omega_2) = P_1(A_1)$, $A_1 \in \mathbf{A}_1$, $A_2 \in \mathbf{A}_2$ are fulfilled, which are called *consistency conditions* of P with P_1 and P_2.

(b) The functions $\eta(\omega_2)$ and $\zeta(\omega_1)$, where

$$\eta(\omega_2) = \int_{\Omega_1} \xi(\omega_1, \omega_2)\, dP_1(\omega_1),$$

$$\zeta(\omega_1) = \int_{\Omega_2} \xi(\omega_1, \omega_2)\, dP_2(\omega_2),$$

are r.v.'s (i.e., $\eta(\omega_2)$ is $\mathbf{A}_2$-measurable, and $\zeta(\omega_1)$ is $\mathbf{A}_1$-measurable) and

$$\int_{\Omega_1\times\Omega_2} \xi(\omega_1, \omega_2)\, dP(\omega_1, \omega_2) = \int_{\Omega_2}\left(\int_{\Omega_1} \xi(\omega_1, \omega_2)\, dP(\omega_2)\right)dP_1(\omega_1) =$$

$$= \int_{\Omega_2}\left(\int_{\Omega_1} \xi(\omega_1, \omega_2)\, dP_1(\omega_1)\right)dP_2(\omega_2).$$

The concepts, section of a set, section of a r.v., as well as Fubini's theorem, are generalized in a natural way for more than two spaces.

Illustrative Examples

Example 14.1. Let $(\Omega, \mathbf{A}, P)$ be a probability space and let $\Omega = \Omega_1 \times \Omega_2$. Let us denote by $\mathbf{A}_1$ the class of such subsets $A' \in \Omega_1$ that $A' \times \Omega_2 \in \mathbf{A}$. Let us consider the function

$$P_1(A') = P(A' \times \Omega_2), \quad A' \in \mathbf{A}_1,$$

which actually is the *projection* of the probability P on the class $\mathbf{A}_1$. Prove the following assertions:
(i) The class $\mathbf{A}_1$ is a σ-algebra of subsets of Ω_1.
(ii) The function $P_1(\cdot)$ is a probability on $\mathbf{A}_1$.

Solution. (i) The following equalities hold for arbitrary A', $\{A'_k\}_{k=1}^{\infty}$, which are elements of $\mathbf{A}_1$:

$$\overline{A'} \times \Omega_2 = \overline{A' \times \Omega_2}$$

and

$$\sum_{k=1}^{\infty} (A'_k \times \Omega_2) = \left(\sum_{k=1}^{\infty} A'_k\right) \times \Omega_2.$$

These relations imply that the class $\mathbf{A}_1$ is a σ-algebra of subsets of Ω_1.

(ii) Obviously $P_1(\Omega_1) = P(\Omega_1 \times \Omega_2) = P(\Omega) = 1$. For arbitrary $A' \in \mathbf{A}_1$,

$0 \leq P_1(A') \leq 1$ as well. Similarly $A'_k \in \mathcal{A}_1$, $k = 1, 2, \ldots$, and $A'_i \cap A'_j = \emptyset$, $i \neq j$ implies the σ-additivity of P_1:

$$P_1\left(\sum_{k=1}^{\infty} A'_k\right) = P\left[\sum_{k=1}^{\infty} (A'_k \times \Omega_2)\right] = \sum_{k=1}^{\infty} P(A'_k \times \Omega_2) = \sum_{k=1}^{\infty} P_1(A'_k).$$

Therefore, P_1 is a probability on $\mathcal{A}_1$.

The σ-algebra $\mathcal{A}_2$ and the projection P_2 can be defined analogously.

Example 14.2. Let $(\Omega_1, \mathcal{A}_1, P_1)$, $(\Omega_2, \mathcal{A}_2, P_2)$ and $(\Omega, \mathcal{A}, P)$ be probability spaces with $(\Omega, \mathcal{A})$ being the product of the measurable spaces $(\Omega_1, \mathcal{A}_1)$ and $(\Omega_2, \mathcal{A}_2)$; i.e., $\Omega = \Omega_1 \times \Omega_2$ and $\mathcal{A} = \mathcal{A}_1 \times \mathcal{A}_2$. Let the probability P satisfy the consistency conditions with P_1 and P_2 (see the Introductory Notes to this Section.) Let us denote

$$\mathcal{F}_1 = \{A \in \mathcal{A} : A = A_1 \times \Omega_2,\ A_1 \in \mathcal{A}_1\}$$

$$\mathcal{F}_2 = \{A \in \mathcal{A} : A = \Omega_1 \times A_2,\ A_2 \in \mathcal{A}_2\};$$

i.e., $\mathcal{F}_1$ and $\mathcal{F}_2$ are the classes of the cylinders with bases in $\mathcal{A}_1$ and $\mathcal{A}_2$, respectively. Show that $\mathcal{F}_1$ and $\mathcal{F}_2$ are independent σ-algebras in $(\Omega, \mathcal{A}, P)$ if and only if P is the product of P_1 and P_2.

Solution. It is easy to see that $\mathcal{F}_1$ and $\mathcal{F}_2$ are sub-σ-algebras of $\mathcal{A}$ (they are isomorphic to $\mathcal{A}_1$ and $\mathcal{A}_2$, respectively). (i) Let P be the product of P_1 and P_2; i.e., $P = P_1 \times P_2$. Then for arbitrary $A_1^* \in \mathcal{F}_1$ and $A_2^* \in \mathcal{F}_2$ we have:

$$P(A_1^*) = P(A_1 \times \Omega_2) = P_1(A_1), \quad A_1 \in \mathcal{A}_1$$

and

$$P(A_2^*) = P(\Omega_1 \times A_2) = P_2(A_2), \quad A_2 \in \mathcal{A}_2.$$

Analogously

$$P(A_1^* \cap A_2^*) = P\{(A_1 \times \Omega_2) \cap (\Omega_1 \times A_2)\} = P(A_1 \times A_2) =$$

$$= P_1(A_1) \cdot P_2(A_2);$$

therefore,

$$P(A_1^* \cap A_2^*) = P(A_1^*)P(A_2^*).$$

This equality shows that $\mathcal{F}_1$ and $\mathcal{F}_2$ are independent σ-algebras.

(ii) Let the σ-algebras $\mathcal{F}_1$ and $\mathcal{F}_2$ be independent. Then it can be proved analogously that the relation $P(A_1 \times A_2) = P_1(A_1)P_2(A_2)$ holds for arbitrary $A_1 \in \mathcal{A}_1$ and $A_2 \in \mathcal{A}_2$. Hence the probability P is the product of P_1 and P_2.

Exercises

14.1. Let $A_1, B_1 \in \mathcal{P}(\Omega_1)$ and $A_2, B_2 \in \mathcal{P}(\Omega_2)$. In the product $\Omega_1 \times \Omega_2$ prove the following equalities: (a) $(A_1 \times A_2) \cap (B_1 \times B_2) = (A_1 \cap B_1) \times (A_2 \cap B_2)$; (b) $(A_1 \times A_2) \setminus (B_1 \times B_2) = A_1 \setminus B_1) \times (A_2 \setminus B_2) + (A_1 \setminus B_1) \times (A_2 \cap B_2) + (A_1 \cap B_1) \times (A_2 \setminus B_2)$.

14.2. Let $\mathcal{C}$ be an algebra of subsets of Ω, and $\mathcal{C}'$ be an algebra of subsets of Ω'. Prove that the set of the rectangles of the type $A \times A'$, $A \in \mathcal{C}$, $A' \in \mathcal{C}$, is a semi-algebra in $\Omega \times \Omega'$ (see Exercise 11.31). Is the assertion true, if $\mathcal{C}$ and $\mathcal{C}'$ are assumed to be only semi-algebras?

14.3. Let $(\Omega_1, \mathcal{A}_1)$ and $(\Omega_2, \mathcal{A}_2)$ be measurable spaces. Prove that the class of all the finite sums of rectangles of the type $A_1 \times A_2$, $A_1 \in \mathcal{A}_1$, $A_2 \in \mathcal{A}_2$, is an algebra.

14.4. Let $(\Omega_1, \mathcal{A}_1)$ and $(\Omega_2, \mathcal{A}_2)$ be arbitrary measurable spaces and let $(\Omega_1 \times \Omega_2, \mathcal{A}_1 \times \mathcal{A}_2)$ be their product. Prove that for an arbitrary $\omega_1 \in \Omega_1$ the section A_{ω_1} in ω_1 of an arbitrary measurable set A from $\mathcal{A}_1 \times \mathcal{A}_2$ is a measurable set in $\mathcal{A}_2$.

14.5. Let $(\Omega_1, \mathcal{A}_1, P_1)$ and $(\Omega_2, \mathcal{A}_2, P_2)$ be given and let A be an event in their product $(\Omega_1 \times \Omega_2, \mathcal{A}_1 \times \mathcal{A}_2, P_1 \times P_2)$. Show that A is a null event if and only if its sections $A_{\omega_1} \doteq \{\omega_2 : \omega_2 \in \Omega_2, (\omega_1, \omega_2) \in A\}$ are a.s. null events in $(\Omega_2, \mathcal{A}_2, P_2)$. The same assertion holds for the sections in the points from Ω_2 as well.

14.6. Let ξ and η be integrable r.v.'s in the spaces $(\Omega_1, \mathcal{A}_1, P_1)$ and $(\Omega_2, \mathcal{A}_2, P_2)$, respectively. Let ζ be a r.v. in $\Omega_1 \times \Omega_2$, defined by means of the equality $\zeta(\omega_1, \omega_2) = \xi(\omega_1)\eta(\omega_2)$. Prove that ζ is integrable, and $\iint_{\Omega_1 \times \Omega_2} \zeta \, dP = \int_{\Omega_1} \xi \, dP_1 \cdot \int_{\Omega_2} \eta \, dP_2$, where $P = P_1 \times P_2$.

*14.7. Let $(\Omega_1, \mathcal{A}_1, P_1)$ and $(\Omega_2, \mathcal{A}_2, P_2)$ be probability spaces. Let $B_1 \in \mathcal{A}_1$ and $B_2 \in \mathcal{A}_2$ be events such that $P_1(B_1) = P_2(B_2) = \frac{1}{2}$. Put $B = (B_1 \times \Omega_2) \Delta (\Omega_1 \times B_2)$, and define the function $\vec{Q}$ in the following way: $Q(A) = 2 \int_A 1_B \, dP$ for $A \in \mathcal{A}_1 \times \mathcal{A}_2$, where $P = P_1 \times P_2$. Show that: (a) Q is a probability; (b) P and Q coincide for each cylindrical set (i.e.,

for sets of the type $A_1 \times \Omega_2$ or $\Omega_1 \times A_2$, where $A_1 \in \mathbf{A}_1$, $A_2 \in \mathbf{A}_2$);
(c) $P \neq Q$ (i.e., there exist events such that $P \neq Q$).

14.8. Let $(\Omega_1, \mathbf{A}_1, P_1)$ and $(\Omega_2, \mathbf{A}_2, P_2)$ be probability spaces and let $(\Omega_1 \times \Omega_2, \mathbf{A}_1 \times \mathbf{A}_2, P_1 \times P_2)$ be their product. Let us denote $P = P_1 \times P_2$. Let A and B be events in $\mathbf{A}_1 \times \mathbf{A}_2$ such that the equality $A_{\omega_1} = B_{\omega_1}$ holds for $\omega_1 \in \tilde{\Omega}_1$, where $\tilde{\Omega}_1 \subset \Omega_1$ and $P_1(\tilde{\Omega}_1) = 1$. Prove that $P(A) = P(B)$.

*14.9. Let $\mathbf{C}_i$ be classes of subsets of $\mathbb{R}_n$, defined in the following way: $\mathbf{C}_1$ is the class of all closed rectangles; i.e., sets of the type $\{(x_1, \ldots, x_n) : a_i \leqslant x_i \leqslant b_i,\ i = 1, \ldots, n\}$; $\mathbf{C}_2$ is the class of all open rectangles; i.e., sets of the type $\{(x_1, \ldots, x_n) : a_i < x_i < b_i,\ i = 1, \ldots, n\}$; $\mathbf{C}_3$ is the class of all closed sets; $\mathbf{C}_4$ is the class of all open sets; $\mathbf{C}_5$ is the class of the rectangles of the type $\{(x_1, \ldots, x_n) : x_i < b_i,\ i = 1, \ldots, n\}$. Prove that

$$\sigma(\mathbf{C}_1) = \sigma(\mathbf{C}_2) = \sigma(\mathbf{C}_3) = \sigma(\mathbf{C}_4) = \sigma(\mathbf{C}_5).$$

This σ-algebra is called *Borel σ-algebra* in $\mathbb{R}_n$ and is denoted by $\mathbf{B}_n$.

*14.10. Show that the Borel σ-algebra $\mathbf{B}_n$ coincides with the product $\mathbf{B}_1 \times \mathbf{B}_1 \times \ldots \times \mathbf{B}_1 = \mathbf{B}_1^{\times n}$.

Chapter 3

CHARACTERISTICS OF RANDOM VARIABLES

15. Distribution Function

Introductory Notes

The function $F(x)$, $x \in \mathbb{R}_1$, is called a *distribution function* (d.f.) if:

(a) F is a non-decreasing function; i.e., $F(x_1) \leqslant F(x_2)$ if $x_1 < x_2$;

(b) F is a left-continuous function; i.e., $F(x - 0) = F(x)$, $x \in \mathbb{R}_1$;

(c) $\lim_{x\to\infty} F(x) = 1$, $\lim_{x\to-\infty} F(x) = 0$.

If ξ is a r.v. defined on the probability space $(\Omega, \mathcal{F}, P)$, then the function $F_\xi(x) = P\{\xi < x\}$, $x \in \mathbb{R}_1$ is a d.f., the so-called d.f. of the r.v. ξ.

Given F_ξ, the distribution $P\{\xi \in B\}$, $B \in \mathcal{B}_1$, of the r.v. ξ is uniquely determined by

$$P\{\xi \in B\} = \int_{\mathbb{R}_1} 1_B(x)\, dF_\xi(x), \tag{15.1}$$

where $1_B(x)$ is the indicator of the Borel set B; i.e., the distribution $P\{\xi \in B\}$, $B \in \mathcal{B}_1$, coincides with the Borel-Stieltjes measure on $\mathcal{B}_1$ generated by F_ξ.

Let $g(x)$, $x \in \mathbb{R}_1$, be a Borel-measurable function and let ξ be a r.v. with known d.f. $F_\xi(x)$, $x \in \mathbb{R}_1$. The d.f. $F_\eta(x)$, $x \in \mathbb{R}_1$, of the r.v. $\eta = g(\xi)$ is given by

$$F_\eta(x) = \int_{\mathbb{R}_1} 1_B(y)\, dF_\xi(y), \tag{15.2}$$

where $B = \{y : g(y) < x\}$.

If there exists a sequence $x_1, x_2, \ldots$, finite or infinite, such that $F(x_n + 0) - F(x_n) = p_n > 0$ and $\sum_n p_n = 1$, then the d.f. F is called a *discrete d.f.* In this case

$$F(x) = \sum_{n:x_n<x} p_n.$$

The d.f. $F(x)$, $x \in \mathbb{R}_1$, is said to be *absolutely continuous* with respect to the Lebesgue measure if there exists a Borel-measurable function $f(y)$, $y \in \mathbb{R}_1$, such that

$$F(x) = \int_{-\infty}^{x} f(y)\,dy \tag{15.3}$$

for all $x \in \mathbb{R}_1$. Every d.f. F is almost everywhere differentiable; i.e., the set S of all points x such that the derivative $F'(x)$ does not exist has a Lebesgue measure $L(S) = 0$. If $\underline{F}$ is an absolutely continuous d.f., then from (15.3) $f(x) = F'(x)$ for $x \bar{\in} S$. On the set S we can define $f(x)$ arbitrarily. The function f is called the *density function* of the d.f. F (with respect to the Lebesgue measure L). If $f(x)$, $x \in \mathbb{R}_1$, and $g(x)$, $x \in \mathbb{R}_1$, are density functions of the d.f. F, then f and g coincide almost everywhere; i.e., $L(\{x : f(x) \neq g(x)\}) = 0$.

The function $f(x)$, $x \in \mathbb{R}_1$, is a density function of some d.f. if $f(x)$ is non-negative almost everywhere on $\mathbb{R}_1$ and $\int_{\mathbb{R}_1} f(x)\,dx = 1$.

If the d.f. $F_\xi(x)$, $x \in \mathbb{R}_1$, of the r.v. ξ has a density $f_\xi(x)$, then the probability of the event $\{\omega : \xi(\omega) \in B\}$, $B \in \mathcal{B}_1$, can be expressed by the density $f_\xi(x)$ as

$$P\{\xi \in B\} = \int_B f_\xi(u)\,du. \tag{15.5}$$

Let the r.v. ξ be absolutely continuous with density f_ξ. Assume $g(x)$, $x \in \mathbb{R}_1$, to be monotonic and differentiable and suppose $g'(x) \neq 0$ for $x \in \mathbb{R}_1$. If $\eta = g(\xi)$, then the r.v. η is absolutely continuous with a density function

$$f_\eta(y) = \begin{cases} f_\xi(g^{-1}(y))/|g'(g^{-1}(y))|, & \text{if } \inf_{x\in\mathbb{R}_1} g(x) < y < \sup_{x\in\mathbb{R}_1} g(x) \\ 0, & \text{otherwise,} \end{cases} \tag{15.6}$$

where $x = g^{-1}(y)$ is the inverse function of $y = g(x)$.

Every real number x_p fulfilling $P\{\xi < x_p\} \leqslant p \leqslant P\{\xi \leqslant x_p\}$ is said to be a p-quantile of the r.v. ξ or, more precisely, of the corresponding d.f. F_ξ. The p-*quantile* x_p is the unique solution of the equation $F_\xi(x) = p$ when $F_\xi(x)$ is continuous and strictly increasing for every x and such that $0 < F_\xi(x) < 1$.

The (1/2)-quantile $x_{1/2}$ is called a *median*.

If the density function $f(x)$ has a local maximum at x_0, then x_0 is called a *mode* of $f(x)$.

The following density functions often appear in the probabilistic literature:

(a) the *uniform distribution* on the interval (a, b):

$$f(x) = \begin{cases} \frac{1}{b-a}, & \text{if } x \in (a, b) \\ 0, & \text{if } x \overline{\in} (a, b); \end{cases} \tag{15.7}$$

(b) the *normal (Gaussian) distribution* with parameters a and σ^2:

$$f(x) = \frac{1}{\sqrt{2\pi}\sigma} \exp(-(x - a)^2/(2\sigma^2)), \quad x \in \mathbb{R}_1;\ a \in R_1,\ \sigma > 0;$$

(c) the *exponential distribution* with parameter λ, $\lambda > 0$:

$$f(x) = \begin{cases} 0, & \text{if } x < 0 \\ \lambda e^{-\lambda x}, & \text{if } x \geqslant 0; \end{cases} \tag{15.8}$$

(d) the *gamma distribution* with parameters α and β:

$$f(x) = \begin{cases} 0, & \text{if } x < 0 \\ \frac{\beta^\alpha}{\Gamma(\alpha)} x^{\alpha-1} e^{-\beta x}, & \text{if } x \geqslant 0;\ \alpha > 0,\ \beta > 0; \end{cases} \tag{15.9}$$

(e) the *beta distribution* with parameters p and q:

$$f(x) = \begin{cases} 0, & \text{if } x \overline{\in} (0, 1) \\ \frac{\Gamma(p+q)}{\Gamma(p)\Gamma(q)} x^{p-1}(1-x)^{q-1}, & \text{if } x \in (0, 1);\ p > 0,\ q > 0; \end{cases} \tag{15.10}$$

(f) the *Cauchy distribution* with parameters a and b:

$$f(x) = \frac{b}{\pi(b^2 + (x-a)^2)}, \quad x \in \mathbb{R}_1,\ a \in \mathbb{R}_1,\ b > 0. \tag{15.11}$$

We shall use the following abbreviated notations for frequently used density functions:

$N(a, \sigma^2)$ for the set of normally distributed r.v.'s with parameters a and σ^2;

$E(\lambda)$ for the set of exponentially distributed r.v.'s with parameter λ.

Illustrative Examples

Example 15.1. Let the r.v. ξ be binomially distributed with parameters $n = 2$ and $p = 1/2$. Find the d.f. F_ξ of the r.v. ξ.

Solution. From the definition of the d.f. it follows that:

(a) $F_\xi(x) = P\{\xi < x\} = 0$ for $x \leqslant 0$;

(b) $F_\xi(x) = P\{\xi < x\} = P\{\xi = 0\} = 1/4$ for $0 < x \leqslant 1$;

(c) $F_\xi(x) = P\{\xi = 0\} + P\{\xi = 1\} = 3/4$ for $1 < x \leqslant 2$;

(d) $F_\xi(x) = P\{\xi = 0\} + P\{\xi = 1\} + P\{\xi = 2\} = 1$ for $x > 2$.

The graph of the function $F_\xi(x)$, $x \in \mathbb{R}_1$ is represented in Figure 15.1.

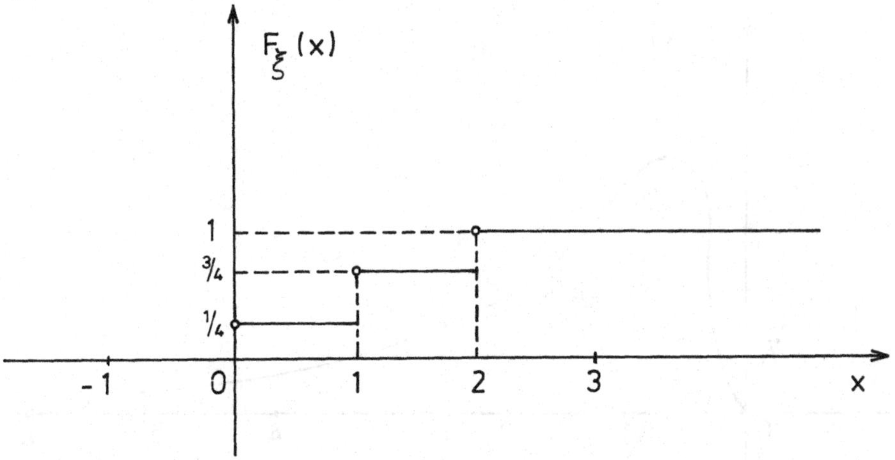

Figure 15.1

Example 15.2. Let the r.v. $\xi \in N(a, \sigma^2)$. Show that the density of the r.v. $\eta = \exp \xi$ exists and is equal to

$$f_\eta(y) \begin{cases} (\sqrt{2\pi}\sigma y)^{-1} \exp(-(\ln y - a)^2/(2\sigma^2)), & \text{if } y > 0 \\ 0, & \text{if } y \leqslant 0. \end{cases} \tag{15.13}$$

A random variable having the density function (15.13) is said to be *lognormal*.

Solution. Put $g(x) = e^x$, $x \in \mathbb{R}_1$. The function g fulfills the conditions for applying the formula (15.6). The inverse function of $g(x) = e^x$, $x \in \mathbb{R}_1$, is $g^{-1}(y) = \ln y$; $g'(x) = g(x)$ and $\inf_{x \in \mathbb{R}_1} \exp x = 0$, $\sup_{x \in \mathbb{R}_1} \exp x = \infty$. Hence by (15.6) $f_\eta(y) = 0$ if $y \leqslant 0$. If $y > 0$, then

$$f_\eta(y) = \frac{1}{\sqrt{2\pi}\sigma}\left(\exp\left(-\frac{(\ln y - a)^2}{2\sigma^2}\right)\right)/|\exp(\ln y)|,$$

which coincides with (15.13).

The lognormal distribution is of great importance in the theory of crushing of materials. The distribution of the grains of a granular material (stone, crystal powder, etc.), produced by a breaking-process, is lognormal under rather general conditions. The density function $f_\eta(y)$, $y \in \mathbb{R}_1$, for $a = 0$ and $\sigma = 1$ is represented by the curve in Figure 15.2.

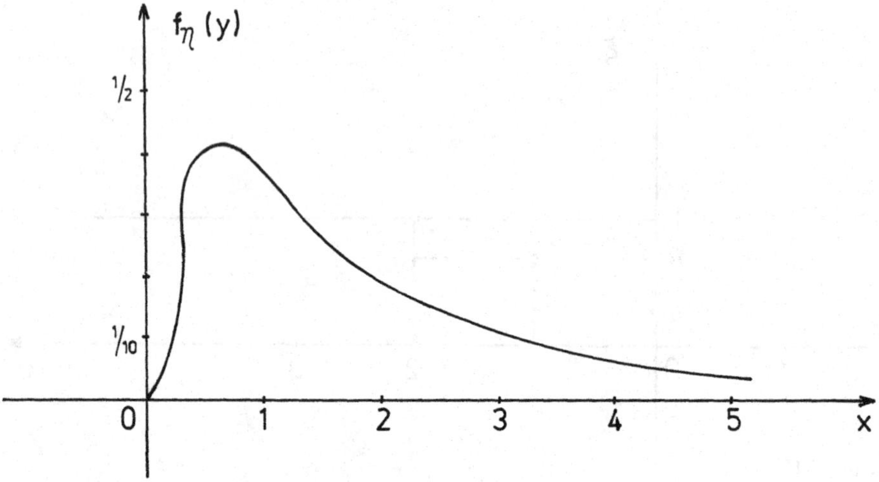

Figure 15.2.

Example 15.3. Let the r.v. ξ have d.f. $F(x)$, $x \in \mathbb{R}_1$. Find the d.f. of the r.v. $\eta = F(\xi)$.

Solution. Let $g(y) = \sup_x \{x : F(x) < y\}$ for $0 < y < 1$. Since F is a non-decreasing left-continuous function we have:

(a) $\{x : F(x) < y\} = \{x : g(y) > x\}$ for $F(g(y)) = y$,

(b) $\{x : F(x) < y\} = \{x, g(y) \geq x\}$ for $F(g(y)) < y$.

Hence in case (a), we have

$$P\{\eta < y\} = P\{F(\xi) < y\} = P\{\xi < g(y)\} = F(g(y)) = y,$$

and in the case (b),

$$P\{\eta < y\} = P\{F(\xi) < y\} = P\{\xi \leq g(y)\} = F(g(y) + 0).$$

When $0 < y < 1$, we had $F_\eta(y) = y$, if $F(g(y)) = y$ and $F_\eta(y) = F(g(y) + 0)$, if $F(g(y)) < y$. It can also be seen that $F_\eta(y) = 0$ when $y \leqslant 0$; $F_\eta(y) = 1$ when $y > 1$ and $F_\eta(1)$ is the limit of $F_\eta(y)$ for $y \uparrow 1$. For instance, if $F(x)$, $x \in \mathbb{R}_1$, is continuous, then $F(g(y)) = y$ for every $0 < y < 1$ and the r.v. $\eta = F(\xi)$ is uniformly distributed on the interval $(0, 1)$.

Exercises

15.1. Show that if the r.v. $\xi \in \mathcal{B}(n, p)$, then the d.f. $F_\xi(x)$, $x \in \mathbb{R}_1$, has the representation:

$$F_\xi(x) = \begin{cases} 0, & \text{if } x \leqslant 0 \\ 1 - \dfrac{1}{B(m, n-m+1)} \displaystyle\int_0^P x^{m-1}(1-x)^{n-m}\,dx, & \text{if } m - x < 1 \leqslant m, \\ & m = 1, \ldots, n, \\ 1, & \text{if } x > n. \end{cases}$$

15.2. Let the r.v. ξ have a gamma distribution (see (15.10)) with parameters: α a positive integer, $\beta = 1$, and the r.v. $\eta \in \mathcal{P}(\lambda)$. Prove that $P\{\xi \geqslant \lambda\} = P\{\eta \leqslant \alpha - 1\}$.

15.3. Let the r.v. $\xi \in \mathcal{N}(2, 4)$ (see (15.8)). Compute: (a) $P\{0 \leqslant \xi < 3\}$; (b) $P\{|\xi| < 1\}$; (c) $P\{-1 \leqslant \xi < 1 | 0 \leqslant \xi < .3\}$.

15.4. Given the function $f(x) = a(s^2 - x^2)^{-1/2}$ for $|x| < s$ and $f(x) = 0$ for $|x| \geqslant s$. Determine the parameter a so that f is the density of a certain r.v., denoted by ξ. Find the d.f. F_ξ of ξ and the probability $P\{0 \leqslant \xi < s\}$.

15.5. Let the r.v. $\xi \in \mathcal{E}(\lambda)$. Find the densities of the following r.v.'s: (a) $\eta = -\xi$; (b) $\eta = 2\xi - 1$; (c) $\eta = \sqrt{\xi}$; (d) $\eta = \xi^\alpha$, $\alpha > 0$; (e) $\eta = c\xi$, $c > 0$.

15.6. Let $\xi \in \mathcal{E}(1)$. Find the probability of the event $A = \{[\xi] \text{ even}\}$, where $[\xi]$ is the integer part of ξ.

*15.7. Let the r.v. ξ assume only non-negative values, have continuous d.f. F and let $P\{\xi < y + x | \xi \geqslant y\} = P\{\xi < x\}$, $x \geqslant 0$, $y \geqslant 0$. Prove that ξ is exponentially distributed.

15.8. Consider the problem of breaking the yarn on a loom. Suppose that the yarn breaks at random at time $\xi \geqslant 0$ with $P\{t \leqslant \xi < t + \Delta t\} = \Delta t/\lambda + o(\Delta t)$ for $\lambda \geqslant 0$ and $t \geqslant 0$, and that the events $\{t \leqslant \xi < t + \Delta t\}$ and $\{\xi < t\}$ are independent for every $t \geqslant 0$. Find the density of ξ.

15.9. Find the density of the r.v. $\eta = c\xi + d$, where c and d are real numbers and $\xi \in \mathcal{N}(a, \sigma^2)$.

15.10. Let ξ and η be r.v.'s with d.f.'s F and G respectively. Prove that $P(\xi > \eta) = 1$ implies $F(x) \leqslant G(x)$ for every $x \in \mathbb{R}_1$. Is the converse statement true if ξ and η are defined on the same probability space?

15.11. Let $F(x)$, $x \geqslant 0$, be a d.f. Show that $G(x) = 0$ for $x \leqslant 0$ and $G(x) = \exp[1 - \alpha(1 - F(x))]$, $x > 0$, $\alpha > 0$, is also a d.f. with a jump at the point $x = 0$. Is is possible that for every x, $f(x) = G(x)$?

15.12. Let the r.v. ξ have a Cauchy distribution with density $[\pi(1 + x^2)]^{-1}$, $x \in \mathbb{R}_1$. Prove that the variables $\eta_1 = 1/\xi$, $\eta_2 = 2\xi/(1 - \xi^2)$ and $\eta_3 = (3\xi - \xi^3)/(1 - 3\xi^2)$ have the same distribution.

15.13. Let ξ have a Laplace distribution with parameters $a > 0$ and $b \in \mathbb{R}_1$; i.e., $f_\xi(x) = (a/2)\exp[-a|x - b|]$, $x \in \mathbb{R}_1$. Find the distribution of $|\xi - b|$.

15.14. Let the point ξ be normally distributed $N(0, 1)$ over the real axis Ox. Find the probability for the segments AB, $A = (0, 1)$, $B = (0, -2)$, to be visible the point ξ at an angle less than $\pi/2$.

15.15. Let the point A from the semicircle $x^2 + y^2 = a^2$, $y \geqslant 0$, $-a \leqslant x \leqslant a$, be chosen at random and let A_1 be the projection of A on the axis $y = 0$. Find the d.f. F_ξ and the density f_ξ of the r.v. ξ defined as the length of the segment AA_1.

15.16. Given the r.v.'s ξ and η and let $P\{|\xi - \eta| \leqslant \delta\} \geqslant 1 - \varepsilon$ hold for any ε and $\delta > 0$. If F and G are the d.f.'s of ξ and η, respectively, then show that for an arbitrary x, we have $|F(x) - G(x)| \leqslant F(x + \delta) - F(x - \delta) + \varepsilon$.

15.17. Let the r.v. ξ be uniformly distributed over the interval $(-\pi/2, \pi/2)$ (see (15.7)). Show that the r.v. $\eta = \cos \xi$ is absolutely continuous and find the density of η.

15.18. Let ξ be an arbitrary r.v. The function $Q_\xi(x)$ defined by

$$Q_\xi(x) = \sup_{u \in \mathbb{R}_1} P\{u \leqslant \xi \leqslant u + x\}, \quad x \in \mathbb{R}_1,$$

is said to be a *concentration function*. Prove the following statements:

$$Q_\xi(\alpha x) = ([\alpha] + 1) Q_\xi(x), \quad x \in \mathbb{R}_1, \tag{a}$$

for any $\alpha > 0$, where $[\alpha]$ is the integer part of α;

$$m \leqslant \binom{n}{2} + 1 \tag{b}$$

where n is the number of the jump points of the d.f. of ξ and m is the number of the jumps of Q_ξ.

15.19. Let us define the function:

$$F(x) = \begin{cases} 0, & \text{if } x \quad 0 \\ \displaystyle\sum_{n=1}^{\infty} \frac{[2^{n-1}x + 0.5]}{2^{2n-1}}, & \text{if } 0 \leqslant x \leqslant 1 \\ 1, & \text{if } x > 0. \end{cases}$$

Show that a discrete r.v. ξ exists for which $P(\xi \leqslant x) = F(x)$ for every $x \in \mathbb{R}_1$.

15.20. The r.v. ξ is called *symmetric* if ξ and $(-\xi)$ have one and the same distribution. Express the symmetry property via the d.f. F and via the density f of ξ. Determine which of the following distributions are symmetric: (a) $N(0, \sigma^2)$; (b) uniform over the interval $(0, 1)$; (c) uniform over the interval $(-1, 1)$; (d) beta distribution (see (15.11)) with $p = q$; (e) Cauchy distribution with $a = 0$; (f) gamma distribution (see (15.10)).

15.21. Let f be the density of the beta distribution with parameters p and q, $p > 0$, $q > 0$. Show that $f(x)$ has only one maximum, reached at the point $x = p/(p + q)$. For what values of p and q does the density f have no inflection points?

15.22. Let ξ be a r.v. with density $f(x) \neq 0$, $x \in \mathbb{R}_1$, having a continuous first derivative. Suppose that for arbitrary real numbers u, v and w the function $f(u - x)f(v - x)f(w - x)$ has maximum at the point $x = \frac{1}{3}(u + v + w)$. Find the 0.95 quantile of the d.f. of the r.v. ξ if $f(0) = 1$.

15.23. Given the function $\phi(x) = \frac{\pi}{4} 2^{-[x]} \sin \frac{\pi}{2}(x - [x])$, $x \geqslant 0$, Show that ϕ is a probability density function with infinitely many modes.

15.24. Prove that $F(x) = (1 - e^{-(ax+b)})^{-1}$, $a > 0$, $x \in \mathbb{R}_1$, is a d.f. (called the *logistic distribution*) and that F and its density f are related by the expression $f(x) = aF(x)(1 - F(x))$, $x \in \mathbb{R}_1$.

15.25. Show that the function $f(x) = 0$ for $x \leqslant b$ and $f(x) = (a/b)(b/x)^{a+1}$ for $x > b$, where $a > 0$, $b > 0$ are constants, is the density of a distribution (called the *Pareto distribution*). Find the d.f. F and its median.

15.26. Let the r.v. ξ have d.f. $F(x) = 1 - \exp(-x^2/2\sigma^2)$ for $x \geqslant 0$ and $F(x) = 0$ for $x < 0$, where $\sigma > 0$ (*Rayleigh distribution*). Find: (a) the median; (b) the density of ξ; (c) the mode.

15.27. Let the r.v. ξ have d.f. $F(x) = 1 - \exp(-x^m/a)$ for $x \geqslant 0$ and $F(x) = 0$ for $x < 0$, where $a > 0$, $m > 0$ (*Weibull distribution*). Find: (a) the p-quantile; (b) the density of ξ; (c) the mode.

16. Multivariate Distributions and Functions of Random Variables

Introductory Notes

Let $\xi_1, \ldots, \xi_n$ be r.v.'s defined on the probability space $(\Omega, \mathcal{F}, P)$. The vector $\vec{\xi} = (\xi_1, \ldots, \xi_n)$ is called *multivariate random variable* or a *random vector*, taking values in the n-dimensional Euclidean space $\mathbb{R}_n$. The function

$$F_{\vec{\xi}}(x_1, \ldots, x_n) = P\{\xi_1 < x_1, \ldots, \xi_n < x_n\},$$

$$(x_1, \ldots, x_n) \in \mathbb{R}_n \tag{16.1}$$

is called the distribution function of the random vector ξ. It satisfies the following properties:

(a) $F_{\vec{\xi}}(x_1, \ldots, x_n)$ is a non-decreasing function in each of the variables;

(b) $F_{\vec{\xi}}(x_1, \ldots, x_n)$ is a left-continuous in each of the variables;

(c) $F_{\vec{\xi}}(x_1, \ldots, x_n) \to 0$ if $x_k \to -\infty$ $(k = 1, \ldots, n)$ at least for one k and $F_{\vec{\xi}}(x_1, \ldots, x_n) \to 1$ if all the variables $x_1, \ldots, x_n$ tend to ∞;

(d) for arbitrary $a_k \leqslant b_k$, $k = 1, \ldots, n$,

$$\Sigma(-1)^{\varepsilon_1+\ldots+\varepsilon_n} F_{\vec{\xi}}(\varepsilon_1 a_1 + (1 - \varepsilon_1)b_1, \ldots, \varepsilon_n a_n + (1 - \varepsilon_n)b_n \geqslant 0,$$

where the summation is taken over all possible $\varepsilon_1, \ldots, \varepsilon_n$, where $\varepsilon_i = 0$ or 1, $i = 1, \ldots, n$.

Every function satisfying conditions (a) - (d) is called a *multivariate distribution function*.

If $F_{\vec{\xi}}(x_1, \ldots, x_n)$ is a d.f. of the random vector $\vec{\xi} = (\xi_1, \ldots, \xi_n)$ and $B \in \mathcal{B}_n$, then

$$P\{\vec{\xi} \in B\} = \int \ldots \int_B 1 \, dF_{\vec{\xi}}(x_1, \ldots, x_n), \tag{16.2}$$

where the integration is in the sense of Lebesgue-Stieltjes.

We say that the d.f. $F(x_1, \ldots, x_n)$, $(x_1, \ldots, x_n) \in \mathbb{R}_n$, of the random vector $\vec{\xi}$ is *absolutely continuous* if there exists an integrable function $f(x_1, \ldots, x_n)$, $(x_1, \ldots, x_n) \in R_n$, such

$$F(y_1, \ldots, y_n) = \int_{-\infty}^{y_1} \ldots \int_{-\infty}^{y_n} f(x_1, \ldots, x_n)dx_1, \ldots, dx_n \tag{16.3}$$

for every $(y_1, \ldots, y_n) \in \mathbb{R}_n$.

The function $f(x_1, \ldots, x_n)$ is called the *density* of the d.f. $F(x_1, \ldots, x_n)$ and also the density of the random vector $\vec{\xi}$.

A necessary and sufficient condition for the function f to be the density of a certain random vector is: $f(x_1, \ldots, x_n) \geqslant 0$ almost everywhere in $\mathbb{R}_n$ and

$$\int_{-\infty}^{\infty} \cdots \int_{-\infty}^{\infty} f(x_1, \ldots, x_n)dx_1 \ldots dx_n = 1.$$

Conditions (a) and (d) imply the existence and the non-negativity of the derivative $\frac{\partial^n F(x_1, \ldots, x_n)}{\partial x_1 \ldots \partial x_n}$ almost everywhere (with respect to the Lebesgue measure in $\mathbb{R}_n$). If the d.f. $F(x_1, \ldots, x_n)$ possesses the density $f(x_1, \ldots, x_n)$, then we have

$$\frac{\partial^n F(x_1, \ldots, x_n)}{\partial x_1 \ldots \partial x_n} = f(x_1, \ldots, x_n) \tag{16.4}$$

almost everywhere.

We say that the d.f. is *singular* with respect to the Lebesgue measure, if F is a continuous function in $\mathbb{R}_n$ and $\frac{\partial^n F(x_1, \ldots, x_n)}{\partial x_1 \ldots \partial x_n} = 0$ almost everywhere in $\mathbb{R}_n$.

Let the random vector $\vec{\xi} = (\xi_1, \ldots, \xi_n)$ have a d.f. $F_{\vec{\xi}}(x_1, \ldots, x_n)$. Denote by $F_{\vec{\eta}}(x_{i_1}, \ldots, x_{i_k})$ the d.f. of the random vector $\vec{\eta} = (\xi_{i_1}, \ldots, \xi_{i_k})$, where $1 \leqslant i_1 < i_2 < \ldots < i_k \leqslant n$. Then for an arbitrary point $(x_{i_1}, \ldots, x_{i_k}) \in \mathbb{R}_k$, we have

$$F_{\vec{\eta}}(x_{i_1}, \ldots, x_{i_k}) = F_{\vec{\xi}}(\infty, \ldots, \infty, x_{i_1}, \infty, \ldots, \infty, x_{i_2}, \ldots, \infty). \tag{16.5}$$

If $f_{\vec{\xi}}(x_1, \ldots, x_n)$, $(x_1, \ldots, x_n) \in \mathbb{R}_n$, is the density of the random vector $\vec{\xi}$ and $A \in \mathcal{B}_n$, then

$$P\{\vec{\xi} \in A\} = \int \underset{A}{\cdots} \int f_{\vec{\xi}}(x_1, \ldots, x_n)dx_1, \ldots dx_n \tag{16.6}$$

The density of the random vector $\vec{\eta} = (\xi_{i_1}, \ldots, \xi_{i_k})$ is given as

$$f_{\vec{\eta}}(x_{i_1}, \ldots, x_{i_k}) = \int_{-\infty}^{\infty} \cdots \int_{-\infty}^{\infty} f_{\vec{\xi}}(x_1, \ldots, x_n)dx_1 \ldots dx_{i_1-1} \cdot \cdot d_{x_{i}+1} \ldots dx_{i_k-1} \cdot dx_{i_k+1} \ldots dx_n. \tag{16.7}$$

The equalities (16.5) and (16.7) are called *consistency conditions* for the multivariate d.f.'s and densities.

A necessary and sufficient condition for the r.v.'s $\xi_1, \ldots, \xi_n$ to be independent is

$$F_{\vec{\xi}}(x_1, \ldots, x_n) = F_{\xi_1}(x_1) \ldots F_{\xi_n}(x_n),$$
$$(x_1, \ldots, x_n) \in \mathbb{R}_n, \tag{16.8}$$

and if $\vec{\xi}$ has a density $f_{\vec{\xi}}$, then this condition is

$$f_{\vec{\xi}}(x_1, \ldots, x_n) = f_{\xi_1}(x_1) \ldots f_{\xi_n}(x_n),$$
$$x_1, \ldots, x_n \in \mathbb{R}_1. \tag{16.9}$$

Let $u_i = u_i(x_1, \ldots x_n)$, $1 \leqslant i \leqslant n$, be Borel-measurable functions from $\mathbb{R}_n$ in $\mathbb{R}_1$. Put $\eta_i = u_i(\xi_1, \ldots, \xi_n)$, $1 \leqslant i \leqslant n$. Then the random vector $\vec{\eta} = (\eta_1, \ldots, \eta_n)$ has the d.f.

$$F_{\vec{\eta}}(y_1, \ldots, y_n) = \int \underset{D}{\ldots} \int dF_{\vec{\xi}}(x_1, \ldots, x_n), \tag{16.10}$$

where $D = \{(x_1, \ldots, x_n) : u_i(x_1, \ldots, x_n) < y_i,\ i = 1, \ldots, n\}$.

Suppose the random vector $\vec{\xi} = (\xi_1, \ldots, \xi_n)$ has a density $f_{\vec{\xi}}$ and the mapping $u_i = u_i(x_1, \ldots, x_n)$, $i = 1, \ldots, n$, of $\mathbb{R}_n$ over the open set $U \subset R_n$ is convertible and continuous. Let $x_i = x_i(u_1, \ldots, u_n)$, $i = 1, \ldots, n$, $(u_1, \ldots, u_n) \in U$, be the inverse mapping. Suppose that the first derivatives $\frac{\partial x_i}{\partial u_j}$ exist and are continuous and the Jacobian $J = \det\left(\frac{\partial x_i}{\partial u_j}\right) \neq 0$ on U. Then the random vector $\vec{\eta} = (\eta_1, \ldots, \eta_n)$ is absolutely continuous and its density is

$$f_{\vec{\eta}}(u_1,\ldots,u_n) = \begin{cases} f_{\vec{\xi}}(x_1(u_1,\ldots,u_n),\ldots,x_n(u_1,\ldots,u_n))\ J\ , & \text{if } (u_1,\ldots,u_n) \in U \\ 0, & \text{if } (u_1, \ldots, u_n) \bar{\in} U. \end{cases} \tag{16.11}$$

It is known from the integral calculus that formula (16.11) holds true in many cases when the conditions imposed on the mapping $u_i = u_i(x_1, \ldots, x_n)$, $i = 1, \ldots, n$, are not satisfied in some points or even over certain hyperplanes in $\mathbb{R}_n$.

Let $(x_1, \ldots, x_n) \in \mathbb{R}_n$ and $Q_k(x_1, \ldots, x_n) = x_{i_k}$, $k = 1, \ldots, n$, where $x_{i_1} \leqslant x_{i_2} \leqslant \ldots \leqslant x_{i_{k-1}} \leqslant x_{i_k} \leqslant \ldots \leqslant x_{i_n}$ are the ordered numbers $x_1, \ldots, x_n$. Let $\xi_1, \ldots, \xi_n$ be n independent r.v.'s. Denote $\xi_{(k)} = Q_k(\xi_1, \ldots, \xi_n)$, $k = 1, \ldots, n$. The r.v. $\xi_{(k)}$ is called the *kth order statistics* of the sample $\xi_1, \ldots, \xi_n$, whereas $\xi_{(1)}, \xi_{(2)}, \ldots, \xi_{(n)}$ are called the *order statistics*.

Let the column vector $\vec{a} \in \mathbb{R}_n$ and the positive definite $n \times n$-matrix Σ be given. The density function of the multivariate normal distribution on $\mathbb{R}_n$ with parameters $\vec{a}$ and Σ is defined by

$$f(\vec{x}) = (2\pi)^{-n/2}(\det \Sigma)^{-1/2}\exp(-\frac{1}{2}(\vec{x} - \vec{a})^T\Sigma^{-1}(\vec{x} - \vec{a})),$$

$$(\vec{x} \in \mathbb{R}_n), \tag{16.12}$$

where $(\vec{x} - \vec{a})^T$ is the transposition of the vector $\vec{x} - \vec{a}$ ($\vec{x}$ is as $\vec{a}$ a column vector), $\vec{a}^T = (a_1, \ldots, a_n)$, $\Sigma^{-1} = (A_{ij})$. If the random vector $\vec{\xi} = (\xi_1, \ldots, \xi_n)$ has the density function f from (16.12), then we say that $\vec{\xi}$ is *normally distributed* and denote this by $\vec{\xi} \in N(\vec{a}, \Sigma)$.

Illustrative Examples

Example 16.1. Let the r.v.'s $\xi_1, \ldots, \xi_n$ be independent with densities $f_1(x_1), f_2(x_2), \ldots, f_n(x_n)$, $x_1, x_2, \ldots, x_n \in \mathbb{R}_1$, respectively. Let $\eta_1 = \xi_1$, $\eta_2 = \xi_1 + \xi_2, \ldots, \eta_n = \xi_1 + \ldots + \xi_n$. Find the density $f_{\vec{\eta}}$ of the random vector $\vec{\eta} = (\eta_1, \ldots, \eta_n)$.

Solution. From (16.9) we have that the density of the random vector $\vec{\xi} = (\xi_1, \ldots, \xi_n)$ is $f_{\vec{\xi}}(x_1, \ldots, x_n) = f_1(x_1) \ldots f_n(x_n)$. Consider the linear mapping $u_1 = x_1$, $u_2 = x_1 + x_2, \ldots, u_n = x_1 + \ldots + x_n$ of $\mathbb{R}_n$ in $\mathbb{R}_n$. One can easily find the inverse mapping: $x_1 = u_1$, $x_2 = u_2 - u_1, \ldots, x_n = u_n - u_{n-1}$, and the Jacobian $J = 1$. Applying (16.11) we obtain

$$f_{\vec{\eta}}(u_1, \ldots, u_n) = f_1(u_1)f_2(u_2 - u_1)f_n(u_n - u_{n-1}).$$

In the special case $n = 2$, for the density of $\eta = (\eta_1, \eta_2)$ we find

$$f_{\vec{\eta}}(u_1, u_2) = f_1(u_1)f_2(u_2 - u_1);$$

hence, applying (16.7) for the density of the r.v. $\eta_2 = \xi_1 + \xi_2$, we get

$$f_{\eta_2}(u_2) = \int_{-\infty}^{\infty} f_1(u_1)f_2(u_2 - u_1)\, du_1.$$

The last equality is called a *convolution* of the density functions f_1 and f_2.

Example 16.2. Let the r.v.'s $\xi_1, \ldots, \xi_n$ be independent and $\xi_i \in N(0, 1)$, $i = 1, \ldots, n$. Find: (a) the density of the r.v. $\eta = \xi_1^2 + \ldots + \xi_n^2$; (b) the density of the r.v. $\sqrt{\eta/n}$.

Solution. (a) Since $\xi_1, \ldots, \xi_n$ are independent, their joint density is

$$f(x_1, \ldots, x_n) = (2\pi)^{-n/2}\exp[-\tfrac{1}{2}(x_1^2 + \ldots + x_n^2],$$

$$(x_1, \ldots, x_n) \in \mathbb{R}_n.$$

Consider the mapping

$$x_1 = \rho \cos \Theta_1 \cos \Theta_2 \ldots \cos \Theta_{n-3} \cos \Theta_{n-2} \cos \Theta_{n-1},$$

$$x_2 = \rho \cos \Theta_1 \cos \Theta_2 \ldots \cos \Theta_{n-3} \cos \Theta_{n-2} \sin \Theta_{n-1},$$

$$x_3 = \rho \cos \Theta_1 \cos \Theta_2 \ldots \cos \Theta_{n-3} \sin \Theta_{n-2},$$

$$\ldots\ldots\ldots\ldots\ldots\ldots\ldots\ldots\ldots\ldots\ldots\ldots\ldots\ldots$$

$$x_n = \rho \sin \Theta_1,$$

where $U = \{\rho, \Theta_1, \ldots, \Theta_{n-1}) : \rho > 0, 0 \leqslant \Theta_i \leqslant \pi, i = 2, 3, \ldots, n - 1, 0 \leqslant \Theta_1 \leqslant 2\pi\}$. We have $\rho^2 = x_1^2 + \ldots + x_n^2$, $J = \rho^{n-1}K(\Theta_1, \ldots, \Theta_{n-1})$, where $K(\Theta_1, \ldots, \Theta_{n-1})$ is a suitable function, depending only on $\Theta_1, \ldots, \Theta_{n-1}$. If we express $\rho, \Theta_1, \ldots, \Theta_{n-1}$ through $x_1, \ldots, x_n$; i.e., $\rho = (x_1^2 + \ldots + x_n^2)^{1/2}$, $\Theta_k = \Theta_k(x_1, \ldots, x_n)$, $k = 1, \ldots, n - 1$ and set $\zeta_1 = (\xi_1^2 + \ldots + \xi_n^2)^{1/2}$, $\zeta_{k+1} = \Theta_k(\xi_1, \ldots, \xi_n)$, $k = 1, \ldots, n - 1$, then, from (16.11) for the random vector $\vec{\zeta} = (\zeta_1, \ldots, \zeta_n)$, we get

$$f_{\vec{\zeta}}(\rho, \Theta_1, \ldots, \Theta_{n-1}) = (2\pi)^{-n/2}\exp[-\rho^2/2]\rho^{n-1}|K(\Theta_1, \ldots, \Theta_{n-1})|$$

for $\rho > 0$, $0 \leqslant \Theta_k \leqslant \pi$ and $k = 2, \ldots, n - 1$, $0 \leqslant \Theta_1 < 2\pi$ and $f_{\vec{\zeta}}(\rho, \Theta_1, \ldots, \Theta_{n-1}) = 0$, otherwise.

For the density of the r.v. ζ_1, according to (16.7), we get

$$f_{\zeta_1}(\rho) = \begin{cases} c(2\pi)^{-n/2}\rho^{n-1}\exp(-\frac{1}{2}\rho^2), & \text{if } \rho > 0 \\ 0, & \text{if } \rho \leqslant 0, \end{cases}$$

where $c = \int_0^{2\pi} d\Theta_1 \int_0^{\pi} d\Theta_2 \ldots \int_0^{\pi} |K(\Theta_1, \ldots, \Theta_{n-1})| d\Theta_{n-1}$. The value of c can be determined from the last integral in conjunction with the equality

$$1 = \int_{\infty}^{\infty} f_{\vec{\zeta}}(\rho)d\rho = c \int_0^{\infty} (2\pi)^{-n/2}\rho^{n-1}\exp(-\tfrac{1}{2}\rho^2)d\rho;$$

hence, we obtain $c = 2\pi^{n/2}/\Gamma(n/2)$. Thus

$$f_{\zeta_1}(\rho) = \begin{cases} \dfrac{2^{(n-2)/2}}{\Gamma(n/2)}\rho^{n-1}\exp(-\tfrac{1}{2}\rho^2), & \text{if } \rho > 0 \\ 0, & \text{if } \rho \leqslant 0. \end{cases}$$

Since $\rho = \zeta_1^2$, we have

$$F_\eta(x) = \begin{cases} P\{\zeta_1^2 < x\} = P\{-\sqrt{x} < \zeta_1 < \sqrt{x}\} = F_{\zeta_1}(\sqrt{x}) - F_{\zeta_1}(-\sqrt{x}), & \text{if } x > 0 \\ 0, & \text{if } x \leqslant 0. \end{cases}$$

Differentiating $F_\eta(x)$ we obtain

$$f_\eta(x) = \begin{cases} \dfrac{1}{\sqrt{4x}}\left(f_{\zeta_1}(\sqrt{x}) + f_{\zeta_1}(-\sqrt{x})\right), & \text{if } x > 0 \\ 0, & \text{if } x \leqslant 0; \end{cases}$$

hence,

$$f_\eta(x) = \begin{cases} \dfrac{(x/2)^{(n-2)/2}}{2\Gamma(n/2)}\exp(-\tfrac{1}{2}x), & \text{if } x > 0 \\ 0, & \text{if } x \leqslant 0. \end{cases} \tag{16.13}$$

The distribution of the r.v. η is called χ^2*-distribution* (*chi-square distribution*) or Pearson distribution with n degrees of freedom.

Comparing the density in (16.13) with that in (15.10), we observe that the r.v. η has a χ^2-distribution with n degrees of freedom, which is in fact a gamma distribution with parameters $\alpha = n/2$ and $\beta = 1/2$.

(b) From $\sqrt{\eta/n} = \zeta_1/\sqrt{n}$, one easily gets

$$f_{\sqrt{\eta/n}}(y) = \begin{cases} 2[\Gamma(n/2)]^{-1}(n/2)^{n/2}\exp(-ny^2/2), & \text{if } y > 0 \\ 0, & \text{if } y \leqslant 0. \end{cases}$$

Example 16.3. Let the random vector $\vec{\xi}$ be normally distributed $N(\vec{a}, \Sigma)$ (see (16.12)), where $\vec{a}^T = (a_1, \ldots, a_n)$, $\vec{\xi}^T = (\xi_1, \ldots, \xi_n)$, $\Sigma^{-1} = (A_{ij})$.

Suppose that the matrix B satisfies $\Sigma^{-1} = B^T B$. Prove that the random vector $\vec{\eta} = B(\vec{\xi} - \vec{a})$ has independent $N(0, 1)$ components; i.e., if $\vec{\eta}^T = (\eta_1, \ldots, \eta_n)$, then the r.v.'s $\eta_1, \ldots, \eta_n$ are independent and identically $N(0, 1)$ distributed.

Solution. Introduce the transformation $\vec{u} = B(\vec{x} - \vec{a})$ of $\mathbb{R}_n$ on $\mathbb{R}_n$, where $\vec{x}^T = (x_1, \ldots, x_n)$, $\vec{u}^T = (u_1, \ldots, u_n)$, $\vec{a}^T = (a_1, \ldots, a_n)$. Since Σ is a positive definite matrix, then $\det \Sigma \neq 0$, and hence $\det(\Sigma^{-1}) \neq 0$ and from $\Sigma^{-1} = B^T B$ we have $\det(B^{-1}) \neq 0$. Since the Jacobian $J = \det(B^{-1}) \neq 0$, we can apply formula (16.11) to find the density of the random vector $\vec{\xi}$. Let us express the density $f_{\vec{\xi}}$ of the random vector $\vec{\xi}$ as (see (16.12)):

$$f_{\vec{\xi}}(\vec{x}) = (2\pi)^{-n/2}(\det \Sigma)^{-1/2}\exp(-\frac{1}{2}(\vec{x} - \vec{a})^T \Sigma^{-1}(\vec{x} - \vec{a})) =$$

$$= (2\pi)^{-n/2}(\det((B^T B)^{-1}))^{-1/2}\exp(-\frac{1}{2}(\vec{x} -$$

$$- \vec{a})^T B^T B(\vec{x} - \vec{a})), \quad \vec{x} \in \mathbb{R}_n.$$

Hence, from (16.11), we get

$$f_{\vec{\eta}}(\vec{u}) = (2\pi)^{-n/2}(\det B^T B)^{1/2}\exp(-\frac{1}{2}\vec{u}^T\vec{u})|\det B^{-1}| =$$

$$= (2\pi)^{-n/2}\exp(-\frac{1}{2}\vec{u}^T\vec{u}) =$$

$$= (2\pi)^{-n/2}\exp(-\frac{1}{2}(u_1^2 + u_2^2 + \ldots + u_n^2) \quad \text{for every } \vec{u} \in \mathbb{R}_n.$$

Thus

$$f_{\vec{\eta}}(u_1, \ldots, u_n) = \prod_{i=1}^{n} (2\pi)^{-1/2}\exp(-\frac{1}{2}u_i^2) \quad (u_1, \ldots, u_n) \in \mathbb{R}_n;$$

therefore, according to (16.9), the r.v.'s $\eta_1, \ldots, \eta_n$ are independent identically $N(0, 1)$ distributed.

Exercises

16.1. Show that the set of numbers $\{p(i, j) = ij(\frac{2}{n(n + 1)})^2, i, j = 1, \ldots, n\}$ defines a two-dimensional discrete distribution.

16.2. Let the random vector (ξ, η) have the distribution

$$P\{\xi = j, \eta = k\} = c\,\frac{\lambda^j \mu^k \nu^{jk}}{j!k!},$$

where $\lambda > 0$, $\mu > 0$, $0 < \nu \leqslant 1$, $j, k = 0, 1, 2, \ldots$ and c is a suitable constant. (This distribution is called a *distribution of Obrechkoff* and also *bivariate Poisson distribution*.) Determine c and find the distributions of ξ and η. Show that ξ and η are independent if and only if $\nu = 1$. Find $P\{\xi = j | \eta = k\}$, $j = 0, 1, \ldots$

*16.3. Let ξ and η be independent r.v.'s whose common distribution is geometric with parameter p, $0 < p < 1$; i.e., $P\{\xi = k\} = P\{\eta = k\} = pq^k$ $k = 0, 1, \ldots$, $q = 1 - p$. Find: (a) the distribution of the r.v. $\zeta = \max[\xi, \eta]$; (b) the distribution of the random vector (ζ, ξ).

16.4. Given the function $F(x, y) = 1$ for $x + y > 0$ and $F(x, y) = 0$, otherwise. Show that it obeys conditions (a), (b) and (c) from the Introductory Notes to this section but does not obey condition (d).

16.5. Let the random vector (ξ, η) have density $f(x, y) = g(x)h(y)$, $x \in \mathbb{R}_1$, $y \in \mathbb{R}_1$. Find the densities of the r.v.'s ξ and η.

16.6. Let the random vector (ξ_1, ξ_2, ξ_3) have density $f(x, y, z) = cg(x, y, z)$, where $x \in \mathbb{R}_1$, $y \in \mathbb{R}_1$, $z \in \mathbb{R}_1$ and c is a suitable constant. Find the unknown constant c. Are the r.v.'s ξ_1, ξ_2 and ξ_3 independent if: (a) $g(x, y, z) = 1$ for $0 \leqslant x \leqslant 1$, $-2 \leqslant y \leqslant 3$, $4 \leqslant z \leqslant 5$ and $g(x, y, z) = 0$, otherwise; (b) $g(x, y, z) = 1$ for $x^2 + y^2 + z^2 \leqslant 1$ and $g(x, y, z) = 0$, otherwise; (c) $g(x, y, z) = x^{k-1}y^{m-1}z^{n-1}$ for $x \geqslant 0$, $y \geqslant 0$, $z \geqslant 0$, $x + y + z \leqslant 1$ and $g(x, y, z) = 0$, otherwise, where $k \geqslant 1$, $m \geqslant 1$, and $n \geqslant 1$ are fixed numbers?

16.7. Let $g(x)$, $x \in \mathbb{R}_1$, be a strictly monotonic function and ξ be a r.v. Express the d.f. of the random vector $\vec{\zeta} = (\xi, \eta)$ in terms of the d.f.'s F_ξ and F_η, where $\eta = g(\xi)$.

*16.8. Let the r.v.'s ξ_1 and ξ_2 be independent and have distribution $E(\lambda)$. Find the density of the random vector $\vec{\zeta} = (\eta_1, \eta_2)$, where $\eta_1 = \max[\xi_1, \xi_2]$, $\eta_2 = \min[\xi_1, \xi_2]$.

16.9. Let $F(x, y)$, $(x, y) \in \mathbb{R}_2$, be the d.f. of the random vector (ξ_1, ξ_2) and $G(x, y)$, $(x, y) \in \mathbb{R}_2$, be the d.f. of the random vector (η_1, η_2), where $\eta_1 = \max[\xi_1, \xi_2]$, $\eta_2 = \min[\xi_1, \xi_2]$. Express G in terms of F.

16.10. Let the r.v.'s ξ and η be independent, $\xi \in N(a_1, \sigma_1^2)$, $\eta \in N(a_2, \sigma_2^2)$ (see (15.8)). Find the density of $\xi + \eta$.

16.11. Given the independent r.v.'s ξ and η, where $\xi \in N(a_1, \sigma_1^2)$, $\eta \in N(a_2, \sigma_2^2)$. Let $\Theta = \xi/\eta$. Is it true that the density $f_\Theta(z)$, $z \in \mathbb{R}_1$, is equal to the function $g(z) = \sigma_1\sigma_2/[\pi(\sigma_1^2 + \sigma_2^2 z^2)]$, $z \in \mathbb{R}_1$? Answer without computing.

16.12. Let the r.v.'s ξ_1 and ξ_2 be independent and identically distributed, $N(0, \sigma^2)$. Find the distribution of ξ_1/ξ_2.

16.13. Given the independent r.v.'s ξ and η, where $\xi \in P(\lambda)$, and η is uniformly distributed over the interval (0, 1). Find the distribution of $\xi + \eta$.

16.14. The point ξ is chosen at random on the axis 0x with distribution $N(0, 1)$. Independently, the point η is chosen on the axis 0y, with the same distribution as ξ. What is the probability that the angle $0\xi\eta$ does not exceed 45 degrees?

16.15. Let the point (ξ_1, ξ_2) be uniformly distributed over the quadrant $\{(x_1, x_2) : 0 \leqslant x_i \leqslant 1, i = 1, 2\}$. For what values of c will the events $A_c = \{|\xi_1 - \xi_2| \geqslant c\}$ and $B_c = \{\xi_1 + \xi_2 \leqslant 3c\}$ be independent?

16.16. Let the r.v.'s ξ_1 and ξ_2 be independent and identically $E(1)$ distributed. Consider $\eta = \xi_1 + \xi_2$ and $\zeta = \xi_1/(\xi_1 + \xi_2)$. Find: (a) the density of η; (b) the density of ζ. Are η and ζ independent?

16.17. The density of the random vector (ξ, η) is given by $f(x, y) = \frac{1}{4}(1 + xy(x^2 - y^2))$ for $|x| < 1$, $|y| < 1$ and $f(x, y) = 0$, otherwise. Show that the density of the r.v. $\xi + \eta$ satisfies the relation $f_{\xi+\eta}(x) = f_\xi(x)f_\eta(x)$. Are the r.v.'s ξ and η independent?

16.18. Let the r.v.'s ξ and η be independent and uniformly distributed over the intervals $(0, a)$ and $(0, \pi/2)$, respectively. Find $P\{\xi < b \cos \eta\}$, where $0 < b < a$.

16.19. Let ξ_1 and ξ_2 be independent r.v.'s uniformly distributed over the interval $(0, 1)$. Find the density of the random vector $\vec{\zeta} = (\eta_1, \eta_2)$, where $\eta_1 = \sqrt{-2\ln\xi_1} \cos(2\pi\xi_2)$, $\eta_2 = \sqrt{-2\ln\xi_1} \sin(2\pi\xi_2)$. Are η_1 and η_2 independent?

16.20. Let a, b and c be r.v.'s, independent and uniformly distributed over the interval $(0, 1)$. What is the probability that the equation $ax^2 + bx + c = 0$ has real roots (see Exercise 7.5)?

16.21. What is the probability that the equation $x^2 - 2bx + c = 0$ has real and different roots, if b and c are independent r.v.'s with distribution $E(\lambda)$?

16.22. Suppose F and G are arbitrary one-dimensional r.v.'s. Let $H_1(x, y) = \max[0, F(x) + G(y) - 1]$, $H_2(x, y) = \min[F(x), G(y)]$ and $H(x, y) = c_1H_1(x, y) + c_2H_2(x, y)$, where c_1 and c_2 are non-negative constants with $c_1 + c_2 = 1$, $x, y \in \mathbb{R}_1$. Prove that: (a) H_1, H_2 and H are two-dimensional d.f.'s and that their marginal d.f.'s are F and G; (b) if $W(x, y)$, $(x, y) \in \mathbb{R}_2$, is an arbitrary d.f. with marginals F and G, respectively, then $H_1(x, y) \leqslant W(x, y) \leqslant H_2(x, y)$, $(x, y) \in \mathbb{R}_2$.

The d.f.'s H_1 and H_2 are known as *Frechet's minimal and maximal d.f.'s* with given marginals F and G.

16.23. Let the r.v.'s ξ and η be independent and have Cauchy distribution with parameters $a = a_1$, $b = b_1$ and $a = a_2$, $b = b_2$, respectively. Find the density of the r.v. $\xi + \eta$.

16.24. Given the independent r.v.'s ξ_1 and ξ_2 each of which have the Cauchy distribution with parameters $a = 0$, $b = 1$. Find the density of the r.v. $\Theta = \xi_1\xi_2$.

16.25. Let the r.v.'s $\xi_1, \ldots, \xi_n$ be independent and exponentially distributed; i.e., ξ_i has the density $f_i(x) = \lambda_i \exp(-\lambda_i(x - \beta_i))$ for

$x > \beta_i$ and $f_i(x) = 0$ for $x \leqslant \beta_i$, $i = 1, \ldots, n$, $\lambda_i > 0$. Find the density of $\eta = \xi_1 + \ldots + \xi_n$ in the following two cases: (a) $\lambda_1 = \lambda_2 = \ldots = \lambda_n = \lambda$, $\beta_1 = \ldots = \beta_n = \beta$; (b) $\beta_1 = \beta_2 = \ldots = \beta_n = 0$.

16.26. Let the random vector $\vec{\xi} = (\xi_1, \ldots, \xi_n)$ have density $f_{\vec{\xi}}(x_1, \ldots, x_n)$, $(x_1, \ldots, x_n) \in \mathbb{R}_n$. Find the density of the random vector $\vec{\eta} = (\alpha_1\xi_1 + \beta_1, \ldots, \alpha_n\xi_n + \beta_n)$, where α_i, β_i, $i = 1, 2, \ldots, n$, are arbitrary real numbers and $\alpha_1\alpha_2 \ldots \alpha_n \neq 0$.

*16.27. Let the r.v.'s $\xi_1, \ldots, \xi_n$ be independent and uniformly distributed over the interval (α, β). Let $\eta_1 = \min[\xi_1, \ldots, \xi_n]$, $\eta_2 = \max[\xi_1, \ldots, \xi_n]$. Find the density of the random vector $\vec{\zeta} = (\eta_1, \eta_2)$.

16.28. Let $I_n(x) = \pi^{-n/2} \int \cdots \int_{\Delta_{n,x}} \exp\left(- \sum_{j=1}^{n} x_j^2\right) dx_1 \ldots dx_n$,

where $\Delta_{n,x} \subset \mathbb{R}_n$ is defined as

$$\Delta_{n,x} = \left\{(x_1, \ldots, x_n) : \left|\frac{x_1}{1} + \frac{x_2}{\sqrt{2}} + \ldots + \frac{x_n}{\sqrt{n}}\right| \leqslant x\right\}, \quad x > 0.$$

Find the limit of $I_n(x)$ when $n \to \infty$. (*W. Luxemburg*).

*16.29. Let $F(x)$, $x \in \mathbb{R}_1$, be a continuous d.f. Suppose the random vector $(\xi_1, \ldots, \xi_n)$ has the d.f.

$$P\{\xi_1 < x_1, \ldots, \xi_n < x_n\} = \frac{1}{n!} F(x_{(1)})(F(x_{(2)}) + 1) \ldots (F(x_{(n)}) + n - 1),$$

where $x_{(1)} \leqslant x_{(2)} \leqslant \ldots \leqslant x_{(n)}$ are the values $x_1, \ldots, x_n$ arranged in a non-decreasing order. Find: (a) the marginal d.f. $P\{\xi_i < x\}$; (b) $P\{\xi_1 < \xi_2 < \ldots < \xi_n\}$. (*G. S. Chobanov and Z. G. Ignatov*)

16.30. Let the r.v.'s ξ and η be independent and distributed as χ^2 with degrees of freedom m and n, respectively, (see Example 16.2). Show that the density of the r.v. $\zeta = \frac{n\xi}{m\eta}$ is given as

$$f_\zeta(x) = \begin{cases} \dfrac{m^{m/2} n^{n/2}}{\Gamma((m+n)/2)\Gamma(m/2)\Gamma(n/2)} u^{m/2-1}(mu + n)^{-\frac{m+n}{2}}, & \text{if } u > 0 \\ 0, \quad \text{if } u \leqslant 0. \end{cases}$$

The distribution of the r.v. ξ is called the *F-distribution* with (m, n) *degrees of freedom* or *Fisher's distribution.*

16.31. Let the r.v.'s ξ and η be independent. Assume that $\xi \in N(0, 1)$ and η has χ^2-distribution with n degrees of freedom. Show that the density of the r.v. $\zeta = \xi/\sqrt{\eta/n}$ is given by

$$f_\zeta(u) = \Gamma((n+2)/2)[\sqrt{\pi n}\,\Gamma(n/2)]^{-1}(1 + u^2/n)^{-(n+1)/2}, \quad u \in \mathbb{R}_1$$

The distribution of the r.v. ζ is called the *t-distribution with n degrees of freedom* or *Student's distribution*.

16.32. Let the r.v. ξ have t-distribution (see Exercise 16.31) with n degrees of freedom. Find the distribution of the r.v. ξ^2.

16.33. Let the r.v. ξ be Cauchy distributed with parameters $a = 0$, $b = 1$ (see (15.12)). Find the distribution of the r.v. ξ^2.

16.34. Let the r.v.'s ξ and η be independent and have d.f.'s $F_\xi(x)$, $F_\eta(x)$ and densities $f_\xi(x)$, $f_\eta(x)$, $x \in \mathbb{R}_1$, respectively. Prove that: (a) $F_{\xi+\eta}(y) = \int_{-\infty}^{\infty} F_\xi(y - x)dF_\eta(x)$ for every $y \in \mathbb{R}_1$; i.e., the d.f. $F_{\xi+\eta}$ is a convolution of F_ξ and F_η ($F_{\xi+\eta} = F_\xi * F_\eta$); (b) $f_{\xi+\eta}(y) = \int_{-\infty}^{\infty} f_\xi(y - x)f_\eta(x)dx$ for every $y \in \mathbb{R}_1$; i.e., $f_{\xi+\eta}$ is a convolution of f_ξ and f_η ($f_{\xi+\eta} = f_\xi * f_\eta$). Find: (c) $(f_\xi * f_\eta)(y)$, if $\xi \in E(\alpha)$, $\eta \in E(\beta)$ and $\alpha \neq \beta$; (d) $(f_\xi^{*n})(y) = (f_\xi * \ldots * f_\xi)(y)$, if $\xi \in E(\lambda)$; (e) $(f_\xi^{*n})(y)$, if ξ is uniformly distributed over the interval (0, 1).

16.35. Given the independent r.v.'s ξ and η with ξ being absolutely continuous and η discrete. Prove that the r.v. $\zeta = \xi + \eta$ is also absolutely continuous. Find its density.

16.36. Let ξ, η and ζ be independent, identically $N(0, 1)$ r.v.'s. Prove that the following non-linear function $\Theta = (\xi + \eta\zeta)/\sqrt{1 + \zeta^2}$ is a r.v. with the same distribution as ξ, η and ζ. (*A. N. Shiryaev*)

16.37. Let ξ, η and ζ be independent: $\xi \in N(0, 1)$, $\eta \in N(0, 1)$, $\zeta \in N(0, 1)$. Find: (a) a number $r > 0$ such that $P\{(\xi, \eta) \in C_r\} = 0.95$, where $C_r = \{(x, y) : x^2 + y^2 \leq r^2\}$; (b) whether the r.v.'s $\xi^2 + \eta^2$ and ξ/η are independent; (c) the density of $\vec{\delta} = (\alpha, \beta, \gamma)$, where $\alpha = (\xi - \eta)/\sqrt{2}$, $\beta = (\xi + \eta - 2\zeta)/\sqrt{6}$, $\gamma = (\xi + \eta + \zeta)/\sqrt{3}$ and show that the r.v.'s α, β and γ are independent; (d) the density of $\vec{\theta} = (\rho, \phi, \Psi)$, where $\xi = \rho \cos\phi \cos\Psi$, $\eta = \rho \cos\phi \sin\Psi$, $\zeta = \rho \sin\phi$, $\rho > 0$, $\phi \in (-\pi/2, \pi/2)$, $\Psi \in (0, 2\pi)$; i.e., ρ, ϕ, Ψ are spherical coordinates in $\mathbb{R}_3$. Are the r.v.'s ρ, ϕ and Ψ independent?

16.38. Let the random vector $\vec{\zeta} = (\xi, \eta)$ have the density $f_{\vec{\zeta}}(x, y) = D\exp(-g(x, y))$, $(x, y) \in \mathbb{R}_2$, where D is a suitable constant and $g(x, y) = A(x - a)^2 + 2B(x - a)(y - b) + C(y - b)^2$ is a positive definite quadratic form. Find (a) the constant D in terms of A, B and C; (b) the density of the r.v.'s ξ and η; (c) the probability $P\{\zeta \in E_k\}$, E_k is the ellipse $E_k = \{(x, y) : g(x, y) \leq k^2\}$, $k > 0$.

16.39. Let the random vector $\vec{\zeta} = (\xi, \eta)$ have density $f_{\vec{\zeta}}(x, y) = (1/\pi)\exp[-(x^2 + y^2)/2]$ for $xy > 0$ and $f_{\vec{\zeta}}(x, y) = 0$, otherwise. Find the distributions of the r.v.'s ξ and η. Does the random vector $\vec{\zeta}$ have a two-dimensional normal density (see Exercise 16.38)?

16.40. The density of the random vector $\vec{\xi} = (\xi_1, \ldots, \xi_n)$ is given as $f_{\vec{\xi}}(x_1, \ldots, x_n) = (2\pi)^{-n/2}\exp[-(x_1^2 + \ldots + x_n^2)/2]$, $(x_1, \ldots, x_n) \in \mathbb{R}_n$. Let $\eta_k = \xi_1 + \ldots + \xi_k$, $k = 1, \ldots, n$. Find the density of the random vector $\vec{\eta} = (\eta_1, \eta_2, \ldots, \eta_n)$.

*16.41. Let ξ and η be independent r.v.'s and let they have joint density $f(x, y) = g(x^2 + y^2)$, where $g(u)$, $u \in \mathbb{R}_1$, is a continuous function. Show that each of the variables is normally distributed.

16.42. If $\xi_1, \ldots, \xi_n$ are uniformly distributed over the interval $(0, 1)$, find: (a) the d.f. and the density of the kth order statistics $\xi_{(k)}$, $1 \leqslant k \leqslant n$; (b) the d.f. and the density of the r.v. $\eta = \xi_{(1)} + \ldots + \xi_{(n)}$.

*16.43. Let $\xi_1, \ldots, \xi_n$ be n independent r.v.'s uniformly distributed over the interval $(0, 1)$ and $\xi_{(1)}, \ldots, \xi_{(n)}$ be the corresponding order statistics. Set: $V_1 = \xi_{(1)}/\xi_{(2)}$, $V_2 = \xi_{(2)}/\xi_{(3)}$, $\ldots$, $V_{n-1} = \xi_{(n-1)}/\xi_{(n)}$, $V_n = \xi_{(n)}$; $U_1 = \xi_{(1)}$, $U_2 = \xi_{(2)} - \xi_{(1)}$, $\ldots$, $U_{n-1} = \xi_{(n-1)} - \xi_{(n-2)}$, $U_n = \xi_{(n)} - \xi_{(n-1)}$, $U_{n+1} = 1 - \xi_{(n)}$. (a) Find the joint density of $U_1, \ldots, U_{n+1}$. The r.v.'s $U_1, \ldots, U_{n+1}$ are known as the *spacings*; (b) Show that the r.v.'s $V_1, \ldots, V_n$ are independent and find their joint density.

*16.44. Let the independent r.v.'s $\xi_1, \ldots, \xi_n$ be uniformly distributed on $(0, 1)$ and let $\xi_{(1)}, \ldots, \xi_{(n)}$ be their order statistics. Let $\eta_i = \xi_{(i)} - \xi_{(i-1)}$, $i = 1, \ldots, n + 1$ assuming that $\xi_{(0)} = 0$ and $\xi_{(n+1)} = 1$. If $\Theta_1, \ldots, \Theta_{n+1}$ are independent r.v.'s identically distributed $E(1)$ and $\zeta_i = \Theta_i/(\Theta_1 + \ldots + \Theta_{n+1})$, then prove that the random vectors $(\eta_1, \ldots, \eta_{n+1})$ and $(\zeta_1, \ldots, \zeta_{n+1})$ have the same distribution.

*16.45. Let $\xi_1, \ldots, \xi_n$ be independent and identically distributed r.v.'s with d.f. F. Suppose that $F(x)$, $x \in \mathbb{R}_1$, is a continuous function. Find the d.f. of the r.v. $\xi_{(k)}$, $1 \leqslant k \leqslant n$, where $\xi_{(1)}, \ldots, \xi_{(n)}$ are the order statistics.

*16.46. Let $\xi_1, \ldots, \xi_n$ be independent and identically distributed r.v.'s with d.f. F and density f. Let $\xi_{(1)}, \ldots, \xi_{(n)}$ be the correspondind order statistics. Find: (a) the d.f. and the density of the r.v. $\xi_{(1)}$; (b) the d.f. and the density of the r.v. $\xi_{(n)}$; (c) the d.f. and the density of the random vector $\vec{\eta} = (\xi_{(1)}, \xi_{(n)})$; (d) the d.f. and the density of the r.v. $\Theta = \xi_{(n)} - \xi_{(1)}$.

16.47. Find the convolutions $(f * f)(y) = (f^{*2})(y)$ and $(f^{*3})(y)$, $y \in \mathbb{R}_1$, if: (a) $f(x) = (1/2)\exp(-|x|)$, $x \in \mathbb{R}_1$; (b) $f(x) = (1/\pi)(b(b^2 + x^2))$, $b > 0$, $x \in \mathbb{R}_1$.

16.48. Let the r.v.'s $\xi_1, \ldots, \xi_n$ be independent and identically distributed with d.f. $F(x)$ continuous in $\mathbb{R}_1$. Define the integer-valued r.v. η as follows: $\eta = k$, for $k = 2, 3, \ldots$, when $\xi_1 \geqslant \xi_2 \geqslant \ldots \geqslant \xi_{k-1} < \xi_k$. What is the distribution of the r.v. η?

16.49. Let $\xi_1, \ldots, \xi_n$ be independent r.v.'s and let ξ_i have continuous and strictly increasing d.f. F_i, $i = 1, \ldots, n$. Set $\eta_i = F(\xi_i)$. Show that $\Theta = -\frac{1}{2}\sum_{i=1}^{n} \ln(1 - \eta_i)$ has χ^2-distribution with 2n degrees of freedom.

16.50. Let $\xi_1, \xi_2, \ldots$ be independent r.v.'s with distribution $N(0, 1)$. Consider the r.v. $\eta_n = \xi_1\left(\frac{1}{n}\sum_{i=1}^{n}\xi_i^2\right)^{-1/2}$. Show that: (a) the event $\{-\sqrt{n} < \eta_n < \sqrt{n}\}$ has probability one; (b) the r.v. $\tau = \eta_n\sqrt{(n - 1)}/(n - \eta_n^2)^{1/2}$ has a t-distribution with n - 1 degrees of freedom Use this fact to find the density of η_n.

*16.51. Let $\xi_1, \ldots, \xi_n$ be independent r.v.'s, $\xi_i \in E((k + i - 1)\lambda)$, $i = 1, \ldots, n$, where k is a fixed non-negative number. Let $S_n = \xi_1 + \ldots + \xi_n$, and f_n be the density of S_n. Show that:

(a) $f_n(x) = n\binom{k + n - 1}{n}\lambda e^{-\lambda kx}(1 - e^{-\lambda x})^{n-1}$, $x > 0$, $n \geqslant 1$;

(b) $P\{S_n \leqslant x < S_{n+1}\} = \int_0^x f_n(u)\exp(-\lambda(n + k)(x - u))du$.

*16.52. Let $\xi_1, \ldots, \xi_n$ be independent r.v.'s with ξ_j being uniformly distributed over the interval $(0, a_j)$ and $0 < a_1 < a_2 < \ldots < a_n$. Let also $\eta = \max[\xi_1, \ldots, \xi_n]$. Express the probability of the event $\{\eta = \xi_i\}$ through the index i and the numbers a_j, $j = 1, \ldots, n$. (*P. Abad and D. Freedman*)

*16.53. Let $\xi_1, \xi_2, \ldots$ be independent and identically distributed with $\xi_i \in E(\lambda)$. Set $S_0 = 0$, $S_n = \xi_1 + \ldots + \xi_n$ and let $\Theta_t = \max\{k : S_t \leqslant t\}$, $t > 0$. Find: (a) $P\{\Theta_t = n\}$, $n = 0, 1, \ldots$; (b) the density f_{η_t} of the r.v. $\eta_t = \xi_{\Theta_t+1}$ and also $\lim_{t\to\infty} f_{\eta_t}(x)$; (c) the density f_{ζ_t} of the r.v. $\zeta_t = S_{\Theta_t+1} - t$; (d) the d.f. $F_{\nu_t}(x)$, $x \in \mathbb{R}_1$ of the r.v. $\nu_t = t - S_{\Theta_t}$ and also $\lim_{t\to\infty} F_{\nu_t}(x)$.

17. Expectation, Variance and Moments of Higher Order

Introductory Notes

Let ξ be a r.v. defined on the probability space $(\Omega, \mathcal{F}, P)$ and let $F(x) = P\{\xi < x\}$, $x \in \mathbb{R}_1$, be its d.f. The *expectation* (*mean value*) of ξ is denoted by $E(\xi)$ and defined by the equality

$$E(\xi) = \int_{-\infty}^{\infty} x \, dF(x) \tag{17.1}$$

assuming that the integral exists; i.e., $E\{|\xi|\} < \infty$.

This definition of $E(\xi)$ is equivalent to that in Section 12 and to that, when ξ is a discrete r.v. (see Section 9). More generally, if $g(x)$, $x \in \mathbb{R}_1$, is a Borel-measurable function, then for the expectation $E(\xi)$ of the r.v. $\eta = g(\xi)$ we have

$$E(\eta) = \int_{-\infty}^{\infty} g(x) dF(x), \tag{17.2}$$

assuming again that the integral exists; i.e. $\int_{-\infty}^{\infty} |g(x)| dF(x) < \infty$ (see also Exercise 12.15).

If the r.v. ξ has density $f(x)$, $x \in \mathbb{R}_1$, then

$$E(\eta) = E\{g(\xi)\} = \int_{-\infty}^{\infty} g(x) f(x) dx. \tag{17.3}$$

Recall some of the properties of the expectation:

(a) If ξ and η are r.v.'s with $E\{|\xi|\} < \infty$ and $E\{|\eta|\} < \infty$, then $E\{|\alpha\xi + \beta\eta|\} < \infty$, α, $\beta \in \mathbb{R}_1$ and

$$E\{\alpha\xi + \beta\eta\} = \alpha E(\xi) + \beta E(\eta). \tag{17.4}$$

(b) If ξ and η are independent r.v.'s, $E\{|\xi|\} < \infty$ and $E\{|\eta|\} < \infty$, then $E\{|\xi\eta|\} < \infty$ and

$$E\{\xi\eta\} = E(\xi)E(\eta). \tag{17.5}$$

The quantities

$$a_k = E\{(\xi - c)^k\} \quad \text{and} \quad \gamma_k = E\{|\xi - c|^k\} \tag{17.6}$$

are called the *moment* and *absolute moment of order* k, respectively, of the r.v. ξ about the constant c. Recall that when $c = 0$ they are called moments (moments about the origin) and when $c = E(\xi)$ they are called *central moments* (moments about the mean). For the variance of the r.v. ξ (i.e. the second central moment) the properties given in Section 9 also hold; i.e.,

(c) $V(\xi)$ can be written as

$$V(\xi) = E\{\xi^2\} - (E(\xi))^2. \tag{17.7}$$

(d) If the r.v. ξ has finite variance, for every $\alpha \in \mathbb{R}_1$, we have

$$V\{\alpha\xi\} = \alpha^2 V(\xi). \tag{17.8}$$

(e) If the r.v.'s ξ and η are independent and have finite variances, then

$$V\{\xi + \eta\} = V(\xi) + V(\eta). \tag{17.9}$$

From (17.2) and (17.3) it is clear how to express the moments of the r.v. ξ through its d.f. F and also through its density f.

Let $\vec{\xi} = (\xi_1, \ldots, \xi_n)$ be a random vector. As in (17.6) we define the *multivariate moments*:

$$a_{k_1,\ldots,k_n} = E\left\{(\xi_1 - c_1)^{k_1} \ldots (\xi_n - c_n)^{k_n}\right\},$$

$$\gamma_{k_1,\ldots,k_n} = E\left\{\left|\xi_1 - c_1\right|^{k_1} \ldots \left|\xi_n - c_n\right|^{k_n}\right\} \tag{17.10}$$

of the random vector $\vec{\xi}$ about the vector $\vec{c} = (c_1, \ldots, c_n)$.

The covariance matrix and the correlation matrix of the random vector $\vec{\xi}$ can be defined in an obvious way following formulas (9.9) and (9.10) in Section 9.

Illustrative Examples

Example 17.1. Let the random vector $\vec{\xi}$ be normally distributed; i.e., $\vec{\xi} \in N(\vec{a}, \Sigma)$ (see (16.12)), where $\vec{\xi}^T = (\xi_1, \ldots, \xi_n)$, $\vec{a}^T = (a_1, \ldots, a_n)$, $\Sigma = (b_{ij})$, $\Sigma^{-1} = (A_{ij})$. Find $\vec{E}(\xi_i)$ and $\mathrm{cov}(\xi_i, \xi_j)$, $i, j = 1, \ldots, n$.

Solution. According to Example 16.3 if $B^TB = \Sigma^{-1}$, then the random vector $\vec{\eta} = B(\vec{\xi} - \vec{a})$ has independent $N(0, 1)$ components. From $\vec{\eta} = B(\vec{\xi} - \vec{a})$ we get $\vec{\xi} = B^{-1} \cdot \vec{\eta} + \vec{a}$. Put $B^{-1} = (c_{ij})$. We have $\xi_i = \sum_{k=1}^{n} c_{ik}\eta_k + a_i$.

Hence, using (17.4) and the fact that $E(\eta_j) = \frac{1}{\sqrt{2\pi}} \int_{-\infty}^{\infty} x \exp(-x^2/2)dx = 0$, we obtain $E(\xi_i) = E\left\{\sum_{k=1}^{n} c_{ik}\eta_k + a_i\right\} = \sum_{k=1}^{n} c_{ik}E(\eta_k) + E(a_i) = a_i$; i.e., the vector $\vec{a}^T = (a_1, \ldots, a_n)$ is the vector of mean values of $\vec{\xi}$.

Similarly, using (17.4), (17.5) and the relation

$$E\{\eta_j^2\} = \frac{1}{\sqrt{2\pi}} \int_{-\infty}^{\infty} x^2 \exp(-\tfrac{1}{2}x^2)dx,$$

we easily get

$$\mathrm{cov}(\xi_i, \xi_j) = E\{(\xi_i - E(\xi_i))(\xi_j - E(\xi_j)) =$$

$$= E\{(\xi_i - a_i)(\xi_j - a_j)\} =$$

$$= E\left\{\left(\sum_{k=1}^{n} c_{ik}\eta_k\right)\left(\sum_{k=1}^{n} c_{jk}\eta_k\right)\right\} =$$

$$= E\left\{\sum_{k=1}^{n} c_{ik}c_{jk}\eta_k^2 + \sum_{\substack{k\neq s\\ 1\leqslant k,s\leqslant n}} c_{ik}c_{js}\eta_k\eta_s\right\} =$$

$$= \sum_{k=1}^{n} c_{ik}c_{jk}E\{\eta_k^2\} + \sum_{\substack{k\neq s\\ 1\leqslant k,s\leqslant n}} c_{ik}c_{js}E(\eta_k)E(\eta_s) =$$

$$= \sum_{k=1}^{n} c_{ij}c_{jk}.$$

From $B^TB = \Sigma^{-1}$ we have $B^{-1}(B^{-1})^T = \Sigma = (b_{ij})$; therefore,

$$b_{ij} = \sum_{k=1}^{n} c_{ij}c_{jk} = \mathrm{cov}(\xi_i, \xi_j);$$

i.e., the matrix Σ in (16.12) is just the covariance matrix of the random vector $\vec{\xi}$.

Example 17.2. Let $\xi_1, \xi_2, \ldots$ be a sequence of independent r.v.'s and let $E(\xi_i) = a_i$ for $i = 1, 2, \ldots$ Suppose $A_1, A_2, \ldots$ is a sequence of independent events such that $A_iA_j = \emptyset$ for $i \neq j$, $\bigcup_{i=1}^{\infty} A_i = \Omega$ and $P(A_i) = p_i$. Let also the r.v.'s $\xi_1, \xi_2, \ldots$ and $1_{A_1}, 1_{A_2} \ldots$ be independent and $V(\xi_i) < \infty$, $i = 1, 2, \ldots$ Prove that

$$V(\eta) = \sum_{i=1}^{\infty} p_iV(\xi_i) + V(\zeta),$$

where $\eta = \sum_{i=1}^{\infty} 1_{A_1} \cdot \xi_i$ and ζ is a r.v. taking values a_i, $i = 1, 2, \ldots$ with probability $P\{\zeta = a_i\} = p_i$, $i = 1, 2, \ldots$

Solution. Denote by $F, F_1, F_2, \ldots$ the d.f.'s of the r.v.'s $\eta, \xi_1, \xi_2, \ldots$, respectively. We have $F(x) = \sum_{i=1}^{\infty} p_iF_i(x)$, $x \in \mathbb{R}_1$, $E(\eta) = \sum_{i=1}^{\infty} p_iE(\xi_i)$ and $E\{\eta^2\} = E\left\{\left(\sum_{i=1}^{\infty} 1_{A_i} \cdot \xi_i\right)^2\right\} = \sum_{i=1}^{\infty} E\left\{\left(1_{A_i} \cdot \xi_i\right)^2\right\} +$

$$\sum_{i\neq j} E\{1_{A_i}1_{A_j}\xi_i\xi_j\} = \sum_{i=1}^{\infty} E\{1^2_{A_i}\}\vec{E}\{\xi_i^2\} + \sum_{i\neq j} E\{1_{A_i}1_{A_j}\}E\{\xi_i\xi_j\} =$$

$$\sum_{i=1}^{\infty} p_i E\{\xi_i^2\}.$$

For the variance of η we get

$$V(\eta) = E\{\eta^2\} - (E(\eta))^2 = \sum_{i=1}^{\infty} p_i E\{\xi_i^2\} - \left(\sum_{i=1}^{\infty} p_i E(\xi_i)\right)^2 =$$

$$= \sum_{i=1}^{\infty} p_i E\{\xi_i^2\} - \sum_{i=1}^{\infty} p_i(\vec{E}(\xi_i))^2 + \sum_{i=1}^{\infty} p_i(E(\xi_i))^2 -$$

$$- \left(\sum_{i=1}^{\infty} p_i E(\xi_i)\right)^2 = \sum_{i=1}^{\infty} p_i(E\{\xi_i^2\} - (E(\xi_i))^2) + \sum_{i=1}^{\infty} p_i a_i^2 -$$

$$- \left(\sum_{i=1}^{\infty} p_i a_i\right)^2 = \sum_{i=1}^{\infty} p_i V(\xi_i) + V(\zeta).$$

Exercises

17.1. Let the r.v. ξ have a binomial distribution with parameters (n, p) (see Section 9). Find: (a) $E\{|\xi - np|\}$; (b) $V\{|\xi - np|\}$.

17.2. Let the r.v. ξ assume the values 0, 1, 2, ... Prove that $E(\xi) = \sum_{n=1}^{\infty} P\{\xi \geqslant n\}$ and also that $E(\xi)$ exists if and only if the series $\sum_{n=1}^{\infty} P\{\xi \geqslant n\}$ converges.

*17.3. Let the r.v.'s $\xi_1, \xi_2, \ldots$ be independent and uniformly distributed over the interval (0, 1) (see (15.7)). Let $\pi_n = (1 + \xi_1)(1 + \xi_2) \ldots (1 + \xi_n)$, $S_n = \xi_1 + \xi_2 + \ldots + \xi_n$, $\tau_c = \inf[n : S_n > c]$, $\nu_c = \inf[n : \pi_n > c]$, where $c > 0$. Compute: (a) $E\{\tau_c\}$; (b) $E\{\nu_c\}$.

17.4. Let ξ be a non-negative r.v. with distribution $\vec{P}\{\xi = 0\} = p_0$, $P\{\xi = k\} = p_k$, $k = 1, 2, \ldots$, where $p_1, p_2, \ldots$ form a geometric progression. Let $E(\xi) = a$ and $E\{\xi^2\} = b$. Express p_k, $k = 0, 1, \ldots$, in terms of a and b. Is it possible that a and b are correspondingly equal to 4 and 25?

17.5. Let $\xi_1, \xi_2, \ldots$ be a sequence of independent r.v.'s, each distributed uniformly on the interval (0, 1). Let $\xi_{1,n} \leqslant \xi_{2,n} \leqslant \ldots \leqslant \xi_{n,n}$ be the order statistics of the sample $\xi_1, \xi_2, \ldots, \xi_n$. (In Exercise 16.42 and in other places we use the standard notation $\xi_{(1)}, \xi_{(2)}, \ldots, \xi_{(n)}$. Writing $\xi_{k,n}$ for the kth order statistics we indicate explicitly the size n of the sample.) Consider the linear combination $\eta_n = \sum_{i=1}^{n} a_i \xi_{n+1-i,n}$, where a_1, a_2, ... are arbitrary positive numbers. (a) Find the density $f_{\eta_n}(x)$, $x \in \mathbb{R}_1$, of η_n. (b) If $\mu_{r,n} = E\{\eta_n^r\}$ is the rth order moment of η_n, r = 0, 1, 2, ..., then prove the recurrence relation:

$$\mu_{r,n} = \binom{n+r}{n}^{-1} \sum_{j=0}^{r} (a_1 + \ldots + a_n)^{r-j} \binom{n+j-1}{j} \mu_{j,n-1},$$

where $\mu_{r,1} = a_1^r/(r + 1)$ for r = 0, 1, 2, ... (c) Prove the following recurrence relation for the densities of the r.v.'s η_n, η_{n-1} and $\zeta_{n-1} = a_1 + \sum_{i=1}^{n-1} a_{i+1}\xi_{n-i,n-i}$:

$$f_{\eta_n}(x) = \frac{n}{n-1}\left[\frac{x}{a_1 + \ldots + a_n} f_{\eta_{n-1}}(x) + \right.$$

$$\left. + \left(1 - \frac{x}{a_1 + \ldots + a_n}\right) f_{\zeta_{n-1}}(x)\right].$$

(d) Derive explicit expressions for the first two moments, $\mu_{1,n}$ and $\mu_{2,n}$ in both cases when a_1, ..., a_n are arbitrary and when $a_1 = a_2 = \ldots = a_n = 1$. (*V. Kaishev and Z. Ignatov*)

17.6. Let ξ be a discrete r.v. with distribution $P\{\xi = k\} = c/[(d + k)(d + k + 1)(d + k + 2)]$, for k = 0, 1, 2, ..., where the constants c and d are such that $E(\xi) = 3/(2 + d)$. Find $E(\xi)$ and $E\{\xi^2\}$.

17.7. Let the r.v.'s ξ_1, ..., ξ_n be independent, identically distributed with zero mean and a finite fourth moment. Set $S_n = \xi_1 + \ldots + \xi_n$ and express $E\{S_n^3\}$ and $E\{S_n^4\}$ in terms of the moments $a_k = E\{\xi_1^k\}$ for k = 1, 2, 3, 4.

*17.8. Let μ_n be the number of successes in a Bernoulli scheme (n, p) (see Section 9, binomial distribution). Find $\lim_{n\to\infty} n^{-1}E\{\eta_n\}$, where $\eta_n = \max[\mu_n, n - \mu_n]$.

17.9. Let $F(x)$, $x \in \mathbb{R}_1$, be a d.f. Prove that if at least one of the integrals

$$I_1(\alpha) = \int_0^\infty x^\alpha \, dF(x) \quad \text{and} \quad I_2(\alpha) = \alpha\int_0^\infty x^{\alpha-1}[1 - F(x)]dx$$

exists, then $I_1(\alpha) = I_2(\alpha)$ for every $\alpha > 0$.

17.10. Let the r.v. ξ have d.f. $F(x)$, $x \in \mathbb{R}_1$. Show that for $\alpha > 0$ the expectation $E\{|\xi|^{\alpha}\}$ exists if and only if the function $|x|^{\alpha-1}[1 - F(x) + F(-x)]$ is integrable on $\mathbb{R}_1$.

*17.11. Let the r.v.'s ξ and η be independent and let their sum $\xi + \eta$ have finite second moment. Prove that ξ and η also have finite second moments.

17.12. Let the r.v. ξ have d.f. $F(x)$, $x \in \mathbb{R}_1$, and $E(\xi)$ exist. Show that $\sum_{n=1}^{\infty} (1/n^2)\int_{-n}^{n} x^2\, dF(x) < \infty$.

17.13. Let ξ be a non-negative r.v. with d.f. $F(x)$, $x \in \mathbb{R}_1$, and let $\lim_{x\to\infty} x[1 - F(x)]$ exist. Show that $\lim_{x\to\infty} x[1 - F(x)] = 0$ is a necessary, but not a sufficient condition for the existence of $E(\xi)$.

17.14. Find the mean value and the variance (if they exist) of the r.v. ξ when: (a) ξ is uniformly distributed on the interval (a, b) (see 15.7)); (b) $\xi \in N(a, \sigma^2)$ (see (15.8)); (c) $\xi \in E(\lambda)$ (see (15.9)); (d) ξ has gamma distribution with parameters $\alpha > 0$ and $\beta > 0$ (see (15.10)); (e) ξ has beta distribution with parameters $p > 0$ and $q > 0$ (see (15.11)); (f) ξ has Cauchy distribution with parameters a and $b > 0$ (see (15.12)); (g) ξ has χ^2-distribution with n degrees of freedom (see Example 16.2); (h) ξ has F-distribution with (m, n) degrees of freedom (see Exercise (16.30)); (i) ξ has t-distribution with n degrees of freedom (see Exercise 16.31).

17.15. Given the independent r.v.'s ξ and η, where $\xi \in E(\lambda)$ and $\eta \in E(\mu)$. Let $\zeta = \max[\xi, \eta]$. Find $E\{\zeta^n\}$.

17.16. Let $\xi_1, \ldots, \xi_{m+n}$ be n + m independent r.v.'s and $\xi_i \in N(0, 1)$ for i = 1, 2, ..., n + m. Find the mean value and the variance of the r.v. η if: (a) $\eta = (\xi_1^2 + \ldots + \xi_{n+m}^2)/(\xi_1^2 + \ldots + \xi_n^2)$; (b) $\eta = (\xi_1^2 + \ldots + \xi_m^2)/(\xi_1^2 + \ldots + \xi_{n+m}^2)$.

17.17. Let the r.v.'s ξ_1 and ξ_2 be independent and $\xi_i \in N(a, \sigma^2)$ for i = 1, 2. Compute $E\{\max[\xi_1, \xi_2]\}$.

17.18. Compute $E\{|\xi - a|\}$ and $V\{|\xi - a|\}$ for the r.v. $\xi \in N(a, \sigma^2)$.

17.19. Let the r.v. ξ be uniformly distributed on the interval (0, 1). Denote $\eta = \min[\xi, 1 - \xi]$. Find: (a) $E(\eta)$; (b) $E\{\eta/(1 - \eta)\}$; (c) $E\{(1 - \eta)/\eta\}$.

17.20. Let the r.v. $\xi_1 \in N(0, 1)$ and $\xi_2 = \xi_1^2 - 1$. Show that the correlation coefficient $\rho(\xi_1, \xi_2) = 0$.

17.21. Calculate the correlation coefficient ρ between the r.v.'s η and ζ in the case when: (a) $\eta = \sin(2\pi\xi)$, $\zeta = \cos(2\pi\xi)$, where ξ is uniformly distributed on (0, 1); (b) $\eta = \xi$, $\zeta = \xi^2$, where ξ is uniformly distributed on (-1, 1). Are the r.v.'s η and ζ independent in each of cases (a) and (b)?

*17.22. Let ξ_1, ξ_2 and ξ_3 be symmetrically dependent r.v.'s; i.e., for every permutation (i_1, i_2, i_3) of the numbers 1, 2 and 3, the following equality holds: $P\{\xi_{i_1} < x_1, \xi_{i_2} < x_2, \xi_{i_3} < x_3\} = P\{\xi_1 < x_1, \xi_2 < x_2, \xi_3 < x_3\}$. Now let $R_2 = \max[\xi_1, \xi_2] - \min[\xi_1, \xi_2]$, $R_3 = \max[\xi_1, \xi_2, \xi_3] - \min[\xi_1, \xi_2, \xi_3]$. Prove that $3E(R_2) = 2E(R_3)$.

17.23. Let the r.v.'s ξ and η be independent and let conditions $E(\xi) = E(\eta) = E\{\xi^3\} = 0$, $E\{\xi^4\} = 3(E\{\xi^2\})^2$, $E\{\eta^4\} = 3(E\{\eta^2\})^2$, hold. Let $\zeta = \xi + \eta$. Prove that $E\{\zeta^4\} = 3(E\{\zeta^2\})^2$. Can the distribution of the r.v. ξ be given by $P\{\xi = p\} = q$ and $P\{\xi = -q\} = p$, where p and q are suitable constants, $p, q \in (0, 1)$, $p + q = 1$?

17.24. Given the r.v. $\xi \in N(a, 1)$. Find the mean value and the variance of the r.v. $\eta = e^{\xi/2}\int_\xi^\infty e^{-u^2/2}\,du$.

*17.25. Let the random vector (ξ, η) have a two-dimensional normal density; i.e., $f(x, y) = D\exp[-g(x, y)]$, where D and $g(x, y)$ are determined as in Exercise 16.38. Find: (a) $E(\xi)$ and $V(\xi)$; (b) $E(\eta)$ and $V(\eta)$; (c) the covariance of the r.v.'s ξ and η; (d) the correlation coefficient ρ between ξ and η expressed in terms of $q = P\{\xi\eta < 0\}$ for $E(\xi) = E(\eta) = 0$ and $V(\xi) = V(\eta) = 1$; (e) the correlation coefficient ρ between ξ and $F(\eta)$, where $F(x)$, $x \in \mathbb{R}_1$, is the d.f. of the r.v. η.

17.26. Let the random vector (ξ, η) have density

$$f(x, y) = \frac{1}{2\pi\sqrt{3}}\left[e^{-\frac{2}{3}(x^2+xy+y^2)} + e^{-\frac{2}{3}(x^2-xy+y^2)}\right], \qquad (x, y) \in \mathbb{R}_2.$$

Find: (a) $E(\xi)$ and $V(\xi)$; (b) $E(\eta)$ and $V(\eta)$; (c) the correlation coefficient ρ between the r.v.'s $\xi + \eta$ and $\xi - \eta$; (d) the correlation coefficient ρ between ξ and η. (e) Are the r.v.'s ξ and η independent?

17.27. Let $\xi_1, \ldots, \xi_n$ be independent r.v.'s with finite second moment and a common symmetric d.f. $F(x)$, $x \in \mathbb{R}_1$; i.e., $F(-x) = 1 - F(x)$ for all $x \geq 0$. Let

$$\bar{\xi}_n = \frac{1}{n}\sum_{k=1}^n \xi_k \quad \text{and} \quad s_n^2 = \frac{1}{n-1}\sum_{k=1}^n (\xi_k - \bar{\xi}_n)^2.$$

Find the covariance $\mathrm{cov}(\bar{\xi}_n, s_n^2)$.

17.28. Let the r.v. ξ have density $f(x)$, $x \in \mathbb{R}_1$, and $E\{|\xi|\} < \infty$. Show that $\min_{c\in\mathbb{R}_1} E\{|\xi - c|\} = E\{|\xi - m|\}$, where m is the median of the r.v. ξ; i.e., determine m so that $\int_{-\infty}^m f(x)\,dx = \frac{1}{2}$.

17.29. Let ξ be a r.v. with finite variance. Show that

$$\min_{c\in\mathbb{R}_1} E\{(\xi - c)^2\} = E\{(\xi - E(\xi))^2\}.$$

17.30. Let ξ and η be r.v.'s whose d.f.'s F_ξ and F_η are absolutely continuous. Show that $E\{F_\xi(\eta)\} + E\{F_\eta(\xi)\} = 1$.

17.31. Let $\xi_1, \ldots, \xi_{k+1}$ be k + 1 independent r.v.'s and $\xi_i \in N(0, 1)$, for i = 1, 2, ..., k + 1. Set $\eta_i = \xi_i^2/(\xi_1^2 + \ldots + \xi_{k+1}^2)$, i = 1, ..., k. Find: (a) the covariance matrix (b_{ij}) of the random vector $(\eta_1, \ldots, \eta_k)$, where $b_{ij} = \mathrm{cov}(\eta_i, \eta_j)$; (b) the multivariate moment $a_{r_1,\ldots,r_k} = E\{\eta_1^{r_1} \ldots \eta_k^{r_k}\}$ (see (17.10)).

17.32. Let the random vector (ξ_1, ξ_2) have two-dimensional normal distribution with $E(\xi_1) = E(\xi_2) = 0$ and $V(\xi_1) = V(\xi_2) = 1$ and correlation coefficient $\rho(\xi_1, \xi_2) = r$. Find the density of $\Theta = \xi_1/\xi_2$.

*17.33. Let the random vector $\vec{\xi}$ be normally distributed $N(\vec{a}, \Sigma)$ (see (16.12)). Here $\vec{\xi}$ is a column vector; i.e., $\vec{\xi}^T = (\xi_1, \ldots, \xi_n)$. Find: (a) $E\{\xi_1\xi_2\xi_3\xi_4\}$ in terms of the entries of the covariance matrix $\Sigma = (b_{ij})$, when n = 4 and $a_i = 0$ for i = 1, 2, 3, 4; (b) $E(\eta)$ and $V(\eta)$, if $\eta = c_1\xi_1 + c_2\xi_2 + \ldots + c_n\xi_n$, where $c_1, \ldots, c_n \in \mathbb{R}_1$; (c) $E(\Theta)$ and $V(\Theta)$, if $\Theta = \frac{1}{2}(\xi - \vec{a})^T . \Sigma^{-1} \cdot (\xi - \vec{a})$.

17.34. Find: (a) $E\{\xi^r\}$ and $E\{|\xi|^r\}$ for r = 1, 2 if the r.v. $\xi \in N(0, 1)$; (b) $E\{\xi^r\}$ for r = 1, 2 if the r.v. ξ has a gamma distribution with parameters $\alpha > 0$ and $\beta > 0$ (see (15.10)); (c) $E\{\xi^r\}$ for r = 1, 2, ... if the r.v. ξ is beta distributed with parameters $p > 0$ and $q > 0$ (see (15.11)); (d) $E\{\xi^r\}$ for r = 1, 2, ..., if the r.v. ξ has a t-distribution with n degrees of freedom (see Exercise 16.31); (e) the multivariate moment $a_{k_1,\ldots k_n}$ about the vector $\vec{c} = (0, 0, \ldots, 0)$ of the random vector $\vec{\xi} = (\xi_1, \ldots, \xi_n)$ (see (17.10)) if the random vector $\vec{\xi}$ has the *Dirichlet distribution* $D(\nu_1, \ldots, \nu_n; \nu_{n+1})$ with parameters $\nu_i > 0$ for i = 1, 2, ..., n + 1; i.e., the random vector $\vec{\xi}$ has density

$$f_{\vec{\xi}}(x_1, \ldots, x_n) = \frac{\Gamma(\nu_1 + \ldots + \nu_{n+1})}{\Gamma(\nu_1) \ldots \Gamma(\nu_{n+1})}$$

$$\cdot\, x_1^{\nu_1 - 1} \ldots x_n^{\nu_n - 1}\left(1 - x_1 - \ldots - x_n\right)^{\nu_{n+1} - 1},$$

where $x_1 + \ldots + x_n \leqslant 1$ and $x_i \geqslant 0$ for i = 1, ..., n and $f_\xi(x_1, \ldots, x_n) = 0$ otherwise (see also the solution of Exercise 16.44).

17.35. Let the point M be uniformly distributed on the triangle $\Delta A_1A_2A_3$, where $A_1 = (0, 0)$, $A_2 = (1, 0)$, $A_3 = (0, 1)$. The straight line A_iM crosses the side opposite to the vertex A_i at the point B_i, i = 1, 2, 3. Find the mean value of the area of the triangle $\Delta B_1B_2B_3$.

17.36. Let the r.v. ξ have density

$$f_\xi(x) = \begin{cases} b\left(1 - \frac{x^2}{a^2}\right)^m, & \text{if } x \in (-a, a) \\ 0, & \text{if } x \overline{\in} (-a, a), \end{cases}$$

where $a > 0$, $m > 0$ and b is a suitable constant. Express the quantities a, b and m in terms of the variance $\sigma^2 = V(\xi)$ and the *coefficient of skewness* $\beta = E\{\xi^4\}/(E\{\xi^2\})^2$.

*17.37. Given the functions $f(x) = k\exp(-\sqrt[4]{x})$ and $g(x) = k(1 + \frac{1}{2}\sin\sqrt[4]{x})\exp(-\sqrt[4]{x})$ for $x > 0$ and $f(x) = g(x) = 0$ for $x \leqslant 0$. Prove that: (a) for a suitable value of the constant k, the functions f and g are densities of certain r.v.'s which we denote by ξ and η; (b) although ξ and η have different densities, their moments of arbitrary order n coincide; i.e., for every natural number n we have $E\{\xi^n\} = E\{\eta^n\}$.

18. Generating Functions and Characteristic Functions

Introductory Notes

Let ξ be an integer-valued non-negative r.v. with distribution $p_k = P\{\xi = k\}$, $k = 0, 1, \ldots$ The function $g(s)$ of the complex variable s, $|s| \leqslant 1$, defined by the formula $g(s) = \sum_k p_k s^k = E\{s^\xi\}$ is called the *probability generating function* (p.g.f.) of the distribution of the r.v. ξ (see also Section 9). It is an analytic function on the closed circle $|s| \leqslant 1$. Moreover, the r.v. ξ has kth moment $E\{\xi^k\}$ if and only if $g(s)$ has kth derivative at the point $s = 1$. In this case $g^{(k)}(1) = E\{\xi(\xi - 1) \ldots (\xi - k + 1)\}$. If ξ and η are indpendent r.v.'s, then the p.g.f. of their sum is given by

$$g_{\xi+\eta}(s) = g_\xi(s)g_\eta(s), \quad |s| \leqslant 1. \tag{18.1}$$

If ξ is an arbitrary r.v., then the functions $h(t) = E\{e^{t\xi}\}$, $m(t) = h(t)e^{-tE(\xi)}$ and $k(t) = \ln m(t)$, $t \in \mathbb{R}_1$, are called respectively the *moment generating function*, the *generating function of the central moments* and the *cumulant generating function*. If ξ is an integer-valued non-negative r.v., then $h(t) = g(e^t)$. If ξ possesses all moments, the necessary and sufficient condition for the representations

$$h(t) = \sum_{t=0}^{\infty} \frac{E\{\xi^r\}}{r!} t^r$$

and

$$m(t) = 1 + \sum_{r=2}^{\infty} \frac{E\{(\xi - E\xi)^r\}}{r!} t^r \quad \text{for } |t| < \delta$$

to hold is that $q = (\lim_{r\to\infty} (r!)^{-1}E\{\xi^r\})^{1/r} < \infty$, where $\delta \leqslant 1/q$. The coefficients k_r, $r = 2, 3, \ldots$, in the representation $k(t) = \sum_{r=2}^{\infty} (k_r t^r/r!)$ are called *cumulants* of the r.v. ξ.

Let ξ be a non-negative r.v. with d.f. F. The function

$$\Psi(\lambda) = E\{e^{-\lambda\xi}\} = \int_0^{\infty} e^{-\lambda x}\, dF(x), \quad \lambda \geqslant 0,$$

is called *Laplace-Stieltjes transformation* of F (and also of ξ).

The function $\Psi(\lambda)$, $\lambda \geqslant 0$, is a Laplace-Stieltjes transformation of a certain d.f. F if and only if $\Psi(0) = 1$ and $\psi(\lambda)$ is absolutely monotonic; i.e., Ψ has derivatives $\Psi^{(k)}(\lambda)$ of an arbitrary order and $(-1)^k\Psi^{(k)}(\lambda) \geqslant 0$, $\lambda \geqslant 0$ (*Bernstein's theorem*).

Now, let ξ be a r.v. with d.f. F. The function $\phi(t)$ of the real variable $t \in \mathbb{R}_1$ defined by the formula

$$\phi(t) = E\{e^{it\xi}\} = \int_{-\infty}^{\infty} e^{itx}\, dF(x), \quad \text{where } i = \sqrt{-1}, \tag{18.2}$$

is called the *characteristic function* (ch.f.) of the r.v. ξ (and also of F).

Recall the most important properties of the ch.f.'s:

(a) $\phi(0) = 1$, $|\phi(t)| \leqslant 1$, $t \in \mathbb{R}_1$;

(b) if ξ has a kth moment, then ϕ has a kth derivative and

$$E\{\xi^k\} = i^{-k}\phi^{(k)}(0); \tag{18.3}$$

(c) if $\phi^{(k)}(0)$ exists and k is even, then $E\{\xi^k\}$ exists, and if k is odd, then $E\{\xi^{k-1}\}$ exists;

(d) if $E\{\xi^k\}$ exists, then for small values of t the ch.f. $\phi(t)$ can be expressed as follows:

$$\phi(t) = \sum_{r=0}^{k} \frac{E\{\xi^r\}}{r!} (it)^r + o(t^k); \tag{18.4}$$

(e) ϕ is a ch.f. if and only if $\phi(0) = 1$ and ϕ is positive definite; i.e., $\sum_{k=1}^{n} \sum_{j=1}^{n} \phi(t_k - t_j)\, z_k\bar{z}_j \geqslant 0$ for arbitrary n real numbers $t_1, \ldots, t_n$ and complex numbers $z_1, \ldots, z_n$;

(f) if $\phi(t)$, $t \in \mathbb{R}_1$, is a real, continuous and even function, convex for $t > 0$, for which $\phi(0) = 1$ and $\lim_{t\to\pm\infty} \phi(t) = 0$, then ϕ is a ch.f. (*Polya criterion*);

(g) if the r.v.'s ξ and η have respectively d.f.'s F_ξ and F_η and ch.f.'s ϕ_ξ and ϕ_η, then F_ξ and F_η coincide on $\mathbb{R}_1$ if and only if ϕ_ξ and ϕ_η coincide on $\mathbb{R}_1$;

(h) if ξ and η are independent r.v.'s, then the corresponding ch.f.'s satisfy the equality

$$\phi_{\xi+\eta}(t) = \phi_\xi(t)\phi_\eta(t), \quad t \in \mathbb{R}_1; \tag{18.5}$$

(i) if the r.v. ξ has ch.f. $\phi(t)$, $t \in \mathbb{R}_1$, and d.f. $F(x)$, $x \in \mathbb{R}_1$, which is continuous at the points a and b for $a < b$, then

$$F(b) - F(a) = \frac{1}{2\pi}\int_{-\infty}^{\infty}\left[\phi(t)\,\frac{e^{-ita} - e^{-itb}}{2it} - \phi(-t)\,\frac{e^{ita} - e^{itb}}{2it}\right]dt. \tag{18.6}$$

(This is called the *inversion formula*.)

Let us note that if $F(x)$ is discontinuous at the point x, then

$$F(x + 0) - F(x - 0) = \lim_{T\to\infty} \frac{1}{2T}\int_{-T}^{T} \phi(t)e^{-itx}\,dt.$$

(j) if the ch.f. ϕ of the r.v. ξ is absolutely integrable on $\mathbb{R}_1$, then ξ has continuous density f and

$$f(x) = \frac{1}{2\pi}\int_{-\infty}^{\infty} e^{-itx}\phi(t)\,dt, \quad x \in \mathbb{R}_1; \tag{18.7}$$

(k) the necessary and sufficient condition for ϕ to be a ch.f. of an absolutely continuous distribution is that $\phi(0) = 1$ and that the function $f(x) = \frac{1}{2\pi}\int_{-\infty}^{\infty} e^{-itx}\phi(t)\,dt$, $x \in \mathbb{R}_1$, is non-negative for every $x \in \mathbb{R}_1$. Then f is a density corresponding to the ch.f. ϕ. (*Bochner criterion*).

Let F be the d.f. of the r.v. ξ and let $E\{\xi^r\}$ exist for r = 0, 1, 2, ... If the series

$$\sum_{r=0}^{\infty} \frac{c^r E\{\xi^r\}}{r!} \tag{18.8}$$

is absolutely convergent for every $c > 0$, then F is the unique d.f. with these moments $E\{\xi^r\}$. In this case the ch.f. ϕ is given as

$$\phi(t) = \sum_{r=0}^{\infty} \frac{E\{\xi^r\}}{r!} (it)^r, \quad t \in \mathbb{R}_1. \tag{18.9}$$

In a similar way we define the p.g.f. of the random vector $\vec{\xi} = (\xi_1, \ldots, \xi_n)$ whose components are integer-valued non-negative r.v.'s:

$$g_{\vec{\xi}}(s_1, \ldots, s_n) = E\left\{s_1^{\xi_1} \ldots s_n^{\xi_n}\right\},$$
$$|s_k| \leqslant 1, \; k = 1, \ldots, n, \tag{18.10}$$

and the ch.f. of an arbitrary vector $\vec{\eta} = (\eta_1, \ldots, \eta_n)$:

$$\phi_{\vec{\eta}}(t_1, \ldots, t_n) = E\{\exp[i(t_1\eta_1 + \ldots +$$
$$+ t_n\eta_n)]\}, \quad t_1, \ldots, t_n \in \mathbb{R}_1. \tag{18.11}$$

Illustrative Examples

Example 18.1. Let ξ be an absolutely continuous r.v. with a density $f(x)$, $x \in \mathbb{R}_1$, and a ch.f. $\phi(t)$, $t \in \mathbb{R}_1$. Prove that

$$\int_{-\infty}^{\infty} f^2(x)dx = \frac{1}{2\pi} \int_{-\infty}^{\infty} |\phi(t)|^2 dt$$

if $f^2(x)$, $x \in \mathbb{R}_1$, is integrable.

Solution. Let ξ and η be independent identically distributed r.v.'s whose common density is f. Then the r.v. $\xi - \eta$ has the density

$$g(x) = \int_{-\infty}^{\infty} f(x + y)f(y)dy.$$

The ch.f. of $\xi - \eta$ is $E\{e^{it(\xi-\eta)}\} = E\{e^{it\xi}\}E\{e^{-it\eta}\} = \phi(t)\bar{\phi}(t) = |\phi(t)|^2$, where $\bar{\phi}(t)$ is the conjugate of $\phi(t)$. Therefore, using (18.7), we get

$$g(x) = \frac{1}{2\pi} \int_{-\infty}^{\infty} e^{-itx} |f(t)|^2 dt.$$

Setting $x = 0$ in the last equality, we obtain

$$\int_{-\infty}^{\infty} f^2(x)\,dx = \frac{1}{2}\int_{-\infty}^{\infty} |\phi(t)|^2 dt. \qquad (18.12)$$

The equality (18.12) is known as the Parseval equality.

Example 18.2. Let ξ be a r.v. with a d.f. $F(x)$, $x \in \mathbb{R}_1$, and a ch.f. $\phi(t)$, $t \in \mathbb{R}_1$. Prove that: (a) if ξ is a discrete r.v. whose d.f. F has jump points $x_1, x_2, \ldots$ and size of jumps $p_1, p_2, \ldots$, respectively, then

$$\lim_{T\to\infty} \frac{1}{2T}\int_{-T}^{T} |f(t)|^2 dt = \sum_{i=1}^{\infty} p_i^2;$$

(b) F is a continuous function if and only if

$$\lim_{T\to\infty} \frac{1}{2T}\int_{-T}^{T} |\phi(t)|^2 dt = 0.$$

Solution. (a) As in the solution of Example 18.1, if η and ξ are independent and identically distributed r.v.'s, then the r.v. $\xi - \eta$ has the ch.f. $|\phi(t)|^2$ and the d.f. $G(x) = \int_{-\infty}^{\infty} F(x + t)\,dF(t)$. Obviously $G(0 + 0) - G(0 - 0) = \sum_{k=1}^{\infty} p_k^2$. According to property (i) given in the Introductory Notes, we have

$$\sum_{k=1}^{\infty} p_k^2 = G(0 + 0) - G(0 - 0) = \lim_{T\to\infty} \frac{1}{2T}\int_{-T}^{T} |\phi(t)|^2 dt.$$

(b) For an arbitrary d.f. $F(x)$, $x \in \mathbb{R}_1$, the jump $G(0 + 0) - G(0 - 0)$ of the d.f. $G(x)$, $x \in \mathbb{R}_1$, introduced in (a), is equal to the sum of squares of all jumps of F. On the other hand,

$$G(0 + 0) - G(0 - 0) = \lim_{T\to\infty} \frac{1}{2T}\int_{-T}^{T} |\phi(t)|^2 dt.$$

Hence $F(x)$, $x \in \mathbb{R}_1$, is continuous if

$$\lim_{T\to\infty} \frac{1}{2T}\int_{-T}^{T} |\phi(t)|^2 dt = 0.$$

Example 18.3. Let the random vector $\vec{\xi}$ be normally distributed $N(\vec{a}, \Sigma)$ (see (16.12)). Find the ch.f. $\phi_{\vec{\xi}}(\vec{t})$, $\vec{t} \in \mathbb{R}_n$.

Solution. Let us first consider the r.v. η normally $N(0, 1)$ distributed. We shall find its ch.f. $\phi_\eta(t)$, $t \in \mathbb{R}_1$, and then use it to derive the ch.f. of the random vector $\vec{\xi}$. We have

$$\phi_\eta(t) = (2\pi)^{-1/2} \int_{-\infty}^{\infty} \exp(ixt - x^2/2)dx =$$

$$= (2\pi)^{-1/2}\exp(-t^2/2) \int_L \exp(-z^2/2)dz,$$

where L is the horizontal line $z = x - it$ $(-\infty < x < \infty)$ in the complex plane. The function $\exp(-z^2/2)$ is an entire function. Therefore its integral is zero along any closed curve and in particular along the rectangular Q_x with the vertices $-x - it$, $x - it$, x and $-x$. The inequality

$$\left| \int_{x-it}^{x} \exp(-z^2/2)dz \right| \leqslant \exp(-x^2/2) \int_0^{|t|} \exp(y^2/2)dy$$

implies that

$$\lim_{|x|\to\infty} \left| \int_{x-it}^{x} \exp(-z^2/2)dz \right| = 0.$$

Hence

$$(2\pi)^{-1/2} \int_L \exp(-z^2/2)dz = (2\pi)^{-1/2} \int_{-\infty}^{\infty} \exp(-y^2/2)dy = 1,$$

and consequently $\phi_\eta(t) = \exp(-\frac{1}{2} t^2)$, $t \in \mathbb{R}_1$. For the ch.f. $\phi_{\vec{\xi}}(\vec{t}) = E\{\exp[i(t^T \cdot \vec{\xi})]\}$ of the vector $\vec{\xi}^T = (\xi_1, \ldots, \xi_n)$, we have

$$\phi_{\vec{\xi}}(\vec{t}) = E\{\exp[i(\vec{t}^T \cdot (\vec{\xi} - \vec{a} + \vec{a}))]\} =$$

$$= \exp\{i(\vec{t}^T \cdot \vec{a})\}E\{\exp[i(\vec{t}^T \cdot (\vec{\xi} - \vec{a}))]\}, \; \vec{t}^T = (t_1,\ldots,t_n).$$

Using the matrix $B = (b_{kj})$ and the random vector η introduced in Example 16.3, we have

$$E\{\exp[i(\vec{t}^T \cdot (\vec{\xi} - \vec{a}))]\} = E\{\exp[i(\vec{t}^T B^{-1} B(\vec{\xi} - \vec{a}))]\} =$$

$$= E\{\exp[i(\vec{t}^T B^{-1}\vec{\eta})]\}.$$

Set $\vec{t}^T B^{-1} = \vec{v} = (v_1, \ldots, v_n)$. Since the coordinates of the random vector $\vec{\eta}^T = (\eta_1, \ldots, \eta_n)$ are independent and identically $N(0, 1)$ distributed, the relation (18.5), combined with the equality $\phi_\eta(t) = \exp(-\frac{1}{2} t^2)$, yields

$$E\{\exp[i(\vec{t}^T B^{-1}\vec{\eta})]\} = E\{\exp[i \sum_{j=1}^{n} v_j \eta_j)]\} =$$

$$= \prod_{j=1}^{n} E\{\exp[iv_j\eta_j]\} = \prod_{j=1}^{n} \exp(-\frac{1}{2} v_j^2) =$$

$$= \exp\left(-\frac{1}{2} \sum_{j=1}^{n} v_j^2\right) = \exp(-\frac{1}{2}(\vec{v} \cdot \vec{v}^T)) =$$

$$= \exp(-\frac{1}{2} \vec{t}^T B^{-1})(\vec{t}^T B^{-1})^T) =$$

$$= \exp(-\frac{1}{2}(\vec{t}^T B^{-1} (B^{-1})^T \vec{t})) =$$

$$= \exp(-\frac{1}{2} \vec{t}^T (B^T B)^{-1} \vec{t}) = \exp(-\frac{1}{2} \vec{t}^T (\Sigma^{-1})^{-1} \vec{t}) =$$

$$= \exp(-\frac{1}{2} \vec{t}^T \Sigma \vec{t}).$$

Thus

$$\phi_{\vec{\xi}}(\vec{t}) = \exp(i(\vec{t}^T \cdot \vec{a}))\exp(-\frac{1}{2} \vec{t}^T \Sigma \vec{t}) = \exp(i(\vec{t}^T \cdot \vec{a}) - \frac{1}{2} \vec{t}^T \Sigma \vec{t}).$$

Exercises

18.1. Let ξ be the sum of the points in rolling a pair of balanced dice. Find the ch.f. ϕ_ξ and the p.g.f. g_ξ of the r.v. ξ.

18.2. Let the non-negative integer-valued r.v. ξ have distribution $P\{\xi = k\} = p_k$, $k = 0, 1, 2, \ldots$, and let $h(s) = \sum_{j=0}^{\infty} q_j s^j$, where $q_j = P\{\xi > j\}$. Express: (a) $h(s)$ in terms of the p.g.f. $g_\xi(s)$ of the r.v. ξ; (b) $E(\xi)$ and $V(\xi)$, if they exist, in terms of $h(s)$.

18.3. Let ξ and η be non-negative integer-valued r.v.'s and let the random vector $\zeta = (\xi, \eta)$ have the distribution $P\{\xi = j, \eta = k\} = p_{jk}$, $j, k = 0, 1, \ldots$ (a) Express the p.g.f.'s $g_\xi(s)$ and $g_\eta(s)$ in terms of the p.g.f. $g_\zeta(s_1, s_2)$ of the random vector ζ. (b) Express the p.g.f. $g_{\xi+\eta}(s)$ of the r.v. $\xi + \eta$ in terms of $g_\zeta(s_1, s_2)$. (c) Prove that the r.v.'s ξ and η are independent if and only if $g_\zeta(s_1, s_2) = g_\xi(s_1)g_\eta(s_2)$ for all $|s_1| \leqslant 1$ and $|s_2| \leqslant 1$.

18.4. Let the r.v.'s $\xi_1, \ldots, \xi_n$ be independent and identically distributed with $P\{\xi_1 = j\} = \frac{1}{m}$ for $j = 0, 1, \ldots, m - 1$. Consider the sum

$S_n = \xi_1 + \ldots + \xi_n$ and denote $q_k = P\{S_n \leqslant k\}$. Find: (a) the p.g.f. $g(s)$ of the r.v. S_n; (b) the function $h(s) = \sum_{k=0}^{\infty} q_k s^k$.

18.5. Let the r.v. $\xi \in P(\lambda)$. Find its: (a) p.g.f.; (b) moment g.f.; (c) g.f. of the central moments; (d) cumulant g.f.; (e) cumulants.

18.6. Let the discrete r.v. ξ have the distribution

$$P\{\xi = 0\} = \left(\frac{a}{1 + a}\right)^{\lambda},$$

$$P\{\xi = k\} = \left(\frac{a}{1 + a}\right)^{\lambda} \cdot \frac{\lambda(\lambda + 1) \ldots (\lambda + k - 1)}{(1 + a)^k k!}, \qquad k = 1, 2, \ldots,$$

where $a > 0$, $\lambda > 0$. Find the ch.f. ϕ, the expectation $E(\xi)$ and the variance $V(\xi)$ of the r.v. ξ.

18.7. Let the r.v. η have the negative binomial distribution with parameters (p, r); i.e., $P\{\eta = k\} = \binom{-r}{k} p^r (-q)^k$, $q = 1 - p$, $k = 0, 1, \ldots$ (see Section 9). Find the ch.f. and the moment g.f. of the r.v. η.

18.8. Find the ch.f. $\phi(t)$ of the r.v. ξ if: (a) ξ has density $f(x) = 0$ for $|x| > 2$ and $f(x) = \frac{1}{2}(1 - \frac{1}{2}|x|)$ for $|x| \leqslant 2$; (b) ξ has density $f(x) = 0$ for $x \leqslant a - c$ and $f(x) = (x - a + c)/c^2$ for $a - c < x < a$, $f(x) = -(x - a - c)/c^2$ for $a \leqslant x < a + c$, and $f(x) = 0$ for $x \geqslant a + c$, where a and $c > 0$ are real parameters; (c) ξ is uniformly distributed over the interval (a, b); (d) ξ has the *Laplace distribution*; i.e., it has density $f(x) = \frac{1}{2\sigma} \exp(-|x - a|/\sigma)$, $x \in \mathbb{R}_1$, where a and $\sigma > 0$ are real parameters; (e) ξ has a beta distribution with parameters $p > 0$ and $q > 0$ (see (15.11)).

18.9. Let the r.v. ξ have ch.f. $\phi(t)$, $t \in \mathbb{R}_1$. Express: (a) $V\{\sin \xi\} + V\{\cos \xi\}$ in terms of $\phi(1)$; (b) the ch.f. ϕ_ζ of the r.v. $\zeta = \alpha\xi + \beta$ in terms of ϕ, where $\alpha, \beta \in \mathbb{R}_1$.

18.10. Let the r.v. ξ have density $f(x) = c[(1 + x^2)\ln(e + x^2)]^{-1}$, $x \in \mathbb{R}_1$. Denote by $\phi(t)$, $t \in \mathbb{R}_1$, its ch.f. (a) Does $\phi'(t)$ for $t = 0$ exist? (b) Does $E(\xi)$ exist?

18.11. Find the moment g.f. for each of the r.v.'s: $\xi_1 \in \mathcal{B}(n, p)$; ξ_2 uniformly distributed over the set $\{0, 1, \ldots, N\}$; $\xi_3 \in N(a, \sigma^2)$; ξ_4 gamma distributed with parameters α and β; ξ_5 uniformly distributed over the interval (a, b).

18.12. Show that the function

$$f(x) = \begin{cases} 0, & \text{if } x \leqslant 0, \\ \int_x^{\infty} u^{-1} e^{-u}\, du, & \text{if } x > 0, \end{cases}$$

is a probability density function. Find its moments and its ch.f.

18.13. Let ξ be a r.v. Express the cumulants k_2, k_3, k_4 of ξ through its central moments.

18.14. Let $\xi_1, \xi_2, \ldots$ be independent and identically distributed r.v.'s, with $\xi_j \in E(\lambda)$, $j = 1, 2, \ldots$ and τ be a geometric r.v. with parameter p. Show that $S_\tau = \xi_1 + \ldots + \xi_{\tau+1}$ is exponentially distributed. Determine the parameter of this distribution.

18.15. Find the ch.f. of $\eta = \xi_1\xi_2 - \xi_3\xi_4$, where ξ_1, ξ_2, ξ_3 and ξ_4 are independent r.v.'s each distributed $N(0, \sigma^2)$.

*18.16. Given the function

$$\phi_1(t) = \frac{2}{\pi}\int_0^\infty \frac{1 - \cos x}{x^2} \cos tx\, dx,$$

$$\phi_2(t) = \frac{1}{2} + \frac{4}{\pi^2} \sum_{n=0}^{\infty} \frac{\cos[(2n + 1)\pi t]}{(2n + 1)^2}, \quad t \in \mathbb{R}_1.$$

Show that ϕ_1 and ϕ_2 are ch.f.'s and find the corresponding distributions.

18.17. Let the r.v. ξ have ch.f. $\phi(t)$, $t \in \mathbb{R}_1$. Find the distribution of ξ if: (a) $\phi(t) = (1/4)(1 + e^{it})^2$; (b) $\phi(t) = 1/(2 - e^{it})$; (c) $\phi(t) = \cos t$; (d) $\phi(t) = \cos^2 t$; (e) $\phi(t) = \sum_{k=0}^{\infty} a_k \cos(kt)$, where $a_k > 0$, $\sum_{k=0}^{\infty} a_k = 1$; (f) $\phi(t) = \exp[\lambda(e^{it} - 1)]$, $\lambda > 0$.

*18.18. Let F be the d.f. of the r.v. ξ and let ϕ be its ch.f. Suppose that $E\{\xi^2\} < \infty$, $E\{\xi^2\} \neq 0$. Prove that the function $\Psi(t) = -\phi''(t)/E\{\xi^2\}$, $t \in \mathbb{R}_1$, is also a ch.f.

18.19. Find the d.f. F if its ch.f. $\phi(t)$, $t \in \mathbb{R}_1$, is given by: (a) $\phi(t) = \exp(-a^2t^2)$, $a > 0$; (b) $\phi(t) = (1 - t^2)\exp(-\frac{1}{2}t^2)$.

18.20. Let $\xi_1, \ldots, \xi_n$ be n independent r.v.'s, each uniformly distributed over the interval $(-1, 1)$. Find: (a) the ch.f. of the r.v. $\eta = \xi_1 + \ldots + \xi_n$; (b) $P\{-a < \eta < a\}$, $a > 0$.

18.21. Let ξ_1 and ξ_2 be independent r.v.'s normally $N(a, \sigma^2)$ distributed. Using the ch.f., show that the r.v.'s $\xi_1 + \xi_2$ and $\xi_1 - \xi_2$ are independent.

18.22. Let ξ and η be independent r.v.'s with densities $f(x)$, $x \in \mathbb{R}_1$, and $g(x)$, $x \in \mathbb{R}_1$, and ch.f.'s $\phi(t)$, $t \in \mathbb{R}_1$, and $\Psi(t)$, $t \in \mathbb{R}_1$, respectively. Express the ch.f. of the r.v. $\xi\eta$ in terms of: (a) f and Ψ; (b) g and ϕ.

*18.23. Let the even function $f(x)$, $x \in \mathbb{R}_1$, be a density of a certain distribution, and let $\phi(t)$, $t \in \mathbb{R}_1$, be its ch.f., which is assumed to be strictly positive. Show that the function $f_a(x) = f(x)(1 - \cos ax)/(1 - \phi(a))$, $a > 0$, $x \in \mathbb{R}_1$, is a density and find its ch.f. $\phi_a(t)$,

$t \in \mathbb{R}_1$.

18.24. Let the random vector $\vec{\zeta} = (\xi, \eta)$ have the density

$$f_{\vec{\zeta}}(x, y) = \begin{cases} \frac{1}{4}(1 + xy(x^2 - y^2)), & \text{if } |x| \leqslant 1 \text{ and } |y| \leqslant 1 \\ 0, & \text{otherwise.} \end{cases}$$

Show that the r.v.'s ξ and η are dependent; nevertheless, for their corresponding ch.f.'s the equality $\phi_{\xi+\eta}(t) = \phi_\xi(t)\phi_\eta(t)$ is fulfilled for all $t \in \mathbb{R}_1$.

*18.25. Let ξ_1 and ξ_2 be independent r.v.'s with ch.f.'s $\phi_1(t)$, $t \in \mathbb{R}_1$, and $\phi_2(t)$, $t \in \mathbb{R}_1$, respectively. Suppose that: the r.v.'s $\eta_1 = a_{11}\xi_1 + a_{12}\xi_2$ and $\eta_2 = a_{21}\xi_1 + a_{22}\xi_2$ are independent for $a_{ij} \neq 0$, $i, j = 1, 2$ and $a_{11}a_{22} - a_{12}a_{21} \neq 0$; $\phi_1''(t)$ and $\phi_2''(t)$ exist for every $t \in \mathbb{R}_1$; ϕ_1 and ϕ_2 are even functions; and $\phi_i(t)$ is not a constant, $i = 1, 2$. Under these conditions, prove that the r.v.'s ξ_1 and ξ_2 are normally distributed.

18.26. Let the random vector $\vec{\zeta} = (\xi, \eta)$ have the density

$$f_{\vec{\zeta}}(x, y) = (\pi^2(1 + x^2)(1 + y^2))^{-1}, \quad (x, y) \in \mathbb{R}_2.$$

Find: (a) the ch.f. $\phi(t, s)$, $(t, s) \in \mathbb{R}_2$, of the vector $\vec{\zeta}$; (b) the ch.f. $\phi_\xi(t)$, $t \in \mathbb{R}_1$, of the r.v. ξ.

18.27. Let $\xi_1, \ldots, \xi_n$ be independent and identically distributed r.v.'s. Show that: (a) if the r.v. ξ_1 has the Cauchy distribution with density $f(x) = (b/\pi)(b^2 + (x - a)^2)^{-1}$, $x \in \mathbb{R}_1$, $a, b \in \mathbb{R}_1$, $b > 0$, then the r.v. $\eta = (\xi_1 + \ldots + \xi_n)/n$ has the same distribution; (b) if $\xi_1 \in N(0, 1)$, then the r.v. $\zeta = (\xi_1 + \ldots + \xi_n)/\sqrt{n}$ has the same distribution; (c) if $\xi_i \in P(\lambda_i)$, $i = 1, \ldots, n$ and $\Theta = \xi_1 + \ldots + \xi_n$, then $\Theta \in P(\lambda)$ with $\lambda = \lambda_1 + \ldots + \lambda_n$.

18.28. Let ξ be a r.v. for which all moments $a_k = E\{\xi^k\}$, $k = 1, 2, \ldots$ exist. Find the density of the r.v. ξ and show that it is uniquely defined by its moments a_k, $k = 1, 2, \ldots$ in the following two cases: (a) $a_k = \delta/(\delta + k)$, $k = 1, 2, \ldots$; (b) $a_k = (n + k)!/k!$, $k = 1, 2, \ldots$, n is a fixed natural number.

18.29. Let the r.v. ξ have gamma distribution with parameters α and β (see (15.10)), and let the r.v. η have a χ^2-distribution with n degrees of freedom (see Example 16.2). Find the ch.f.'s of ξ and η.

18.30. Let ξ_1, ξ_2, ξ_3 and ξ_4 be independent r.v.'s with $N(0, 1)$ distributions. Prove that the r.v. $\eta = \xi_1\xi_2 + \xi_3\xi_4$ has Laplace distribution (i.e., the density of η is $\frac{1}{2}\exp(-|y|)$, $y \in \mathbb{R}_1$).

*18.31. If ξ and η are bounded r.v.'s and $E\{\xi^k\eta^j\} = E\{\xi^k\}E\{\eta^j\}$ holds, where k and j are arbitrary natural numbers, then $\vec{\xi}$ and η are independent. (*M. Kac*)

*18.32. Prove that the function $\phi(t)$, $t \in \mathbb{R}_1$, is a ch.f. and find the corresponding density if: (a) $\phi(t) = (\text{ch}t)^{-1}$; (b) $\phi(t) = (\text{ch}t)^2$; (c) $\phi(t) = (\text{sh}t)^{-1}$.

*18.33. Let ξ_1 and ξ_2 be independent r.v.'s having Student's t distribution with 1 and 3 degrees of freedom, respectively. Let $\eta = \frac{1}{2}\xi_1\sqrt{3} + \frac{1}{2}\xi_2$. Show that the densities of ξ_1, ξ_2 and η satisfy the equality $f_\eta(y) = \frac{1}{2} f_{\xi_1\sqrt{3}}(y) + \frac{1}{2} f_{\xi_2}(y)$, $y \in \mathbb{R}_1$. (*I. Kotlarsky*)

*18.34. For which values of the parameter α, is the function

$\phi_\alpha(t) = e^{-\varepsilon|t|^\alpha}$, $t \in \mathbb{R}_1$, $\varepsilon > 0$, a ch.f.?

*18.35. Let ξ be a non-negative r.v. with d.f. $F(x)$, $x \geqslant 0$, $F(0) = 0$ and $\Psi_\xi(\lambda)$ be its Laplace-Stieltjes transformation. Show that: (a) if the r.v. $\eta \in E(\lambda)$ and is independent of ξ then $\Psi_\xi(\lambda) = P\{\eta > \xi\}$; (b) if ξ_1 and ξ_2 are non-negative and independent r.v.'s, then $\Psi_{\xi_1+\xi_2}(\lambda) = \Psi_{\xi_1}(\lambda)\Psi_{\xi_2}(\lambda)$; (c) if the independent and non-negative r.v.'s ξ_1 and ξ_2 have Laplace-Stieltjes transformations $\Psi_{\xi_1}(\lambda)$ and $\Psi_{\xi_2}(\lambda) = e^{-\lambda^\alpha}$, respectively, then for the r.v. $\lambda = \xi_1^{1/\alpha}\xi_2$ we have $\Psi_\eta(\lambda) = \Psi_{\xi_1}(\lambda^\alpha)$.

*18.36. Let $\Psi(\lambda)$ be the Laplace-Stieltjes transformation of the non-negative r.v. ξ with $E(\xi) = a$. Show that: (a) from the functional equation $\beta(\lambda) = \Psi(\lambda + c - c\beta(\lambda))$, $c > 0$, $ac \leqslant 1$, we can uniquely determine the function $\beta(\lambda)$, $\beta(\lambda) \leqslant 1$, $\lambda \geqslant 0$; (b) the function $\beta(\lambda)$ is a Laplace- Stieltjes transformation of a certain r.v.

18.37. Let $\phi(t)$, $t \in \mathbb{R}_1$, be a ch.f. of the r.v. ξ and let for some $t_0 \neq 0$ the equality $|\phi(t_0)| = 1$ hold. Show that ξ has a lattice distribution with step size $h = 2\pi/t_0$; i.e.,

$$\sum_{n=-\infty}^{\infty} P\{\xi = a + nh\} = 1,$$

where a is some constant.

19. Infinitely Divisible and Stable Distributions

Introductory Notes

Let ξ be a r.v. defined on some probability space $(\Omega, \mathcal{F}, P)$ and $F(x)$, $x \in \mathbb{R}_1$, and $\phi(t)$, $t \in \mathbb{R}_1$, be its d.f. and ch.f., respectively.

The r.v. ξ, as well as its d.f. F and its ch.f. ϕ, are said to be *infinitely divisible* if for each $n \geqslant 1$ there exist independent and identically distributed r.v.'s $\xi_{n1}, \ldots, \xi_{nn}$ such that

$$\xi \stackrel{d}{=} \xi_{n1} + \ldots + \xi_{nn},$$

or equivalently, if

$$F = [F_n]^{*n}, \quad \phi = (\phi_n)^n,$$

where F_n is a d.f. and ϕ_n a ch.f. (Recall the symbol $\xi \stackrel{d}{=} \eta$ means that the r.v.'s ξ and η are equivalent in distribution: $F_\xi(x) = F_\eta(x)$ for all $x \in \mathbb{R}_1$.)

Let $\phi(t)$, $t \in \mathbb{R}_1$, be an infinitely divisible ch.f. of some r.v. ξ with finite variance. Then the logarithm of η admits the following representation:

$$\ln\phi(t) = i\gamma t + \int_{-\infty}^{\infty} (e^{itx} - 1 - itx) \frac{dG(x)}{x^2} \tag{19.1}$$

(*Kolmogorov's formula*).

In (19.1) γ is a real number and G is a left-continuous non-decreasing and bounded function. This representation is unique; i.e., for every infinitely divisible ch.f. ϕ there exists just one pair (γ, G) for which (19.1) holds. The converse statement is also true: if γ and G satisfy the above conditions, then the right-hand side of (19.1) is logarithm of an uniquely defined infinitely divisible ch.f.

In the general case (when one relaxes the requirement of a finite variance) for the logarithm of an infinitely divisible ch.f. ϕ the following representation is valid:

$$\ln\phi(t) = i\gamma t + \int_{-\infty}^{\infty} \left(e^{itx} - 1 - \frac{itx}{1 + x^2}\right) \frac{1 + x^2}{x^2}\, dN(x) \tag{19.2}$$

(*Lévy-Khintchine's formula*).

In the last formula γ is a real number and $N(x)$ is a left-continuous non-decreasing and bounded function. It is assumed that the integral equal $(-t^2/2)$ when $x = 0$. If $N(x) \to 0$ as $x \to \infty$, the representation (19.2) is unique.

The function $L(x)$ defined for $x \in \mathbb{R}_1 \setminus \{0\}$ as

$$L(x) = \begin{cases} \displaystyle\int_{-\infty}^{x} \frac{1 + u^2}{u^2}\, dN(u), & \text{if } -\infty < x < 0; \\ \displaystyle -\int_{x}^{\infty} \frac{1 + u^2}{u^2}\, dN(u), & \text{if } 0 < x < \infty, \end{cases}$$

possesses the following properties:

(a) $L(x)$ is non-decreasing in the intervals $(-\infty, 0)$ and $(0, \infty)$;

(b) $L(\infty) = 0$, $L(-\infty) = 0$;

(c) $L(x)$ is continuous at all points (in its domain) at which $N(x)$ is continuous;

(d) for arbitrary $\delta > 0$ we have $\displaystyle \mathop{{\int\!\!\!\!\!-}}_{-\delta}^{\delta} x^2\, dL(x) < \infty$ (here the integral $\int\!\!\!\!\!-$ indicates that the point $x = 0$ is omitted from $\mathbb{R}_1 = (-\infty, \infty)$).

With $v = N(+0) - N(-0)$ and the function L, introduced above, (19.2) can be written equivalently as

$$\ln\phi(t) = i\gamma t - \frac{vt^2}{2} + \mathop{{\int\!\!\!\!\!-}}_{-\infty}^{\infty} \left(e^{itx} - 1 - \frac{itx}{1 + x^2}\right) dL(x) \qquad (19.3)$$

(*Lévy's formula*).

The distribution function F is said to belong to the class of *stable distributions* if for arbitrary $b_1 > 0$, $b_2 > 0$ and real c_1 and c_2 one can find $b > 0$ and real c such that

$$F\left(\frac{x - c_1}{b_1}\right) * F\left(\frac{x - c_2}{b_2}\right) = F\left(\frac{x - c}{b}\right). \qquad (19.4)$$

The corresponding ch.f. is also called stable. If ϕ is the ch.f. corresponding to F, then (19.4) can be written as

$$\phi(b_1 t)\phi(b_2 t) = \phi(bt)e^{i\gamma t}, \qquad (19.5)$$

where $\gamma = c - c_1 - c_2$.

Note that the logarithm of every stable ch.f. also admits canonical representation of the type (19.3).

Illustrative Examples

Example 19.1. Show that any r.v. $\xi \in N(a, \sigma^2)$ is infinitely divisible. Find the corresponding γ and G in the representation (19.1).

Solution. Denote the ch.f. of ξ by $\phi(t)$, $t \in \mathbb{R}_1$. Then $\phi(t) = \exp\left(iat - \dfrac{\sigma^2 t^2}{2}\right)$. For $\phi_n(t) = \exp\left(i\dfrac{a}{n}t - \dfrac{\sigma^2}{n}\dfrac{t^2}{2}\right)$ we have $\phi(t) = [\phi_n(t)]^n$, which means that ξ is infinitely divisible. Therefore, for arbitrary n we have $\xi = \xi_{n1} + \ldots + \xi_{nn}$, where $\xi_{nk} \in N\left(\dfrac{a}{n}, \dfrac{\sigma^2}{n}\right)$, $k = 1, \ldots, n$.

Consider the function $G(x) = 0$, if $x \leq 0$ and $G(x) = \sigma^2$, if $x > 0$. Then

$$\int_{-\infty}^{\infty} (e^{itx} - 1 - itx)\frac{dG(x)}{x^2} =$$

$$= \lim_{x\to 0} \frac{e^{itx} - 1 - itx}{x^2} [G(+0) - G(-0)] = -\frac{\sigma^2 t^2}{2}.$$

Because of the uniqueness of (19.1) we have that G(x) in (19.1) coincides with G(x) defined above and we also have that γ in (19.1) coincides with the constant a.

Example 19.2. Show that every stable distribution is infinitely divisible but the converse statement is not always true.

Solution. From (19.5) we find that for every stable ch.f. ϕ

$$\phi(b_1't)\phi(b_2't) \ldots \phi(b_n't) = \phi(b't)e^{i\gamma' t}$$

where $b_1', \ldots, b_n' > 0$, $b' > 0$ and $\gamma' \in \mathbb{R}_1$. Then for $b_1' = 1, \ldots, b_n' = 1$, $b' = b_n$, $\gamma' = \gamma$ we obtain

$$\phi(t) = \left[\phi\left(\frac{t}{b_n}\right)\exp\left(-\frac{i\gamma t}{nb_n}\right)\right]^n;$$

hence ϕ is infinitely divisible.

Assume now that ϕ is the ch.f. of the Poisson distribution with parameter λ. Then $\phi(t) = \exp[\lambda(e^{it} - 1)]$ and it suffices to show that the last function does not satisfy any relation of the type (19.5). On the other hand ϕ is infinitely divisible (see Exercise 19.1). Thus we have described a distribution which is infinitely divisible, but not stable.

Exercises

19.1. Prove that any r.v. $\xi \in P(\lambda)$ is infinitely divisible. The same holds for any r.v. of the form $\eta = c_1 + c_2\xi$, where $c_1, c_2 \in \mathbb{R}_1$. In both cases find the constant γ and the function G.

19.2. Prove the infinite divisibility of the following distributions: (a) the Cauchy distribution (see (15.12)); (b) the gamma distribution (see (15.10)); (c) the χ^2-distribution with n degrees of freedom (see Exercise 18.29); (d) the Laplace distribution (see Exercise 18.8); (e) the exponential distribution (see (15.9)); (f) the negative binomial distribution (see Section 9 and Exercise 18.7); (g) the degenerate distribution concentrated in one point.

19.3. Let the non-negative integer-valued r.v. ξ be infinitely divisible. Express the property of ξ in terms of its probability generating function g(s), $|s| \leq 1$. (In this case g is called an *infinitely divisible probability generating function*).

19.4. Let ξ be an infinitely divisible r.v. whose distribution is not concentrated in one point. Prove then that the range of ξ cannot be contained in any finite interval.

19.5. Prove that $\exp[i\gamma t - \frac{\sigma^2 t^2}{2} + \sum_{k=1}^{N} (e^{itc_k} - 1)\lambda_k]$, with $0 < \lambda_k < \infty$,

$c_k \in \mathbb{R}_1$, is an infinitely divisible ch.f.

19.6. The distribution F is called the *generalized Poisson distribution* if it can be written as

$$F(x) = e^{-p} \sum_{k=0}^{\infty} \frac{p^k}{k!} H^{*k}(x), \quad x \in \mathbb{R}_1,$$

where $0 < p < 1$ and H is some d.f. Prove that any distribution of this type is infinitely divisible. Find the explicit form of the logarithm of the ch.f. of F.

19.7. Let $\xi_1, \xi_2, \ldots$ be i.i.d. r.v.'s with ch.f. ϕ and let $\tau \in P(\lambda)$ be independent of all ξ_k. The r.v. η is defined by

$$\eta = \begin{cases} 0, & \text{if } \tau = 0; \\ \xi_1 + \xi_2 + \ldots + \xi_\tau, & \text{if } \tau \geq 1. \end{cases}$$

Prove that η is infinitely divisible.

19.8. Prove that $\phi^{1/n}$ is a ch.f. for any positive integer n if and only if ϕ is an infinitely divisible ch.f.

19.9. If ϕ is an infinitely divisible ch.f., then ϕ has no real roots. Prove this statement and use it to show that the uniform distribution on any interval is not infinitely divisible.

*19.10. Let $\gamma(t)$, $t \in \mathbb{R}_1$, be any ch.f. and let $\lambda > 0$. Show that $\phi_\lambda(t) = \lambda/(\lambda + 1 - \gamma(t))$, $t \in \mathbb{R}_1$, is an infinitely divisible ch.f.

*19.11. Let ξ be a Γ-distributed r.v. with parameters (α, β). Then ξ is infinitely divisible, according to Exercise 19.2. Find the constant γ and the function N in the representation (19.2) of the logarithm of the ch.f. of ξ.

19.12. Assume that $\phi(t)$ is an infinitely divisible ch.f. which admits the representation (19.3) and let $\int_{|x|>1} x^2 \, dL(x) < \infty$. Show that

$$\ln\phi(t) = i\tilde{\gamma}t - \frac{vt^2}{2} + \int_{-\infty}^{\infty} (e^{itx} - 1 - itx)\,dL(x)$$

and find the constant $\tilde{\gamma}$.

*19.13. Let $\phi(t)$, $t \in \mathbb{R}_1$, be a ch.f. of some distribution which is symmetric with respect to the origin. Prove that $\ln\phi(t)$, $t \in \mathbb{R}_1$, can be written as

$$\ln\phi(t) = -\frac{vt^2}{2} + \int_{0+}^{\infty} (\cos tx - 1)\,dQ(x),$$

where $v \geq 0$, $Q(x) \geq 0$ for $x > 0$ and $\int_{0+}^{\infty} \frac{x^2}{1 + x^2}\,dQ(x) < \infty$.

19.14. Show that for any $\alpha \in (0, 2]$, the function $\phi(t) = \exp(-|t|^\alpha)$, $t \in \mathbb{R}_1$, is a stable ch.f.

19.15. Show that the normal distribution and the Cauchy distribution are stable.

19.16. Show that the distribution with density

$$f(x) = \begin{cases} 0, & \text{if } x \leqslant 0 \\ (2\sqrt{\pi})^{-1} e^{-\frac{3}{2}} e^{-\frac{1}{4}x}, & \text{if } x > 0, \end{cases}$$

is stable. (*P. Lévy*)

19.17. Let F be a d.f. with mean 0 and variance 1 and let F satisfy the relationship of the type (19.4):

$$F\left(\frac{x}{\sigma_1}\right) * F\left(\frac{x}{\sigma_2}\right) = F\left(\frac{x}{(\sigma_1^2 + \sigma_2^2)^{1/2}}\right).$$

Show that F is the standard normal distribution.

*19.18. Let $\Psi_\alpha(\lambda) = \exp(-\lambda^\alpha)$, $0 < \lambda < 1$, $\lambda > 0$. Show that: (a) Ψ_α is a Laplace-Stieltjes transformation of some d.f. F_α; (b) $F_\alpha(x)$, $x \in \mathbb{R}_1$, is *strongly stable*; i.e., there exists a sequence $\{c_n\}$ of positive numbers, such that $F_\alpha^{*n}(x) = F_\alpha(c_n x)$, $x \in \mathbb{R}_1$. Determine c_n, $n = 1, 2, \ldots$ Note that this property is stronger than (19.4).

*19.19. Let $\xi_{n,k} \in P(\lambda_{n,k})$ be independent r.v.'s, where $\lambda_{n,k} = \frac{cn^\alpha}{|k|^{1+\alpha}}$, $c > 0$, $0 < \alpha \leqslant 2$, $n = 1, 2, \ldots$, $k = 1, 2, \ldots$ Put $\eta_n = \sum_{k=-n^2}^{n^2} k\xi_{n,k}$. Show that $\eta_n \xrightarrow{d} \eta$ as $n \to \infty$ for some r.v. η with a stable distribution. Also show that the constant c can be chosen such that $\phi_\eta(t) = \exp(-|t|^\alpha)$, $t \in \mathbb{R}_1$.

19.20. Let F_n, $n = 1, 2, \ldots$ be a sequence of infinitely divisible distribution functions with $F_n \xrightarrow{d} F$ as $n \to \infty$, where F is a d.f. Show that F is also infinitely divisible.

19.21. Show that $\phi(t) = \dfrac{1 - b}{1 - be^{it}}$, $t \in \mathbb{R}_1$, $b \in (0, 1)$, is a ch.f. of an infinitely divisible distribution.

*19.22. Let $\zeta(z) = \zeta(s + it)$ be the Rieman zeta-function which is defined for any real t and s, $s > 1$; i.e., $\zeta(z)$ is either given by the series $\zeta(z) = \sum_{n=1}^{\infty} n^{-z}$ or by the Euler product $\zeta(z) = \prod_p (1 - p^{-z})^{-1}$, taken over all the prime numbers p. Prove that for any fixed $s > 1$, the function $\phi(t) = \zeta(s + it)/\zeta(s)$, $t \in \mathbb{R}_1$, is a ch.f. of an infinitely divisible distribution.

20. Conditional Distributions and Conditional Expectation

Introductory Notes

The definition and the main properties of conditional probability and conditional expectation are given in Section 13. Here we give concrete realizations of some of the general ideas developed in Section 13.

Let ξ and η be r.v.'s defined on the probability space $(\Omega, \mathcal{F}, P)$ and ξ be integrable. Denote by P_η the measure on $\mathcal{B}_1$ generated by η; i.e., $P_\eta(B) = P(\eta^{-1}(B))$, $B \in \mathcal{B}_1$. Denote also by $\mathcal{F}_\eta$ the σ-algebra generated by η; i.e., $\mathcal{F}_\eta = \{\eta^{-1}(B) : B \in \mathcal{B}_1\}$.

In Section 13 (see (13.3)) the conditional expectation of ξ given η, notationally $E\{\xi|\eta\}$, is defined. Here we shall introduce the conditional expectation of ξ given $\eta = y$ and denote it by $E\{\xi|\eta = y\}$. The *conditional expectation* $E\{\xi|\eta = y\}$, $y \in \mathbb{R}_1$, is any $\mathcal{B}_1$-measurable function $\phi(y)$, $y \in \mathbb{R}_1$, satisfying the relation

$$\int_B \phi(y)\,dP_\eta(y) = \int_{\{\omega:\eta(\omega)\in B\}} \xi(\omega)\,dP(\omega)$$

for all $B \in \mathcal{B}_1$.

Any two versions $\phi_1(y)$ and $\phi_2(y)$ of $E\{\xi|\eta = y\}$ are equal almost surely with respect to the measure P_η on $\mathcal{B}_1$ (abbr. P_η-a.s.). Thus $E\{\xi|\eta = y\}$, $y \in \mathbb{R}_1$, is a class of P_η-equivalent r.v.'s (see Section 12) defined on the probability space $(\mathbb{R}_1, \mathcal{B}_1, P_\eta)$. The difference between $E\{\xi|\eta = y\}$, $y \in \mathbb{R}_1$, and $E\{\xi|\eta\}$ introduced in (13.3) is that $E\{\xi|\eta\}$ is a class of P-equivalent r.v.'s defined on $(\Omega, \mathcal{F}_\eta, P)$ but not on $(\mathbb{R}_1, \mathcal{B}_1, P)$ with the property

$$\int_A \Psi(\omega)\,dP(\omega) = \int_A \xi(\omega)\,dP(\omega)$$

for all $A \in \mathcal{F}_\eta$ and for every version $\Psi(\omega)$ of $E\{\xi|\eta\}$.

If $\phi(y)$ is a version of $E\{\xi|\eta = y\}$, then a version $\Psi(\omega)$ of the conditional expectation $E\{\xi|\eta\}$ or $E\{\xi|\mathcal{F}_\eta\}$ (see (13.3)) could be $\phi(\eta(\omega))$; i.e.,

$$E\{\xi|\eta\} = \phi(\eta(\omega)) \quad (P\text{-a.s.}).$$

Analogously to $E\{\xi|\eta = y\}$ we can introduce the notation $P\{A|\eta = y\}$ for the *conditional probability* of the event A given $\eta = y$. By definition we put

$$P\{A|\eta = y\} = E\{1_A|\eta = y\} \quad (P_\eta\text{-a.s.}).$$

Thus $P\{A|\eta = y\}$ is an equivalence class under P_η of positive $\mathcal{B}_1$-measurable functions such that if $\phi(y)$ belongs to this class, then

$$\int_B \phi(y)dF_\eta(y) = \int_{\{\omega:\eta(\omega)\in B\}} 1_A dP(\omega) = P\{A \cap (\eta^{-1}(B))\}$$

for all $B \in \mathcal{B}_1$.

It can be proved that for every $A \in \mathcal{F}$ one can select a version $\phi(A, y)$ of $P\{A|\eta = y\}$ such that the function $\phi(A, y)$ defined on the product $\mathcal{F} \times \mathbb{R}_1$ is a probability distribution on $\mathcal{F}$ for every fixed $y \in \mathbb{R}_1$. The function $\phi(A, y)$ is called a *regular conditional probability* and is denoted by $P^*\{A|\eta = y\}$. If in $P^*\{A|\eta = y\}$ we substitute A by $\{\omega : \xi(\omega) \in B\}$, then we obtain the regular conditional distribution $P^*\{\xi \in B|\eta = y\}$ of the r.v. ξ given $\eta = y$. This means that if y is fixed, then $P^*\{\xi \in B|\eta = y\}$, $B \in \mathcal{B}_1$, is a distribution on $\mathcal{B}_1$, and moreover, when B is fixed, $P^*\{\xi \in B|\eta = y\}$, $y \in \mathbb{R}_1$, is a version of the conditional expectation $E\{1_{[\xi\in B]}|\eta = y\}$, $y \in \mathbb{R}_1$. In particular, if $B = (-\infty, x)$, then $P^*\{\xi < x|\eta = y\}$, $x \in \mathbb{R}_1$, is called a *regular conditional distribution* of ξ given $\eta = y$. In the sequel we drop the sign * assuming that the corresponding conditional d.f., conditional distribution, etc., are regular.

Let the random vector (ξ, η) be absolutely continuous (with respect to the Lebesgue measure in $\mathbb{R}_2$) with density $f(x, y)$, $(x, y) \in \mathbb{R}_2$.

Consider the following functions:

$$f(x|y) = \frac{f(x, y)}{\int_{-\infty}^{\infty} f(u, y)du} \tag{20.1}$$

and

$$F(x|y) = \int_{-\infty}^{x} f(u|y)du = \frac{\int_{-\infty}^{x} f(u, y)du}{\int_{-\infty}^{\infty} f(u, y)du}. \tag{20.2}$$

Then $\int_A f(x|y)dx$ is a version of $P\{\xi \in A|\eta = y\}$; i.e.,

$$P\{\xi \in A|\eta = y\} = \int_A f(x|y)dx \quad (P_\eta\text{-a.s.}). \tag{20.3}$$

The function $f(x|y)$ (denoted also by $f_{\xi|\eta}(x|y)$, is called the *conditional density* of the r.v. ξ given $\eta = y$. Through it we can express the following conditional probability:

$$P\{\xi \in A|\eta \in B\} = \int_B\int_A f(x|y)dx\,dy, \quad \text{for } A, B \in \mathcal{B}_1. \tag{20.4}$$

If ξ is integrable, then $\int_{-\infty}^{\infty} xf(x|y)dx$ and $\int_{-\infty}^{\infty} x\,dF(x|y)$ are versions of $E\{\xi|\eta = y\}$; i.e.,

$$E\{\xi|\eta = y\} = \int_{-\infty}^{\infty} xf(x|y)dx \quad (P_\eta\text{-a.s.}) \tag{20.5}$$

and

$$E\{\xi|\eta = y\} = \int_{-\infty}^{\infty} x\,dF(x|y) \quad (P_\eta\text{-a.s.}). \tag{20.6}$$

The notions of conditional distribution, conditional d.f. and conditional density of one r.v. with respect to another are naturally generalized to the case of multivariate r.v.'s.

Illustrative Examples

Example 20.1. Let the random vector (ξ, η) be defined on the probability space $(\Omega, \mathcal{F}, P)$. Suppose that it is normally $N(\vec{a}, \Sigma)$ distributed (see (16.12)) with $a_1 = a_2 = 0$ and $\Sigma = \begin{pmatrix}\sigma_1^2 & \sigma_1\sigma_2\rho \\ \sigma_1\sigma_2\rho & \sigma_2^2\end{pmatrix}$, where $\sigma_1 > 0$, $\sigma_2 > 0$, $0 < \rho < 1$. Find the conditional expectation $E\{\xi|\eta\}$.

Solution. Obviously $\det \Sigma = \sigma_1^2\sigma_2^2(1 - \rho^2)$ and

$$\Sigma^{-1} = (1 - \rho^2)^{-1}\begin{pmatrix}\sigma_1^{-2} & -\rho\sigma_1^{-1}\sigma_2^{-1} \\ -\rho\sigma_1^{-1}\sigma_2^{-1} & \sigma_2^{-2}\end{pmatrix}$$

Therefore, the density $f_{\xi,\eta}(x, y)$, $x, y \in \mathbb{R}_1$ of the random vector (ξ, η) is (see (16.12))

$$f_{\xi,\eta}(x, y) = \frac{1}{2\pi\sigma_1\sigma_2\sqrt{1 - \rho^2}} \exp\left(-\frac{1}{2(1 - \rho^2)}\left(\frac{x^2}{\sigma_1^2} - \frac{2\rho xy}{\sigma_1\sigma_2} + \frac{y^2}{\sigma_2^2}\right)\right).$$

From $f_{\xi,\eta}(x, y)$, using (16.7), we obtain the marginal density

$$f_\eta(y) = \frac{1}{\sqrt{2\pi}\sigma_2} \exp\left(-\frac{y^2}{2\sigma_2^2}\right), \quad y \in \mathbb{R}_1.$$

From (20.1) we have

$$f(x|y) = \frac{f_{\xi,\eta}(x, y)}{f_\eta(y)} =$$

$$= \frac{1}{\sqrt{2\pi}\sigma_1\sqrt{1-\rho^2}} \exp\left(-\frac{1}{2(1-\rho^2)}\left(\frac{x^2}{\sigma_1^2} - \frac{2\rho xy}{\sigma_1\sigma_2} + \frac{\rho^2 y^2}{\sigma_2^2}\right)\right) =$$

$$= \frac{1}{\sqrt{2\pi\sigma_1^2(1-\rho^2)}} \exp\left(-\frac{\left(x - \rho\frac{\sigma_1}{\sigma_2}y\right)^2}{2\sigma_1^2(1-\rho^2)}\right).$$

The latter means that the conditional distribution of ξ given $\eta = y$ is normal $N(\rho\sigma_1\sigma_2^{-1}y, \sigma_1\sqrt{1-\rho^2})$; i.e., $\rho\sigma_1\sigma_2^{-1}y$ is a version of $E\{\xi|\eta = y\}$. Therefore,

$$E\{\xi|\eta\} = \rho\sigma_1\sigma_2^{-1}\eta(\omega) \qquad \text{(P-a.s.)}.$$

Example 20.2. Let the point (ξ, η) be uniformly distributed over the domain of the plane defined by $|x^2 - y^2| \leqslant 1$. Find the conditional distribution of the r.v. ξ given $\eta = y$.

Solution. Set $D = \{(x, y) : |x^2 - y^2| < 1\}$; i.e., D is the domain closed between the hyperboles $x^2 - y^2 = 1$ and $x^2 - y^2 = -1$. It can easily be found that the Lebesgue measure of D is equal to 1. Therefore, the density of the random vector (ξ, η) is

$$f_{\xi,\eta}(x, y) = \begin{cases} 1, & \text{if } (x, y) \in D \\ 0, & \text{otherwise.} \end{cases}$$

From $f_{\xi,\eta}(x, y)$, using (16.7), we obtain the marginal density

$$f_\eta(y) = \begin{cases} 2(\sqrt{y^2+1} - \sqrt{y^2-1}), & \text{if } |y| > 1 \\ 2\sqrt{y^2+1}, & \text{otherwise} \end{cases}$$

From (20.1) we have $f(x|y) = f_{\xi,\eta}(x, y)/f_\eta(y)$ and hence

$$f(x|y) = \begin{cases} (2(\sqrt{y^2+1} - \sqrt{y^2-1}))^{-1}, & \text{if } |y| > 1,\ \sqrt{y^2-1} < |x| < \sqrt{y^2+1} \\ (2\sqrt{y^2+1})^{-1}, & \text{if } |y| \leqslant 1,\ 0 \leqslant |x| < \sqrt{y^2+1} \\ 0, & \text{otherwise} \end{cases}$$

Hence $P\{\xi \in B|\eta = y\} = \int_B f(x|y)dx$ (P_η-a.s.) for $B \in \mathcal{B}_1$.

Comparing $f(x|y)$ with (15.7), we observe that the conditional distribution of the r.v. ξ given $\eta = y$ is uniform over the interval $(-\sqrt{y^2+1}), \sqrt{y^2+1})$ if $|y| \leqslant 1$ and is uniform over the set $\{x : \sqrt{y^2-1} < |x| < \sqrt{y^2+1}\}$ when $|y| > 1$.

Exercises

20.1. Let the r.v.'s ξ and η be independent with $\xi \in B(n_1, p)$ and $\eta \in B(n_2, p)$. Find the conditional distribution of the random vector (ξ, η) given $\xi + \eta = k$, $0 \leqslant k \leqslant n_1 + n_2$.

20.2. Let ξ and η be independent negative binomially distributed r.v.'s with parameters (λ_1, p) and (λ_2, p) (see Section 9). Find the condiotional distribution of the random vector (ξ, η) given $\xi + \eta = k$, where k is a non-negative integer.

*20.3. Let ξ_j, $j = 0, 1, \ldots$, be independent r.v.'s and $\xi_j \in P(\lambda_j)$, $\lambda_j = xz^j/j!$, $x > 0$, $z > 0$. Let $\eta_1 = \sum_{j=0}^{\infty} j\xi_j$ and $\eta_2 = \sum_{j=0}^{\infty} \xi_j$. Find $\vec{P}\{\xi_j = m_j, j = 0, 1, \ldots | \eta_1 = n, \eta_2 = N\}$, where m_j are non-negative integers with n and N being natural numbers. (*A. Obretenov*)

20.4. Let ξ and η be independent r.v.'s and $\zeta = \xi - \eta$. Find $P\{|\zeta| \leqslant 1 | \xi = 1\}$, $F_{\zeta|\xi}(0, 1)$ and $f_{\zeta|\xi}(0, 1)$ if: (a) ξ and η are uniformly distributed over the interval (0, 2); (b) ξ and η are exponentially distributed $E(1)$.

20.5. Let ξ_1 and ξ_2 be independent r.v.'s, $\xi_i \in N(0, 4)$, $i = 1, 2$. With $\eta = \xi_1 + \xi_2$, $\zeta = \xi_1 - \xi_2$, find: (a) $P\{|\zeta| \leqslant 1 | \eta = 1\}$; (b) $F_{\zeta|\eta}(0, 1)$; (c) $f_{\zeta|\eta}(0, 1)$; (d) $\vec{P}\{\eta \geqslant 0 | |\zeta| \leqslant 1\}$.

20.6. Let $\xi_1, \xi_2, \ldots, \xi_n$ be *exchangeable r.v.'s*; i.e., for every permutation π of the indices the random vector $(\xi_{\pi(1)}, \xi_{\pi(2)}, \ldots, \xi_{\pi(n)})$ is distributed as $(\xi_1, \xi_2, \ldots, \xi_n)$. If $\xi_1, \xi_2, \ldots, \xi_n$ are also non-negative integer-valued r.v.'s, find

$$P\{\xi_1 + \ldots + \xi_r < r,\ r = 1, 2, \ldots, n | \xi_1 + \xi_2 + \ldots + \xi_n = k\}.$$

20.7. Prove that if $\phi(t)$, $t \in \mathbb{R}_1$, is a ch.f., then $\exp(\lambda\phi(t) - \lambda)$, $t \in \mathbb{R}_1$, $\lambda > 0$, is also a ch.f. (see Exercise 19.7). (*B. de Finetti*)

20.8. Let ξ_1 and ξ_2 be independent r.v.'s with corresponding densities $f_1(x)$ and $f_2(x)$, $x \in \mathbb{R}_1$. For $\eta = \xi_1 + \xi_2$, find: (a) the conditional density $f(x|y)$ of the r.v. ξ_1 given $\eta = y$; (b) $f(x|y)$, when ξ_1 and ξ_2 are exponential $E(\lambda)$.

20.9. Let $\tau, \xi_1, \xi_2, \ldots$ be independent r.v.'s with

$$P\{\xi_i = k\} = -\frac{1}{\ln(1-q)} \cdot \frac{q^k}{k}, \quad 0 < q < 1,\ k = 1, 2, \ldots,\ i = 1, 2, \ldots$$

and $\tau \in P(\lambda)$, where $\lambda = -\ln(1 - q)$. Define the r.v. ζ as $\zeta = 0$ for $\tau = 0$ and $\zeta = \xi_1 + \ldots + \xi_\tau$ for $\tau > 0$. What is the distribution of the r.v. ζ?

20.10. Introduce the probability space $(\Omega, \mathcal{F}, P)$, where $\Omega = \{z \in (-1, +1)\}$, $\mathcal{F} = \{B : B \in \mathcal{B}_1, B \subset (-1, 1)\}$ and $P(A) = \int_A \frac{1}{2}\,dz$ for $A \in \mathcal{F}$.

Let the r.v. ξ be defined on $(\Omega, \mathcal{F}, P)$ as $\xi(z) = z^2$. Find a version of the conditional distribution $P\{A|\xi\}$.

20.11. Let the random vector $\vec{\xi} = (\xi_1, \ldots, \xi_n)$ be normally $N(\vec{a}, \Sigma)$ distributed (see (16.12)) with $a_1 = a_2 = \ldots = a_n = 0$. Find $E\{\xi_n | \xi_1, \xi_2, \ldots, \xi_{n-1}\}$.

20.12. Let ξ and η be r.v.'s, where $V(\eta) < \infty$ and let $\mathbb{H}$ be the set of all measurable functions $g(x)$, $x \in \mathbb{R}_1$ for which $V\{g(\xi)\} < \infty$. Is there a function $g^*(x) \in \mathbb{H}$ such that

$$E\{(\eta - g^*(\xi))^2\} = \min_{g \in \mathbb{H}} E\{(\eta - g(\xi))^2\}?$$

20.13. Let ξ and η be integrable r.v.'s and let $E\{\xi|\eta\}$ be a constant with probability 1. Find $\mathrm{cov}(\xi, \eta)$.

20.14. A point ξ_1 is chosen at random from the interval (0, 1). Similarly, a second point ξ_2 from the interval $(\xi_1, 1)$ is chosen, and so on, until the point $\xi_n \in (\xi_{n-1}, 1)$ is chosen. Find: (a) $E\{\xi_n | \xi_{n-1}\}$; (b) $E(\xi_n)$.

20.15. Let the variance of the r.v.'s ξ and η be finite and $E\{\eta|\xi\} = a\xi + b$ (P-a.s.), $a, b \in \mathbb{R}_1$. Express a, b and $V(\zeta)$ in terms of $E(\xi)$, $E(\eta)$, $V(\xi)$, $V(\eta)$ and $\mathrm{cov}(\xi, \eta)$, where $\zeta = \eta - E\{\eta|\xi\}$.

20.16. Let $\xi_1, \ldots, \xi_n$ be independent r.v.'s uniformly distributed over the interval (0, 1) and $\xi_{(1)}, \ldots, \xi_{(n)}$ be their order statistics. Find the conditional density of the random vector $(\xi_{(1)}, \ldots, \xi_{(k)})$ given $\xi_{(k+1)} = y$, if $k < n$.

20.17. Let $\xi_1, \xi_2, \ldots$ be independent and identically distributed r.v.'s, $E(\xi_i) = a$, $E\{|\xi_i|\} = c < \infty$, $i = 1, 2, \ldots$ and τ be an integer-valued r.v. independent of the r.v.'s ξ_i, $i = 1, \ldots$ with $E(\tau) < \infty$. Prove that $E\{\xi_1 + \xi_2 + \ldots + \xi_\tau\} = aE(\tau)$. (*A. Wald*)

20.18. Let $\xi_0, \xi_1, \ldots$ be independent and identically distributed r.v.'s with a ch.f. $\phi(t)$, $t \in \mathbb{R}_1$, and let also τ be an integer-valued non-negative r.v. with a p.g.f. $g(s)$. Prove that the ch.f. of the r.v. $\eta = \xi_0 + \xi_1 + \ldots + \xi_\tau$ is $g(\phi(t))$, $t \in \mathbb{R}_1$.

21. Inequalities for Random Variables

Introductory Notes

Inequalities, involving moments of random variables, are widely used in probability theory and sometimes even in the standard analysis.

In general these inequalities consist of two groups. The first group includes the so-called *Chebyshev-type inequalities*, in which upper and lower bounds are found for probabilities of some events, related to r.v.'s. For example we have:

Chebyshev's inequality. For any r.v. ξ with finite mean and variance and for any $\varepsilon > 0$, the following relation holds:

$$P\{|\xi - E(\xi)| \geqslant \varepsilon\} \leqslant \frac{1}{\varepsilon^2} V(\xi).\square \tag{21.1}$$

Various generalizations and corollaries of the Chebyshev's inequality are given in the exercises in this section.

The second group are the so-called *moment inequalities*. In these inequalities upper and lower bounds are obtained for the moments of r.v.'s of some fixed order. For example we have:

Hölder's inequality (see Exercise 21.21). For arbitrary r.v.'s ξ and η, we have

$$E\{|\xi\eta|\} \leqslant (E\{|\xi|^r\})^{1/r} \cdot$$

$$\cdot\ (E\{|\eta|^s\})^{1/s}, \quad \text{where } r > 1,\ \frac{1}{r} + \frac{1}{s} = 1, \tag{21.2}$$

assuming that all the above moments are finite.□

Note that when inequalities of this type are considered, it will always be assumed that the moments involved are finite.

Of course, not all inequalities can be classified into one of the above groups. Inequalities of other kinds are also included in this section.

Illustrative Examples

Example 21.1. Let ξ be a r.v. with a finite expectation and a finite variance. Prove that for any $x > 0$ we have

$$P\{\xi \leqslant -x\} \leqslant \frac{V(\xi)}{V(\xi) + x^2} \tag{21.3}$$

and

$$P\{\xi \geqslant x\} \leqslant \frac{V(\xi)}{V(\xi) + x^2} \tag{21.4}$$

Solution. Without loss of generality we can assume that $E(\xi) = 0$, $V(\xi) = 1$. Denote by F the d.f. of ξ. Then for any fixed $x > 0$ and any $a \geqslant 0$, we have the following chain of relations:

$$1 + a^2 = \int_{-\infty}^{\infty} (z - a)^2 dF(z) \geqslant \int_{-\infty}^{-x} (z - a)^2 dF(z) \geqslant$$

$$\geqslant (x + a)^2 F(-x).$$

It follows that

$$F(-x) \leqslant (1 + a^2)/(x + a)^2.$$

Since $a \geqslant 0$ is arbitrary, we can choose $a = 1/x$. Then $F(-x) \leqslant 1/(1 + x^2)$ and hence (21.3) is proved.

Similar reasoning leads to the inequality $F(x) \geqslant x^2/(1 + x^2)$. Thus we establish the validity of (21.4).

The general case when $E(\xi) \neq 0$ and $V(\xi) \neq 1$ is left to the reader as a useful exercise.

Example 21.2. Let ξ and η be r.v.'s with zero means, unit variances and correlation coefficient ρ. Prove that

$$E\{\max[\xi^2, \eta^2]\} \leqslant 1 + \sqrt{1 - \rho^2}.$$

Solution. We shall use the following well-known fact: if a and b are arbitrary real numbers, then

$$\max[a, b] = \frac{1}{2}(|a + b| + |a - b|).$$

This and the Cauchy-Bunyakovski-Schwarz inequality (see Exercise 21.21 (a)) allow us to obtain the following relations:

$$E\{\max[\xi^2, \eta^2]\} = \frac{1}{2} E\{\xi^2 + \eta^2\} + \frac{1}{2} E\{|\xi^2 - \eta^2|\} =$$

$$= \frac{1}{2}(1 + 1) + \frac{1}{2} E\{|\xi - \eta||\xi + \eta|\} \leqslant$$

$$\leqslant 1 + \frac{1}{2}(E\{(\xi - \eta)^2\}E\{(\xi + \eta)^2\})^{1/2} =$$

$$= 1 + \frac{1}{2}(E\{\xi^2\} + E\{\eta^2\} - 2E\{\xi\eta\})^{1/2}(E\{\xi^2\} +$$

$$+ E\{\eta^2\} + 2E\{\xi\eta\})^{1/2}.$$

Since each of the variables ξ and η has zero mean and unit variance, then $E\{\xi\eta\} = \rho$ and we easily arrive at the desired inequality.

Example 21.3. Consider n independent r.v.'s, say $\xi_1, \ldots, \xi_n$, each distributed symmetrically with respect to the origin 0. Let $S_k = \xi_1 + \ldots + \xi_k$, $k = 1, \ldots, n$. Prove that for any real number x the following inequality holds:

$$P\{\max_{1\leqslant k\leqslant n} S_k > x\} \leqslant 2P\{S_n > x\}.$$

Solution. Let us consider the following events:

$$A_k = \{S_1 \leqslant x, \ldots, S_{k-1} \leqslant x, S_k > x\}, \quad k = 1, \ldots, n \quad \text{and}$$

$$B = \{S_n > x\}.$$

Then obviously $A_k A_j = \emptyset$ if $k \neq j$. Moreover,

$$P(A_k B) \geqslant P(A_k \cap [S_n \geqslant S_k]) = P(A_k)P(S_n \geqslant S_k) =$$

$$= P(A_k)P(\xi_{k+1} + \ldots + \xi_n \geqslant 0).$$

For the equality we have used the independence of the variables $\xi_1, \ldots, \xi_n$. Again using this property we see that the ch.f. of the sum $\xi_{k+1} + \ldots + \xi_n$ is equal to the product of the ch.f.'s of $\xi_{k+1}, \ldots, \xi_n$. However, let us recall that a r.v. is symmetric if its ch.f. is real. So the ch.f. of $\xi_{k+1} + \ldots + \xi_n$ is real, which means that $\xi_{k+1} + \ldots + \xi_n$ is a symmetric r.v. (with respect to 0). Therefore

$$P(\xi_{k+1} + \ldots + \xi_n \geqslant 0) \geqslant \frac{1}{2}.$$

This implies that $P(A_k B) \geqslant \frac{1}{2} P(A_k)$. Further we find that

$$P(B) \geqslant \sum_{k=1}^{n} P(A_k B) \geqslant \frac{1}{2} \sum_{k=1}^{n} P(A_k) \geqslant \frac{1}{2} P(\bigcup_{k=1}^{n} A_k).$$

Hence

$$2P(B) \geqslant P(\bigcup_{k=1}^{n} A_k),$$

which is equivalent to the inequality

$$2P\{S_n > x\} \geqslant P\{\max_{1\leqslant k\leqslant n} S_k > x\}.$$

Thus the desired inequality is proved.

Exercises

21.1. Let $f(x)$, $x \in \mathbb{R}_1$, be an even, non-negative function which is non-decreasing for $x > 0$. Prove that for any r.v. ξ and for any choice of the constant $c > 0$, we have

$$P\{|\xi| \geqslant c\} \leqslant \frac{E\{f(\xi)\}}{f(c)}$$

(*Markov's inequality*).

21.2. Let ξ and η be r.v.'s with finite variances and let ρ be their correlation coefficient. Prove the following two-dimensional analogue of the Chebyshev's inequality:

$$P\{|\xi - E(\xi)| \geqslant \varepsilon\sqrt{V(\xi)} \quad \text{or} \quad |\eta - \vec{E}(\eta)| \geqslant \varepsilon\sqrt{V(\eta)}\} \leqslant$$

$$\leqslant \frac{1}{\varepsilon^2}(1 + \sqrt{1 - \rho^2}).$$

21.3. Let the function $f(x)$, $x \in \mathbb{R}_1$, be defined as in Exercise 21.1, let ξ be any r.v. and let $c > 0$ be a constant. Prove that: (a) if $|f(x)| \leqslant K$, $x \in \mathbb{R}_1$, then

$$P\{|\xi| \geqslant c\} \geqslant \frac{E\{f(\xi)\} - f(c)}{K} ;$$

(b) if $|\xi| \leqslant M$, then

$$P\{|\xi| \geqslant c\} \geqslant \frac{E\{f(\xi)\} - f(c)}{f(M)} .$$

(*Kolmogorov's inequalities*)

21.4. Prove that for any r.v. ξ and for arbitrary $\varepsilon > 0$ and $r > 0$, the following inequalities hold:

$$E\left\{\frac{|\xi|^r}{1 + |\xi|^r}\right\} - \frac{\varepsilon^r}{1 + \varepsilon^r} \leqslant P\{|\xi| \geqslant \varepsilon\} \leqslant \frac{1 + \varepsilon^r}{\varepsilon^r} E\left\{\frac{|\xi|^r}{1 + |\xi|^r}\right\} .$$

21.5. Let ξ be a r.v. for which $P\{\xi = 0\} = 1 - \sigma^2\varepsilon^{-2}$, $P\{\xi = \varepsilon\} = P\{\xi = -\varepsilon\} = \sigma^2/(2\varepsilon^2)$. Compare $P\{|\xi| \geqslant \varepsilon\}$ with its bound, given by the Chebyshev's inequality.

21.6. Let $\xi_1, \ldots, \xi_n$ be i.i.d. r.v.'s for which $P\{\xi_1 > 0\} = 1$ and $E\{(\log_b \xi_1)^2\} < \infty$ for some $b > 1$. Prove that for any $\varepsilon > 0$,

$$P\{b^{n(a-\varepsilon)} < \xi_1\xi_2 \dots \xi_n < b^{n(a+\varepsilon)}\} \geqslant 1 - \frac{1}{n\varepsilon^2} V\{\log_b \xi_1\},$$

where $a = E\{\log_b \xi_1\}$.

21.7. Let $\xi \in P(\lambda)$. Prove that: (a) $P\{\xi \geqslant 1\} \leqslant \lambda$; (b) $P\{\xi \geqslant 2\} \leqslant \frac{1}{2}\lambda^2$.

21.8. Consider the r.v.'s $\xi_1 \in P(\lambda_1)$ and $\xi_2 \in P(\lambda_2)$, where $\lambda_2 > \lambda_1$. Prove that for arbitrary non-negative integer n, we have $P\{\xi_1 \leqslant n\} > P\{\xi_2 \leqslant n\}$.

21.9. Let the r.v. ξ have a Γ-distribution with parameters $m + 1$ and 1 (see (15.10)). Show that $P\{0 < \xi < 2m + 2\} > m/(m + 1)$.

21.10. Let ξ be a r.v. and let $a = E(\xi)$, $\sigma^2 = V(\xi)$, $\gamma_r = E\{|\xi - a|^r\}$ and $\sigma = \gamma_1/\sigma$. Prove the following Chebyshev-type inequalities:
(a) $P\{|\xi - a| \geqslant \lambda\gamma_r^{1/r}\} \leqslant \lambda^{-r}$; (b) $P\{|\xi - a| \geqslant \lambda\sigma\} \leqslant \gamma_r(\sigma\lambda)^{-r}$;
(c) $P\{|\xi - a| \geqslant \lambda\} \geqslant \gamma_r\lambda^{-r}$; (d) $P\{\xi - a \leqslant \lambda\} \leqslant \sigma^2(\sigma^2 + \lambda^2)^{-1}$, if $\lambda \leqslant 0$, and $P\{\xi - a \leqslant \lambda\} \geqslant \lambda^2(\sigma^2 + \lambda^2)^{-1}$, if $\lambda > 0$ (*Cantelli's inequality*);
(e) $P\{|\xi - a| \geqslant \lambda\sigma\} \leqslant (1 - \sigma^2)(\lambda^2 - 2\lambda\sigma + 1)^{-1}$ if $\lambda \geqslant \sigma$ (*Pick's inequality*).

21.11. Prove that for arbitrary r.v.'s ξ and η, we have

$$P\{\xi + \eta < x + y\} \leqslant P\{\xi < x\} + P\{\eta < y\}, \quad x, y \in \mathbb{R}_1.$$

21.12. Let $F(x_1, \dots, x_n)$, $(x_1, \dots, x_n) \in \mathbb{R}_n$, be any n-variate d.f. and let $F_1(x_1), \dots, F_n(x_n)$ be the corresponding marginal d.f.'s. Prove that

$$F(x_1, \dots, x_n) \leqslant (F_1(x_1) \dots F_n(x_n))^{1/n}.$$

(*H. Robbins*)

21.13. For the standard normal d.f. Φ, prove that

$$\Phi(-a - b) \leqslant 2\Phi(-a)\Phi(-b), \quad \text{where } a, b \geqslant 0.$$

(*K. G. Esseen*)

21.14. Let $\xi \in N(0, 1)$. Show that for any $x > 0$,

$$\frac{1}{1 + x^2} \exp(-x^2/2) \leqslant \sqrt{2\pi} P\{\xi \geqslant x\} \leqslant \frac{1}{x} \exp(-x^2/2).$$

21.15. Let $\xi \in N(0, \sigma^2)$. For arbitrary $x > 0$ and $c > 0$, show that

$$P\{\xi - x > \frac{c}{x} \mid \xi > x\} < e^{-c/\sigma^2}.$$

*21.16. Let $\xi \in N(0, 1)$. Then prove that for arbitrary real numbers

$c > 0$ and x the following inequality holds:

$$|P\{\xi < x\} - \vec{P}\{\xi < cx\}| \leqslant (2\pi e)^{-\frac{1}{2}}|c - 1|\max[1, c^{-1}].$$

21.17. Let $\xi_1, \xi_2, \ldots$ be independent and identically $N(0, 1)$-distributed r.v.'s. Find the minimum value of n for which $P\{\max[|\xi_1|, \ldots, |\xi_n|] \geqslant 2\} \geqslant \frac{1}{2}$.

21.18. Consider a r.v. ξ with density $f(x)$, $x \in \mathbb{R}_1$, and a finite mean $a = E(\xi)$. Let $H(x) = -\int_{-\infty}^{x} (u - a)f(u)du$. (Obviously $H(x) \geqslant 0$, $H(-\infty) = H(\infty) = 0$). Prove that if for some $c > 0$ and for any $x \in \mathbb{R}_1$, $H(x) \leqslant cf(x)$, then for any smooth function $g(x)$, $x \in \mathbb{R}_1$, one has

$$V\{g(\xi)\} \leqslant cE\{[g'(\xi)]^2\}.$$

(*A. Borovkov, S. Utev*)

21.19. Let ξ be a non-negative integer r.v. $P\{\xi = k\} = p_k$, $k = 0, 1, \ldots$ Let the function $g(k)$, defined for $k = 0, 1, \ldots$, be such that $V\{g(\xi)\} < \infty$. Denote $\Delta g(k) = g(k + 1) - g(k)$. Prove that

$$V\{g(\xi)\} \leqslant \sum_{k=0}^{\infty} [\Delta g(k)]^2 \sum_{j=k+1}^{\infty} jp_j.$$

(*T. Cacoullos*)

21.20. Let $\xi \in N(0, 1)$ and let g be a smooth function with $E\{|g'(\xi)|\} < \infty$. Then prove that

$$V\{g(\xi)\} \leqslant E\{[g'(\xi)]^2\}.$$

(*H. Chernoff*)

*21.21. Let ξ and η be r.v.'s for which the moments below exist. Prove the following relations:

$$(E\{|\xi\eta|\})^2 \leqslant E\{\xi^2\}E\{\eta^2\} \qquad \text{(a)}$$

(*Cauchy-Bunyakovski-Schwarz inequality*);

(b) If $g(x)$, $x \in \mathbb{R}_1$, is a continuous and convex downwards function, for which $E(\xi)$ and $E\{g(\xi)\}$ exist, then

$$g(E(\xi)) \leqslant \vec{E}\{g(\xi)\}$$

(*Jensen's inequality*);

(c) $\ln(E\{|\xi|^r\}^{1/r}$, $r > 0$, is downwards convex function of r;

(d) $(E\{|\xi|^r\}^{1/r}$, $r > 0$, is non-decreasing function of r. In particular, for $0 < r < s$, we have

$$(E\{|\xi|^r\})^{1/r} \leqslant (E\{|\xi|^s\})^{1/s}$$

(*Lyapunov's inequality*);

$$E\{|\xi\eta|\} \leqslant (E\{|\xi|^r\})^{1/r}(E\{|\xi|^s\})^{1/s}, \ r \geqslant 1, \ r^{-1} + s^{-1} = 1, \quad \text{(e)}$$

(*Hölder's inequality*);

$$(E\{|\xi + \eta|^r\}^{1/r} \leqslant (E\{|\xi|^r\}^{1/r} + (E\{|\eta|^r\})^{1/r}, \ r \geqslant 1, \quad \text{(f)}$$

(*Minkovski's inequality*);

$$E\{|\xi + \eta|^r\} \leqslant c_r E\{|\xi|^r\} + c_r E\{|\eta|^r\}, \quad \text{(g)}$$

where $c_r = 1$ for $r \leqslant 1$ and $c_r = 2^{r-1}$ for $r > 1$.

21.22. Let ξ and η be uniformly distributed r.v.'s over the interval (0, 1). Show that, whatever the dependence between ξ and η, the following inequality holds:

$$E\{|\xi - \eta|\} \leqslant 1/2.$$

21.23. Let $\xi_1, \ldots, \xi_n$ be r.v.'s with zero means and unit variances. Show that if these r.v.'s are equally correlated; i.e., if $\rho(\xi_1, \xi_j) = c$ for arbitrary $i, j = 1, \ldots, n$, $i \neq j$, then $c \geqslant -1/(n - 1)$.

21.24. Let F be any d.f., with corresponding ch.f. ϕ and let $c > 0$ be arbitrarily chosen. Then

$$\int_{-c}^{c} x^2 \, dF(x) \leqslant 3t^{-2}|1 - \phi(t)|, \quad \text{for } t \in [-\tfrac{1}{c}, \tfrac{1}{c}].$$

21.25. Let ξ and η be given r.v.'s with finite second moments and let η be symmetrically distributed (see Exercise 15.20). Show that $E\{|\xi + \eta|^r\} \leqslant E\{|\xi|^r\} + E\{|\eta|^r\}$ for arbitrary $r \in [1, 2]$.

21.26. Prove that, for arbitrary r.v.'s $\xi_1, \ldots, \xi_n \in \mathbf{L}_r$, $r \geqslant 1$, the following inequality holds:

$$E\{|\xi_1 + \ldots + \xi_n|^r\} \leqslant n^{r-1}(E\{|\xi_1|^r\} + \ldots + E\{|\xi_n|^r\}).$$

*21.27. (a) Let $\xi_1, \ldots, \xi_n$ be independent r.v.'s with $E(\xi_i) = 0$, $E\{\xi_i^2\} = \sigma_i^2 < \infty$, $i = 1, \ldots, n$, and let $S_n = \xi_1 + \ldots + \xi_n$. Then for arbitrary $\varepsilon > 0$, we have

$$P\{\max_{1 \leqslant k \leqslant n} |S_k| \geqslant \varepsilon\} \leqslant \frac{1}{\varepsilon^2} E\{S_n^2\}.$$

(b) Let ξ_i be defined as in (a). Assume further that all ξ_i are uniformly bounded by some constant c; i.e., $P\{|\xi_i| \leqslant c\} = 1$, $i \leqslant n$. Then prove that

$$P\{\max_{1\leqslant k\leqslant n} |S_k| \geqslant \varepsilon\} \geqslant 1 - \frac{(c+\varepsilon)^2}{E\{S_n^2\}}$$

(*Kolmogorov's inequalities*).

21.28. Suppose $\xi_1, \xi_2, \ldots$ are independent r.v.'s with $E(\xi_i) = 0$ and $V(\xi_i) = \sigma_i^2$, $i = 1, 2, \ldots$ Prove that if $c_1, c_2, \ldots$ is a non-decreasing sequence of positive numbers, then for arbitrary naturals m and n with $m < n$ and for arbitrary $\varepsilon > 0$, we have

$$P\{\max_{m\leqslant k\leqslant n} c_k|\xi_1 + \ldots + \xi_k| \geqslant \varepsilon\} \leqslant \frac{1}{\varepsilon^2}\left[c_m^2 \sum_{i=1}^{m} \sigma_i^2 + \sum_{i=m+1}^{n} c_i^2\sigma_i^2\right]$$

(*J. Hájek, A. Rényi*).

21.29. Let $A_1, \ldots, A_n$ be arbitrary events. If $P(A_1) = p_1, \ldots, P(A_n) = p_n$, then show that

$$P\left(\bigcap_{i=1}^{n} A_i\right) \geqslant 1 - \sum_{i=1}^{n} (1 - p_i).$$

(*S. Bonferoni*)

21.30. Consider an n-dimensional random vector $(\xi_1, \ldots, \xi_n)$ and let $B_1, \ldots, B_n$ be arbitrary Borel sets in $\mathbb{R}_1$. Let $C_i = \{\xi_i \bar{\in} B_i\}$, $q_i = P(C_i)$, $Q_1 = \sum_{i=1}^{n} q_i$, $q_{ij} = P(C_iC_j)$, $i, j = 1, \ldots, n$, $Q_2 = \sum_{i<j} q_{ij}$. Then the following inequalities hold:

$$1 - Q_1 \leqslant P(\bigcap_{i=1}^{n} \{\xi_i \in B_i\}) \leqslant 1 - \frac{Q_1^2}{Q_1 + 2Q_2}.$$

(*K. L. Chung, P. Erdös*)

21.31. For arbitrary events $A_1, \ldots, A_n$, show that

$$P(A_1 \ldots A_n) \geqslant \max\left[0, \sum_{i=1}^{n} P(A_i) - n + 1\right].$$

(*D. Galin*)

21.32. Let ξ be a r.v. with mean a and variance σ^2, $0 < \sigma^2 < \infty$. If the numbers x_1 and x_2 are such that $x_1 < a < x_2$ and if $\sigma^2 < (a - x_1)(x_2 - a)$, then

$$P\{x_1 < \xi < x_2\} \geqslant 4[(a - x_1)(x_2 - a) - \sigma^2]/(x_2 - x_1)^2.$$

(*K. Ferentinos*)

21.33. Let ξ_1 and ξ_2 be independent and $N(0, 1)$-distributed r.v.'s and let $\xi_3 = (\xi_1 + \xi_2)/\sqrt{2}$. For arbitrary $x > 0$ and sufficiently small $\varepsilon > 0$, prove that

$$P\{|\xi_1| \geqslant x, |\xi_2| \geqslant x, |\xi_3| \leqslant \varepsilon\} <$$

$$< P\{|\xi_1| \geqslant x\}P\}|\xi_2| \geqslant x\}P\{|\xi_3| \geqslant \varepsilon\}$$

(*Zb. Šidak*)

*21.34. Let $\xi_1, \xi_2, \ldots$ be a sequence of independent r.v.'s and let $S_k = \xi_1 + \ldots + \xi_k$, $k = 1, 2, \ldots$ If for every k it holds that $P\{|S_n - S_k| \leqslant x\} \geqslant \alpha$, then prove that

$$P\{\max_{1\leqslant k\leqslant n} |S_k| > 2x\} \leqslant \frac{1 - \alpha}{\alpha}.$$

(*A. Skhorohod*)

21.35. Let ξ be a non-negative r.v. and let $E\{\xi^k\} < \infty$ for some $k > 2$. Show that, for arbitrary $r \in [2, k)$, we have

$$E\{\xi^k\} \geqslant \left(E\left\{\xi^{\frac{k}{r}}\right\}\right)^r \geqslant (E(\xi))^k + (E\{\xi^{k/r}\} - (E(\xi))^{k/r})^r.$$

In particular, for $r = 2$: $E\{\xi^k\} \geqslant (E(\xi))^k + (V(\xi))^{k/2}$. (*Y. Tong*)

21.36. Let $\xi_1, \ldots, \xi_n$ be i.i.d. r.v.'s with $E(\xi_1) = 0$, $V(\xi_1) = 1$ and let $S_k = \xi_1 + \ldots + \xi_k$. Then

$$P\{\max_{1\leqslant k\leqslant n} S_k > x\} \leqslant \frac{4}{3} P\{S_n > x - 2\sqrt{n}\}.$$

21.37. Let $(\xi_n, \mathcal{F}_n)$, $n = 1, 2, \ldots$ be any submartingale. Then for arbitrary $\varepsilon > 0$, show that the following inequality holds:

$$P\{\max_{1\leqslant k\leqslant n} \xi_n \geqslant \varepsilon\} \leqslant \frac{1}{\varepsilon} E\{|\xi_n|\}.$$

(*J. L. Doob*)

Chapter 4

LIMIT THEOREMS

22. Types of Convergence for Sequences of Random Variables

Introductory Notes

Let $\{\xi_n\}$, $n = 1, 2, \ldots$ be a sequence of random variables (r.v.'s). We shall be concerned with their limiting behaviour when $n \to \infty$.

We say that:

1. $\{\xi_n\}$ converges *almost surely* (a.s.) to the r.v. ξ, or equivalently that ξ_n converges to ξ *with probability 1* ($\xi_n \xrightarrow{a.s.} \xi$, $\xi_n \to \xi$ a.s.), if

$$P\{\omega : \lim_{n\to\infty} \xi_n(\omega) = \xi(\omega)\} = 1. \tag{22.1}$$

2. $\{\xi_n\}$ converges to ξ in *r-th mean*, or *in* L_r*- sense*, $r > 0$ ($\xi_n \xrightarrow{L_r} \xi$) if $\xi, \xi_n \in \mathbb{L}_r$, $n = 1, 2, \ldots$ and

$$\lim_{n\to\infty} E\{|\xi_n - \xi|^r\} = 0 \tag{22.2}$$

($\mathbb{L}_r$ denotes the space of all r.v.'s ξ with $E\{|\xi|^r\} < \infty$).

The convergence in r-th mean for $r = 2$ plays a particular important role in probability theory and statistics. It is called *convergence in mean square* or convergence in the mean of order two. The corresponding notation is $\xi_n \xrightarrow{m.sq.} \xi$ or $\underset{n\to\infty}{\mathrm{l.i.m.}}\, \xi_n = \xi$.

3. $\{\xi_n\}$ converges to ξ *in probability* ($\xi_n \xrightarrow{P} \xi$ or $\xi = \underset{n\to\infty}{P\text{-}\lim}\, \xi_n$) if for any $\varepsilon > 0$,

$$\lim_{n\to\infty} P\{\omega : |\xi_n(\omega) - \xi(\omega)| \geqslant \varepsilon\} = 0. \tag{22.3}$$

4. ξ_n, with d.f. $F_n(x)$, $x \in \mathbb{R}_1$, converges to ξ, with d.f. $F(x)$, $x \in \mathbb{R}_1$, *in distribution* ($\xi_n \xrightarrow{d} \xi$, $F_n \xrightarrow{d} F$) if

$$\lim_{n\to\infty} F_n(x) = F(x) \tag{22.4}$$

for every point x at which $F(x)$ is continuous.

It turns out that if ξ and ξ_n assume their values in $\mathbb{R}_1$, then (22.4) is equivalent to the following condition:

$$\lim_{n\to\infty} \int_{-\infty}^{\infty} g(x)\,dF_n(x) = \int_{-\infty}^{\infty} g(x)\,dF(x) \tag{22.5}$$

for every bounded and continuous function $g(x)$, $x \in \mathbb{R}_1$.

We note that (22.5) defines the so-called *weak convergence*.

In some solutions the following lemma will be used (see also Exercise 11.26):

Borel-Cantelli lemma. Let $A_1, A_2, \ldots$ be an arbitrary sequence of events. Put $A^* = \limsup_n A_n$ (see Section 11).

(a) If the events $\{A_n\}$ are mutually independent, then $P(A^*) = 0$ or $P(A^*) = 1$ depending on whether $\sum_{n=1}^{\infty} P(A_n)$ is finite or infinite.

(b) If $\sum_{n=1}^{\infty} P(A_n) < \infty$, then $P(A^*) = 0$ for an arbitrary sequence of events $A_1, A_2, \ldots$

Illustrative Examples

The next example clarifies the relation between the a.s. convergence (convergence with probability 1) and the convergence in probability.

Example 22.1. Show that a.s. convergence implies convergence in probability; i.e., if $\xi_n \xrightarrow{a.s.} \xi$, then $\xi_n \xrightarrow{P} \xi$ as $n \to \infty$.

Solution. One can write the event $A = \{\omega : \lim_{n\to\infty} \xi_n(\omega) = \xi(\omega)\}$ in the form:

$$A = \bigcap_{k=1}^{\infty} \bigcup_{N=1}^{\infty} \bigcap_{n=N}^{\infty} [|\xi_n - \xi| < \tfrac{1}{k}], \quad P(A) = 1.$$

Obviously $P\left\{\bigcup_{N=1}^{\infty} \bigcap_{n=N}^{\infty} [|\xi_n - \xi| < \varepsilon]\right\} = 1$ for any $\varepsilon > 0$. It is also evident that events $\bigcap_{n=N}^{\infty} [|\xi_n - \xi| < \varepsilon]$ monotonically increase as N increases. From the Continuity axiom (see Section 11) it follows that $\lim_{N\to\infty} P\left\{\bigcap_{n=N}^{\infty} [|\xi_n - \xi| < \varepsilon]\right\} = 1$. The following inclusion is obvious: $[|\xi_N - \xi| < \varepsilon] \subset \bigcap_{n=N}^{\infty} [|\xi_n - \xi| < \varepsilon]$, and it then follows that

$$\lim_{N\to\infty} P\{|\xi_N - \xi| < \varepsilon\} = 1.$$

This relation means that $\xi_N \xrightarrow{P} \xi$ as $N \to \infty$.

Example 22.2. Consider the sequence $\{\xi_n, n = 1, 2, \ldots\}$, where ξ_n is an absolutely continuous r.v. with density

$$f_n(x) = \frac{n}{\pi} \frac{1}{1 + n^2x^2}, \quad x \in \mathbb{R}_1.$$

Prove that $\{\xi_n\}$ is convergent in probability but not in L_r-sense for any $r \geqslant 1$.

Solution. For an arbitrary $\varepsilon > 0$ we have the relations:

$$P\{|\xi_n| \geqslant \varepsilon\} = 1 - P\{|\xi_n| < \varepsilon\} =$$

$$= 1 - \int_{-\varepsilon}^{\varepsilon} f_n(x)\,dx = 1 - \int_{-\varepsilon}^{\varepsilon} \frac{n}{\pi} \frac{1}{1 + n^2x^2}\,dx =$$

$$= 1 - \frac{2}{\pi} \operatorname{arc\,tg}(n\varepsilon) \to 0 \quad \text{as } n \to \infty.$$

Hence $\xi_n \xrightarrow{P} 0$ as $n \to \infty$.

Further, it is clear that ξ_n, as defined, has a Cauchy distribution. But we know that this distribution has no mean value; i.e., for each $n \geqslant 1$, $E\{|\xi_n|\} = \infty$. Then the Lyapunov inequality (see Exercise 21.21) implies that $E\{|\xi_n|^r\} = \infty$ for any $r \geqslant 1$. Thus the variables ξ_n do not belong to the space $\mathbf{L}_r$ of the r-integrable r.v.'s. The conclusion is obvious: $\xi_n \overset{L_r}{\not\to} 0$ as $n \to \infty$ despite the fact that $\xi_n \xrightarrow{P} 0$.

It will be useful for the reader to clarify what happens if $0 < r < 1$.

Note that in this example the variables ξ_n are absolutely continuous. Another sequence of discrete r.v.'s obeying similar properties (converge in probability but not in L_r-sense) is considered in Exercise 22.13.

Exercises

22.1. Let a non-negative r.v. f_n^m correspond to every pair of natural numbers m and n and let $f_n^m \leqslant f_l^m + f_n^l$ for $m \leqslant l \leqslant n$.

Suppose that $f_n^m \xrightarrow{P} 0$ as $m \to \infty$ and $n \to \infty$. Prove that every sequence of natural numbers converging to ∞, contains a subsequence $\{m_k\}$, $m_k \to \infty$ as $k \to \infty$ such that $f_{m_l}^{m_k} \xrightarrow{a.s.} 0$ as $k \to \infty$ and $l \to \infty$.

22.2. Let $\xi_n \xrightarrow{P} \xi$. Prove the existence of a subsequence $\{\xi_{n_k}\}$ for which $\xi_{n_k} \xrightarrow{a.s.} \xi$.

22.3. Let $\{\xi_n\}$ be a sequence of r.v.'s and let $\varepsilon > 0$ be arbitrary chosen. Prove that: (a) $\{\xi_n\}$ is convergent a.s. if and only if

$$\lim_{k\to\infty} P\{\sup_{n\geqslant k} |\xi_n - \xi_k| \geqslant \varepsilon\} = 0;$$

(b) $\{\xi_n\}$ is convergent in probability if and only if

$$\lim_{k\to\infty} \sup_{n\geqslant k} P\{|\xi_n - \xi_k| \geqslant \varepsilon\} = 0;$$

(c) if $\xi_n \in \mathbf{L}_r$, $r > 0$, then $\{\xi_n\}$ is convergent in $\mathbf{L}_r$-sense if and only if

$$\lim_{n,k\to\infty} E\{|\xi_n - \xi_k|^r\} = 0.$$

22.4. Let us assume that $\xi, \eta, \xi_n, \eta_n \in \mathbf{L}_2$, $n = 1, 2, \ldots$ and $\xi_n \xrightarrow{L_2} \xi$, $\eta_m \xrightarrow{L_1} \eta$ as $n \to \infty$ and $m \to \infty$. Show that: (a) $\xi_n \xrightarrow{L_1} \xi$; (b) $\xi_n\eta_m \xrightarrow{L_4'} \xi\eta$ as $n, m \to \infty$; (c) $E(\xi_n) \to E(\xi)$; (d) $E\{\xi_n^2\} \to E\{\xi^2\}$.

22.5. If $\xi, \xi_n \in \mathbf{L}_r$, $r > 0$, and $\xi_n \xrightarrow{L_1} \xi$, then $\xi_n \xrightarrow{P} \xi$.

22.6. The sequence $\{\xi_n\}$ is such that $\xi_n - \xi_m$ has a Cauchy distribution with density $\dfrac{1}{\pi\sigma_{mn}} \dfrac{1}{1 + x^2/\sigma_{mn}}$, where $\sigma_{mn}^2 = \dfrac{|m - n|}{mn}$. Does the sequence $\{\xi_n\}$ converge in probability?

22.7. Prove that $\xi_n \xrightarrow{P} \xi$ if and only if, for some $r > 0$, we have $E\{|\xi_n - \xi|^r/(1 + |\xi_n - \xi_m|^r)\} \to 0$ as $n \to \infty$.

22.8. If $\xi_n \xrightarrow{P} \xi$ and if in addition $\{\xi_n\}$ is monotonic, then $\xi_n \xrightarrow{a.s.} \xi$.

22.9. Let $\xi_n \xrightarrow{d} \xi$, where $P\{\xi = c \text{ (constant)}\} = 1$. Then $\xi_n \xrightarrow{P} \xi$.

*22.10. Let $\{\xi_n\}$ and $\{\eta_n\}$ be sequences of r.v.'s. Prove that: (a) if $|\xi_n - \eta_n| \xrightarrow{P} 0$ and $\eta_n \xrightarrow{d} \eta$, then $\xi_n \xrightarrow{d} \eta$; (b) if $\xi_n \xrightarrow{d} \xi$ and $\eta_n \xrightarrow{P} c$, then $\xi_n + \eta_n \xrightarrow{d} \xi + c$, $\xi_n\eta_n \xrightarrow{d} \xi c$, $\xi_n/\eta_n \xrightarrow{d} \xi/c$ $(c \neq 0)$; (c) if $\xi_n \xrightarrow{d} \xi$ and $\eta_n \xrightarrow{P} 0$, then $\xi_n\eta_n \xrightarrow{P} 0$. (*H. Cramér*)

22.11. Let the function $g(x)$, $x \in \mathbb{R}_1$, be continuous and bounded and let $\{\xi_n\}$ be an arbitrary sequence of r.v.'s. Show that: (a) if $\xi_n \xrightarrow{d} \xi$, then $g(\xi_n) \xrightarrow{d} g(\xi)$; (b) if $\xi_n \xrightarrow{P} \xi$, then $g(\xi_n) \xrightarrow{P} g(\xi)$. (*E. Slutsky*)

22.12. Let $\{\xi_n\}$ be a sequence of r.v.'s for which, with probability

1, $|\xi_n| \leq c < \infty$, c = constant $< \infty$. Prove that $\xi_n \xrightarrow{P} 0$ if and only if $\lim_{n\to\infty} E\{|\xi_n|\} = 0$.

22.13. For every integer $n \geq 1$ a r.v. ξ_n is given, which assumes the values $n^{\alpha/r}$ and 0 with probability $\frac{1}{n}$ and $1 - \frac{1}{n}$, respectively ($\alpha > 1$, $r > 0$). Prove that $\xi_n \xrightarrow{P} 0$ but ξ_n does not converge to 0 in L_r-sense.

22.14. Prove that convergence in probability implies convergence in distribution. The converse is in general not true. Find a counter-example.

22.15. Let F_n be the d.f. and let ϕ_n be the ch.f. of the r.v. ξ_n, $n = 1, 2, \ldots$ Prove the equivalence of the following three conditions: (a) $\xi_n \xrightarrow{P} c$, c = constant; (b) $F_n(x) \to 0$ for $x \leq c$ and $F_n(x) \to 1$ for $x > c$; (c) $\lim_{n\to\infty} \phi_n(t) = e^{itc}$ for every $t \in \mathbb{R}_1$.

22.16. Let the r.v.'s ξ_n, $n = 1, 2, \ldots$, be such that $P\{\xi_n = \pm\frac{1}{n}\} = \frac{1}{2}$. Show that $\{\xi_n\}$ is convergent in distribution, in probability, a.s., and in L_r-sense.

22.17. Let $\{\xi_n\}$ be a sequence of r.v.'s for which $P\{\xi_n = -n - 4\} = 1/(n + 4)$, $P\{\xi_n = -1\} = 1 - 4/(n + 4)$, $P\{\xi_n = n + 4\} = 3/(n + 4)$. Show that $\{\xi_n\}$ is convergent in probability but $E\{P\text{-}\lim_{n\to\infty} \xi_n\} \neq \lim_{n\to\infty} E(\xi_n)$.

22.18. Let ξ_n, $n = 1, 2, \ldots$, be r.v.'s with the same variance σ^2 and non-positive correlation coefficient $\rho(\xi_i, \xi_j)$, $i \neq j$. Show that $\frac{1}{n}\sum_{k=1}^{n} (\xi_k - E(\xi_k)) \xrightarrow{L_2} 0$ as $n \to \infty$.

22.19. Let $\xi_1, \xi_2 \ldots$ be independent r.v.'s which are uniformly distributed over the interval (0, 1). Denote $\eta_n = \max[\xi_1, \ldots, \xi_n]$ and $\zeta_n = n(1 - \eta_n)$. Prove that $\zeta_n \xrightarrow{d} \zeta$ as $n \to \infty$, where $\zeta \in E(1)$.

22.20. Let $\eta_n = \max[\xi_1, \ldots, \xi_n]$, where $\xi_1, \ldots, \xi_n$ are i.i.d. r.v.'s with d.f. $F(x)$, $x \in \mathbb{R}_1$. Assume that $\lim_{x\to\infty} e^x(1 - F(x)) = b > 0$. Take $\zeta_n = \eta_n - \ln(nb)$ and then prove that $\zeta_n \xrightarrow{d} \zeta$, where ζ is a r.v. whose d.f. is $G(x) = \exp(-e^{-x})$, $x \in \mathbb{R}_1$.

22.21. Let $\xi_1, \xi_2, \ldots$ be i.i.d. r.v.'s with d.f. F, given by $F(x) = 1 - x^{-\alpha}$ for $x > 1$ and $F(x) = 0$ for $x \leq 1$, with $\alpha > 0$. Let $\eta_n = \max[\xi_1, \ldots, \xi_n]$. Show that $n^{-1/\alpha}\eta_n$ is weakly convergent to some r.v. ζ as $n \to \infty$ and find the distribution of ζ.

22.22. Let $\xi_n \xrightarrow{P} \xi$ and for some $r > 0$ let $|\xi_n|^r \leq \eta$, where $E(\eta) < \infty$. Show that $\xi_n \xrightarrow{L_r} \xi$.

22.23. Let the r.v.'s $\xi_1, \ldots, \xi_n$ be independent and uniformly distributed over the interval (0, 1) and let $\xi_{(k)}$ be the corresponding kth order statistic (see Exercise 16.42). Find the asymptotic behaviour of the r.v. $n\xi_{(k)}$ for fixed k and $n \to \infty$.

22.24. Let ξ and $\eta_\sigma \in N(0, \sigma^2)$ be independent r.v.'s and let F and Φ_σ be their d.f.'s, respectively. Denote by F_σ the convolution of F and Φ_σ: $F_\sigma = F * \Phi_\sigma$. Prove that $F_\sigma \xrightarrow{d} F$ as $\sigma \to 0$.

22.25. Let $\xi_1, \xi_2, \ldots$ be a sequence of independent r.v.'s with $E(\xi_n) = 0$, $V(\xi_n) = \sigma_n^2$ and $\sum_{n=1}^{\infty} \sigma_n^2 < \infty$. Prove that the series $\sum_{n=1}^{\infty} \xi_n$ is convergent: (a) with probability 1; (b) in mean square.

22.26. The sequence of r.v.'s $\{\xi_n\}$ satisfies the following condition: $\sum_{n=1}^{\infty} P\{|\xi_n| \geqslant c\} \leqslant \infty$ for every $c > 0$. Show that $\xi_n \to 0$ with probability 1 as $n \to \infty$.

*22.27. Let $\sum_{n=1}^{\infty} \xi_n$ be a series of independent r.v.'s. In order that it convergences with probability 1, both $\sum_{n=1}^{\infty} E(\xi_n)$ and $\sum_{n=1}^{\infty} V(\xi_n)$ must be convergent. If ξ_n, n = 1, 2, ... are uniformly bounded; i.e., $P\{|\xi_n| \leqslant c\} = 1$, then this condition is also necessary. (This exercise is a special case of the *Kolmogorov theorem of the three series*, see the references [17], [19], [32], [37] cited at the end of this Manual.)

22.28. The r.v.'s ξ_n, n = 1, 2, ... are independent and $P\{\xi_n = 1\} = p_n = 1 - P\{\xi_n = 0\}$. Develop conditions for the probabilities p_n, n = 1, 2, ... which imply the convergence a.s. of the series $\sum_{n=1}^{\infty} (\xi_n - p_n)$.

22.29. Let $\{\xi_n\}$ be a sequence of independent r.v.'s with $P\{\xi_n = 1\} = p_n = 1 - P\{\xi_n = 0\}$. Choose the probabilities p_n in such a way that $\{\xi_n\}$ will be convergent in L_r-sense but not with probability 1.

*22.30. Let us assume that the functions f(x) and g(x) are continuous on the closed interval [0, 1] and satisfy the relation $0 \leqslant f(x) \leqslant cg(x)$ for some $c > 0$. Prove that

$$\lim_{n\to\infty} \int_0^1\int_0^1 \cdots \int_0^1 \frac{f(x_1) + f(x_2) + \ldots + f(x_n)}{g(x_1) + g(x_2) + \ldots + g(x_n)} dx_1\, dx_2 \ldots dx_n = \frac{\int_0^1 f(x)dx}{\int_0^1 g(x)dx}.$$

22.31. Let ξ and η be independent r.v.'s, $\xi, \eta \in E(1)$ and let $\Theta =$

$\xi - \eta$. Let $\Theta_1, \Theta_2, \ldots$ be a sequence of independent copies of Θ. Prove that $\sum_{n=1}^{\infty} (\frac{1}{n}\Theta_n)$ is convergent a.s.

22.32. Let $\xi_1, \xi_2, \ldots$ be independent r.v.'s, which are uniformly distributed over the interval (0, 1) and let $\eta_n = \max[\xi_1, \ldots, \xi_n]$. Let $g(x)$, $x \in (0, \infty)$, be any increasing and differentiable function and let g^{-1} be its inverse function. Put $\zeta_n = g(n(1 - \eta_n))$. Prove that when $n \to \infty$, then $\zeta_n \xrightarrow{d} \zeta$, where ζ is a r.v. with density

$$f_\zeta(x) = \begin{cases} \exp[-g^{-1}(x)]\, \frac{d}{dx} g^{-1}(x), & \text{if } x \in (g(0), g(\infty)); \\ 0, & \text{otherwise.} \end{cases}$$

*22.33. The r.v.'s $\xi_1, \xi_2, \ldots$ are independent and such that $S_n = \xi_1 + \ldots + \xi_n$ converges in probability to some r.v. S as $n \to \infty$. Prove that the series $\sum_{n=1}^{\infty} \xi_n$ is convergent a.s.

*22.34. Let $\xi_1, \xi_2, \ldots$ be a sequence of independent identically distributed r.v.'s for which $E(\xi_1)$ exists. For some fixed $c > 0$ denote $\eta_n = \max[\xi_1, \ldots, \xi_n] - cn$, $n = 1, 2, \ldots$, and $\zeta = \sup_n \eta_n$. Prove that (a) $P\{\zeta < \infty\} = 1$; (b) $P\{\lim_{n\to\infty} \eta_n = -\infty\} = 1$. (*M. DeGroot*).

23. Laws of Large Numbers

Introductory Notes

Assume that all r.v.'s in the sequence $\{\xi_n\}$ are defined on some fixed probability space $(\Omega, \mathcal{F}, P)$ and introduce the following notations:

$$a_k = E(\xi_k), \quad S_n = \xi_1 + \ldots + \xi_n, \quad A_n = a_1 + \ldots + a_n.$$

If $\frac{1}{n}(S_n - A_n) \xrightarrow{P} 0$ as $n \to \infty$; i.e., for every $\varepsilon > 0$,

$$\lim_{n\to\infty} P\left\{ \left| \frac{\xi_1 + \ldots + \xi_n}{n} - \frac{a_1 + \ldots + a_n}{n} \right| < \varepsilon \right\} = 1, \tag{23.1}$$

then we say that $\{\xi_n\}$ obeys the *Weak law of large numbers* (WLLN).

If $\frac{1}{n}(S_n - A_n) \xrightarrow{a.s.} 0$ as $n \to \infty$; i.e., if

$$P\left\{\lim_{n\to\infty}\left(\frac{\xi_1 + \dots + \xi_n}{n} - \frac{a_1 + \dots + a_n}{n}\right) = 0\right\} = 1, \tag{23.2}$$

then we say that $\{\xi_n\}$ obeys the *Strong law of large numbers* (SLLN).

It follows from Example 22.1 that if $\{\xi_n\}$ obeys the SLLN then it also obeys the WLLN.

For the sequence $\{\xi_n\}$ we distinguish the following two cases:

First case: $\xi_1, \xi_2, \dots$ are identically distributed.

Khintchine's theorem. Let $\{\xi_n\}$ be a sequence of independent and identically distributed r.v.'s with $E\{|\xi_1|\} < \infty$. Then $\{\xi_n\}$ obeys the WLLN; i.e., $\frac{1}{n} S_n \xrightarrow{P} a$ as $n \to \infty$, where $a = E(\xi_1)$. □

Kolmogorov's theorem. Let $\{\xi_n\}$ be a sequence of independent identically distributed r.v.'s. Then the existence of the mean $E\{|\xi_1|\}$ is a necessary and sufficient condition for $\{\xi_n\}$ to obey the SLLN; i.e., $\frac{1}{n} S_n \xrightarrow{a.s.} a$ as $n \to \infty$, where $a = E(\xi_1)$. □

Second case: $\xi_1, \xi_2, \dots$ are not identically distributed.

Chebyshev's theorem. Let $\{\xi_n\}$ be a sequence of pairwise independent r.v.'s with uniformly bounded variances; i.e., for arbitrary $n = 1, 2, \dots$ we have $V(\xi_n) \leqslant c < \infty$. Then $\{\xi_n\}$ obeys the WLLN. □

Markov's theorem. Let $\{\xi_n\}$ be a sequence of r.v.'s for which the following relation holds:

$$\lim_{n\to\infty} \frac{1}{n^2} V\{\xi_1 + \dots + \xi_n\} = 0. \tag{23.3}$$

Then $\{\xi_n\}$ obeys the WLLN. □

Kolmogorov's theorem. Let $\{\xi_n\}$ be a sequence of independent r.v.'s with finite variances $V(\xi_n) = \sigma_n^2$, $n = 1, 2, \dots$ If

$$\sum_{n=1}^{\infty} \frac{\sigma_n^2}{n^2} < \infty, \tag{23.4}$$

then $\{\xi_n\}$ obeys the SLLN. □

Illustrative Examples

Example 23.1. Let $\{\xi_n\}$ be a sequence of r.v.'s and let $f_n(x) = (\sqrt{\pi}\sqrt[4]{n})^{-1}\exp\left[-\frac{(x-c^n)^2}{\sqrt{n}}\right]$, $x \in \mathbb{R}_1$, be the density of ξ_n, $n = 1, 2, \ldots$, where $c \in (0, 1)$. Does the sequence $\{\xi_n\}$ obey: (a) the WLLN; (b) the SLLN?

Solution. It follows (see Section 15) that each r.v. ξ_n is normally distributed with parameters $E(\xi_n) = c^n$ and $V(\xi_n) = \frac{1}{2}\sqrt{n}$. Since $\xi_1, \xi_2, \ldots$ are not identically distributed, we have to check conditions (23.3) and (23.4).

(a) We have that

$$\frac{1}{n^2}V\{\xi_1 + \ldots + \xi_n\} = \frac{1}{n^2}\left(\frac{\sqrt{1}}{2} + \frac{\sqrt{2}}{2} + \ldots + \frac{\sqrt{n}}{2}\right) <$$

$$< \frac{1}{n^2}\frac{n\sqrt{n}}{2} = \frac{1}{2\sqrt{n}}.$$

Since $1/(2\sqrt{n}) \to 0$ as $n \to \infty$, from Markov's theorem we conclude that $\{\xi_n\}$ obeys the WLLN.

(b) We have

$$\sum_{n=1}^{\infty}\frac{\sigma_n^2}{n^2} = \frac{1}{2}\sum_{n=1}^{\infty}\frac{\sqrt{n}}{n^2} = \frac{1}{2}\sum_{n=1}^{\infty} n^{-\frac{3}{2}} < \infty.$$

Now we conclude from Kolmogorov's theorem that $\{\xi_n\}$ obeys the SLLN.

Note. Since in general the convergence with probability 1 implies convergence in probability (see Example 21.1), we conclude that the SLLN implies the WLLN.

Example 23.2. Let $\{\xi_n\}$ be a sequence of i.i.d. r.v.'s with $E(\xi_1) = a$ and $V(\xi_1) = \sigma^2 < \infty$. Put $\zeta_n = \sum_{i<j}^{n} \xi_i\xi_j$. Show that $\{\zeta_n\}$ obeys the WLLN in the following form:

$$\binom{n}{2}^{-1}\zeta_n \xrightarrow{P} a^2 \quad \text{as } n \to \infty.$$

Solution. It is easily seen that ζ_n is a sum of $\binom{n}{2}$ terms of the form $\xi_i\xi_j$. Because of the independence, the mean of each term is $E(\xi_i\xi_j) = E(\xi_i)E(\xi_j) = a^2$. Applying Chebyshev's inequality to the r.v. $\binom{n}{2}^{-1}\zeta_n$, we obtain the desired result.

Exercises

23.1. Let $\{\xi_n\}$, $\{\eta_n\}$, $\{\zeta_n\}$, $\{\Theta_n\}$ be sequences of independent r.v.'s, where: ξ_n assumes the values $\sqrt{n}$, $-\sqrt{n}$ and 0 with probabilities $\frac{1}{n}$, $\frac{1}{n^2}$ and $1 - \frac{2}{n^2}$; η_n assumes the values n and $-\frac{1}{n}$ with probabilities $1/(1 + n^2)$ and $n^2/(1 + n^2)$; ζ_n equals $\pm\sqrt{\ln n}$ each with probability $\frac{1}{2}$ for $n \geq 3$; Θ_n assumes the values 1, -1, 2^n and -2^n with probability $\frac{1}{2}(1 - \frac{1}{2^n})$, $\frac{1}{2}(1 - \frac{1}{2^n})$, $\frac{1}{2^{n+1}}$ and $\frac{1}{2^{n+1}}$, respectively, n = 1, 2, ... Do the above sequences obey the WLLN and the SLLN?

23.2. A sequence $\{\xi_n\}$ of i.i.d. r.v.'s is given and it is known that $E(\xi_1) = 0$ and $V(\xi_1) = 1$. Two new sequences $\{\eta_n\}$ and $\{\zeta_n\}$ are formed, where $\eta_n = a + \alpha^n\xi_n$ and $\zeta_n = a + n^\alpha\xi_n$ for some fixed $a \in \mathbb{R}_1$. (a) Find the values of α for which $\{\eta_n\}$ obeys the WLLN; the SLLN. (b) Do the same for the sequence $\{\zeta_n\}$.

23.3. The following three sequences of independent r.v.'s are given: $\{\xi_n\}$, $\{\eta_n\}$, $\{\zeta_n\}$. Here ξ_n has a χ^2-distribution with n degrees of freedom, $\eta_n \in P(\lambda_n)$ are such that $\frac{1}{N}(\lambda_1 + \ldots + \lambda_N) \to 0$ as $N \to \infty$ and $\zeta_n \in E(2^{-n/2})$. Do the above three sequences obey the WLLN?

23.4. Let $\{\xi_n\}$ be a sequence of i.i.d. r.v.'s with finite variance. Then (see Section 24) $\{\xi_n\}$ satisfies the Central limit theorem. Use this fact to show that $\{\xi_n\}$ obeys the WLLN.

23.5. Let $\{\xi_n\}$ be a sequence of r.v.'s with uniformly bounded variances, every pair of which is negatively correlated. (An example of such a sequence is given in Exercise 9.30.) Prove that $\{\xi_n\}$ obeys the WLLN.

23.6. Let $\{\xi_n\}$ be a sequence of r.v.'s with uniformly bounded variances. For each n, the variable ξ_n depends only on ξ_{n-1} and ξ_{n+1}. Then prove that $\{\xi_n\}$ obeys the WLLN.

*23.7. Suppose $\{\xi_n\}$ is a sequence of r.v.'s whose variances are uniformly bounded ($V(\xi_n) \leq c$ for all n and some $c < \infty$) and for the correlation coefficients we have $\rho(\xi_i, \xi_j) \to 0$ as $|i - j| \to \infty$. Show that $\{\xi_n\}$ obeys the WLLN. (*S. N. Bernstein*)

23.8. Consider the sequence $\{\xi_n\}$ of independent r.v.'s with $\frac{1}{n} V(\xi_n) \to 0$ as $n \to \infty$. Show that the WLLN is satisfied. (*A. Ya. Khintchine*)

23.9. Let $\{\xi_n\}$ be a sequence of independent r.v.'s such that ξ_n takes on the values n^β, $-n^\beta$ and 0 with probability $1/(2n^\alpha)$, $1/(2n^\alpha)$ and $1 - 1/n^\alpha$, respectively, where $\alpha > 0$ and $\beta > 0$ are fixed. For what relation between α and β, will the sequence $\{\xi_n\}$ obey: (a) the WLLN; (b) the SLLN?

*23.10. Let $f(x)$, $x \in [0, 1]$ be a continuous function and let

$$B_n(f; x) = \sum_{k=0}^{n} f(\tfrac{k}{n}) \binom{n}{k} x^k (1 - x)^{n-k}$$

be the nth *Bernstein polynomial* for the function f. Show that $B_n(f, x) \to f(x)$ uniformly w.r.t. $x \in [0, 1]$ as $n \to \infty$.

23.11. Let $f(x)$, $x \in [0, 1]$, be a continuous function such that f(0) and f(1) both are integers. Show that one can approximate f(x) uniformly by polynomials, whose coefficients are integer numbers.

23.12. Let $\xi_1, \xi_2, \ldots$ be i.i.d. r.v.'s with ch.f. $\phi(t)$, which is differentiable at $t = 0$ and such that $\phi'(0) = ia$, where $i = \sqrt{-1}$, $a \in \mathbb{R}_1$. Show that $\frac{1}{n}(\xi_1 + \ldots + \xi_n) \xrightarrow{P} a$ as $n \to \infty$.

*23.13. Using the result of Exercise 22.15, show that any sequence of i.i.d. r.v.'s for which $E\{|\xi_1|\} < \infty$, obeys the WLLN. (*Khintchine theorem*, see the Introductory Notes of this section)

23.14. Let $\xi_1, \xi_2, \ldots$ be i.i.d. r.v.'s with d.f. $F(x)$, $x \in \mathbb{R}_1$. If: (a) $\lim_{n\to\infty} \int_{-n}^{n} x\, dF(x) = 0$ and (b) $\lim_{x\to-\infty} xF(x) = \lim_{x\to\infty} x(1 - F(x)) = 0$, then

$$\frac{1}{n}(\xi_1 + \ldots + \xi_n) \xrightarrow{P} 0 \quad \text{as } n \to \infty.$$

(*A. N. Kolmogorov*)

23.15. Let $\{\xi_n\}$ be a sequence of independent r.v.'s for which $P\{\xi_n = \sqrt{n}\} = P\{\xi_n = -\sqrt{n}\} = \frac{1}{2}$. Show that $\{\xi_n\}$ does not obey the WLLN; i.e., $\frac{1}{n}(\xi_1 + \ldots + \xi_n)$ does not converge in probability to 0 as $n \to \infty$, although $E\{\frac{1}{n}(\xi_1 + \ldots + \xi_n)\} = 0$ for each n.

23.16. Let $\{\xi_n\}$ be a sequence of r.v.'s and let $S_n = \xi_1 + \ldots + \xi_n$. Assume that for every $n \geq 1$ we have $|S_n| < cn$ and $V(S_n) > \alpha n^2$, where $c > 0$ and $\alpha > 0$ are fixed constants. Prove that $\{\xi_n\}$ does not obey the WLLN.

23.17. Let $\xi_i \in N(0, 1)$, $i = 1, 2, \ldots$, be i.i.d. r.v.'s and let $\xi^{(n)}$ denote the random vector $(\xi_1, \ldots, \xi_n)$. Consider the following two regions in $\mathbb{R}_n$:

$$B_{r,\varepsilon} = \{(x_1, \ldots, x_n) \in \mathbb{R}_n : (1 - \varepsilon)r \leq (x_1^2 + \ldots + x_n^2)^{1/2} \leq (1 + \varepsilon)r\};$$

$$C_{r,\varepsilon} = \{(x_1, \ldots, x_n) \in \mathbb{R}_n : (1 - \varepsilon)r \leq |x_1| + \ldots + |x_n| \leq (1 + \varepsilon)r\}.$$

Show that for every $\varepsilon > 0$, we have

$$\lim_{n\to\infty} P\{\xi^{(n)} \in B_{\sqrt{\pi},\varepsilon}\} = 1, \quad \lim_{n\to\infty} P\{\xi^{(n)} \in C_{n\sqrt{2/\pi},\varepsilon}\} = 1.$$

23.18. Let $\{\xi_n\}$ be a sequence of i.i.d. r.v.'s for which $E(\xi_1) = a$ for some a with $0 < a < \infty$. Put $S_n = \xi_1 + \ldots + \xi_n$. Then show that: (a) $P\{\sup_n S_n = \infty\} = 1$; (b) $P\{\inf_n S_n > -\infty\} = 1$.

*23.19. Let $\xi_1, \xi_2, \ldots$ be i.i.d. r.v.'s with $E\{|\xi_1|\} = a < \infty$ and $E\{|\xi_1 - a|^4\} = b < \infty$. Without using condition (23.4) in Kolmogorov's theorem, show that $\{\xi_n\}$ obeys the SLLN. As a corollary prove Borel's version of SLLN: In the Bernoulli scheme (n, p) we have that $\tilde{p}_n = \mu_n/n \xrightarrow{a.s.} p$ as $n \to \infty$.

*23.20. Show that any sequence $\{\xi_n\}$ of identically distributed and pairwise uncorrelated r.v.'s, with $E(\xi_n) = 0$ and $V(\xi_n) \leqslant c < \infty$, obeys the SLLN.

*23.21. Consider the sequence $\{\xi_n\}$: $P\{\xi_n = +1\} = P\{\xi_n = -1\} = \frac{1}{2}(1 - 2^{-n})$, $P\{\xi_n = 2^n\} = P\{\xi_n = -2^n\} = \frac{1}{2} 2^{-n}$. Show that (23.4) does not hold for the sequence $\{\xi_n\}$ and nevertheless $\{\xi_n\}$ obeys the SLLN.

*23.22. Let $\xi_1, \xi_2, \ldots$ be i.i.d. r.v.'s with $E\{|\xi_1|\} = +\infty$. Prove that

$$P\left\{\limsup_n \left|\frac{\xi_1 + \ldots + \xi_n}{n}\right| = +\infty\right\} = 1.$$

(A. N. Kolmogorov)

*23.23. Let $\{\xi_n\}$ be a sequence of i.i.d. r.v.'s, such that for every $n \geqslant 1$ we have

$$P\{\xi_n = (-1)^{k-1}k\} = \frac{\sigma}{\pi^2 k^2}, \quad k = 1, 2, \ldots$$

Put $S_n = \xi_1 + \ldots + \xi_n$. Then prove that $\frac{1}{n} S_n \xrightarrow{P} \frac{1}{\pi^2} \sigma \ln 2$ as $n \to \infty$. Nevertheless, $\{\xi_n\}$ does not obey the SLLN.

23.24. Let ρ_n be the distance between two randomly and independently chosen points in the unit cube in $\mathbb{R}_n$. Find

$$\lim_{n\to\infty} n^{-\frac{1}{2}} E(\rho_n).$$

(J. van de Lun)

24. Central Limit Theorem and Related Topics

Introductory Notes

Let $\{\xi_n\}$ be a sequence of independent r.v.'s defined on the probability space $(\Omega, \mathcal{F}, P)$. The following notations will be frequently used:

$$E(\xi_k) = a_k, \quad V(\xi_k) = \sigma_k^2, \quad S_n = \xi_1 + \ldots + \xi_n,$$

$$A_n = a_1 + \ldots + a_n, \quad B_n = (\sigma_1^2 + \ldots + \sigma_n^2)^{1/2}.$$

The sequence $\{\xi_n\}$ will be said to obey the *Central limit theorem* (CLT) (or equivalently, the CLT holds for the sequence $\{\xi_n\}$) if for every $x \in \mathbb{R}_1$ the following relation holds:

$$\lim_{n\to\infty} P\left\{\frac{S_n - A_n}{B_n} < x\right\} = \Phi(x) = \frac{1}{\sqrt{2\pi}}\int_{-\infty}^{x} e^{-u^2/2}\, du. \tag{24.1}$$

Recall that Φ is the standard normal d.f. and it is d.f. of any r.v. $\xi \in N(0, 1)$. In other words, if $\xi \in N(0, 1)$ and $\eta_n = (S_n - A_n)/B_n$, then $\eta_n \xrightarrow{d} \xi$ as $n \to \infty$. If the last relation holds, we shall say that the r.v. η_n is asymptotically normal $N(0, 1)$.

Lindeberg's theorem. Let $\{\xi_n\}$ be a sequence of independent r.v.'s and ξ_n has a d.f. F_n, $n = 1, 2, \ldots$ Suppose for every $\varepsilon > 0$ the following condition (called *Lindeberg's condition*) is fulfilled:

$$\lim_{n\to\infty} \frac{1}{B_n^2} \sum_{k=1}^{n} \int_{|x-a_k|>\varepsilon B_n} (x - a_k)^2 dF_k(x) = 0 \tag{24.2}$$

Then the sequence $\{\xi_n\}$ obeys the CLT. □

Lyapunov's theorem. Suppose $\{\xi_n\}$ is a sequence of independent r.v.'s such that $E\{|\xi_k - a_k|^{2+\delta}\}$ exists for every $k = 1, 2, \ldots$ and some fixed $\delta > 0$. Let the following condition (called *Lyapunov's condition*) be fulfilled:

$$\lim_{n\to\infty} \frac{1}{B_n^{2+\delta}} \sum_{k=1}^{n} E\{|\xi_k - a_k|^{2+\delta}\} = 0. \tag{24.3}$$

Then the sequence $\{\xi_n\}$ obeys the CLT. □

Notice that the most common application of Lyapunov's theorem is for $\delta = 1$. In this case we put $\rho_n^3 = \sum_{k=1}^{n} E\{|\xi_k - a_k|^3\}$ and then condition (24.3) can be written as

$$\lim_{n\to\infty} \frac{\rho_n}{B_n} = 0. \tag{24.4}$$

Many problems concerning the CLT can be easily solved with the aid of characteristic functions. The following theorem plays an important role when this method is used. It clarifies the relationship between ch.f.'s and d.f.'s.

Continuity theorem. (a) Let F and F_n be d.f.'s and let ϕ and ϕ_n be the corresponding ch.f.'s, n = 1, 2, ... If $F_n \xrightarrow{d} F$ as $n \to \infty$, then $\phi_n(t) \to \phi(t)$ uniformly w.r.t. t on every finite interval. (b) Let for n = 1, 2, ..., F_n be a d.f. and let ϕ_n be the corresponding ch.f. Assume that for every $t \in \mathbb{R}_1$ the limit $\phi(t) = \lim_{n\to\infty} \phi_n(t)$ exists and assume that $\phi(t)$ is continuous at t = 0. Then there exists a d.f. F, whose ch.f. is exactly ϕ and $F_n \xrightarrow{d} F$ as $n \to \infty$. □

Illustrative Examples

Example 24.1. Let μ_n be the number of successes in the Bernoulli scheme (n, p). Prove that

$$\lim_{n\to\infty} P\left\{\frac{\mu_n - np}{\sqrt{np(1-p)}} < x\right\} = \Phi(x), \quad x \in R_1. \tag{24.5}$$

Solution. Consider the i.i.d. r.v.'s $\xi_1, \ldots, \xi_n$, chosen such that each ξ_k assumes the values 0 or 1 with probability q = 1 - p and p, respectively. Then $\mu_n = \xi_1 + \ldots + \xi_n$, $E(\mu_n) = np$, $V(\mu_n) = npq$. Writing (24.2) for the discrete case, it is easily seen that Lindeberg's condition holds. Thus we conclude that (24.5) is fulfilled.

Now we shall show how the same result can be derived using the Continuity theorem. Put $\eta_n = (\mu_n - np)/\sqrt{npq}$. The variables ξ_k and η_n have ch.f.'s given respectively by $\phi_{\xi_k}(t) = pe^{it} + q$ and $\phi_{\eta_n}(t) = [q\exp(-it\sqrt{p/nq}) + p\exp(it\sqrt{q/np})]^n$. From the expansion of e^z we get $\phi_{\eta_n}(t) = \left[1 - \frac{t^2}{2n} + o(\frac{1}{n})\right]^n$; therefore, $\lim_{n\to\infty} \phi_{\eta_n}(t) = e^{-t^2/2}$. However $e^{-t^2/2}$, $t \in \mathbb{R}_1$, is the ch.f. of a r.v. $\xi \in N(0, 1)$. Thus $\eta_n \xrightarrow{d} \xi$ as $n \to \infty$, which is exactly (24.5).

Example 24.2. Let $\xi_1, \xi_2, \ldots$ be a sequence of i.i.d. r.v.'s with zero mean and finite but unknown variance $\sigma^2 = V(\xi_1)$. Suppose $S_n = \xi_1 + \dots + \xi_n$ and we know that $\lim_{n\to\infty} P\{S_n/\sqrt{n} > 1\} = \frac{1}{3}$. Determine the unknown variance σ^2.

Solution. It is easy to check (see Exercise 24.1) that the CLT holds for the sequence $\{\xi_n\}$; i.e.,

$$\lim_{n\to\infty} P\left\{\frac{S_n}{\sigma\sqrt{n}} < x\right\} = \Phi(x), \quad x \in \mathbb{R}_1.$$

However, $P\{S_n/\sqrt{n} > 1\} \to \frac{1}{3}$ as $n \to \infty$, and it follows that

$$P\left\{\frac{S_n}{\sqrt{n}} > 1\right\} = P\left\{\frac{S_n}{\sigma\sqrt{n}} > \frac{1}{\sigma}\right\} \to 1 - \Phi\left(\frac{1}{\sigma}\right) = \frac{1}{3}.$$

Hence σ is the solution of the equation $\Phi\left(\frac{1}{\sigma}\right) = \frac{2}{3}$. Using Table 1 (or more precise tables), we find that $\frac{1}{\sigma} \approx 0.43$ and $\sigma^2 \approx 5.41$.

Example 24.3. Consider the sequence $\xi_1, \xi_2, \dots$ of i.i.d. r.v.'s with zero mean and unit variance. Let $\nu_1, \nu_2, \dots$ be another sequence of r.v.'s which are integer-valued and such that for each k, ν_k does not depend on $\{\xi_n, n = 1, 2, \dots\}$. Prove that if $S_n = \xi_1 + \dots + \xi_n$ and $\nu_n \xrightarrow{P} \infty$ as $n \to \infty$, then $S_{\nu_n}/\sqrt{\nu_n} \xrightarrow{d} \Theta$ as $n \to \infty$, where $\Theta \in N(0, 1)$.

Solution. The formula for the total probability yields

$$P\{S_{\nu_n}/\sqrt{\nu_n} < x\} = \sum_{k=1}^{\infty} P\{\nu_n = k\}P\{S_k/\sqrt{k} < x\}.$$

The sequence $\{\xi_n\}$ satisfies the CLT and since $E(S_k) = 0$ and $V(S_k) = k$, we get

$$P\{S_k/\sqrt{k} < x\} \to \Phi(x) \quad \text{as } k \to \infty.$$

This implies that for an arbitrary $\varepsilon > 0$ there exists an index k_ε such that $|P\{S_k/\sqrt{k} < x\} - \Phi(x)| < \frac{\varepsilon}{2}$ for all $k \geq k_\varepsilon$. However $\nu_n \xrightarrow{P} \infty$ as $n \to \infty$, implying the existence of an index n_ε such that for each $n \geq n_\varepsilon$ we have $P\{\nu_n < k_\varepsilon\} < \frac{\varepsilon}{2}$. Therefore

$$|P\{S_{\nu_n}/\sqrt{\nu_n} < x\} - \Phi(x)| = \left|\sum_{k=1}^{\infty} P\{\nu_n = k\}(P\{S_k/\sqrt{k} < x\} - \Phi(x))\right| \leq$$

$$\leq P\{\nu_n < k_\varepsilon\} + \sum_{k=k_\varepsilon}^{\infty} P\{\nu_n = k\}|P\{S_k/\sqrt{k} < x\} - \Phi(x)|.$$

Thus we arrive at the relation

$$|P\{S_{\nu_n}/\sqrt{\nu_n} < x\} - \Phi(x)| \leq \varepsilon,$$

which is valid for each $n \geq n_\varepsilon$. Obviously this means that for some r.v. $\Theta \in N(0, 1)$ we have $S_{\nu_n}/\sqrt{\nu_n} \xrightarrow{d} \Theta$ as $n \to \infty$.

Exercises

24.1. Prove that any sequence $\{\xi_n\}$ of i.i.d. r.v.'s with finite variances obeys the CLT.

24.2. Let $\{\xi_n\}$ be a sequence of i.i.d. r.v.'s, $\xi_n \in P(\lambda)$, $n = 1, 2, \ldots$, for some fixed $\lambda > 0$ and let $\eta_n = (S_n - n\lambda)/\sqrt{n\lambda}$. Prove, using only the method of ch.f.'s (i.e., not general theorems), that $\eta_n \xrightarrow{d} \eta \in N(0, 1)$ as $n \to \infty$.

24.3. Let $\xi \in P(\lambda)$, $\lambda > 0$. Prove that $\eta_\lambda = (\xi - \lambda)/\sqrt{\lambda}$ is asymptotically normal $N(0, 1)$ as $\lambda \to \infty$.

24.4. Prove the convergence of the Binomial distribution to the Poisson distribution, using the method of ch.f.'s (see the Poisson theorem in Section 10).

24.5. Let F and F_n be d.f.'s of integer-valued r.v.'s and let g and g_n be the corresponding p.g.f.'s. Prove that $F_n \xrightarrow{d} F$ as $n \to \infty$ if and only if $g_n(s) \to g(s)$, $s \in [0, 1)$ and $\lim_{s \uparrow 1} g(s) = 1$.

24.6. Prove the convergence of the Binomial distribution to the Poisson distribution, using the method of p.g.f.'s (see Exercise 24.4).

24.7. For $n = 1, 2, \ldots$ let ξ_n be a r.v. which takes on the values $\frac{k}{n}$ for $k = 1, 2, \ldots, n$, with the same probability $\frac{1}{n}$. Prove that there exists a r.v. ξ such that $\xi_n \xrightarrow{d} \xi$ as $n \to \infty$. What is the distribution of ξ?

*24.8. For $n = 1, 2, \ldots$ let ξ_n be a r.v. which assumes the values $\frac{k}{n^2}$ for $k = 1, \ldots, n^2$, with the same probability $\frac{1}{n^2}$. Prove that there is no r.v. ξ for which $\xi_n \xrightarrow{d} \xi$ as $n \to \infty$.

24.9. Let $\{\xi_n\}$ be a sequence of r.v.'s, where $\xi_n \in N(a_n, \sigma_n^2)$, $n = 1, 2, \ldots$ and let $\xi_n \xrightarrow{L_2} \xi$ for some r.v. ξ with $0 < V(\xi) < \infty$. Prove that $\xi \in N(a, \sigma^2)$, where $a = \lim_{n\to\infty} a_n$ and $\sigma^2 = \lim_{n\to\infty} \sigma_n^2$.

24.10. Let $\xi_1, \xi_2, \ldots$ be independent r.v.'s with $\xi_n \in N(0, n^{-2})$. Prove that the series $\sum_{n=1}^{\infty} \xi_n$ is convergent with probability 1 and if ξ

denotes its sum, then $\xi \in N(0, \frac{\pi^2}{\sigma_\infty})$. Prove also the following more general assertion: If $\xi_n \in N(0, \sigma_n^2)$ and $\sum_{n=1}^{\infty} \sigma_n^2 = \sigma^2 < \infty$, then $\xi = \sum_{n=1}^{\infty} \xi_n \in N(0, \sigma^2)$.

24.11. Let $\xi_1, \ldots, \xi_n$ be i.i.d. r.v.'s which are uniformly distributed over the interval (0, 1) and let $\eta_n = (S_n - \frac{1}{2}n)/\sqrt{n}$. By the method of ch.f.'s prove that η_n is asymptotically normal $N(0, 1)$.

24.12. Let $\xi_1, \ldots, \xi_n$ be i.i.d. r.v.'s which have gamma-distribution with parameters α and β (see Section 15). Put $\eta_n = (S_n - n\alpha\beta)/(\beta\sqrt{n\alpha})$. By method of ch.f.'s prove that $\eta_n \xrightarrow{d} \eta \in N(0, 1)$ as $n \to \infty$.

24.13. For each $n \geqslant 1$ let the r.v. ξ_n have a χ^2-distribution with n degrees of freedom (see Exercise 16.29). Put $\eta_n = (\xi_n - n)/n\sqrt{2})$. Prove that η_n is asymptotically normal $N(0, 1)$ as $n \to \infty$.

24.14. Let ξ be a gamma-distributed r.v. with parameters α and β. Prove that the r.v. $\eta_\alpha = (\beta\xi - \alpha)/\sqrt{\alpha}$ is asymptotically normal $N(0, 1)$ as $\alpha \to \infty$.

24.15. Suppose that $\{\xi_n\}$ and $\{\eta_n\}$ are two sequences of independent r.v.'s for which $P\{\xi_n = n^{-\alpha}\} = P\{\xi_n = -n^{-\alpha}\} = p$, $P\{\xi = 0\} = 1 - 2p$, for $0 < p < \frac{1}{2}$, $\frac{1}{3} < \alpha < \frac{1}{2}$, and $P\{\eta_n = \sqrt{n}\} = P\{\eta_n = -\sqrt{n}\} = \frac{1}{2}$. Prove that each of these two sequences obeys the CLT.

24.16. Suppose $\xi_1, \xi_2, \ldots$ are independent r.v.'s and for some fixed $\alpha \in (0, 1)$, ξ_n assumes the values n^α, $-n^\alpha$ and 0 with probability $1/(2n^\alpha)$, $1/(2n^\alpha)$ and $1 - 1/n^\alpha$, respectively. Prove the validity of Lyapunov's theorem for the sequence $\{\xi_n\}$.

24.17. Let $\xi_1, \ldots, \xi_n$ be i.i.d. r.v.'s each with variance equal to 5. (a) For n = 4,500 find the probability that the absolute value of the difference between the sample mean $\eta_n = \frac{1}{n} S_n$ and the mean $a = E(\xi_1)$ will not exceed 0.04. (b) For which n will $P\{|\eta_n - a| \leqslant 0.2\} \geqslant 0.8$?

24.18. Let $\xi_1, \ldots, \xi_n$ be i.i.d. r.v.'s. Find $\lim_{n\to\infty} P\{S_n < \sqrt{n}\}$ in the following two cases: (a) ξ_1 has density $f(x) = 3x^2/2$ for $x \in (-1, 1)$ and $f(x) = 0$, otherwise; (b) $P\{\xi_1 = 1\} = P\{\xi_1 = -1\} = \frac{1}{2}$.

24.19. Let $\{\xi_n\}$ be a sequence of independent r.v.'s such that ξ_n has the density

$$p_n(x) = \begin{cases} 2^n, & \text{if } -2^{-n-2} \leqslant x \leqslant 2^{-n-3} \quad \text{or} \quad ||x| - 1| < 2^{-n-3}; \\ 0, & \text{otherwise.} \end{cases}$$

Does the CLT hold for the sequence $\{\xi_n\}$?

24.20. Let $\xi_1, \xi_2, \ldots$ be independent r.v.'s and let $P\{\xi_n = n^\delta\} = P\{\xi_n = -n^\delta\} = \frac{1}{2}$. Prove that $\{\xi_n\}$ obeys the CLT for every fixed $\delta > -\frac{1}{3}$.

24.21. Let $\xi_1, \xi_2, \ldots$ be independent r.v.'s where ξ_n assumes the values: (a) 2^n and -2^n; (b) 2^n, -2^n and 0; (c) n, -n and 0, respectively with probability: (a) $\frac{1}{2}$ and $\frac{1}{2}$; (b) $2^{-(2n+1)}$, $2^{-(2n+1)}$ and $1 - 2^{-2n}$; (c) $\frac{1}{2\sqrt{n}}$, $\frac{1}{2\sqrt{n}}$ and $1 - \frac{1}{\sqrt{n}}$. In which of these cases is condition (24.3) in Lyapunov's theorem valid for the sequence $\{\xi_n\}$?

24.22. Let $\{\xi_n\}$ be a sequence of independent r.v.'s for which $\xi_n \in N(0, c^{-\frac{n}{2}})$ for $c > 0$ and n = 1, 2, ... Is it possible to apply Lindeberg's theorem to the sequence $\{\xi_n\}$?

24.23. Suppose $\xi_1, \xi_2, \ldots$ are i.i.d. r.v.'s with density

$$f(x) = \begin{cases} \frac{1}{3}, & \text{if } x \in (-1, 0), \\ \frac{2}{3}, & \text{if } x \in [0, 1), \\ 0, & \text{if } x \overline{\in} (-1, 1). \end{cases}$$

For n = 60 find approximately the value of $P\{S_n < 13\}$.

24.24. Let $\{\xi_n\}$ be a sequence of independent r.v.'s and let $P\{\xi_n = n\alpha\} = P\{\xi_n = -n\alpha\} = \frac{1}{4}$, $P\{\xi_n = 0\} = \frac{1}{2}$, with α = constant. Prove the validity of Lindeberg's theorem for the sequence $\{\xi_n\}$.

24.25. Let $\{F_n(x), x \in \mathbb{R}_1\}$ be a family of d.f.'s, where

$$F_n(x) = \begin{cases} 0, & \text{if } x < -n, \\ \frac{1}{2n}(n + x), & \text{if } -n \leqslant x < n, \\ 1, & \text{if } x \geqslant n. \end{cases}$$

and n = 1, 2, ... Prove that $\lim_{n\to\infty} F_n(x)$ exists but it is not a d.f.

*24.26. Let $G_n(\lambda) = e^{-n\lambda} \sum_{k=0}^{n} (n\lambda)^k/k!$. Apply the CLT to an appropriately chosen sequence of r.v.'s and prove that $\{G_n(\lambda)\}$ is a convergent sequence for every $\lambda > 0$. Let $G_0(\lambda)$ denote its limit. Prove the following relations: $G_0(\lambda) = 1$ for $0 < \lambda < 1$, $G_0(\lambda) = \frac{1}{2}$ for $\lambda = 1$ and $G_0(\lambda) = 0$ for $\lambda > 1$.

24.27. Prove that if the kth moment of the r.v. ξ_n converges as $n \to \infty$ to the kth moment of the r.v. $\xi \in N(0, 1)$ for k = 1, 2, ..., then $\xi_n \xrightarrow{d} \xi$.

24.28. Let $\xi_1, \ldots, \xi_n$ be i.i.d. r.v.'s such that $P\{\xi_k = 1\} = P\{\xi_k = -1\} = \frac{1}{2}$, $1 \leqslant k \leqslant n$, and let $\eta_n = \sum_{k=1}^{n} (\xi_k/2^k)$. Show that $\eta_n \xrightarrow{d} \eta$ as $n \to \infty$ for some r.v. η which is uniformly distributed over the interval $(-1, 1)$.

24.29. Let $\xi_1, \xi_2, \ldots$ be independent r.v.'s such that

$$P\{\xi_n = n^{\alpha}\} = P\{\xi_n = -n^{\alpha}\} = \frac{1}{6n^{2(\alpha-1)}}, \quad P\{\xi_n = 0\} = \frac{1}{3n^{2(\alpha-1)}}.$$

Show that condition (24.2) in Lindeberg's theorem holds for the sequence $\{\xi_n\}$ only if $\alpha < \frac{3}{2}$.

*24.30. Assume that $\xi_1, \ldots, \xi_n$ are i.i.d. r.v.'s with $E(\xi_1) = 0$ and $V(\xi_1) = 1$. Prove that each of the r.v.'s

$$\gamma_n = \sqrt{n}\,\frac{\xi_1 + \ldots + \xi_n}{\xi_1^2 + \ldots + \xi_n^2} \quad \text{and} \quad \delta_n = \frac{\xi_1 + \ldots + \xi_n}{(\xi_1^2 + \ldots + \xi_n^2)^{1/2}}$$

is asymptotically normal $N(0, 1)$ as $n \to \infty$.

24.31. Let $\xi_1, \ldots, \xi_n$ be i.i.d. r.v.'s with expectation $a > 0$ and variance σ^2. Prove that $\eta_n = 2\sqrt{a}\sigma^{-1}(|S_n|^{\frac{1}{2}} - (an)^{\frac{1}{2}})$ is asymptotically $N(0, 1)$ as $n \to \infty$.

*24.32. Let $\xi_1, \xi_2, \ldots$ be i.i.d. and non-negative r.v.'s with $E(\xi_1) < \infty$ and let the r.v. τ have a geometric distribution with parameter p (see Section 9). Put $\eta = 0$ when $\tau = 0$ and $\eta = \xi_1 + \ldots + \xi_n$ when $\tau > 0$ and let $\Theta = \tau/E(\eta)$. Prove that $\Theta \xrightarrow{d} \zeta$ as $E(\tau) \to 0$, where $\zeta \in E(1)$.

24.33. Let $\xi_1, \xi_2, \ldots$ be i.i.d. r.v.'s with expectation a and variance σ^2. According to the CLT if $\overline{\xi}_n = (\xi_1 + \ldots + \xi_n)/n$, then $\eta_n = \frac{\sqrt{n}}{\sigma}(\overline{\xi}_n - a)$ converges in distribution to some r.v. $\eta \in N(0, 1)$. Show that $\{\xi_n\}$ is not convergent in mean square, a.s., or in probability.

*24.34. Let the r.v.'s ξ_n and ξ have densities $f_n(x)$ and $f(x)$, $x \in \mathbb{R}_1$. Assume that $f_n(x) \to f(x)$ as $n \to \infty$ for almost all $x \in \mathbb{R}_1$. Then show that $\xi_n \xrightarrow{d} \xi$. (*H. Scheffé*).

*24.35. Let $\{\xi_n\}$ be a sequence of integer-valued r.v.'s for which

$$P\{\xi_n = r\} = p_n(r), \; r = 0, \pm 1, \pm 2, \ldots, \quad \sum_{r=-\infty}^{\infty} p_n(r) = 1, \; n = 1, 2, \ldots$$

Let ξ be a r.v. with density $f(x)$, $x \in \mathbb{R}_1$. Assume that there exists a sequence $\{c_n\}$ of real numbers, such that for almost all $x \in \mathbb{R}_1$ the following relation holds:

$$\lim_{n\to\infty} \sqrt{n}p_n([c_n + x\sqrt{n}]) = f(x). \tag{24.6}$$

Show that $(\xi_n - c_n)/\sqrt{n} \xrightarrow{d} \xi$ as $n \to \infty$. (*M. Okamoto*)

24.36. Let ξ_N have a hypergeometric distribution (see Section 9); i.e., $\vec{P}\{\xi_N = k\} = \binom{M}{k}\binom{N-M}{n-k}/\binom{N}{n}$, $k = 0, 1, \ldots, \min[M, n]$, $M < N$, $n < N$, and let p and $\tilde{p}$ be fixed numbers in the interval $(0, 1)$. Put $M = pN + o(\sqrt{N})$, $N = \tilde{p}N + o(\sqrt{N})$. Prove that if $q = 1 - p$ and $\tilde{q} = 1 - \tilde{p}$, then the r.v. $(\xi_N - Np\tilde{p})/\sqrt{Np\tilde{p}q\tilde{q}}$ is asymptotically normal $N(0, 1)$ as $N \to \infty$.

24.37. Let $\xi_1, \ldots, \xi_n$ be i.i.d. r.v.'s, which are normally distributed. Put

$$\eta_n = \frac{S_n}{n}, \quad \Theta_n = \left(\frac{1}{n-1}\sum_{k=1}^{n} (\xi_k - \eta_n)^2\right)^{\frac{1}{2}}, \quad \zeta_n = \frac{\eta_n\sqrt{n}}{\Theta_n}.$$

Prove that $\zeta_n \xrightarrow{d} \zeta \in N(0, 1)$ as $n \to \infty$.

*24.38. For $n = 1, 2, \ldots$, let $\xi_1^{(n)}, \ldots, \xi_n^{(n)}$ be i.i.d. r.v.'s such that $P\{\xi_j^{(n)} = \sqrt{n}\} = P\{\xi_j^{(n)} = -\sqrt{n}\} = \frac{1}{2n}$, $P\{\xi_j^{(n)} = 0\} = \frac{n-1}{n}$, $j = 1, \ldots, n$. Find $E(\xi_j^{(n)})$ and $V(\xi_j^{(n)})$. Find also the limit distribution of the r.v. $\eta_n = (nV(\xi_1^{(n)}))^{-1}(\xi_1^{(n)} + \ldots + \xi_n^{(n)})$ as $n \to \infty$.

*24.39. Prove *Stirling's formula*:

$$n! \approx n^n e^{-n}\sqrt{2\pi n} \quad \text{for large } n$$

using the CLT.

*24.40. Let $\{\xi_n\}$ be a sequence of i.i.d. r.v.'s with the following symmetric density:

$$f(x) = \begin{cases} |x|^{-3}, & \text{if } |x| > 1; \\ 0, & \text{if } |x| \leqslant 1. \end{cases}$$

Show that $(n \ln n)^{-\frac{1}{2}} \sum_{k=1}^{n} \xi_k \xrightarrow{d} \Theta$ as $n \to \infty$ for some $\Theta \in N(0, 1)$.

24.41. Let $\xi_1, \xi_2, \ldots$ be i.i.d. r.v.'s which assume two values: 1 and -1 with probability $\frac{1}{2}$. Put $\eta_n = \sqrt{15}\xi_n/4^n$, $n = 1, 2, \ldots$ Show that $\{\eta_n\}$ does not obey the CLT.

*24.42. Let $\xi_1, \xi_2, \ldots$ be independent r.v.'s and let ξ_n have density

$$f_n(x) = \begin{cases} 2^n, & \text{if } |x| \leqslant 2^{-(n+2)} \quad \text{or} \quad ||x| - 1| < 2^{-(n+3)}; \\ 0, & \text{otherwise,} \end{cases}$$

Let $F_n(x)$, $x \in \mathbb{R}_1$, and $p_n(x)$, $x \in \mathbb{R}_1$, be respectively the d.f. and the density of the r.v. $\eta_n = (S_n - A_n)/B_n$, where $S_n = \xi_1 + \ldots + \xi_n$, $A_n = E(S_n)$, $B_n^2 = V(S_n)$. Prove that when $n \to \infty$, then the d.f. $F_n(x)$ converges to the standard normal d.f. $\Phi(x)$ for every $x \in \mathbb{R}_1$, but $p_n(x)$ does not converge to the standard normal density $\phi_0(X) = \frac{1}{\sqrt{2\pi}} e^{-x^2/2}$ for all $x \in \mathbb{R}_1$. (This means that $\{\xi_n\}$ obeys the integral CLT but does not obey the local CLT.)

SOLUTIONS, HINTS, AND ANSWERS

1. Combinatorics

1.1. The result follows from the definition of the union of disjoint sets.

1.2. The elements of AB are counted both in $\nu(A)$ and $\nu(B)$.

1.3. *Method 1*: Mathematical induction.

Method 2 (method of inclusion and exclusion): We have to prove that each element of $\bigcup_{i=1}^{n} A_i$ is counted exactly once in the sum on the right-hand side of the considered equality. Let x be a common element of exactly k of the sets A_i; hence, x does not belong to the remaining n - k sets. (The number k may be any one of the numbers 1, 2, ..., n.) For definiteness, let $x \in A_1A_2 \ldots A_k$. Then in the sum $\sum_{i=1}^{n} \nu(A_i)$ the element x is counted $\binom{k}{1}$ times, in $\sum_{i<j} \nu(A_iA_j)$ it is counted $\binom{k}{2}$ times, in $\sum_{i<j<k} \nu(A_iA_jA_k)$ it is counted $\binom{k}{3}$ times, ..., in $\sum_{i_1<\ldots<i_k} \nu(A_{i_1} \ldots A_{i_k})$ it is counted $\binom{k}{k} = 1$ time. Thus we have $\binom{k}{1} - \binom{k}{2} + \binom{k}{3} - \ldots + (-1)^{k-1}\binom{k}{k} = 1 - \left[\binom{k}{0} - \binom{k}{1} + \binom{k}{2} - \ldots + (-1)^k\binom{k}{k}\right] = 1 - (1 - 1)^k = 1.$

1.4. After substituting the data in formula $\nu(A \cup B \cup C) = \nu(A) + \nu(B) + \nu(C) - \nu(AB) - \nu(AC) - \nu(BC) + \nu(ABC)$, we obtain $\nu(ABC) = -13$. This is impossible since $\nu(\cdot) \geqslant 0$.

1.5. Let A, B and C be the sets of those cubes which do not have red, blue and green face, respectively, and D be the set of the cubes which do not have at least one of the three colours. We have $\nu(D) + \nu(\overline{D}) = 100$, $\nu(\overline{D}) = 100 - \nu(D) = 100 - \nu(A \cup B \cup C) = 100 - \nu(A) - \nu(B) - \nu(C) + \nu(AB) + \nu(AC) + \nu(BC) - \nu(ABC) \geqslant 100 - \nu(A) - \nu(B) - \nu(C) - \nu(ABC) = 100 - 20 - 15 - 25 - 0 = 40$. The equality is attained only if $AB = AC = BC = \emptyset$.

1.6. (a) $M_n = K_0^n + K_1^n + \ldots + K_n^n$. Since $K_i^nK_j^n = \emptyset$ for $i \neq j$, we have $\nu(M_n) = \binom{n}{0} + \binom{n}{1} + \ldots + \binom{n}{n} = (1 + 1)^n = 2^n$. (b) Let n = 0; i.e., $M = \emptyset$. Then $M_0 = \{\emptyset\}$ and $\nu(M_0) = 1 = 2^0$. Suppose that $\nu(M_{n-1}) = 2^{n-1}$ for each set M_{n-1}, for which $\nu(M_{n-1}) = n - 1$. Let $A(n)$ and $A(\overline{n})$ denote, respectively, the set of those elements formed from M_n which contain the element n and which do not contain this element. It is easy to establish a one-to-one correspondence between $A(n)$ and $A(\overline{n})$, for instance, if we compare the element $\{i_1, \ldots, i_k, n\} \in A(n)$ with the element $\{i_1, \ldots, i_k\} \in A(\bar{n})$. However, according to the above assumption we have $\nu(A(\bar{n})) = \nu(M_{n-1}) = 2^{n-1}$. Hence $\nu(M_n) = 2 \cdot 2^{n-1} = 2^n$. (c) Consider the ordered n-tuples of 1's and 0's, e.g. {0, 1, ..., 1, 1}, and let us introduce, between this set and the set M_n the following correspondence, which is a one-to-one correspondence: to each n-tuple we assign that subset of $M_n = \{a_1, a_2, \ldots, a_n\}$ which consists of the elements a_i at whose place there is 1 in the ordered n-tuple. For instance, to the n-tuple {0, 0, ..., 0}, consisting of only 0's, we assign the set $\emptyset$, while to the n-tuple {1, 1, ..., 1}, consisting of only 1's, we assign the set M_n. Then $\nu(M_n) = \nu(\{0, 1\}^{\times n}) = 2^n$.

1.7. $2^{m+n} - 2^m - 2^n + 1$.

1.8. (a) $\binom{n}{k}$; (b) $\binom{n+k-1}{k}$; (c) $n!/(n-k)!$; (d) n^k.

1.9. $\binom{n}{k}$.

1.10. $\binom{n+k-1}{k}$.

1.11. n^k.

1.12. $n!/(n-k)!$.

1.13. $\nu(\Omega) = 3^n$. (a) 3^{n-1}; (b) $\binom{n-2}{k}2^{n-2-k}$; (c) $\binom{n}{k}2^{n-k}$; (d) $n!/(k_1!k_2!k_3!)$.

1.14. (a) Let $M = \{a_1, a_2, \ldots, a_n\}$. Consider the sets $\mathbf{K}_k^n$ and $\mathbf{K}_{n-k}^n$. To each element of $\mathbf{K}_k^n$, for example $a_{i_1}a_{i_2} \ldots a_{i_k}$, we assign that element of $\mathbf{K}_{n-k}^n$ which consists of the remaining elements of M. This is a one-to-one correspondence. (b) Consider the set $\mathbf{K}_k^n$ and select an arbitrary element of M, for example a_n. Split the set $\mathbf{K}_k^n$ into two subsets A(n) and $A(\bar{n})$, where A(n) consists of those elements of $\mathbf{K}_k^n$ which contain the element a_n, while $A(\bar{n})$ contains the remaining elements of $\mathbf{K}_k^n$. Obviously, $A(n) \cap A(\bar{n}) = \emptyset$ and $\mathbf{K}_k^n = A(n) + A(\bar{n})$, but $\nu(A(n)) = \binom{n-1}{k-1}$ and $\nu(A(\bar{n})) = \binom{n-1}{k}$.

1.15. Let $M = \{a_1, a_2, \ldots, a_n\}$. Denote by A_i the set of samples containing exactly i times a fixed element, for example a_n. (a) $\mathbf{B}_k^n = A_0 + A_1$; (b) $\tilde{\mathbf{B}}_k^n = A_0 + \ldots + A_k$; (c) $\tilde{\mathbf{K}}_k^n = A_0 + \ldots + A_k$; (d) Let B_i be the set of samples which contain the element a_i in a fixed place, say, in the first place. Then $\tilde{\mathbf{B}}_k^n = B_1 + \ldots + B_k$; (e) Let $\mathbf{K}_k^n$ be a set of combinations of the elements of the set M = {1, 2, ..., n} taken k at a time, and let r be a natural number, $1 \leqslant r \leqslant n$. Any one of these combinations, say L, may be written in an increasing order of its elements: $L = l_1l_2 \ldots l_{r-1}l_rl_{r+1} \ldots l_k$, where $1 \leqslant l_1 < \ldots < l_r < \ldots < l_k \leqslant n$. The number l_r being the rth largest element of L cannot be less than r; i.e., $l_r \geqslant r$.

If $l_r = r$, then the only possible values of $l_1, l_2, \ldots, l_{r-1}$ are the numbers $1, 2, \ldots, r - 1$. On the other hand, $l_r \leqslant n - (k - r)$, because otherwise there will be no other $k - r$ elements of M following l_r in the k-tuple L. If $l_r = n - k + r$, then the only possible values of l_{r+1}, $l_{r+2}, \ldots, l_k$ are the numbers $n - k + r + 1, n - k + r + 2, \ldots, n - 1$, n. Let $l_r = i$, $r \leqslant i \leqslant n - k + r$. Denote by A_i the set of those k-tuples in which $l_r = i$. Obviously, $A_iA_j = \emptyset$ for $i \neq j$, and $\mathbf{K}_k^n = A_r + A_{r+1} + \ldots + A_{n-k+r}$. Each element of A_i, except the number i, contains another $r - 1$ numbers, less than i, which are chosen from the numbers $1, 2, \ldots, i - 1$ in $\binom{i-1}{r-1}$ ways, and $k - r$ numbers, greater than i, which are chosen from the numbers $i + 1, i + 2, \ldots, i + n$ in $\binom{n-i}{k-r}$ ways; i.e., $\nu(A_i) = \binom{i-1}{r-1}\binom{n-i}{k-r}$. The required equality follows from the additivity of the function ν.

1.16. (a) $\nu(A_i) = \binom{m}{i}\binom{n}{k-i}$, $\nu(B_i) = \binom{m}{k-i}\binom{n}{i}$. (b) Follows from (a). (c) Use the representation $\mathbf{K}_k^{m+n} = A_0 + A_1 + \ldots + A_k$; (d) $A = \mathbf{K}_k^{m+n} \setminus A_0$ and $\nu(A) = \binom{m+n}{k} - \binom{n}{k}$; (e) $\nu(D) = \binom{m+n}{k} - \binom{n}{k} - \binom{m}{k}$.

1.17. (a) $k!/(k_1!k_2! \ldots k_n!)$; (b) 1.

1.18. (a) $P(n - 1, k - 1)$; (b) $P(n - 2, k - 2)$; (c) $2P(n - 2, k - 2)$.

1.19. (a) $P(10, 6) = 151\ 200$; (b) $\tilde{P}(5, 6) = 15\ 625$.

1.20. $8! \cdot 3! = 241\ 920$.

1.21. $\binom{4}{3} + \binom{4}{2} = 10$.

1.22. (a) $6^2 = 36$; (b) $\binom{6+2-1}{2} = 21$; (c) 11.

1.23. (a) $3^{13} = 1\ 594\ 323$; (b) $\binom{49}{6} = 13\ 983\ 816$.

1.24. $28^3 = 21\ 952$.

1.25. $\binom{6+3-1}{6} = 28$.

1.26. If $x_1, x_2, \ldots, x_m$ are non-negative integers and

$$x_1 + x_2 + \ldots + x_m = n, \tag{1.6}$$

then we may form an n-tuple by taking x_1 elements of the 1st type, x_2 elements of the 2nd type, ..., x_m elements of the mth type. Conversely, if we form the n-tuple of x_1 elements of the 1st type, ..., x_m elements of the mth type, then we may get a solution of (1.6). Therefore, between the set of combinations of m elements taken n at a time, with repetition allowed, and the set of non-negative solutions of (1.6) there is a one-to-one correspondence. The required number of solutions of (1.6) is equal to $\tilde{C}(m, n) = \binom{m+n-1}{n}$.

1.27. The partial derivatives of nth order of the function $f(x_1, \ldots, x_m)$

do not depend on the order of differentiating, but only on the following fact how many times we have differentiated with respect to each of the arguments. If we have differentiated k_1 times with respect to x_1, ..., k_m times with respect to x_m, and if $k_1 + \dots + k_m = n$, then we obtain the partial derivative $\partial^n f/\partial x_1^{k_1} \dots \partial x_m^{k_m}$. Hence, the number of the different partial derivatives of nth order of f is equal to the number of the different non-negative solutions of the equation $k_1 + \dots + k_m = n$. The answer is $\tilde{C}(m, n) = \binom{n + m - 1}{n}$.

1.28. $\binom{9}{3} \cdot \binom{6}{3} \cdot (3!)^{-1} = 280$.

1.29. $\binom{6}{2}\binom{4}{2} \frac{1}{3!} 3! = 90$.

1.30. $\binom{m}{k}\binom{n}{k}k!$.

1.31. (a) $\binom{m}{n}\binom{m - n}{r}$; (b) $P(m, n)P(m - n, r)$.

1.32. $\binom{n}{2k}\binom{2k}{k}k!$.

1.33. Choose j cells in $\binom{m}{j}$ ways. Place in each cell exactly one ball. The remaining n - j balls can be distributed into the selected cells in $N_{a-1}(n - j, j)$ different ways so that each cell to contain at most a - 1 balls. Thus the number of the different distributions, such that n balls are placed in j arbitrary chosen cells in such a way that in each cell there are at most a balls, is equal to $\binom{m}{j}N_{a-1}(n - j, j)$. After summing with respect to j from 1 to n we obtain $N_a(n, m)$.

1.34. If the number of the distributions of the balls is N, then

$$N = \begin{cases} \binom{k}{s} \dfrac{(n - 1)!}{(n - k + s - 1)!}, & \text{if } k \geqslant s \text{ and } n - 1 \geqslant k - s, \\ 0, & \text{otherwise;} \end{cases} \tag{a}$$

$$N = \begin{cases} \displaystyle\sum_{j=\max[0,k-n+1]}^{\min[s,k]} \frac{(n - 1)!}{(n - k + j - 1)!}, & \text{if } \max[0,k-n+1] \leqslant \min[s,k] \\ 0, & \text{otherwise;} \end{cases} \tag{b}$$

(c) only the case $k \geqslant s$ is reasonable:

$$N = \begin{cases} \displaystyle\sum_{j=s}^{k} \binom{k}{j} \frac{(n - 1)!}{(n - k + j - 1)!}, & \text{if } n - 1 \geqslant k - s, \\ \displaystyle\sum_{j=k-n+1}^{k} \binom{k}{j} \frac{(n - 1)!}{(n - k + j - 1)!}, & \text{if } n - 1 < k - s. \end{cases}$$

1.35. (a) In this case the number of the different k-tuples $i_1 i_2 \dots i_k$

(they are ordered and with repetition) is equal to N^k, the number of the permutations of N elements taken k at a time, with repetitions permitted. The minimum, $\min[i_1, \ldots, i_k]$, can take any value from 1 to N inclusive. The number of k-tuples in which the number 1 does not appear is equal to $(N-1)^k$. Then for the remaining k-tuples, which are $N^k - (N-1)^k$ in number, we have $\min[i_1, \ldots, i_k] = 1$. Further, $\min[i_1, \ldots, i_k] = 2$ exactly $(N-1)^k - (N-2)^k$ times, etc. Hence

$$\sum_{i_1,\ldots,i_k=1}^{N} \min[i_1, \ldots, i_k] = \sum_{j=1}^{N} j[(N-j+1)^k - (N-j)^k] =$$

$$= \sum_{j=1}^{N} (N-j+1)[j^k - (j-1)^k].$$

An analogous reasoning show that

$$\sum_{i_1,\ldots,i_k=1}^{N} \max[i_1, \ldots, i_k] = \sum_{j=1}^{N} j[j^k - (j-1)^k].$$

Combining the two equalities we obtain

$$\Sigma^{(1)} (\min[i_1, \ldots, i_k] + \max[i_1, \ldots, i_k]) = N^k(N+1).$$

(b) The proof in this case is carried out by the same scheme as in case (a) only here permutations without repetitions are used.

1.36. Let $\mathbf{K}_k^n$ be the set of the combinations of n elements taken k at a time formed of the elements of the set $\{1, 2, \ldots, n\}$. We know that $\nu(\mathbf{K}_k^n) = \binom{n}{k}$. It is easy to realize that the number of different mappings of the set $\{1, 2, \ldots, n\}$ onto any set with k elements is equal to k^m. Let E(k) be the set of the pairs (X, f), where X is an arbitrary k-tuple of $\mathbf{K}_k^n$, and f is a mapping of the set $\{1, 2, \ldots, n\}$ onto X. Then $\nu(E(k)) = \binom{n}{k}k^m$.

Let Ψ be the mapping of $\{1, 2, \ldots, m\}$ onto $\{1, 2, \ldots, n\}$ and $R(\Psi)$ be its image. Denote by $E(k, \Psi)$ the subset of E(k) containing those pairs (X, f) for which $f(i) = \Psi(i)$, $i = 1, \ldots, m$. Since $E(k, \Psi) = \emptyset$ for $k < \nu(R(\Psi))$ and $E(K) = \bigcup_{\Psi} \{E(k, \Psi) : \nu(R(\Psi)) \leqslant k\}$, then

$$\nu(E(k)) = \sum_{\Psi} \nu(E(k, \Psi)).$$

Hence it follows that

$$\sum_{k=0}^{n} (-1)^{n-k}\binom{n}{k}k^m = \sum_{\Psi} \sum_{k=0}^{n} (-1)^{n-k}\nu(E(k, \Psi)). \tag{1.7}$$

On the right-hand side of (1.7) the external sum is calculated through all mappings Ψ of $\{1, \ldots, n\}$ onto $\{1, \ldots, n\}$. It is evident that for each Ψ the number of the pairs in $E(k, \Psi)$ is equal to the number of k-tuples of $\{1, \ldots, n\}$ which contain $R(\Psi)$. This implies that if $R(\Psi)$ is a t-tuple, then

$$\nu(E(k, \Psi)) = \begin{cases} 0, & \text{if } k < t, \\ \binom{n - t}{k - t}, & \text{if } k \geqslant t. \end{cases}$$

Substituting $\nu(E(k, \Psi))$ in (1.7) and noting that $\sum_{i=0}^{n} (-1)^i\binom{n}{i} = 0$, we obtain

$$\sum_{k=0}^{n} (-1)^{n-k}\nu(E(k, \Psi)) = \sum_{k=t}^{n} (-1)^{n-k}\binom{n - t}{k - t} =$$

$$= (-1)^{n-t}\sum_{s=0}^{n-t} (-1)^s\binom{n - t}{s} =$$

$$= \begin{cases} 0, & \text{if } t < n \\ 1, & \text{if } t = n. \end{cases} \tag{1.8}$$

However, if $0 \leqslant m < n$, then $\nu(R(\Psi)) < n$ for each mappings Ψ of $\{1, \ldots, m\}$ onto $\{1, \ldots, n\}$ and it follows from (1.7) and (1.8) that

$$\sum_{k=0}^{n} (-1)^{n-k}\binom{n}{k}k^m = 0.$$

Now consider the case when $m = n$. Here for each mapping Ψ whose image $R(\Psi) \subset \{1, \ldots, n\}$ (the inclusion is strong), the corresponding term on the right-hand side of (1.7) is equal to 0. However, the number of permutations of the set $\{1, \ldots, n\}$ is equal to $n!$, and for each mapping $\Psi : \{1, \ldots, n\} \to R(\Psi) = \{1, \ldots, n\}$ the corresponding term on the right-hand side of (1.7) is equal to 1, according to (1.8). Hence, we have shown that for $m = n$

$$\sum_{k=0}^{n} (-1)^{n-k}\binom{n}{k}k^m = n!.$$

2. Events and Relations among Them

2.1. $B = A_6$, $C = A_5$.

2.2. (a) The certain event Ω. (b) The impossible event $\emptyset$. (c) The event B. (d) The event A.
2.3. (a) At least one book is taken. (b) At least one volume of each edition is taken. (c) Either at least one book of the first edition is taken or three books of the second edition are taken. (d) Two volumes of the first and two volumes of the second edition are taken. (e) One volume of the first edition and three volumes of the second edition are taken, or the converse.
2.4. $A \setminus B = A\bar{B}$ = {the chosen number ends in 5}.
2.5. (a) 0; (b) 0; (c) 4; (d) 2; (e) 1; (f) 1; (g) 0.
2.6. (a) $A\bar{B}\bar{C}$; (b) $AB\bar{C}$; (c) ABC; (d) $A \cup B \cup C$; (e) $AB \cup AC \cup BC$; (f) $A\bar{B}\bar{C} + \bar{A}B\bar{C} + \bar{A}\bar{B}C$; (g) $\bar{A}BC + A\bar{B}C + AB\bar{C} = (AB \cup AC \cup BC) \setminus ABC$; (h) $\bar{A}\bar{B}\bar{C}$; (i) $\overline{ABC}$.
2.7. Hint: Apply the method used in Example 2.2.
2.8. (a) Let x be an arbitrary element of A. From $x \in A$ and $A \subset B$ it follows that $x \in B$. From $x \in B$ and $B \subset C$ follows $x \in C$. (b) Let $x \in AB$. Then $x \in A$. (c) Let $x \in A$. Then $x \in A \cup B$.
2.9. (a) $A \subset BC$; i.e., the event BC occurs every time when the event A occurs; (b) $B \subset A$ and $C \subset A$; i.e., A occurs whenever either B or C occurs.
2.10. Using the distributive properties (Example 2.2) and formulas of Exercise 2.7 we have: (a) $(A \cup B)(B \cup C) = [A(B \cup C)] \cup [B(B \cup C)] = (AB \cup AC) \cup B = (AB \cup B) \cup AC = B \cup AC$; (b) A; (c) AB; (d) Ω; (e) $\emptyset$.
2.11. (a) A; (b) B; (c) AC; (d) $B \cup C$.
2.12. $A(\bar{A}B) = (A\bar{A})B = \emptyset$; $A(\overline{A \cup B}) = A(\bar{A}\bar{B}) = (A\bar{A})\bar{B} = \emptyset$; $\bar{A}B(\overline{A \cup B}) = (\bar{A}B)(\bar{A}\bar{B}) = (\bar{A}\bar{A})(B\bar{B}) = \bar{A}\emptyset = \emptyset$. Besides $A \cup \bar{A}B \cup \overline{A \cup B} = A \cup [\bar{A}B \cup \bar{A}\bar{B}] = A \cup [\bar{A}(B \cup \bar{B})] = A \cup [\bar{A}\Omega] = A \cup \bar{A} = \Omega$.
2.13. $(A \cup B)(\bar{A} \cup B)(A \cup \bar{B}) = [(A \cup B)\bar{A} \cup B](A \cup \bar{B}) = [\bar{A}B \cup B] \cap (A \cup \bar{B}) = B(A \cup \bar{B}) = AB$. The required property is $AB = \emptyset$.
2.14. (a) Yes; (b) not always; (c) not always.
2.15. We have: (a) $A = A(A \cup C) = A(B \cup C) = AB \cup AC = AB \cup BC = B(A \cup C) = B(B \cup C) = B$; (b) $A = A \cup \emptyset = A \cup C\bar{C} = (A \cup C)(A \cup \bar{C}) = (B \cup C)(B \cup \bar{C}) = [B(B \cup \bar{C})] \cup [C(B \cup \bar{C})] = B \cup BC = B$.
2.16. We have $B = \overline{(X \cup \bar{A})} \cup \overline{(X \cup A)} = (\bar{X}A) \cup (\bar{X}\bar{A}) = \bar{X}(\bar{A} \cup A) = \bar{X}$; $X = \bar{B}$.
2.17. We have: (a) $AX = A[AB \cup \bar{A}D] = A(AB) \cup A(\bar{A}D) = (AA)B \cup (A\bar{A})D = AB$; (b) $A \cup Y = A \cup [\bar{A}B \cup AD] = \bar{A}B \cup [A \cup AD] = \bar{A}B \cup A = (\bar{A} \cup A)(B \cup A) = (A \cup B)\Omega = A \cup B$; (c) $AB \cup Z = AB \cup [\bar{A}BC \cup ABD] = \bar{A}BC \cup [(AB) \cup (AB)D] = (\bar{A}BC \cup AB = (\bar{A}B \cup AB)(C \cup AB) = \Omega(C \cup AB) = C \cup AB = (C \cup A)(C \cup B)$.
2.18. A cut-off of the chain will happen when either the element a or all the three elements b_k $(k = 1, 2, 3)$ will go out of order. That are the events A and $B_1B_2B_3$ respectively. Then $C = A \cup B_1B_2B_3$;
$\bar{C} = \overline{A \cup B_1B_2B_3} = \bar{A}(\overline{B_1B_2B_3}) = \bar{A}(\bar{B}_1 \cup \bar{B}_2 \cup \bar{B}_3)$
2.19. $D = A(B_1 \cup B_2 \cup B_3 \cup B_4)(C_1 \cup C_2)$; $\bar{D} = \bar{A} \cup \bar{B}_1\bar{B}_2\bar{B}_3\bar{B}_4 \cup \bar{C}_1\bar{C}_2$.
2.20. $C = (A_1 \cup A_2)(B_1B_2 \cup B_1B_3 \cup B_2B_3)$.

3. Classical Definition of Probability

3.1. 1/12.
3.2. 15/27 = 5/9.
3.3. 0.3.

3.4. $(n + 1)/6^n$.
3.5. $2(k - 1)(n - k)/[n(n - 1)]$.
3.6. 4/9.
3.7. (a) 1/8; (b) 25/216; (c) 1/2.
3.8. The number of possible outcomes is equal to $P(5, 3) = 5 \cdot 4 \cdot 3 = 60$. Among these, the favourable ones are ordered triplets with last digit 2 (their number is $P(4, 2) = 4 \cdot 3 = 12$) as well as those with last digit 4 (also 12 in number). Thus we find $P(A) = 0.4$.
3.9. $(3^2 + 6 \cdot 3)6^{-2} = 0.75$.
3.10. $(n - 2)/n$.
3.11. $1 \cdot 3 \cdot 5 \cdot \ldots \cdot (2n - 1)(n!2^{n-1})^{-1}$.
3.12. $\binom{m + n - k - s - 2}{m - k - 1}/\binom{m + n}{m}$ when $m > k$ and $n > s$; $1/\binom{m + n}{m}$ when $m = k$ and $n = s$.
3.13. $P(A) = \dfrac{n - k}{m + n - k}$.
3.14. Consider the model of placing five distinguishable particles (the disks) into six different cells (the sectors on the disks) without restrictions about the number of particles in a cell. Then $P(A) = 6^{-5} \approx 0.00013$.
3.15. 1/360.
3.16. (a) $\binom{N}{k}/N^k$; (b) $1/k!$.
3.17. $1 - \binom{m}{2}/\binom{m + n - b}{2}$.
3.18. (a) 1/15. (b) $(10 - k + 1)!k!(10!)^{-1}$.
3.19. $10 \cdot 9 \cdot 8 \cdot 7 \cdot 6(10^5 - 1)^{-1} \approx 0.302$.
3.20. (a) $\dfrac{9!}{(3!)^3 3^9}$; (b) $\dfrac{9!}{4!3!2!3^9} \cdot \dfrac{1}{3!}$.
3.21. (a) $(n!)/\dfrac{(2n)!}{(2!)^n} = \dfrac{1}{(2n - 1)!!}$; (b) $(n!)^2/\dfrac{(2n)!}{(2!)^n} = \dfrac{n!}{(2n - 1)!!}$.
(Recall that $(2n - 1)!! = 1 \cdot 3 \cdot 5 \cdot \ldots \cdot (2n - 1)$.)
3.22. $\dfrac{n!}{n_1!n_2! \ldots n_6!6^n}$.
3.23. (a) $\dfrac{1}{6^n}$; (b) $\dfrac{n}{6^n}$; (c) $\dfrac{n!}{6^n} \Sigma \dfrac{1}{n_1!n_2! \ldots n_6!}$, where summation is taken over all possible integers $n_1, n_2, \ldots, n_6$, for which $n_1 + 2n_2 + 3n_3 + \ldots + 6n_6 = s$.
3.24. $6^{-m} \cdot 2^{-n}$.
3.25. (a) $12!12^{-12} \approx 0.000054$; (b) $\binom{12}{2}(2^6 - 2)/12^6 \approx 0.000132$.
3.26. Let A_r = {the birthdays of at least two of r people coincide}. The total number of the possible outcomes is 365^r. Among them the favourable for the occurrence of $\overline{A}_r$ are $\binom{365}{r}!$. We have $P_r = P(A_r) = 1 - P(\overline{A}_r) = 1 - \dfrac{365!}{365^r(365 - r)!}$. The sequence $\{P_r\}$ increases monotonically. Actuallu $(1 - P_{r+1})/(1 - P_r) = (365 - r)/365 < 1$; hence $P_{r+1} > P_r$. Using loga-

rithmic tables and tables for the factorials of the integer numbers, we obtain $P_{22} \approx 0.4757$ and $P_{23} \approx 0.5073$. Therefore, the smallest number r, for which $P_r > 1/2$, is $r = 23$.

3.27. $2(n - r - 1)(n - 2)!(n!)^{-1} = 2(n - r - 1)[n(n - 1)]^{-1}$.

3.28. (a) If one of the two persons has chosen a place, the other can choose one of the remaining n - 1 places. Two of them will be neighbouring the seat of the first person; hence, $P(A) = 2/(n - 1)$;
(b) $(n - 3)/(n - 1)$.

3.29. $\dfrac{2(m - 1)n(mn - 2)!}{(mn)!} = \dfrac{2(m - 1)}{m(mn - 1)}$.

3.30. $\binom{n}{m}\binom{n - k - m}{n - m}/\binom{n + k}{n}$.

3.31. (a) $10 \cdot 9 \cdot 8 \cdot 7 \cdot 10^{-4} = 0.504$; (b) $10\binom{9}{2}\dfrac{4!}{2!} \cdot 10^{-4} = 0.432$; (c) $10\binom{9}{1}\dfrac{4!}{3!1!}10^{-4} = 0.036$; (d) $\binom{10}{2}\dfrac{4!}{2!2!}10^{-4} = 0.027$; (e) $[2(1^2 + 2^2 + \dots + 9^2) + 10^2]10^{-4} = 0.067$; (f) 0.001.

3.32. (a) $\binom{2N}{N}^2/\binom{4N}{2N}$; (b) $\binom{4N - 4}{2N - 4}/\binom{4N}{2N}$; (c) $\binom{4}{2}\binom{4N - 4}{2N - 2}/\binom{4N}{2N}$.

3.33. Let A_i = {both partners receive i diamonds}. Then

$$P(A_0)/P(A_{13}) = \left[\binom{13}{0}\binom{39}{26}/\binom{52}{26}\right] : \left[\binom{13}{13}\binom{39}{13}/\binom{52}{26}\right] = 1.$$

3.34. The number of possible outcomes is equal to $\binom{n}{1} + \binom{n}{2} + \dots + \binom{n}{n} = 2^n - 1$. The number of the outcomes favourable for drawing an odd number of balls is $\nu_{odd} = \binom{n}{1} + \binom{n}{3} + \binom{n}{5} + \dots = \binom{n}{0} + \binom{n}{2} + \binom{n}{4} + \dots = 2^n - \nu_{even}$. We have $p_{odd} = \dfrac{2^{n-1}}{2^n - 1} > p_{even} = \dfrac{2^{n-1} - 1}{2^n - 1}$.

3.35. (a) $1 - \binom{5}{1}\binom{3}{1}\binom{2}{1}/\binom{10}{3} = 0.75$; (b) $\dfrac{20 + 15}{120} = \dfrac{7}{24}$.

3.36. (a) $\sum_{k=0}^{2r} \binom{n}{k}\binom{n - k}{2r - k}/\binom{2n}{2r} = 2^{2r}\binom{n}{2r}/\binom{2n}{2r}$;

(b) $n \sum_{k=0}^{2r-2} \binom{n - 1}{k}\binom{n - k - 1}{2r - k - 2}/\binom{2n}{2r} = n \cdot 2^{2r-2}\binom{n - 1}{2r - 2}/\binom{2n}{2r}$;

(c) $\binom{n}{2} \sum_{k=0}^{2r-4} \binom{n - 2}{k}\binom{n - k - 2}{2r - k - 4}/\binom{2n}{2r} = \binom{n}{2}\binom{n - 2}{2r - 4}2^{2r-4}/\binom{2n}{2r}$.

3.37. We can describe the result of n tossings of the coin with a sequence of n letter H (head) and T (tail), say TTHHH ... T. The number of the possible outcomes containing exactly k letters H is equal to the number of the different ways in which we may select k of all n places in the sequence; i.e., $\binom{n}{k}$. Thus the number of all outcomes is equal to $\binom{n}{0} + \binom{n}{1} + \dots + \binom{n}{n} = 2^n$. The number of favourable outcomes of the event A = {odd number of heads} is $\binom{n}{1} + \binom{n}{3} + \dots = 2^{n-1}$. Hence $P(A) = 2^{n-1}/2^n = 0.5$.

3.38. The number of possible outcomes is $\binom{2n + 1}{3}$. Consider the triangles which do not contain the centre of the polygon. These triangles are obtuse (why?). If we select an arbitrary vertex M of the polygon, then there exist exactly n vertices of the polygon such that in choosing two of them we obtain an obtuse triangle whose obtuse angle vertex is at the point M. Hence, the number of outcomes favourable for the event $\bar{A}$ is $(2n + 1)\binom{n}{2}$. Thus $P(A) = 1 - P(\bar{A}) = (n + 1)/[2(2n - 1)]$.

3.39. $2(n + 1)\binom{n^2 - n}{k}\Big/\binom{n^2}{n + k}$.

3.40. If N is the chosen number, then N can be written in the form $N = 10a + b$, where a and b are non-negative integer numbers. Since $N^3 = 10^3 \cdot a^3 + 3 \cdot 10^2 \cdot a^2b + 3 \cdot 10 \cdot ab^2 + b^3$, we easily find that the last two digits of N^3 will be both one only if $b = 1$ and $a = 7$. Thus any number ending by 71 is a favourable outcome for the event A = {a randomly chosen natural number has a cube ending by 11}. Obviously among any 100 consecutive numbers, there is only one which is favourable for A and if N is chosen from {1, 2, ..., M}, then [M/100] is the number of the favourable outcomes. This implies that

$$P(A) = \lim_{M\to\infty} \frac{[M/100]}{M} = 0.01.$$

3.41. (a) Represent N as $N = a + 10b + \dots$, where each of a and b can take arbitrarily one of the values 0, 1, 2, ..., 9. Thus $N^2 = a^2 + 100b^2 + 2 \cdot 10ab + \dots$ It is easy to see that the last digit of N^2 depends only on the value of a. Among the ten possible values of a, those which are favourable, are $a = 1$ and $a = 9$. Thus, $P(A) = 0.2$. (b) $N^4 = a^4 + \dots$ We can see that the last digit of N^4 depends only on the value of a. The favourable values for a are 1, 3, 7 and 9. $P(A) = 0.4$. (c) Denote by $M = x + 10y + \dots$, the other integer chosen at random. Then $M \cdot N = ax + 10(ay + bx) + \dots$ The last digit of the product depends only on the units digits of M and N. There are 10^2 possible outcomes of ordered pairs among which (1, 1), (3, 7), (7, 3) and (9, 9) are favourable. Thus $P(A) = 0.04$.

3.42. (a) The difference $x^2 - y^2$ is divisible by 2 only when x and y are with equal parity; $x^2 - y^2$ is divisible by 3 only when either both numbers x and y are divisible by 3 or both are not divisible by 3. (Verify that!) The number of the elements of M which are divisible by 2 is equal to $[\frac{n}{2}]$ (the integer part of $\frac{n}{2}$), while the number of those which are divisible by 3 is $[\frac{n}{3}]$. Let $n_2 = [\frac{n}{2}]$ and $n_3 = [\frac{n}{3}]$. Then

$$P(A_2) = \frac{1}{n^2}\left(n_2^2 + (n - n_2)^2\right) = 1 - \frac{2n_2}{n} + \frac{2n_2^2}{n^2};$$

$$P(A_3) = 1 - \frac{2n_3}{n} + \frac{2n_3^2}{n^2};$$

(b) It is easy to show that $P(A_3) > P(A_2)$.

3.43. $\binom{M-1}{m-1}\binom{N-M}{n-m}/\binom{N}{n}$.

3.44. $6^{-10}\sum_{i=0}^{5}\binom{10}{i}\binom{10-i}{i}\cdot 4^{10-2i} \approx 0.219$.

3.45. $5!\binom{10}{2}\binom{8}{2}\binom{6}{2}\binom{4}{2}\binom{2}{2}\Big/\left[\binom{15}{3}\binom{12}{3}\binom{9}{3}\binom{6}{3}\binom{3}{3}\right] = \frac{81}{1001}$.

3.46. Let the initial 2^n positions be divided into two halves; the first one consisting of the first 2^{n-1} positions and the second of the remaining ones. Obviously player B will reach the final only if he will not meet player A before. This event will occur only if A is in the first half of the initial list and B is in the second one, or conversely. Hence the desired probability is equal to

$$\frac{2\times 2^{n-1}\times 2^{n-1}\times(2^n-2)!}{(2^n)!} = \frac{2^{n-1}}{2^n-1}.$$

3.47. (a) $\binom{n+r-k-2}{r-k}/\binom{n+r-1}{r}$; (b) $\binom{n}{m}\binom{r-1}{r-n+m}/\binom{n+r-1}{r}$; (c) $\binom{r-n-1}{r-2n}/\binom{n+r-1}{r}$.

3.48. $(n-r)/n$.

3.49. (a) $\frac{r!}{k_1!k_2!\ldots k_n!}\,\frac{1}{n^r}$; (b) $\frac{n!}{n^n}$; (c) $\frac{n!}{n^n}\binom{n}{2}$;

(d) $\binom{n}{m}\frac{r!}{n^r}\sum\frac{1}{k_1!k_2!\ldots k_{n-m}!}$, where the summation is over all possible positive integers $k_1, k_2, \ldots, k_{n-m}$, for which $k_1 + k_2 + \ldots + k_{n-m} = r$.

3.50. Denote the faces of the three dice by a, b and c, where a, b, c can be 1, 2, 3, 4, 5, 6. Then from the equality $a + b + c = abc$ it follows that

$$\frac{1}{ab} + \frac{1}{ac} + \frac{1}{bc} = 1.$$

On the left-hand side of this equation either all three fractions are equal to 1/3 or at least one of them is not 1/3. Suppose that $\frac{1}{ab} = \frac{1}{ac} = \frac{1}{bc} = \frac{1}{3}$, then $ab = ac = bc = 3$; i.e., at least one of these numbers should satisfy the equality $x^2 = 3$ and this is impossible. Therefore, at least one of the fractions above is not 1/3. Since all three fractions cannot be smaller than 1/3, there will be at least one greater than 1/3. For instance, let $\frac{1}{ab} > \frac{1}{3}$; hence, either $ab = 1$ or $ab = 2$. The case ab =

1 is not possible. If ab = 2, then a = 1, b = 2 (or conversely) and hence c = 3. Thus the ordered triple (1, 2, 3) is a favourable outcome for the event. The total number of favourable outcomes is 3! Since the total number of possible outcomes is 6^3, the required probability is $P(A) = 3!/6^3 = 1/36$.

3.51. The space of outcomes is $\Omega = \tilde{B}^n_N$, and $\nu(\Omega) = N^n$. Denote the set of favourable outcomes by A, and let the ordered n-tuple

$$(x_1, x_2, \ldots, x_n) \in A; \tag{3.11}$$

i.e., if we arrange the elements of (3.11) in a non-decreasing sequence

$$x_{i_1} \leqslant x_{i_2} \leqslant \ldots \leqslant x_{i_m} \leqslant \ldots \leqslant x_{i_n}, \tag{3.12}$$

then we have $x_{i_m} = M$.

We shall describe now a two-step procedure which will allow us to calculate $\nu(A)$.

First step: (1) Display in a table all the orderings with repetitions of the numbers 1, 2, ..., M taken m - 1 at a time. These will be the first m - 1 terms of (3.12); (2) On the right-hand side of each element of this table add M at the mth position; i.e., $x_{i_m} = M$. (3) On the right of these ordered m-tuples add to each exactly one of the orderings (repetition allowed) of the numbers M, M + 1, ..., N. These will be the last n - m elements of (3.12). The so obtained new table contains $M^{m-1}(N - M + 1)^{n-m}$ ordered n-tuples of the type

$$(x_1^*, x_2^*, \ldots, x_{m-1}^*, M, x_{m+1}^*, \ldots, x_n^*). \tag{3.13}$$

Denote their set by A*. We have seen that

$$\nu(A^*) = M^{m-1}(N - M + 1)^{n-m}. \tag{3.14}$$

Obviously, $A^* \subset A$, but this is not the whole set of elements of A because the numbers $x_1^*, \ldots, x_{m-1}^*$ in (3.13) do not exceed M, $x_m^* = M$ and the numbers $x_{m+1}^*, \ldots, x_n^*$ are not smaller than M, while for the elements (3.12) of A such restrictions do not exist.

Second step: (4) Let (3.13) be a fixed element of A*. There are exactly n different elements of A which contain the same numbers $x_1^*, \ldots, x_{m-1}^*, x_{m+1}^*, \ldots, x_n^*$, in the same order. These are obtained by placing the number M in each of the n positions in (3.13). A typical element with M in the first position is given by

$$(M, x_1^*, x_2^*, \ldots, x_{m-1}^*, x_{m+1}^*, \ldots, x_n^*).$$

(5) With M in the first position, there are exactly $\binom{n-1}{m-1}$ different elements of A which are obtained by filling the remaining n - 1 positions with $x_1^*, \ldots, x_{m-1}^*$ in that order and the remaining positions with $x_{m+1}^*, \ldots, x_n^*$ in that order. Then $\nu(A) = n\binom{n-1}{m-1}\nu(A^*)$; hence, in view of (3.14) the required probability is

$$n\binom{n-1}{m-1}M^{m-1}(N - M + 1)^{n-m}N^{-n}.$$

3.52. Let $A_n = \{ad - bc \neq 0\}$. Then $p_n = P(A_n) = 1 - P(\bar{A}_n)$. Since every one of the elements of the ordered quadruple (a, b, c, d) is some number from the set $M = \{0, \pm 1, \pm 2, \ldots, \pm n\}$, the number of the possible outcomes is $m = (2n + 1)^4$. Let us consider those which are favourable for the event $\bar{A}_n = \{ad - bc = 0\}$. If ad = bc = 0, then the pair (a, d) with ad = 0 can be selected in 2(2n + 1) different ways. The same is true for the pair (b, c) with bc = 0. Thus, the number of the ordered quadruples (a, b, c, d) with ad = bc = 0 is $m_1 = 4(2n + 1)^2$. If ad = bc ≠ 0, then every one of the numbers a, b, c and d must be different from 0. When the choice of numbers of M is an arbitrary one, then for any 3 of these numbers the value of the fourth one is uniquely determined through the conditions ad = bc ≠ 0. However, the value of the fourth does not always belong to M (why?). Therefore, the number m_2 of cases in which ad = bc ≠ 0 satisfies the inequality $m_2 \leqslant \binom{4}{3}(2n)^3 = 32n^3$. Then

$$p_n = 1 - P(\bar{A}) = 1 - \frac{m_1 + m_2}{m} \geqslant 1 - \frac{32n^3 + 4(2n + 1)^2}{(2n + 1)^4}.$$

Hence we obtain $\lim_{n\to\infty} p_n = 1$.

3.53. Let the line of ticket buyers be represented by the sequence $\varepsilon_1, \varepsilon_2, \ldots, \varepsilon_{n+m}$, where

$$\varepsilon_i = \begin{cases} +1, & \text{if the ith buyer has only one-lev notes,} \\ -1, & \text{if the ith buyer has only two-lev notes.} \end{cases}$$

Now consider the sequence $s_1, s_2, \ldots, s_{n+m}$, where $s_k = \varepsilon_1 + \varepsilon_2 + \ldots + \varepsilon_k$. Clearly $s_{n+m} = n - m$. The problem can be visualized more easily with the help of a t×s rectangular plot. On the t-axis we plot the consecutive number of buyers: 1, 2, ..., n + m; the values of s_k for k = 1, 2, ..., n + m are plotted on the s-axis. Thus any possible realization of the queue will be represented by a polygonal line starting at (0, 0) and terminating at (n + m, n - m). For each i the line makes a jump of +1 if $\varepsilon_i = +1$ and -1 if $\varepsilon_i = -1$ (Figure 3.2).

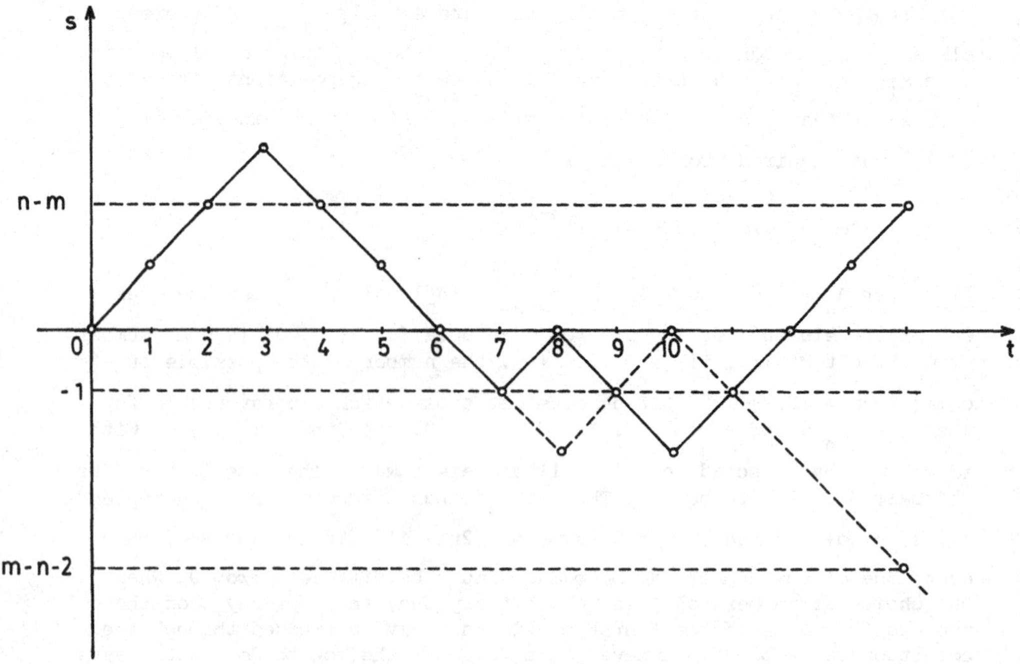

Figure 3.2.

Since each path is composed of n + m jumps of which n are positive and m are negative, every path is uniquely determined by considering on the positive jumps. Then the total number of paths (outcomes) is $\binom{n+m}{n}$. Let A be the event that no ticket buyer has to wait for a change. In order to determine the number of outcomes favourable to A, we will count those paths which are strictly positive; i.e., those lying entirely above the line s = -1. Only these are favourable outcomes since, if $s_k = -1$ for some k, this implies that the kth buyer will have to wait for change.

To count the number of these outcomes we shall use the *reflection principle*, first noted by Bertrand in 1887 but attribted to D. André by probabilitists. With each path which touches or crosses the line s = -1, we associate a single path which coincides with the initial one up the point at which it first touches the line s = -1. To the right of this point, it is a symmetric reflection on the initial line with respect to s = -1. We call the points with coordinates (s, t_1) and (s, t_2) symmetric with respect to the line s = c when $(t_1 + t_2)/2 = c$. On Figure 3.2. the dotted line shows a new path after first touching the line s = -1. The new path ends at the point with coordinates (n + m, m - n - 2), which is symmetrical to the point (n + m, n - m) with respect to the straight line s = -1. For both the number of the positive jumps x and the number of the negative jumps y, which the new path makes in connecting the points (0, 0) and $(n + m, m - n - 2)$, we have respectively $x + y = n + m$ and $x - y = m - n - 2$; hence we find that $x = m - 1$, $y = n + 1$. Thus the number of all paths from (0, 0) to $(m + n, n - m)$, touching or crossing the line $s = -1$, is equal to $\binom{n+m}{m-1}$, and the number of the favourable outcomes is $\binom{m+n}{n} - \binom{m+n}{n+1}$. Therefore,

$$P(A) = 1 - \binom{n+m}{n+1}/\binom{n+m}{n} = \frac{n - m + 1}{n + 1} .$$

3.54. $\frac{n - m}{n + m}$. *Hint*: Apply the method of Exercise 3.53. Favourable outcomes are those paths for which $s_1 = +1$, and which do not touch the t-axis.

3.55. (a) Consider the first N natural numbers 1, 2, ..., N and denote by $q(N)$ the number of the even numbers which do not exceed N. We have $N = 2q + r$, where $q = q(N) = [N/2]$ and $r = 0$ or 1. Then

$$P(A) = \lim_{N\to\infty} \frac{q(N)}{N} = \lim_{q\to\infty} \frac{q}{2q + r} = \frac{1}{2} ;$$

(b) $\frac{1}{5}$.

Note. If we arrange the set of natural numbers in a different way (not in their natural order) an application of definition (3.10), given in the note at the end of Exercise 3.55, may lead to a different result. Consider for example the sequence

$$1, 2, 3, 5, 4, 7, 9, 11, 13, 6, 15, 17, 19, 21, 23, 25,$$
$$27, 29, 8, \ldots, 2, \ldots \tag{3.15}$$

which we obtain by rearranging the sequence of natural numbers 1, 2, ..., n, ... so that the odd numbers appear consecutively with the successive even numbers inserted in such a way that in front of the first even number 2 to stand the first odd number 1; behind the first even number and in front of the second even number 4 to stand the next two odd numbers; ...; and finally, behind the $2(n - 1)$ even number and in front of the 2n even number to stand the next 2^{n-1} odd numbers, etc. It is not difficult to see that the sequence (3.15) contains all natural numbers.

Let $a_1, a_2, \ldots, a_N$ be the first N terms of the sequence (3.15) and let A = {an arbitrary chosen number of (3.15) is even}. Then

$$P(A) = \lim_{N\to\infty} \frac{q(N)}{N} ,$$

if this limit exists.

In the sequence (3.15) in front of each even number 2n there are $n - 1$ even and $1 + 2 + 2^2 + \ldots + 2^{n-1} = 2^n - 1$ odd numbers. Therefore, the serial number of the number 2n in the sequence (3.15) is equal to

$1 + (n - 1) + (2^n - 1) = 2^n + n - 1$.

Consider the numerical sequence $\{b(N)\} = \{\frac{q(N)}{N}\}$, $N = 1, 2, \ldots$ and form its subsequence $\{b(2^k + k - 1)\}$, $k = 1, 2, \ldots$ Then

$$\lim_{k\to\infty} b(2^k + k - 1) = \lim_{k\to\infty} \frac{q(2^k + k - 1)}{2^k + k - 1} = \lim_{k\to\infty} \frac{k}{2^k + k - 1} = 0.$$

Hence, it follows that the limit (3.14) exists and is equal to 0; i.e., it is not equal to $\frac{1}{2}$ as obtained above. The reason for that is the special way that the set of natural numbers has been arranged.

3.56. The set S_m contains $\nu(S_m) = 9 \cdot 10^{m-1}$ numbers. The greatest $(m - 1)$-digit number divisible by k is $k[(10^{m-1} - 1)/k]$, and the greatest m-digit number divisible by k is $k[(10^m - 1)/k]$. (Here $[\cdot]$ is the integer part of the number in the brackets.) Thus the number of the elements of S_m which are divisible by k is equal to $[(10^m - 1)/k] - [(10^{m-1} - 1)/k]$. Hence the required probability is

$$P_m(k) = \left(\left[\frac{10^m - 1}{k}\right] - \left[\frac{10^{m-1} - 1}{k}\right]\right)/(9 \cdot 10^{m-1}).$$

Since $[x]$ and x differ at most with 1 for each x, then

$$\lim_{m\to\infty} P_m(k) = \lim_{m\to\infty} \frac{\frac{10^m - 1}{k} - \frac{10^{m-1} - 1}{k}}{9 \cdot 10^{m-1}} = \frac{1}{k}.$$

3.57. Consider a plane with the coordinate system (x, y). As the chosen numbers ξ and η have to satisfy the condition $\xi^2 + \eta^2 \leqslant N^2$, it is clear that the point with coordinates (ξ, η) must lie in that part of the circle with radius N which is situated in the first quadrant. However, the number of all integer-valued points in a circle with radius N is approximately equal to πN^2, and therefore the number of favourable outcomes of the considered event is approximately equal to $\frac{1}{4}\pi N^2$. Since the total number of outcomes is N^2 the required probability is $p_N \approx \pi N^2/(4N^2)$ and its limit is $\lim_{N\to\infty} p_N = \pi/4$.

4. Conditional Probability. Independence of Events.

4.1. 0.251.

4.2. 0.75.

4.3. $\frac{1}{6} \cdot \frac{2}{5} \cdot \frac{1}{4} \cdot \frac{1}{3} \cdot \frac{1}{2} \cdot 1 = \frac{1}{360}$.

4.4. Let A = {the visible face of the chip is white} and B = {the non-visible face of the chip is white}. Then $P(B|A) = P(AB)/(P(A) = 2/3$.

4.5. $\frac{1}{2} \cdot \frac{3}{4} \cdot \frac{5}{6} \cdots \frac{99}{100} = \frac{100!}{2^{100}(50!)^2} \approx 0.08$.

4.6. $\frac{1}{n} \cdot \frac{1}{n-1} \cdots \frac{1}{n-k+1} = \frac{(n-k)!}{n!}$.

4.7. $2 \cdot \frac{n}{2n} \cdot \frac{n}{2n-1} \cdot \frac{n-1}{2n-2} \cdot \frac{n-1}{2n-3} \cdots \frac{1}{2} \cdot 1 = \frac{2(n!)^2}{(2n)!}$.

4.8. $\frac{mn}{\binom{m+n}{2}} \cdot \frac{(m-1)(n-1)}{\binom{m+n-2}{2}} \cdots \frac{(m-n+1)(n-n+1)}{\binom{m-n+2}{2}} = \frac{2^n m! n!}{(m+n)!}$.

4.9. If A = {k unicoloured balls}, B = {k black balls}, then $P(A) = [\binom{m}{k} + \binom{n}{k}]/\binom{m+n}{k}$, $P(B) = \binom{n}{k}/\binom{m+n}{k}$ and since AB = B we find

$$P(B\ A) = P(BA)/P(A) = P(B)/P(A) = \binom{n}{k}/[\binom{n}{k} + \binom{m}{k}].$$

4.10. (a) $1 - \frac{9}{10} \cdot \frac{8}{9} \cdot \frac{7}{8} = 0.3$; (b) $1 - \frac{4}{5} \cdot \frac{3}{4} \cdot \frac{2}{3} = 0.6$.

4.11. $P(A) = 1 - P(\bar{A}) = 1 - \prod_{k=1}^{n} (1 - p_k)$.

4.12. $1 - (1-p)^4 = 0.5$; $p \approx 0.159$.

4.13. $P(A) = P(\bar{A}_1\bar{A}_2 \ldots \bar{A}_k A_{k+1}) = P(A_{k+1}) \prod_{i=1}^{k} P(\bar{A}_i) = (1-p)^k p$.

4.14. Since $P(K \cup K_1K_2) = 1 - P(\overline{\bar{K}(\overline{K_1K_2})}) = 1 - P(\bar{K})P(\overline{K_1K_2}) = 1 - P(\bar{K})(1 - P(K_1)P(K_2))$, then $1 - (1 - 0.3)(1 - 0.2^2) = 0.328$.

4.15. $P(A_1) = 1 - (\frac{5}{6})^4$, $P(A_2) = 1 - (\frac{35}{36})^{24}$. Then to show $P(A_1) > P(A_2)$ we have $(\frac{35}{36})^{24} - (\frac{5}{6})^4 = \frac{5^4}{6^4} \cdot \left(\frac{5^{10}7^{12}}{6^{22}} + 1\right) \cdot \left(\frac{5^5 7^6}{6^{11}} + 1\right) \cdot \left(\frac{5^5 7^6}{6^{11}} - 1\right)$.

But $\frac{5^5 7^6}{6^{11}} - 1 = (1 - \frac{1}{6})^5 (1 + \frac{1}{6})^6 - 1 = (1 - \frac{1}{36})^5 \cdot \frac{7}{6} - 1 \geqslant (1 - \frac{5}{36})\frac{7}{6} - 1$
$= \frac{217}{216} - 1 = \frac{1}{216} > 0$.
(We applied above the *inequality of Bernoulli*: $(1+x)^n \geqslant 1 + nx$ which holds for every n and $x \geqslant -1$.) Therefore,

$$(\frac{35}{36})^{24} - (\frac{5}{6})^4 > 0.$$

4.16. n = 253. Let N be the number of the equally probable birthdays and suppose you are inquiring n people. If A_k = {the birthday of the kth person coincides with your own} and A = {at least one of the n inquired people has the same birthday as yours}, then $P(A) = 1 - P(\bar{A}) = 1 - P(\bar{A}_1\bar{A}_2 \ldots \bar{A}_n) = 1 - ((N-1)/N)^n = P_n$. From the condition $P_n > 1/2$ with N = 365 we find $n > \ln 2/(\ln 365 - \ln 364) \approx 252.9$.

4.17. Use the inequality $e^{-x} \geqslant 1 - x$.

4.18. (a) $1 - (\frac{5}{6})^n > 0.5$, $n \geq 4$; (b) $n \geq 9$; (c) $n \geq 13$.
4.19. $n \geq 4$.
4.20. $1 - (\frac{35}{36})^n > \frac{1}{2}$, $n > 25$.
4.21. $n \geq \frac{\ln(1 - r)}{\ln(1 - p)}$.
4.22. $P(B|A) = \frac{P(AB)}{P(A)} = \frac{P(AB)}{P(B)} \cdot \frac{P(B)}{P(A)} = \frac{P(A|B)}{P(A)} P(B) > P(B)$.
4.23. The events are not independent: $P(A|B) = P(AB)/(P(B) = 0 \neq P(A)$.
4.24. $P(AB) = P(A)P(B)$. But since $A \subset B$ we have $P(AB) = P(A)$. Thus

$$P(A) = P(A)P(B), \quad P(A)[1 - P(B)] = 0.$$

4.25. The events W, G and R are pairwise independent, but they are not mutually independent.
4.26. See Exercise 4.25.
4.27. We find that: $A_1A_2A_3 = \{\omega_1\}$, $A_1A_2 = \{\omega_1, \omega_2\}$, $P(A_1) = 1/2$, $P(A_2) = 1/2$, $P(A_3) = 1/2$, $P(A_1A_2A_3) = 1/8 = (1/2)^3 = P(A_1)P(A_2)P(A_3)$, $P(A_1A_2) = 5/16 \neq (1/2)^2 = P(A_1)P(A_2)$.
4.28. $P(A_1) = P(A_2) = P(A_3) = 0.5$; $P(A_1A_2) = P(A_1A_3) = P(A_2A_3) = 0.25$; $P(A_1A_2A_3) = 0.5 \cdot 0.5 \cdot 0.5 = 0.125$.
4.29. In both cases the events are independent.
4.30. In the first k trials some k numbers are drawn and they can be arranged (permuted) in k! ways. Among the drawn numbers there will be one which will be the greatest and will occupy the kth place in $(k - 1)!$ of the cases. Therefore, $P(A_k) = (k - 1)!/k! = 1/k$, with $k = 1, \ldots, n$. It is interesting that $P(A_k)$ do not depend on n. Let now $1 \leq i_1 < i_2 < \ldots < i_s \leq n$, $s \leq n$. Then $P(A_{i_1} A_{i_2} \ldots A_{i_k}) = P(A_{i_1})P(A_{i_2}|A_{i_1}) \ldots P(A_{i_s}|A_{i_1} \ldots A_{i_{s-1}})$. The conditional probabilities are well-defined because $P(A_{i_1} \ldots A_{i_{s-1}}) > 0$. Thus $P(A_{i_1}) = 1/i_1$. It is easy to show, further, that $P(A_{i_2}|A_{i_1}) = 1/i_2$, $P(A_{i_3}|A_{i_1}A_{i_2}) = 1/i_3$, etc.; hence the independence of $A_1, \ldots, A_n$ follows.
4.31. Let the conditions of the exercise be fulfilled for the events $A_1, \ldots, A_n$. We shall find a lower bound of the number of elements of Ω. For an arbitrary n-tuple $(\varepsilon_1, \ldots, \varepsilon_n)$ with $\varepsilon_i = 0$ or 1 we have $P\left(\bigcap_{i=1}^{n} A_i^{(\varepsilon_i)}\right) > 0$, where $A_i^{(0)} = A_i$ and $A_i^{(1)} = \overline{A}$. This implies that $\bigcap_{i=1}^{n} A_i^{(\varepsilon_i)} \neq \emptyset$. Since, when $(\varepsilon_1, \ldots, \varepsilon_n) = (\varepsilon_1', \ldots, \varepsilon_n')$ the events $\bigcap_{i=1}^{n} A_i^{(\varepsilon_i)}$ and

$\bigcap_{i=1}^{n} A_i^{(\varepsilon_i')}$ are mutually exclusive and the number of different n-tuples $(\varepsilon_1, \ldots, \varepsilon_n)$, formed by the symbols 0 and 1, is equal to 2^n, then $N \geqslant 2^n$. Hence $n \leqslant \log_2 N$. The following case is extremal: $N = 2^n$; i.e., Ω consists of 2^n elements, $\Omega = \{(\varepsilon_1, \ldots, \varepsilon_n), \varepsilon_i = 0$ or 1, $i = 1, \ldots, n\}$, every elementary event $(\varepsilon_1, \ldots, \varepsilon_n)$ has probability $P((\varepsilon_1, \ldots, \varepsilon_n)) = 2^{-n}$, and the events $A_1, \ldots, A_n$ are defined as: $A_i = \{\varepsilon_i = 0\}$.

4.32. We have: A = {HHH, HHT, HTH, HTT}, B = {HHH, HHT, HTH, THH}, C = {HHH, TTT}, AB = {HHH, HHT, HTH}, AC = {HHH}, BC = {HHH}. With P(H) = P(T) = 1/2 we find according to formula (3.3) that P(A) = P(B) = 1/2, P(C) = 1/4, P(AB) = 3/8, P(AC) = P(BC) = 1/8. Thus A and C are independent, B and C are independent, but A and B are not independent.

4.33. If h_j and t_j denote respectively a head and a tail at the jth trial, then the space of elementary outcomes is $\Omega = \{h_1, t_1h_2, t_1t_2h_3, t_1t_2t_3\}$. It is easy to verify that if $P(h_j) = P(t_j) = 1/2$, $j = 1, 2, 3$, then $\sum_{k=1}^{4} P(\omega_k) = 1$, where $\omega_k \in \Omega$. Let A = {the coin is tossed three times} and B = {at the first trial a tail occurs}. According to formula (3.3) we find

$$P(A) = \frac{1}{8} + \frac{1}{8} = \frac{1}{4}, \quad \vec{P}(B) = \frac{1}{4} + \frac{1}{8} + \frac{1}{8} = \frac{1}{2}.$$

Since AB = A, the desired conditional probability is

$$P(A|B) = P(AB)/P(B) = P(A)/P(B) = \frac{1}{2}.$$

4.34. As in Exercise 4.33 we write down the sample space Ω:

$$\Omega = \{h_1, t_1h_2, t_1t_2h_3, \ldots, t_1t_2 \ldots t_{k-1}h_k, t_1t_2 \ldots t_{k-1}t_k\}.$$

Then we find the probabilities of the events A = {the coin is tossed k times} and B = {at the first two trials a tail occurs}: $P(A) = 1/2^{k-1}$ and $P(B) = 1/2^2$. Since AB = A, the desired probability is

$$P(A|B) = P(AB)/P(B) = P(A)/P(B) = 1/2^{k-3}.$$

4.35. If we denote by a and b the winning and by $\bar{a}$ and $\bar{b}$ the losing moves of the players A and B respectively, then $\Omega = \{a, \bar{a}b, \bar{a}\bar{b}a, \bar{a}\bar{b}\bar{a}\}$. Thus we find that P(A) = 0.44, P(B) = 0.35 and P(O) = 0.21.

4.36. The probability that the numerator and the denominator of a fraction will not be divisible by k is equal to $1 - 1/k^2$. Then the desired probability is

$$\left(1 - \frac{1}{2^2}\right)\left(1 - \frac{1}{3^2}\right)\left(1 - \frac{1}{5^2}\right)\left(1 - \frac{1}{7^2}\right) \cdots = \prod_k \left(1 - \frac{1}{k^2}\right),$$

where the multiplication is taken over all indices k which are prime numbers. This infinite product is convergent and is equal to $\frac{6}{\pi^2} \approx 0.608$.

This is a generalization of Exercise 3.55 for an arbitrary natural number n.

Note. The reader could find a more detailed solution of this exercise in the book [45, Problem 90] cited at the end of this Manual .

4.37. According to Exercise 4.31, it is possible to construct a probability space Ω and n independent events $A_1, A_2, \ldots, A_n$ such that $\vec{P}(A_1) = x_1, \ldots, P(A_n) = x_n$.

From the inequality $P^2(A_k) > 0$ it follows that $1 - P^2(A_k) < 1$ or $(1 - P(A_k))(1 + P(A_k)) < 1$, which we can write as

$$P(\bar{A}_k) < \frac{1}{1 + P(A_k)} .$$

From this and from the independence of the events $A_1, A_2, \ldots, A_n$, we find

$$\prod_{k=1}^{n} (1 - x_k) = P(\bar{A}_1, \ldots, \bar{A}_n) = \prod_{k=1}^{n} P(\bar{A}_k) < \prod_{k=1}^{n} \frac{1}{1 + P(A_k)} =$$

$$= \prod_{k=1}^{n} \frac{1}{1 + x_k} < \left(1 + \sum_{k=1}^{n} x_k\right)^{-1};$$

thus the right-hand equality given in (a) is proved.

Using De Morgan's formulas, the semi-additive property of the probability P and the independence of the events $A_1, \ldots, A_n$, we find

$$\prod_{k=1}^{n} (1 - x_k) = P(\bar{A}_1\bar{A}_2 \ldots \bar{A}_n) = 1 - P(\overline{\bar{A}_1\bar{A}_2 \ldots \bar{A}_n}) =$$

$$= 1 - P(A_1 \cup \ldots \cup A_n) > 1 - \sum_{k=1}^{n} P(A_k) =$$

$$= 1 - \sum_{k=1}^{n} x_k,$$

which is the left-hand inequality in (a). Notice that the two inequalities in (a) are strong because of $P(A_iA_j) = P(A_i)P(A_j) = x_ix_j > 0$.

If $\sum_{k=1}^{n} x_k < 1$, then the inequalities in (b) are equivalent to the following ones:

$$1 - \sum_{k=1}^{n} x_k < \prod_{k=1}^{n} \frac{1}{1 + x_k} < \frac{1}{1 + \sum_{k=1}^{n} x_k} .$$

To prove these inequalities it is sufficient to notice that

$$1 - \sum_{k=1}^{n} x_k < P(\bar{A}_1 \ldots \bar{A}_n)$$

and to repeat some of the above reasoning.

Note. Another proof (not a probabilistic one) of inequalities more general than those treated here is due to I. Baičev and M. Petkov and can be found in *Phys.-Math. Journal* (Bulg. Acad. of Sci.) 5 (1962), pp. 142-143.

5. Probability of a Sum of Events. Formula for Total Probability. Bayes' Formula

5.1. $\left[\sum_{k=0}^{m-1} \binom{M}{k}\binom{N - M}{n - k}\right] / \binom{N}{n}$.

5.2. 0.323.

5.3. $p + (1 - p)p = p(2 - p)$.

5.4. $P(A) = 1 - \sum_{k=0}^{2} \binom{6}{k}\binom{43}{6 - k} / \binom{49}{6} \approx 0.018638$.

5.5. $P(A) = 1 - P(S_n < n + 4) = 1 - \sum_{k=0}^{3} P(S_n = n + k) = 1 - 9^{-n}\{\binom{n}{n} + \binom{n}{n - 1} + [\binom{n}{n - 1} + \binom{n}{n - 2}] + [\binom{n}{n - 1} + 2\binom{n}{n - 2} + \binom{n}{n - 3}]\}$.

Let B_2 = {at least one even number is chosen}, B_5 = {a 5 is chosen at least once}, B_7 = {a 7 is chosen at least once}. Then

$$P(B) = P(B_2B_5B_7) = 1 - P(\overline{B_2B_5B_7}) = 1 - P(\bar{B}_2 \cup \bar{B}_5 \cup \bar{B}_7) =$$

$$= 1 - [P(\bar{B}_2) + P(\bar{B}_5) + P(\bar{B}_7) - P(\bar{B}_2\bar{B}_5) - P(\bar{B}_2\bar{B}_7) -$$

$$- P(\bar{B}_5\bar{B}_7) + P(\bar{B}_2\bar{B}_5\bar{B}_7)] =$$

$$= 1 - [(\tfrac{5}{9})^n + (\tfrac{8}{9})^n + (\tfrac{8}{9})^n - (\tfrac{4}{9})^n - (\tfrac{4}{9})^n - (\tfrac{7}{9})^n + (\tfrac{3}{9})^n].$$

5.6. The first player wins the match in one of the following n cases:

(1) when he wins the first m individual games; (2) in the first m individual games he loses one and wins the (m + 1)st one; (3) from the first m + 1 individual games he loses two but wins the (m + 2)nd; ...; (n) from the first m + n - 2 individual games he loses n - 1 but wins the (m + n + 1)st one. Hence, the probability that the first player wins is

$$p^m[1 + \binom{m}{1}q + \binom{m+1}{2}q^2 + \dots + \binom{m+n-2}{n-1}q^{n-1}].$$

5.7. The stake should be divided proportionally to the quotient p_1/p_2 of the probabilities of the first and the second player for winning the whole game. According to Exercise 5.6 we have

$$p_1 = \frac{1}{2^m}\left[1 + \binom{m}{1}\frac{1}{2} + \binom{m+1}{2}\frac{1}{2^2} + \dots + \binom{m+n-2}{n-1}\frac{1}{2^{n-1}}\right],$$

$$p_2 = \frac{1}{2^n}\left[1 + \binom{n}{1}\frac{1}{2} + \binom{n+1}{2}\frac{1}{2^2} + \dots + \binom{m+n-2}{m-1}\frac{1}{2^{m-1}}\right].$$

5.8. Let H_{i1} = {a white ball is drawn from the ith urn}, H_{i2} = {a black ball is drawn from the ith urn}. Then

$$P(H_{11}) = \frac{m}{m+n}, \quad P(H_{12}) = \frac{n}{m+n},$$

$$P(H_{21}) = \frac{m}{m+n}\cdot\frac{m+1}{m+n+1} + \frac{n}{m+n}\cdot\frac{m}{m+n+1} = \frac{m}{m+n},$$

$$P(H_{22}) = \frac{n}{m+n}.$$

Show by induction that $P(H_{i1}) = \frac{m}{m+n}$ and $P(H_{i2}) = \frac{n}{m+n}$ for every i. Then the required probability is $p = m/(m+n)$.

5.9. Assume that the elementary outcomes of $\Omega = \{2, 3, 4, 5, 6, 7, 8, 9, 10, 11, 12\}$ are equally probable. Hence we should have $r = 1/11$. Denote by p_j and q_j ($j = 1, 2, \dots, 6$) the probabilities that j dots will occur on the face of the first and of the second die respectively. According to the assumption we should have $p_1q_1 = p_6q_6$ or $q_1 = p_6q_6/p_1$. Substitute this value of q_1 in $I = (p_1 - p_6)(q_1 - q_6) \Rightarrow I = -q_6/p_1(p_6 - p_1)^2 \leqslant 0$. Therefore, $p_1q_1 + p_6q_6 \leqslant p_1q_6 + p_6q_1$; i.e., $r_2 + r_{12} \leqslant p_6q_1 + p_1q_6$. But $r_7 = (p_6q_1 + p_1q_6) + (p_5q_2 + p_2q_5) + (p_4q_3 + p_3q_4) \geqslant p_6q_1 + p_1q_6 \geqslant r_2 + r_{12}$; i.e., $r_7 \geqslant r_2 + r_{12}$ or $r \geqslant 2r$, which is not true because of $r = 1/11$.

5.10. Let ξ be the number marked on the face on which one of the dice is fallen. Denote respectively by x, y, z the probabilities: ξ is divisible by 3, it has a remainder 1 when divided by 3, it has a remainder 2 when divided by 3. Thus $p_3 = P\{S_3$ is a multiple of $3\} = x^3 + y^3 + z^3 + 6xyz$.

From $x + y + z = 1$ it follows that either $x^3 = y^3 = z^3 = 1/3$ (and then $p_3 = 1/3$ and $p_3 > 1/4$), or at least for one of the numbers, say for x, $x \geqslant 1/3$. Then

$$p_3 = x^3 + (y + z)^3 + 6xyz - 3y^2z - 3yz^2 =$$

$$= x^3 + (y + z)^3 + 3yz(2x - y - z) =$$

$$= x^3 + (1 - x)^3 + 3yz(3x - 1) = 3x^2 - 3x + 1 + 3yz(3x - 1) =$$

$$= 1/4 + 3(x - 1/2)^2 + 3yz(3x - 1) \geqslant 1/4.$$

5.11. (a) 1/3; (b) 5/6. See Exercise 3.55.

5.12. Let $p_1 = P(A_i)$, $i = 1, 2, 3$. We have $A_1A_2 = \bar{A}_3$, $A_2A_3 = \bar{A}_1$, $A_3A_1 = \bar{A}_2$; hence $P(A_1A_2) = 1 - p_3$, $P(A_2A_3) = 1 - p_1$, $P(A_3A_1) = 1 - p_2$. Taking into account that $P(A_1A_2A_3) = P(\bar{A}_1\bar{A}_2\bar{A}_3) = 0$ we obtain $0 = P(\bar{A}_1\bar{A}_2\bar{A}_3) = 1 - P(A_1 \cup A_2 \cup A_3) = 1 - P(A_1) - P(A_2) - P(A_3) + P(A_1A_2) + P(A_2A_3) + P(A_1A_3) = 2(2 - p_1 - p_2 - p_3)$.

5.13. Let A_k = {the ball with number k appears at the kth draw in the sample} for $k = 1, 2, \ldots, n$. Then, by formula (5.2), P(A) can be obtained. Here

$$P(A_k) = \frac{(n-1)!}{n!}, \quad P(A_kA_j) = \frac{(n-2)!}{n!},$$

$$P(A_kA_jA_i) = \frac{(n-3)!}{n!}, \ldots, P\left(\bigcap_{k=1}^{n} A_k\right) = \frac{1}{n!};$$

$$P(A) = \binom{n}{1}\frac{(n-1)!}{n!} - \binom{n}{2}\frac{(n-2)!}{n!} + \binom{n}{3}\frac{(n-3)!}{n!} - \ldots +$$

$$+ (-1)^{n-1}\frac{1}{n!} = 1 - \frac{1}{2!} + \frac{1}{3!} - \ldots + (-1)^{n-1}\frac{1}{n!} =$$

$$= 1 - \sum_{k=0}^{n} \frac{(-1)^{k-1}}{k!}.$$

For large enough values of n this probability is close to $1 - e^{-1} \approx 0.63212$. It is interesting to note that it approaches its boundary value even for unexpectedly small values of n, which can be seen from the following table:

n	3	4	5	6	7
P(A)	0.66667	0.62500	0.63333	0.63196	0.63214

5.14. $P_r = S_r - \binom{r+1}{1}S_{r+1} + \binom{r+2}{2}S_{r+2} - \ldots + (-1)^{n-r}\binom{n}{n-r}S_n$,

$Q_r = S_r - \binom{r}{1}S_{r+1} + \binom{r+1}{2}S_{r+2} - \ldots + (-1)^{n-r}\binom{n-1}{n-r}S_n$.

5.15. (a) $1/n!$; (b) 0; (c) $\sum_{k=1}^{n} (-1)^{k-1}/k!$; (d) P_r from Exercise 5.14.

5.16. $\lim_{n\to\infty} p_n(m) = e^{-m}$. (In Exercise 5.15 $m = 1$.)

5.17. Let A_k = {no passenger chooses the kth carriage}, $k = 1, 2, \ldots, n$. We have

$$P(A_k) = \frac{(n-1)^k}{n^k}, \quad P(A_kA_j) = \frac{(n-2)^k}{n^k}, \ \ldots, \ P(A_1 \ldots A_n) = 0.$$

Then

$$P(A) = P\left(\bigcap_{k=1}^{n} \overline{A}_k\right) = 1 - P\left(\bigcup_{k=1}^{n} A_k\right) = 1 - \binom{n}{1}\left(1 - \frac{1}{n}\right)^k +$$

$$+ \binom{n}{2}\left(1 - \frac{2}{n}\right)^k + \ldots + (-1)^{n-1}\left(1 - \frac{n-1}{n}\right)^k.$$

5.18. Consider a rectangular table with n rows and m columns. Suppose that the two outcomes for tossing a coin H (head) and T (tail) have probabilities x and $1 - x$ respectively; i.e., $x \in [0, 1]$. We shall associate with each cell of the table one of the letters H or T based on a tossing a coin. Consider an arbitrary series. Then x^m is the probability that the whole series consists of letters H only. The probability of having at least one 'T' in this series is equal to $1 - x^m$. However, this is true for each of the n rows. Therefore, in each of the rows there would be at least one letter T with probability $(1 - x^m)^n$. Quite analogously we find that $(1 - (1 - x)^n)^m$ is the probability that in each column we shall have at least one letter H. Consider the events: A = {at least one H in each column} and B = {at least one T in each row}. Obviously $A \cup B = \Omega$ is the certain event and therefore

$$1 = P(\Omega) = P(A \cup B) \leqslant P(A) + P(B) = (1 - x^m)^n + (1 - (1 - x)^n)^m.$$

Here the equality is attained only if $AB = \emptyset$. However, that is possible only in the two extreme cases: $x = 1$ and $x = 0$. If $0 < x < 1$, then $AB = \emptyset$. and the given inequality is strict.

5.19. Use the equality $P\left(\bigcap_{k=1}^{n} A_k\right) = 1 - P\left(\bigcup_{k=1}^{n} \overline{A}_k\right)$ and the formula of Poincaré (5.2) from the Introductory Notes to this section.

5.20. Since AB is the complement of $\overline{A} \cup \overline{B}$ by De Morgan's laws, then

$$P(AB) = 1 - P(\overline{A} \cup \overline{B}) = 1 - P(\overline{A}) - P(\overline{B}) + P(\overline{A}\overline{B});$$

hence,

$$P(AB) \geq 1 - P(\bar{A}) - P(\bar{B}).$$

5.21. $P\{A(B + C)\} = P(AB + AC) = P(AB) + P(AC) = P(A)P(B) + P(A)P(C) = P(A)[P(B) + P(C)] = P(A)P(B + C)$.
5.22. $P(A\bar{B}) = P(A) - P(AB)$.
5.23. $1 = P[(A \cup B) + (\overline{A \cup B})] = P(A) + P(B) - P(AB) + P(\bar{A}\bar{B})$.
5.24. $P(B) = P(AB) + P(\bar{A}B) = P(A)P(B|A) + P(\bar{A})P(B|\bar{A}) = [P(A) + P(\bar{A})]P(B|A) = P(B|A)$.
5.25. (a) $P(A\bar{B}) = P(A) - P(AB) = P(A) - P(A)P(B) = P(A)P(\bar{B})$. Case (b) and case (c) are checked analogously.
5.26. $A = A(B + \bar{B}) = AB + A\bar{B} = \emptyset + A\bar{B} = A\bar{B}$; i.e., $A \subset \bar{B}$ and $P(A) \leqslant P(\bar{B})$.
5.27. $P(A|B) = P(AB)/P(B) = [P(B) - P(\bar{A}B)]/P(B) \geqslant [P(B) - P(\bar{A})]/P(B) = (a + b - 1)/b$.
5.28. $P(A|B) = 0.98$.
5.29. $A_2 = A_2\Omega = A_2(A_1 \cup \bar{A}_1) = A_1A_2 + \bar{A}_1A_2 = A_1 + \bar{A}_1A_2$. Hence $P(\bar{A}_1A_2) = P(A_2) - P(A_1) = 1/6$. By analogy $P(\bar{A}_1A_3) = 1/3$, $P(\bar{A}_2A_3) = 1/6$, $P(A_1\bar{A}_2\bar{A}_3) = P(A_1\bar{A}_3) = 0$, $P(\bar{A}_1\bar{A}_2\bar{A}_3) = P(\bar{A}_3) = 5/12$.
5.30. (a) $p_1(1 - p_3)p_4$; (b) $p_1 + p_2 - p_1p_2$; (c) $(p_1 + p_2 - p_1p_2)[1 - p_3 + 1 - p_4 - (1 - p_3)(1 - p_4)]$.
5.31. (a) $P(A) \geqslant P(A_1A_2) = 1 - P(\overline{A_1A_2}) = 1 - P(\bar{A}_1 \cup \bar{A}_2) = 1 - P(\bar{A}_1) - P(\bar{A}_2) + P(\bar{A}_1\bar{A}_2) = P(A_1) + P(A_2) - 1 + P(\bar{A}_1\bar{A}_2) \geqslant P(A_1) + P(A_2) - 1$;
(b) $P(A_1A_2) = P(A_1)P(A_2|A_1) = P(A_1)[1 - P(\bar{A}_2|A_1)] = P(A_1)[1 - P(\bar{A}_2A_1)/P(A_1)] = P(A_1) - P(\bar{A}_2)P(A_1|\bar{A}_2) \geqslant P(A_1) - P(\bar{A}_2)$.
5.32. $P(D) \geqslant P(ABC) = 1 - P(\bar{A} \cup \bar{B} \cup \bar{C}) = 1 - P(\bar{A}) - P(\bar{B}) - P(\bar{C}) + P(\bar{A}\bar{B}) + P(\bar{A}\bar{C}) + P(\bar{B}\bar{C}) - P(\bar{A}\bar{B}\bar{C}) = P(A) + P(B) + P(C) - 2 + P(\bar{A}\bar{B}) - P(\bar{A}\bar{B})P(\bar{C}|\bar{A}\bar{B}) + P(\bar{A}\bar{C}) + P(\bar{B}\bar{C}) \geqslant P(A) + P(B) + P(C) - 2$.
5.33. (a) $AB \neq \emptyset$, $AC \neq \emptyset$, $BC \neq \emptyset$. Therefore, the events A, B and C do not form a partition of the space Ω. (b) From $AC = C$ we have $P(AC) = P(C) > P(A)P(C)$ and the events A and C are not independent. (c) From $A \cup B = \Omega$ we have $P(C(A \cup B)) = P(C\Omega) = P(C) = P(C)P(A \cup B)$ and therefore C and $A \cup B$ are independent. (d) $P(A|C) = P(AC)/P(C) = P(C)/P(C) = 1$.
(e) 10/23.
5.34. We have $P(A(B \cup C)) = P(A)P(B \cup C) = P(A)[P(B) + P(C) - P(BC)]$. On the other hand $P(A(B \cup C)) = P(AB) + P(AC) - P(ABC)$. Hence

$$P(A)[P(B) + P(C)] - P(A)P(BC) = P(AB) + P(AC) - P(ABC). \quad (5.8)$$

According to the conditions of the exercise, $P(A(BC)) = P(A)P(BC)$ and (5.8) becomes

$$P(A)[P(B) + P(C)] = P(AB) + P(AC). \quad (5.9)$$

Furthermore, $P(ABC) = P(C)P(AB) = P(B)P(AC)$, whence $P(AC) = P(C)P(AB)/P(B)$ (by condition $P(B) > 0$). After substituting in (5.9): $P(A)[P(B) + P(C)] = P(AB) + P(C)P(AB)/P(B) = P(AB)[P(B) + P(C)]/P(B)$, hence $P(AB) = P(A)P(B)$. Therefore $P(ABC) = P(AB)P(C) = P(A)P(B)P(C)$. From that and from the independence of the pairs of events B and AC, A and BC, it follows that P(AC) = P(A)P(C) and P(BC) = P(B)P(C). Hence the events A, B and C are independent.

5.35. Let $P(A_i) = p_i$, $i = 1, \ldots, n$. Thus $p_i > 0$ and since $P\left(\bigcup_{i=1}^{n} A_i\right) = 1$, then $p_1 + \ldots + p_n = 1$. Referring to the property of independence, we find $P(A_1 \ldots A_n) = p_1 \ldots p_n$. But $(p_1 + \ldots + p_n)/n \geqslant \sqrt[n]{p_1 \ldots p_n}$, where the equality is attained only if $p_1 = \ldots = p_n$, and therefore $P(A_1 \ldots A_n) \leqslant 1/n^n$. The equality is possible only if $P(A_1) = \ldots = P(A_n) = 1/n$.

5.36. We have $1 = P(\bar{A}\bar{B}\bar{C} + \overline{\bar{A}\bar{B}\bar{C}}) = P(\bar{A}\bar{B}\bar{C}) + P(A \cup B \cup \bar{C}) = P(\bar{A}\bar{B}\bar{C}) + P(A) + P(B) + P(\bar{C}) - P(AB) - P(A\bar{C}) - P(B\bar{C}) + P(AB\bar{C})$. But $AB\bar{C} = AB\bar{C} + ABC = AB$. Therefore

$$P(\bar{A}\bar{B}\bar{C}) = 1 - 2x - (1 - x) + x^2 + 2x(1 - x) - x^2 = x - 2x^2 \geqslant 0;$$

i.e., $x \leqslant \frac{1}{2}$. On the other hand $A \cup B \cup C = A + B\bar{A} + C\bar{A}\bar{B} \subset \Omega$, $P(A) + P(\bar{A}B) + P(\bar{A}\bar{B}C) = \alpha \leqslant 1$; hence $P(\bar{A}\bar{B}C) = \alpha - x - x(1 - x)$. It follows that x satisfies the equation $x - 2x^2 = \alpha - 2x + x^2$, or $3x^2 - 3x + \alpha = 0$. Its roots are $x_1 = \frac{1}{6}(3 + \sqrt{9 - 12\alpha})$ and $x_2 = \frac{1}{6}(3 - \sqrt{9 - 12\alpha})$. However, x is a real non-negative number and as we have seen above $x \leqslant \frac{1}{2}$. Obviously if $\alpha = \frac{3}{4}$ we have $x_1 = x_2 = \frac{1}{2}$; i.e., x attains the value $\frac{1}{2}$ and this is its largest possible value.

5.37. Let A = {A tells the truth} and D = {D tells the truth}. The event D occurs only when (i) every one of the four tells the truth, (ii) any two of them tell the truth, while the remaining two lie, (iii) all four lie. Thus

$$P(D) = \left(\frac{1}{3}\right)^4 + \binom{4}{2}\left(\frac{1}{3}\right)^2\left(\frac{2}{3}\right)^2 + \left(\frac{2}{3}\right)^4 = 41/3^4.$$

Hence

$$P(D|A) = \left(\frac{1}{3}\right)^3 + \binom{3}{2}\left(\frac{2}{3}\right)^2\frac{1}{3} = 13/3^3.$$

Provided that B has received a true information from A, the event D occurs only if every one of the players B, C and D tells the truth or some two of them lie. Then $P(AD) = P(A)P(D|A) = 13/3^4$. For the required conditional probability we obtain

$$P(A|D) = P(AD)/P(D) = 13/41.$$

5.38. (a) $1733/7425 \approx 0.233$; (b) $4193/44\,550 \approx 0.094$.

5.39. $[1^n + 2^n + \ldots + N^n]/[N^n(N + 1)]$.

5.40. Let H_{ij} = {one of the two students is in the ith year and the other is in the jth one}, i, j = 1, 2, 3, A = {both are in different years}, B = {the senior student is in the third year}. If $i \neq j$, then $P(H_{ij}) = 2n_i n_j/[n(n-1)]$. Thus $P(B|A) = P(AB)/P(A) = [P(H_{13}) + P(H_{23})]/[P(H_{13}) + P(H_{23}) + P(H_{12})] = (\frac{1}{n_1} + \frac{1}{n_2})/(\frac{1}{n_1} + \frac{1}{n_2} + \frac{1}{n_3})$.

5.41. Let A = {the first item is non-defective}, B = {the second item is defective}, H_1 = {the lot containing defective items is chosen}, H_2 = {the lot containing only non-defective items is chosen}. Then

$$P(H_1) = P(H_2) = 1/2, \quad P(A|H_1) = 3/4, \quad P(A|H_2) = 1,$$

$$P(A) = (1/2)(1 + 3/4) = 7/8,$$

$$P(AB) = P(H_1)P(AB|H_1) + P(H_2)P(AB|H_2) = \frac{1}{2}\left(\frac{3}{4} \cdot \frac{1}{4} + 1 \cdot 0\right) = \frac{3}{32}.$$

However, A and B are not independent and therefore $P(B|A) = P(AB)/P(A) = \frac{3}{32}/\frac{7}{8} = \frac{3}{28}$.

5.42. Let A_k = {a winning ticket appears at the (k + 1)st draw}. The results of the previous k trials can be described by the following k + 1 hypotheses: H_{ks} = {s of the k tickets are winning}, s = 0, 1, ..., k. But $P(H_{ks}) = \binom{m}{s}\binom{n-m}{k-s}/\binom{n}{k}$, s = 0, 1, ..., k, where $P(H_{ks}) = 0$ if $s > m$. Furthermore, $P(A_k|H_{ks}) = (m - s)/(n - k)$. According to the formula for the total probability, we obtain for each k

$$P(A_k) = \sum_{s=0}^{k} \frac{\binom{m}{s}\binom{n-m}{k-s}}{\binom{n}{k}} \frac{m-s}{n-k} = \frac{m}{n} \sum_{s=0}^{k} \frac{\binom{m-1}{s}\binom{n-m}{k-s}}{\binom{n-1}{k}} = \frac{m}{n}.$$

Thus the chance of winning is the same for each of the individuals and does not depend on the position at which he draws.

5.43. Let A = {the marked ball is drawn}, H_i = {the ball is in the ith urn}, i = 1, 2. By condition $P(H_1) = \alpha$, $P(H_2) = 1 - \alpha$. Suppose that m balls are drawn from the first urn and n - m from the second one. Hence $P(A|H_1) = 1 - (1-p)^m$, $P(A|H_2) = 1 - (1-p)^{n-m}$. Then $P(A) = \alpha[1 - (1-p)^m] + (1-\alpha)[1 - (1-p)^{n-m}]$. We have to determine m so as $\vec{P}(A)$ to be maximal. We assume that m is a continuous variable and differentiate P(A) with respect to m. Then putting $dP(A)/dm = 0$ we obtain $(1-p)^{2m-n} = \frac{1-\alpha}{\alpha}$. Hence $m \approx \frac{n}{2} + (\ln \frac{1-\alpha}{\alpha})/[2\ln(1-p)]$.

5.44. $P(A_j|A) = j(5-j)/20$, max $P(A_j|A) = 3/10$ for j = 2 and j = 3.

5.45. Let the events H_i = {the urn Y_i is chosen}, i = 1, ..., 13, form a partition. By the condition $P(H_i) = \lambda i$, i = 1, ..., 13, and since $P(H_1) + \ldots + P(H_{13}) = 1$, then $\lambda = 1/91$ and $P(H_i) = i/91$. Let A = {two

balls with different colours are drawn}. Then $P(H_i|A) = 12i^2(13 - i)/(13^2(13^2 - 1))$ and $P(H_i|A)$ is maximum for $i = 9$.
5.46. $n^k/(1^k + 2^k + \ldots + n^k)$.
5.47. (a) $\Omega = \{\omega_*, \omega_1, \omega_2, \ldots, \omega_n, \ldots\}$, where $\omega_n = \underbrace{TT \ldots T}_{n-1}H$, and ω_* contains infinitely many symbols 'T'. We have $P(\omega_*) = \lim_{n\to\infty} (0.5)^n = 0$,

$$P(\omega_*) + \sum_{i=1}^{\infty} P(\omega_i) = 0 + \sum_{i=1}^{\infty} P(\omega_i) = 0.5\,\frac{1}{1 - 0.5} = 1.$$

(b) P{the game finishes after a finite number of moves} $= 1 - P(\omega_*) = 1$.
(c) $P(A) = P(\omega_1) + P(\omega_3) + \ldots + P(\omega_{2k-1}) + \ldots = 2/3$, $P(B) = 1/3$. An alternative solution for condition (c) is: Since $P(A) + P(B) = 1$, and $P(B) = (1/2)P(A)$, then $P(A) = 2/3$, $P(B) = 1/3$.
5.48. If p_k is the probability for the kth player to win, $k = 1, 2, 3$, then $p_1 + p_2 + p_3 = 1$, $p_2 = \frac{1}{2} p_1$, $p_3 = \frac{1}{4} p_1$; hence $p_1 = \frac{4}{7}$, $p_2 = \frac{2}{7}$, $p_3 = \frac{1}{7}$.
5.49. (a) If $P_1 = P(A)$ and $P_2 = P(\bar{A})$, then $P_1 = \frac{n}{n + m + 1} c$, $P_2 = \frac{m}{n + m + 1} c$, where $c = 1 + \frac{1}{n + m + 1 - 1} + \frac{1(1 - 1)}{(n + m - 1 - 1)(n + m - 1 - 2)} + \ldots$ Therefore, $P_1 : P_2 = n : m$ and since $P_1 + P_2 = 1$ we find that $P_1 = n/(n + m)$. (b) again $P_1 = n/(n + m)$.
5.50. Let $p(x)$ be the probability that a player will be ruined if at the beginning of the game he has x levs. Then the probability that he will be ruined, provided he has won at the first move, is $p(x + 1)$, while the probability of his being ruined, provided he has lost his first move, is $p(x - 1)$. Denote by B_1 and B_2 the events: the player 'wins' and, respectively, 'loses' at the first move, and by A the event the player 'is ruined', then $P(B_1) = P(B_2) = 1/2$, $P(A|B_1) = p(x + 1)$, $P(A|B_2) = p(x - 1)$. By the formula for the total probability, $p(x) = (1/2)[p(x + 1) + p(x - 1)]$, $0 \leqslant x \leqslant a$, where $p(0) = 1$, $p(a) = 0$. A solution of this equation is the linear function $p(x) = c_1 + c_2x$, where c_1 and c_2 can be determined from the boundary conditions $p(0) = c_1 = 1$, $p(a) = c_1 + c_2a = 0$. Hence $p(x) = 1 - x/a$, $0 \leqslant x \leqslant a$.

6. Urn Models. Polya Urn Model

6.1. (a) 10/23. (b) 7/23. (c) 6/23.
6.2. By simultaneous drawing of two balls the probability of the two balls being white is $8 \cdot 7/(12 \cdot 11)$. The probability for the two balls to be of different colour is $2 \cdot 8 \cdot 4/(12 \cdot 11)$.
6.3. (a) 33/95. (b) 14/95. (c) 48/95-
6.4. (a) $w^2/(w + b)^2$; (b) $b^2/(w + b)^2$; (c) $wb/(w + b)^2$.
6.5. $(2k)!(2l)![(k + l)!]^2/[(k!)^2(l!)^2(2k + 2l)!]$.
6.6. Let H = {a white ball is drawn from the first urn}, and A = {the ball drawn from the second urn is white}. According to the formula for

total probability $P(A) = P(A|H)P(H) + P(A|\bar{H})P(\bar{H}) = (w_2w_1 + w_1 + w_2r_1)/((w_1 + r_1)(w_2 + b_2 + 1))$.

6.7. Let us denote the event "a white ball appears for the first time on the ith drawing" with H_i, $i = 1, 2, \ldots, b + 1$. Indeed, as the drawing is without replacement, a white ball would appear on the (b + 1)st trial at the latest. Therefore $\{H_i\}$ forms a partition. If we denote by A the event "the player, who starts the game is the winner", then $P(A) = P(H_1) + P(H_3) + P(H_5) + \ldots + P(H_{b+1})$ when b is odd; $P(A) = P(H_1) + P(H_3) + P(H_5) + \ldots + P(H_b)$ when b is even. For the probabilities of $\{H_i\}$ we have $P(H_1) = \dfrac{w}{w + b}$; $P(H_i) = \dfrac{b(b - 1) \ldots (b - i + 2)w}{(w + b)(w + b - 1)..(w + b - i + 1)}$, $i > 1$.

6.8. Consider an urn with a balls, b of which are white. Find the probabilities p_i of a white ball appearing for the first time on the ith trial, if the drawing is without replacement; use the fact that $\sum_i p_i = 1$.

6.9. It is supposed that initially in the urn there are w white and b black balls. (a) $2wb/((w + b)(w + b + c))$.

(b) $\dfrac{w(w + c)(w + 2c) + b(b + c)(b + 2c)}{(w + b)(w + b + c)(w + b + 2c)}$.

6.10. Let $B_1 = \{$the first ball is black$\}$ and $B_2 = \{$the second ball is black$\}$. Since $P(B_1) = \dfrac{b}{w + b}$; $P(\bar{B}_1) = \dfrac{w}{w + b}$; $P(B_2|B_1) = \dfrac{b + c}{w + b + c}$; $P(B_2|\bar{B}_1) = \dfrac{b}{w + b + c}$ and using Bayes' formula we find $P(B_1|B_2) = (b + c)/(w + b + c)$. It is interesting that $P(B_1|B_2) = P(B_2|B_1)$ (see Exercise 6.13).

6.11. Let $H_i = \{$exactly i white balls are drawn on the first n - 1 trials$\}$, $i = 0, \ldots, n - 1$, $W_n = \{$drawing a white ball on the nth trial$\}$, $W_{n+1} = \{$drawing a white ball on the (n + 1)th trial$\}$. Obviously $\{H_i\}$ form a partition and $P(W_n) = \sum_i P(W_n|H_i)P(H_i)$. As $W_{n+1} = W_{n+1}W_n + W_{n+1}\bar{W}_n$, then $P(W_{n+1}) = \sum_i P(W_{n+1}W_n|H_i)P(H_i) + \sum_i P(W_{n+1}\bar{W}_n|H_i)P(H_i)$. It is easy to find that $P(W_n|H_i) = \dfrac{w + ic}{w + b + (n - 1)c}$, $P(W_{n+1}W|H_i) = \dfrac{(w + ic)(w + ic + c)}{(w + b + nc - c)(w + b + nc)}$, $P(W_{n+1}\bar{W}_n|H_i) = \dfrac{(b + nc - ic - c)(w + ic)}{(w + b + nc - c)(w + b + nc)}$. Then $P(W_{n+1}W_n|H_i) + P(W_{n+1}\bar{W}|H_i) = P(W_n|H_i)$, which, substituted in the formula for $P(W_{n+1})$, yields $P(W_{n+1}) = P(W_n)$. Since the probability of a white ball being drawn on the first trial is $w/(w + b)$, the probability of a white ball being drawn on the nth trial is also $w/(w + b)$. It is interesting that this probability depends neither on n nor on c.

6.12. (a) Let us consider a series of n trials in the Polya urn model. To each outcome in this series we can juxtapose a sequence of n symbols (w or b), e.g. wbbw ... wbw; the symbol w at the ith position means drawing a white ball on the ith trial, the symbol b at the jth position means drawing a black ball on the jth trial. The probability of each simple event depends only on the number of the letters w and b, but not on their positions (see the Introductory Notes to this section). To find the probability sought we have to add up the probabilities of all the simple events in which w stands on mth place and b stands on the nth place. We shall obtain the same result if we add up the probabilities of all the simple events in which w stands in the first position and b stands in the second position. But this is the probability for a white ball to appear on the first trial, and a black one to appear on the second; $wb/((w + b)(w + b + c))$. (b) Solved analogously to (a).

6.13. (a) According to Exercise 6.11 $P(A_m) = P(A_n)$. But $P(A_m|A_n) = P(A_mA_n)/P(A_n)$ and $P(A_n|A_m) = P(A_mA_n)/P(A_m)$; hence the desired equality follows. (b), (c) and (d) are proved analogously.

6.14. It is easy to see that the probability would be the greatest when one white ball is placed in one urn and the remaining 19 balls are placed in the other.

6.15. The number of all possible outcomes is $\binom{2n}{2m}$. The number of the outcomes in which there are exactly k pairs, $0 \leqslant k \leqslant m$, numbered with the same integer is $\binom{n}{k}\binom{n - k}{2m - 2k}2^{2m-2k}$. The factor 2^{2m-2k} results from the fact that any ball, whose number can be found only once among the drawn ones, can be either white or black; hence it is easily seen that $S = \binom{2n}{2m}$.

6.16. Let $n = w + b$. Construct an n-tuple of numbers $\xi_1, \xi_2, \ldots, \xi_n$ at each drawing of the balls from the urn according to the following rule: $\xi_i = 1$, if the ith ball drawn is white, and $\xi_i = -1$, if it is black. It is clear that w of the numbers ξ_i would be equal to 1, and the remaining b would be equal to (-1). For these n-tuples we can use a geometrical representation with broken lines (trajectories) and the reflection principle (see the solution of Exercise 3.53). Any simple event is represented by such a line with endpoints at (0, 0) and (w + b, w - b); the number of such trajectories is $\binom{w + b}{w}$. Trajectories, which intersect the x-axis at a point different from the origin (0, 0), correspond to the favourable outcomes. If $w > b$, then the number of these trajectories is $2\binom{w + b - 1}{w}$ and the desired probability is $2b/(w + b)$; if $w < b$, the probability is $2w/(w + b)$. When $w = b$, the probability is one.

6.17. Suppose that we have dropped one white ball into the urn. Let the events of interest be denoted by H_i = {in the urn there are exactly i white balls}, $i = 1, 2, \ldots, n + 1$ and A = {the ball drawn is white}. It is clear that

$$P(H_i) = 1/(n + 1) \quad \text{and} \quad P(A|H_i) = i/(n + 1).$$

Using the formula of total probability, we find $P(A) = (n + 2)/(2(n + 1))$.

Note. Consider the case when $n = 2k$. Then $P(A) = (k + 1)/(2k + 1)$. But if an urn contains $2k + 1$ balls, then the probability of drawing a white ball from it is $(k + 1)/(2k + 1)$ only if exactly $k + 1$ balls are white. Therefore, before dropping the white ball into the urn, it must have contained k white balls. This contradicts the assumption that all possible compositions concerning the number of the white balls are equally likely.

A particular case of this paradox when n = 2 is given in the book: Carrol, L. (1958): *Mathematical Recreations, Vol. 2. Pillow Problems and a Tangled Tale*. New York, Dover Publ. Co.

7. Geometric Probability

7.1. In the condition of the exercise the meaning of "a randomly drawn chord" is not defined exactly. On account of this the exercise allows different solutions.

Solution 1. Let us choose the chord with a given direction. Then it is defined by the location of its midpoint X on the diameter MN, perpendicular to the direction of the chord (Figure 7.1, a).

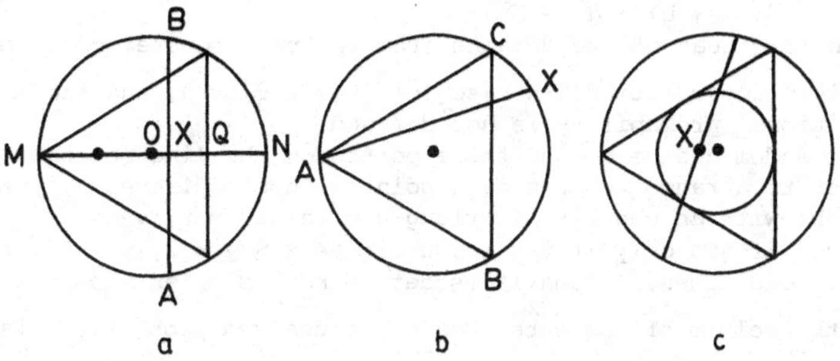

Figure 7.1

The length of the chord AB will exceed the side of the inscribed regular triangle, if the distance OX is less than R/2. If OP = OQ = R/2, then the locations of X on PQ are unfavourable for the event under consideration. Therefore, the desired probability is $R/(2R) = \frac{1}{2}$.

Solution 2. We can suppose that one end A of the chord is fixed, and the other one, X, can fall on any of the arcs $\widehat{AB}$, $\widehat{BC}$ or $\widehat{CA}$ (Figure 7.1, b). Only the locations of X on $\widehat{BC}$ are favourable. Therefore, the probability of interest is $\frac{1}{3}$.

Solution 3. We can regard the chord as being defined through the location of its midpoint in the circle (Figure 7.1, c). The length of the chord will exceed the side of the regular triangle, if its midpoint X lies inside the circumference inscribed in the triangle. Since the surface of the inscribed circle is $\pi R^2/4$, and that of the given one is πR^2, the desired probability is $\frac{1}{4}$.

Note. Only Solution 1 conforms the kinematic measure (see Introductory Notes).

7.2. The simple events (the times of occurrence of X and Y) can be represented by the points $M(t_x, t_y)$ in the square $0 \leqslant t_x \leqslant T$, $0 \leqslant t_y \leqslant T$ (Figure 7.2).

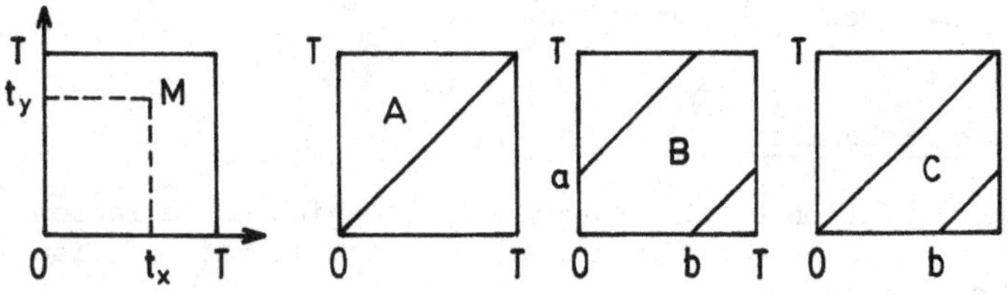

Figure 7.2

The events A = {X occurs before Y}, B = {X and Y coincide} and $C = \overline{A}B$ are also shown in Figure 7.2. Then the desired probabilities are given by: (a) $P(A) = \frac{1}{2}$; (b) $P(B) = (a + b)/T - (a^2 + b^2)/(2T^2)$; (c) $P(\overline{A}|B) = (2bT - b^2)/[2T(a + b) - a^2 - b^2]$.

7.3. Find the areas of the domains from the unit square, corresponding to the given inequalities. In case (c), $P\{x = \frac{1}{2}\} = 0$, and therefore, the conditional probability is not defined.

7.4. The random choice of the three points on the line segment AB is equivalent to a random choice of a point with coordinates x, y and z from a cube with an edge AB. A triangle with sides x, y and z can be constructed if and only if the inequalities $x < y + z$, $y < z + x$, $z < x + y$ hold. These inequalities determine a simplex whose volume is half of the volume of the cube. Hence the desired probability is $\frac{1}{2}$.

7.5. $1 - \sqrt{s}/(3k)$ when $s \leqslant k^2$; $\frac{1}{2} + k^2/(6s)$ when $s \geqslant k^2$.

7.6. Construct an equilateral triangle of height one. The three segments,

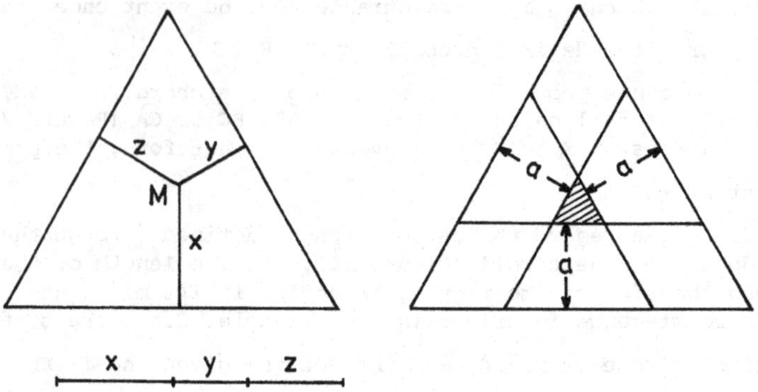

Figure 7.3

x, y and z, are the perpendicular distances from a point M to each of the sides (Figure 7.3). The conditions are $x \geqslant a$, $y \geqslant a$, $z \geqslant a$. The desired probability is equal to the area of the shaded part divided by the area of the whole triangle: $(1 - 3a)^2$.

7.7. $\frac{1}{4}$. Use the representation from Exercise 7.6.

7.8. We can consider only the rectangle with sides a and b, in which the centre of the coin has fallen (the coin is regarded as a circle). The coin will intersect none of the lines if its centre is located at a distance greater than r, from each side of the rectangle. The desired probability is

$$(a - 2r)(b - 2r)(ab)^{-1}.$$

7.9. $(1 - 2r\sqrt{3}/a)^2$. See Exercise 7.8.

7.10. $\frac{1}{2}$.

7.11. Consider only three adjacent straight lines and examine the possible locations of the circle's centre. The desired probability is 6/19.

7.12. $(1/\pi) \arctan(h/s)$. It is interesting that the answer does not depend upon R.

7.13. $2x - x^2$. A representation as in Exercise 7.2 is used.

7.14. $\frac{3}{4}$. The representation from Exercise 7.2 is used.

7.15. $\frac{1}{4}$. The three points A, B and C divide the circumference to three arcs. The triangle is acute-angled, if none of these arcs exceeds half of the circumference. The division of the circumference into three arcs is equivalent to the division of a line segment to three parts by choosing two points at random. The representation from Exercise 7.6 is extended.

7.16. $2s/(\pi L)$. The needle's location is determined by the distance x, $0 \leqslant x \leqslant L$, from its midpoint to the nearest line, and by the angle α, $0 \leqslant \alpha < \pi$ between the needle and the direction of the line. The condition of intersection is $x \leqslant s \sin \alpha$.

7.17. $(1 - 2s)/(\pi a))(1 - 2s/(\pi b))$. Exercise 7.16 should be used.

7.18. $k\sqrt{k}/(5s)$ when $k^3 \leqslant s^2$; $\frac{1}{2} - 3\sqrt[3]{s^2}/(10k)$ when $k^3 \geqslant s^2$.

7.19. Let $MB = x$ and $BN = y$. Then the lengths of the three segments are $a - x$, $x + y$ and $b - y$, respectively. A triangle can be constructed from them, if the inequalities $x + y < a + b - x - y$, $x + y > a - b + y - x$, $x + y > b - a + x - y$ are fulfilled; they are equivalent to the following ones: $x + y < (a + b)/2$, $x > (a - b)/2$, $y > (b - a)/2$. A point from the rectangle $\{(x, y) : 0 \leqslant x \leqslant a, 0 \leqslant y \leqslant b\}$, whose area is ab, corresponds to the choice of the two points M and N. The region of this triangle, determined by the above three inequalities, is a triangle with an area $a^2/2$. Therefore, the desired probability is $a/(2b)$.

7.20. $(\sqrt{a^2 + b^2} - \sqrt{a^2 - b^2})/(2b)$.

7.21. $(1 - a^3/R^3)^N$; $\exp(-4\pi a^3 \lambda/3)$.

7.22. Consider the events: $B = \{A$ appears in the interval $(0, T)\}$, $C =$

{A appears in the interval (t, T)}. We are to find $P(C|\overline{B})$. We know that $P(B + C) = p$ and $P(B)/P(C) = t/(T - t)$; hence $P(B) = pt/T$ and $P(C) = p(T - t)/T$. Therefore, $P(C|\overline{B}) = (T - t)/(T - pt)$.

7.23. 0.6.

7.24. (a) $2/\pi \approx 0.6366$; (b) $5!(\frac{1}{4} - 1/(2\pi))^4 \cdot 2/\pi \approx 0.0052$.

7.25. (a) $\frac{1}{4}$; (b) $\frac{1}{2} - 1/(2\sqrt{2})$; (c) $\frac{1}{8} - 1/(8\sqrt{2})$.

7.26. The aggregate of the two points A and B can be considered as a point in the four-dimensional space (x, y, z, t), x and y being the coordinates of A, z and t being the coordinates of B. The hypervolume of this region is given by the integral

$$\iint\limits_{x^2 + y^2 \leqslant R^2,\ z^2 + t^2 \leqslant R^2}\iint dx\, dy\, dz\, dt = \pi^2 R^4.$$

The condition "the circumference of the circle with centre A and radius AB will lie inside the circle" is equivalent to "B will lie into the circumference with centre A that is tangent to the circumference given". This condition, expressed by means of the coordinates, is

$$\sqrt{(x - z)^2 + (y + t)^2} \leqslant R - \sqrt{x^2 + y^2}.$$

If $D_1 = \{(x, y, z, t) : x^2 + y^2 \leqslant R^2, \sqrt{(x - z)^2 + (y - t)^2} \leqslant R - x^2 - y^2\}$, $D_2 = \{(x, y) : x^2 + y^2 \leqslant R^2\}$, then the hypervolume of the region, corresponding to the event under consideration, is calculated from the integral

$$\iiiint\limits_{D_1} dx\, dy\, dz\, dt = \iint\limits_{D_2} \pi(R - \sqrt{x^2 + y^2})^2\, dx\, dy = \pi^2 R^4/6.$$

Therefore, the desired probability is 1/6.

7.27. 1/20. Solved analogously to Exercise 7.26.

7.28. If we fix one of the points and denote the sizes of the arcs between it and the other two points by x and y, $0 \leqslant x < 2\pi$, $0 \leqslant y < 2\pi$, then an arc of size α, containing all the three points, exists only in the following four cases:

(a) $x < \alpha$, $y < \alpha$;
(b) $x > 2\pi - \alpha$, $y > 2\pi - \alpha$;
(c) $x - y > 2\pi - \alpha$;
(d) $y - x > 2\pi - \alpha$.

The cases are shown on Figure 7.4 by the shaded area when $\alpha \leqslant \pi$. The desired probability is $3\alpha^2/(4\pi^2)$, when $\alpha \leqslant \pi$; $(24\pi\alpha - 9\alpha^2 - 12\pi^2)/(4\pi^2)$ when $\pi \leqslant \alpha \leqslant 4\pi/3$, and one, when $\alpha \geqslant 4\pi/3$.

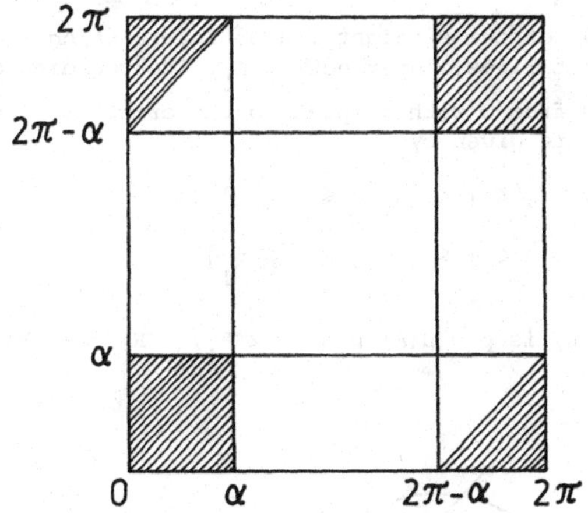

Figure 7.4.

7.29. Consider an arbitrary interval I of length ε. Let p_1 be the probability that the first point is not in I, let p_2 be the probability that the first two points are not in I, ..., let p_n be the probability that the first n points are not in I, etc. We have $p_1 = 1 - \varepsilon$, $p_2 = (1 - \varepsilon)^2$, ..., $p_n = (1 - \varepsilon)^n$, ... Obviously $\lim_{n\to\infty} p_n = 0$.

Let all the subintervals of the interval (0, 1), which have rational ends, be arranged in a sequence. Let us consider the events A_1, A_2, ..., where A_i = {none of the random points is to be found in the ith subinterval with rational ends}. From the above reasoning it follows that $P(A_i) = 0$ for any i. The event $\bar{A}$ means that there exists a subinterval of (0, 1), which contains none of the random points. Then it is clear that $\bar{A} \subset \bigcup_{i=1}^{\infty} A_i$. However, $P\left(\bigcup_{i=1}^{\infty} A_i\right) \leqslant \sum_{i=1}^{\infty} P(A_i)$ and since $P(A_i) = 0$ for any i, then $P(\bar{A}) = 0$. Therefore, $P(A) = 1$.

7.30. Let $P = P\{m_{AB} < a\}$. According to (7.2) we have

$$dP = 2(P_1 - P)\frac{dS}{S}.$$

Here S is the surface of the hemisphere, and $P_1 = \lambda/S$, where λ is half of the surface of a spherical segment with central angle 2a/R. After intergrating the above equation we obtain $S^2P = 2S\lambda + c$. When $S = \lambda$, the probability $P = 1$. Then $c = -\lambda^2$ and $P = (2S\lambda - \lambda^2)/S^2$. For the surface of the segment we have $\lambda = \pi R^2(1 - \cos(a/R))$, and $S = 2\pi R^2$. Finally,

$P = 1 - \cos^4(a/(2R))$. It is easy to see that if $a = 2R \text{ arc cos}(1/\sqrt{2})$, then $P = P\{m_{AB} < a\} = \frac{1}{2}$.

7.31. Let Ω be the set of the straight lines, intersecting K, and A be the set of the straight lines, intersecting K_1. The coordinates (ρ, ϕ) of the straight lines from Ω with respect to the origin O (see for instance Figure 7.5) are given by:

$$\Omega = \{(\rho, \phi) : 0 \leqslant \phi < 2\pi, \ 0 \leqslant \rho \leqslant R\};$$

$$A = \{(\rho, \phi) : 0 \leqslant \phi \leqslant 2\pi, \ 0 \leqslant \rho \leqslant R_1\}.$$

The desired probability is $p = \mu(A)/\mu(\Omega) = 2\pi R_1/(2\pi R) = R_1/R$.

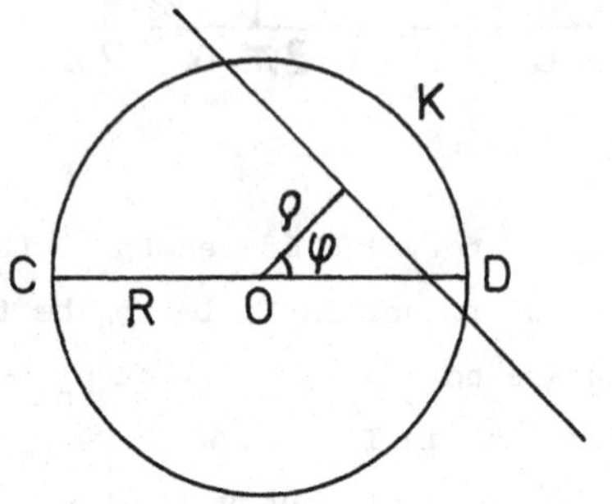

Figure 7.5.

7.32. The probability sought does not depend on the position of the square Q_1 in Q. The two figures can be placed in such a way that they will be homothetic. With homothetic figures, the ratio of the measure of the sets of straight lines which intersect each of the figures is equal to the coefficient of homothety. In our case the coefficient is $\sqrt{2}$ and hence the desired probability will be $\sqrt{2}/2$. (See also Exercise 7.31.)

7.33. Let Ω be the set of the straight lines, intersecting K and A be the set of the straight lines, intersecting CD. The coordinates (ρ, ϕ) of the straight lines from Ω with respect to the origin O are determined by the inequalities $0 \leqslant \rho \leqslant R$, $0 \leqslant \phi < 2\pi$, and the coordinates of the straight lines from A by the inequalities $0 \leqslant \rho \leqslant R|\cos \phi|$, $0 \leqslant \phi < 2\pi$ (see Figure 7.5). For the desired probability we find $\mu(A)/\mu(\Omega) = 2/\pi$. It is interesting to compare this result with the result from Exercise 7.16 when $L = 1$.

7.34. $1/\pi$. The exercise is solved analogously to Exercise 7.33. The same result is also obtained for an arbitrary segment from the circle, whose length is equal to the radius (see Exercise 7.32).

7.35. We shall apply Crofton's formula (7.2) regarding d as fixed and considering s as a parameter. Let us denote $f_n(s) = P(A)$, $s \geqslant (n - 1)d$ and $f^*_{n-1}(s) = P_1$ for the sake of convenience. Then for the sequence of functions $\{f_n(s)\}$ we obtain the differential equations

$$f_n'(s) = f_{n-1}^*(s) - f_n(s)/s, \quad n = 2, 3, \ldots$$

It is easy to see that $f_{n-1}^*(s) = f_{n-1}(s - d)(1 - d/s)^{n-1}$, and these equations are of the form

$$f_n'(s) + nf_n(s)/s = nf_{n-1}(s - d)(1 - d/s)^{n-1}/s, \quad n = 2, 3, \ldots$$

We directly find (see Exercise 7.6) that

$$f_1(s) = 1, \quad f_2(s) = (1 - d/s)^2, \quad s \geqslant d.$$

We shall now proceed by induction. Let us suppose that for some $n \geqslant 2$ we have $f_{n-1}(s) = (1 - (n - 2)d/s)^{n-1}$, $s \geqslant (n - 2)d$. Then for $f_n(s)$ we obtain the linear differential equation

$$f_n'(s) + nf_n(s)/s = n(1 - (n - 2)d/s)^{n-1}/s.$$

The general solution of this equation is

$$f_n(s) = s^{-n}(c + (s - (n - 1)d)^n),$$

where c is a constant. From the condition $f_n((n - 1)d) = 0$ we find that $c = 0$. Therefore,

$$f_n(s) = (1 - (n - 1)d/s)^n.$$

8. Bernoulli Trials: Binomial and Multinominal Distributions

8.1. P{the player does not receive an ace on one deal of the cards} = $p = \binom{48}{13}/\binom{52}{13} = 0.028$. Since the observed event has probability equal to $(0.028)^3$, the player has a reason to complain!

8.2. (a) $\binom{13}{2}\frac{1}{2^{13}} \approx 0.00952$; (b) $\binom{26}{2}\binom{26}{11}/\binom{52}{13} \approx 0.003954$.

8.3. 0.1705.

8.4. (a) $P_4(3) = \frac{1}{4} > P_8(5) = \frac{7}{32}$. (b) $\sum_{k=3}^{k} P_4(k) = \frac{5}{16} < \sum_{k=5}^{8} P_8(k)$.

8.5. $P_{2n}(n) = \frac{(2n)!}{(n!)^2 2^{2n}} = \frac{1 \cdot 3 \cdot 5 \cdot \ldots \cdot (2n - 1)}{2 \cdot 4 \cdot 6 \cdot \ldots \cdot (2n)} < \frac{2}{3} \cdot \frac{4}{5} \cdot \frac{6}{7} \cdot \ldots \cdot \frac{2n}{2n + 1} = \frac{2 \cdot 4 \cdot \ldots \cdot (2n)}{1 \cdot 3 \cdot \ldots \cdot (2n - 1)} \cdot \frac{1}{2n + 1} = \frac{1}{P_{2n}(n)} \cdot \frac{1}{2n + 1}$.

This implies the inequality on the right and the one on the left follows from the fact that

$$2P_{2n}(n) > \frac{2}{3} \cdot \frac{4}{5} \cdot \ldots \cdot \frac{2n-2}{2n-1} = \frac{1}{2n} \cdot \frac{1}{P_{2n}(n)} .$$

8.6. (a) $\tilde{P}_4(0) = 0.3024$, $\tilde{P}_4(1) = 0.4404$, $\tilde{P}_4(2) = 0.2144$, $\tilde{P}_4(3) = 0.0404$, $\tilde{P}_4(4) = 0.0024$. (b) 0.25952.

8.7. (a) The required probability is equal to the coefficient of s^m in the polynomial $\sum_{m=0}^{n-1}\left[\sum_{j=m+1}^{n} \tilde{P}_n(j)\right]s^m$. According to formula (8.2), we have

$$\sum_{m=0}^{n-1}\left[\sum_{j=m+1}^{n} \tilde{P}_n(j)\right]s^m = \sum_{j=1}^{n}\sum_{m=0}^{j-1} s^m\tilde{P}_n(j) = \sum_{j=1}^{n}\left[\tilde{P}_n(j)\sum_{m=0}^{j-1} s^m\right] =$$

$$= \sum_{j=1}^{n}\frac{1-s^j}{1-s}\tilde{P}_n(j) = \frac{1}{1-s}\left[1 - \tilde{P}_n(0) - \right.$$

$$\left. - \left(\sum_{j=0}^{n} s^j\tilde{P}_n(j) - \tilde{P}_n(0)\right)\right] = g_1(s).$$

(b) Show that $\sum_{m=0}^{n}\left[\sum_{j=0}^{m}\tilde{P}_n(j)\right]s^m = g_2(s)$.

8.8. See Exercise 8.7.

8.9. $0.8^n[1 + \frac{n}{4} + \frac{n(n-1)}{32}] \leqslant 0.1$; $n \geqslant 25$.

8.10. $0.99 \cdot 5^{10} \leqslant 4^{10} + \binom{10}{1}4^9 + \ldots + \binom{10}{n}4^{10-n}$; $n = 5$.

8.11. $m_+ = 3$, $m_- = 1$, $p_+ = p_- = 32/81$.

8.12. 79 or 80.

8.13. $82/91 \leqslant p \leqslant 83/91$.

8.14. Let $P = \binom{n}{m}p^m q^{n-m}$ and $P' = \binom{n+1}{m}q^{m'}q^{n+1-m'}$, where $(n+1)p - 1 \leqslant m \leqslant (n+1)p$ and $(n+2)p - 1 \leqslant m' \leqslant (n+2)p$. We have $P/P' = p^{m-m'}q^{m'-m-1}[(m')!(n+1-m')!]/[m!(n-m)!(n+1)!]$.

(1) If $(n+1)p$ is an integer, let $(n+1)p = k$. Then $P = P_n(k-1) = P_n(k)$ and $P' = P_{n+1}(k)$, and the quotient

$$P/P' = P_n(k)/P_{n+1}(k) = (n+1-k)/((n+1)q) =$$

$$= (n+1-(n+1)p)/((n+1)q) = 1.$$

(2) If $(n+1)p$ is not an integer, let $(n+1)p = k + \varepsilon$, where $0 < \varepsilon < 1$. When $0 < \varepsilon + p \leqslant 1$, we have $n = k$, $m' = k$ and, as in case (1), we find that $P = P'$. If $1 < \varepsilon + p < 2$, we have $m = k$, $m' = k + 1$, and then

$$(P/P') - 1 = [k + 1 - (n + 1)p]/[(n + 1)p] =$$

$$= [(n + 1)p - \varepsilon + 1 - (n + 1)p]/[(n + 1)p] > 0.$$

8.15. (a) $P(A_0B_0 + A_1B_1 + A_2B_2 + A_3B_3) = 0.32076$. (b) $P(A_1B_0 + A_2(B_1 + B_0) + A_3(B_0 + B_1 + B_2)) = 0.243$.

8.16. Denote by A the event in which we are interested and let B = {at the first k + r - 1 trials r - 1 successes have occurred}, C = {at the (k + r)th trial success occurs}. Since the trials are independent, then

$$P(A) = P(B)P(C) = \binom{k + r - 1}{r - 1}p^{r-1}q^k p = \binom{k + r - 1}{k}p^r q^k.$$

However, $\binom{k + r + 1}{k} = (-1)\binom{-r}{k}$ and finally $P(A) = \binom{-r}{k}p^r(-q)^k$.

8.17. $\binom{2n - k}{n}/2^{2n-k}$. The reasoning is the same as that in Exercise 8.16.

8.18. The model is equivalent to a Bernoulli scheme (a + b - 1, p) and the aim is to find the probability that at least a successes (with at most b - 1 failures) will occur. The desired probability is equal to

$$\sum_{k=a}^{a+b-1} \binom{a + b - 1}{k}p^k(1 - p)^{a+b-1-k}.$$

8.19. Let A = {the first success occurs after the fifth but before the tenth trial}, B = {failure at the first two trials} and A_i = {success at the ith trial}. Then $B = \bar{A}_1\bar{A}_2$, $A = \bar{A}_1\bar{A}_2\bar{A}_3\bar{A}_4\bar{A}_5A_6 + \bar{A}_1 \ldots \bar{A}_6A_7 + \bar{A}_1 \ldots \bar{A}_7A_8 + \bar{A}_1 \ldots \bar{A}_8A_9$, AB = A since $A \subset B$. Then we easily find that $P(B) = (1 - p)^2$, $P(A) = (1 - p)^5[1 - (1 - p)^4]$; hence the desired probability is

$$P(A|B) = P(AB)/P(B) = P(A)/P(B) = (1 - p)^3[1 - (1 - p)^4].$$

8.20. The events A and A_k are independent when $P(A_k|A) = P(A_k)$. However, $P(A_k) = \binom{n}{k}p^k q^{n-k}$, $P(A_k|A) = \binom{n - 1}{k - 1}p^{k-1}q^{n-k}$. From the equality $\binom{n}{k}p^k q^{n-k} = \binom{n - 1}{k - 1}p^{k-1}q^{n-k}$ we find k = np. Therefore, A and A_k are independent only if np is an integer and k = np.

8.21. $\sum_{i=0}^{n} \binom{n}{i}p^i q^{n-i}[1 - (1 - 1/c)^i] = 1 - (1 - p/c)^n$.

8.22. In a Bernoulli scheme (200, p) let us consider the hypotheses $H_1 = \{p = 1/2\}$ and $H_2 = \{p = 2/3\}$, with $P(H_1) = P(H_2) = 1/2$, and let A = {116 successes in 200 trials}. Then $P(H_1|A)/P(H_2|A) = 3^{200}/2^{316} \approx 1.98$; i.e., $P(H_1|A) \approx 2P(H_2|A)$.

8.23. (a) $p_1 = \dfrac{k^{k_1} m^{m_1} n^{n_1}}{(k+m+n)^{k_1+m_1+n_1}}$. (b) $p_2 = 6p_1$. (c) $p_3 = \dfrac{(k_1+m_1+n_1)!}{k_1!m_1!n_1!} p_1$.

8.24. $P_3(1, 1, 1) + P_3(2, 1, 0) + P_3(1, 2, 0) = 0.24543$.

8.25. Let A_i = {at least one tube burns in the ith block}, i = 1, 2, 3, 4 and A = {the device gets out of order}. Then

$$P(A) = P(A_1 \cup A_2A_3) = 1 - P(\overline{A}_1\overline{A}_2 \cup \overline{A}_1\overline{A}_3) = 1 - P(\overline{A}_1\overline{A}_2) -$$

$$- P(\overline{A}_1\overline{A}_3) + P(\overline{A}_1\overline{A}_2\overline{A}_3) = 1 - 2P(\overline{A}_1\overline{A}_2) + P(\overline{A}_1\overline{A}_2\overline{A}_3) =$$

$$= 1 - 2[P_4(0, 0, 4, 0) + P_4(0, 0, 3, 1) + P_4(0, 0, 2, 2) +$$

$$+ P_4(0, 0, 1, 3) + P_4(0, 0, 0, 4)] + P_4(0, 0, 0, 4) =$$

$$= 1 - 2[p_3^4 + \binom{4}{1}p_3^3p_4 + \binom{4}{2}p_3^2p_4^2 + \binom{4}{3}p_3p_4^3 + p_4^4] + p_4^4 \approx$$

$$\approx 0.983.$$

8.26. We shall show preliminarily that the probability (8.6) (see the Introductory Notes to this section) is maximal if and only if for each pair (i, j) the inequality

$$p_i\tilde{k}_j \leqslant p_j(\tilde{k}_i + 1) \qquad (8.12)$$

holds.

(a) *Necessity*. Let $n!(\tilde{k}_1!\tilde{k}_2! \ldots \tilde{k}_r!)^{-1} p_1^{\tilde{k}_1} p_2^{\tilde{k}_2} \ldots p_r^{\tilde{k}_r}$ be the maximal probability from those given in (8.6). Then, if (i, j) is an arbitrary pair of indices, we have

$$n!(\tilde{k}_1! \ldots \tilde{k}_i! \ldots \tilde{k}_j! \ldots \tilde{k}_r!)^{-1} p_1^{\tilde{k}_1} \ldots p_i^{\tilde{k}_i} \ldots p_j^{\tilde{k}_j} \ldots p_r^{\tilde{k}_r} \geqslant$$

$$\geqslant n!(\tilde{k}_1! \ldots (\tilde{k}_i + 1)! \ldots (\tilde{k}_j - 1)! \ldots \tilde{k}_r!)^{-1} p_1^{\tilde{k}_1} \ldots p_i^{\tilde{k}_i+1}$$

$$\ldots p_j^{\tilde{k}_j-1}; \quad \text{i.e.,} \quad p_j/\tilde{k}_j \geqslant p_i/(\tilde{k}_i + 1).$$

(b) *Sufficiency*. This part of the proof follows in reverse order from that in proving necessity. Summing (8.12) over all i for i ≠ j, we obtain that $\tilde{k}_j(1 - p_j) \leqslant p_j(n - \tilde{k}_j + r - 1)$; i.e.,

$$\tilde{k}_j \leqslant p_j(n + r - 1), \quad j = 1, 2, \ldots, r. \tag{8.13}$$

Summing (8.12) over all j, $j \neq i$ we obtain $p_i(n - \tilde{k}_i) \leqslant (\tilde{k}_i + 1)(1 - p_i)$, i.e. $np_i + p_i - 1 \leqslant \tilde{k}_i$. Therefore

$$np_i - 1 < \tilde{k}_i, \quad i = 1, 2, \ldots, r. \tag{8.14}$$

Hence the required inequalities follow from (8.13) and (8.14).

8.27. Choosing the first three digits of the number is analogous to three independent trials. Each of the digits can be one of 0, 1, ..., 9 with probability $p_i = \frac{1}{10}$, $i = 0, 1, \ldots, 9$. Consider the p.g.f. $g(s_0, s_1, \ldots, s_9) = 10^{-3}(s_0 + s_1 + \ldots + s_9)^3$, where the index k of s_k implies the appearance of the number k. Let $s_k = s^k$. Then the coefficient of s^N in the function $g_1(s) = 10^{-3}(s^0 + s^1 + \ldots + s^9)^3 = 10^{-3}(1 - s^{10})^3(1 - s)^{-3}$ is equal to the probability that the sum of the first three digits of the number is equal to N. By analogy, the coefficient s^{-N} in the function $g_2(s) = 10^{-3}(1 - s^{-10})^3(1 - s^{-1})^{-3}$ is equal to the probability that the sum of the last three digits of the number is equal to N. Thus the coefficient s^0 in the function $g(s) = g_1(s)g_2(s) = 10^{-6}s^{-27}(1 - s^{10})^6(1 - s)^{-6}$ is equal to the probability p that the sum of the first three digits of the number is equal to the sum of the last three digits. Expanding $g(s)$ in a series in powers of s, we obtain

$$p = 10^{-6}\left[\binom{32}{5} - \binom{6}{1}\binom{22}{5} + \binom{6}{2}\binom{12}{5}\right] = 0.05525.$$

8.28. The required probability p is equal to the doubled coefficient of s^4 in the expansion of the function

$$g(s) = 5^{-20}(s + s^{-1} + 3)^{20} =$$

$$= 5^{-20} \sum_{m=0}^{20} \sum_{n=0}^{20-m} \frac{20!}{m!\,n!\,(20 - m - n)!} s^{m-n} \cdot 3^{20-m-n};$$

i.e.,

$$p = 2\,\frac{20!}{5^{20}} \sum_{k=0}^{8} \frac{3^{16-2k}}{k!\,(k + 4)!\,(16 - 2k)!} \approx 0.104.$$

8.29. The required probability p is equal to the sum of coefficients of the powers of s, which do not exceed m, in the expansion of the function

$$g(s) = \left(\frac{1}{16}s^2 + \frac{1}{4}s + \frac{3}{8} + \frac{1}{4}s^{-1} + \frac{1}{16}s^{-2}\right)^n = (4s)^{-2n}(1 + s)^{4n}.$$

We find $p = 2^{-4n} \sum_{k=2n+m}^{4n} \binom{4n}{k}$.

8.30. (a) $P(k)$ is equal to the coefficient of s^k in the expansion of the function $g(s) = s^n m^{-n}(1 - s)^{-n}(1 - s^m)^n$, namely

$$P(k) = m^{-n}\left[\binom{k-1}{n-1} - \binom{n}{1}\binom{k-m-1}{n-1} + \binom{n}{2}\binom{k-2m-1}{n-1} - \ldots\right].$$

(b) $p = 1 + P(k) - \sum_{i=1}^{k} P(i) = 1 + P(k) - m^{-n}\left[\binom{k}{n} - \binom{n}{1}\binom{k-m}{n} + \binom{n}{2}\binom{k-2m}{n} - \ldots\right]$.

8.31. Let p_k be the probability of an even number of successes in carrying out the first k trials. Before the kth trial there are two possibilities: in the first $k - 1$ trials there were either an even or an odd number of successes with probabilities p_{k-1} and $1 - p_{k-1}$, respectively. Then $p_k = p_{k-1}(1 - p) + (1 - p_k - 1)p$; i.e., $p_k = p + p_{k-1}(1 - 2p)$, which is the required recurrence relation. The last equation can be written in the form $p_k - 1/2 = (1 - 2p)(p_{k-1} - 1/2)$ for $k = 1, 2, \ldots, n$. Multiplying termwise these n equations we obtain, after cancelling the common factors, $p_n - \frac{1}{2} = (1 - 2p)^n(p_0 - \frac{1}{2})$. Since $p_0 = 1$, the desired probability is

$$p_n = \frac{1}{2}(1 + (1 - 2p)^n).$$

8.32. (a) The number of the moves of the procedure can be only even. Denote by q_{2k} the probability that it has not finished at the (2k)th move. Then we have $q_0 = 1$ and $q_{2k+2} = \frac{1}{2} q_{2k}$ for $k = 0, 1, \ldots$ It follows that $q_{2k} = \frac{1}{2^k}$. Since $p_{2k} = q_{2k-2}(\frac{1}{2} \cdot \frac{1}{2} + \frac{1}{2} \cdot \frac{1}{2}) = \frac{1}{2^k}$, then the desired probability is

$$p_n = \begin{cases} 0, & \text{if } n \text{ is odd,} \\ \frac{1}{2^k}, & \text{if } n = 2k, \text{ where } k = 0, 1, \ldots \end{cases}$$

(b) Here the number n can only be odd. Let q_{2k+1} be the probability that the procedure will not end at the $(2k + 1)$th move. Then $q_1 = 1$ and $q_{2k+1} = q_{2k-1}(1 - \frac{1}{2} \cdot \frac{1}{2}) = \frac{3}{4} q_{2k-1}$, where $k = 1, 2, \ldots$ Hence $q_{2k+1} = (\frac{3}{4})^k$, $p_{2k+1} = q_{2k-1} \cdot \frac{1}{2} \cdot \frac{1}{2} = \frac{1}{4}(\frac{3}{4})^{k-1}$. Thus the desired probability is

$$p_n = \begin{cases} 0, & \text{if } n \text{ is even} \\ \frac{1}{4}(\frac{3}{4})^{k-1}, & \text{if } n = 2k + 1, \text{ where } k = 1, 2, \ldots \end{cases}$$

8.33. Let p_k be the probability that player A is ruined when he has k levs. Then $p_k = p \cdot p_{k+1} + q \cdot q_{k-1}$, $p + q = 1$, $p_0 = 1$, $p_{m+n} = 0$. Hence we obtain

$$q(p_k - p_{k-1}) = p(p_{k+1} - p_k). \tag{8.15}$$

If $p = q$, we obtain $\vec{P}(A) = \frac{n}{m + n}$ and $\vec{P}(B) = \frac{m}{m + n}$ (see also Exercise 5.50). If $p \neq q$, from (8.15) we obtain $p_k - p_{k-1} = (p/q)^{k-1}(p_1 - 1)$. When summing these equations for $k = 1, \ldots, m$ and for $k = 1, \ldots, m + n$, we obtain

$$p_{m-1} - 1 = (p_1 - 1)\,\frac{1 - (q/p)^m}{1 - q/p}$$

and

$$p_{n+m} - 1 = (p_1 - 1)\,\frac{1 - (q/p)^{n+m}}{1 - q/p}.$$

After eliminating p_1 we finally get

$$P(A) = p_m = \frac{1 - (p/q)^n}{1 - (p/q)^{n+m}}$$

and

$$P(B) = 1 - P(A) = \frac{1 - (q/p)^m}{1 - (q/p)^{n+m}}.$$

8.34. (a) $\sum_{k=0}^{r} P_{n_1}(k)P_{n_2}(r - k) = \sum_{k=0}^{r} \binom{n_1}{k} p^k q^{n_1-k} \binom{n_2}{r - k} p^{r-k} q^{n_2-r+k} =$
$p^r q^{n_1+n_2-r} \sum_{k=0}^{r} \binom{n_1}{k}\binom{n_2}{r - k} = \binom{n_1 + n_2}{r} p^r q^{n_1+n_2-r} = P_{n_1+n_2}(r)$.

(b) Let $A^i_{n_1}$ = {i successes in a Bernoulli scheme (n_1, p)}, $i = 0, 1, \ldots, r$, $B^j_{n_2}$ = {j successes in a Bernoulli scheme (n_2, p)}, $j = 0, 1, \ldots, r$, and $D^r_{n_1+n_2}$ = {r successes in a Bernoulli scheme $(n_1 + n_2, p)$}. Then

$$D^r_{n_1+n_2} = A^0_{n_1} B^r_{n_2} + A^1_{n_1} B^{r-1}_{n_2} + \dots + A^r_{n_1} B^0_{n_2}.$$

Since the trials are independent, we obtain the required equation by using the formula for total probability.

8.35. Suppose that each trial results in one of the events S and F (success and failure) with μ_n as the number of the successes in n independent trials. Let $A^k_n = \{\mu_n = k\}$. Then $A^k_{n+1} = A^k_n F + A^{k-1}_n S$, $k \geqslant 1$, and

$$\{\mu_{n+1} \leqslant r\} = \bigcup_{k=0}^{r} A^k_{n+1} = A^0_{n+1} + \bigcup_{k=1}^{r} [A^k_n F + A^{k-1}_n S] =$$

$$= [A^0_n F + A^1_n F + A^0_n S + \dots + A^r_n F + A^{r-1}_n S +$$

$$+ A^r_n S] \setminus A^r_n S = [(A^0_n + A^1_n + \dots + A^r_n)F + (A^0_n +$$

$$+ A^1_n + \dots + A^r_n)S] \setminus A^r_n S = \left(\sum_{k=0}^{r} A^k_n\right) \setminus A^r_n S =$$

$$= \{\mu_n \leqslant r\} \setminus A^r_n S.$$

Thus

$$P\{\mu_{n+1} \leqslant r\} = P\{\mu_n \leqslant r\} - P(A^r_n S),$$

which implies the first inequality.

A similar argument shows that $\{\mu_{n+1} \leqslant r + 1\} = \{\mu_n \leqslant r\} + A^{r+1}_{n+1} F$, which implies the second equality.

8.36. We have to prove the inequality

$$\sum_{k=0}^{m} \binom{n}{k} p^k q^{n-k} > \sum_{k=m+1}^{n} \binom{n}{k} p^k q^{n-k}. \tag{8.16}$$

Let $B_r = \binom{n}{m-r} p^{m-r} q^{n-m+r}$ for $r = 0, 1, \dots, m$ and $C_r = \binom{n}{m+r} p^{m+r} q^{n-m-r}$ for $r = 0, 1, \dots, n - m$. Then (8.16) could be written in the form

$$\sum_{r=0}^{m} B_r > \sum_{r=0}^{n-m} C_r. \tag{8.17}$$

Let $D_r = B_r/C_r$. Then

$$\frac{D_{r+1}}{D_r} - 1 = \frac{(p - q)(r^2 + r - npq)}{(n - m - r)(n - m + r - 1)p^2} .$$

Hence it follows that for small values of r we have $D_{r+1}/D_r - 1 > 0$. When r increases, the value $\frac{D_{r+1}}{D_r} - 1$ decreases and becomes negative for $r \geqslant s$, where s is the smallest integer for which $s(s + 1) > npq$. Since $D_0 = 1$ and $D_1 = B_1/C_1 = (npq + q)/(npq + p) > 1$, then there exists an integer k such that $B_r/C_r \geqslant 1$ for $r = 0, 1, \ldots, k - 1$ and $B_r/C_r < 1$ for $r = k, \ldots, n - m$. It is easy to verify that for this value k the inequalities

$$\sum_{r=0}^{k-1} (k - r - 1)B_r > \sum_{r=0}^{k-1} (k - r - 1)C_r \qquad (8.18)$$

and

$$\sum_{r=k}^{m} (r - k + 1)B_r > - \sum_{r=k}^{n-m} (r - k + 1)C_r \qquad (8.19)$$

are true. From the identity $\sum_{k=0}^{n} (k - m)\binom{n}{k}p^k q^{n-k} = 0$, it follows that

$$\sum_{r=0}^{m} rB_r = \sum_{r=0}^{n-m} rC_r . \qquad (8.20)$$

Finally, from (8.16), (8.19) and (8.20) we find that

$$(k - 1) \sum_{r=0}^{m} B_r > (k - 1) \sum_{r=0}^{n-m} C_r ,$$

which is, in fact, the required inequality (8.17).

9. Discrete Random Variables and Their Characteristics

9.1. (a) Comparing the coefficients in the both sides of the equality $(1 + x)^M(1 + x)^{N-M} = (1 + x)^N$ with $M < N$, we get

$$\binom{N}{n} = \sum_{k=0}^{\min[M,n]} \binom{M}{k}\binom{N - M}{n - k} .$$

Then it follows that $\Sigma p_i = 1$.

(b) $\sum_{k=0}^{n} \binom{n}{k} p^k q^{n-k} = (p+q)^n = 1.$

(c) $\sum_{k_1+\ldots+k_r=n} \frac{n!}{k_1! \ldots k_r!} p_1^{k_1} \ldots p_r^{k_r} = (p_1 + \ldots + p_r)^n = 1.$

(d) $\sum_{k=0}^{\infty} q^k p = p \sum_{k=0}^{\infty} q^k = \frac{p}{1-q} = 1.$

(e) $\sum_{k=0}^{\infty} \binom{-r}{k} (-q)^k p^r = p^r \sum_{k=0}^{\infty} \binom{-r}{k} (-q)^k = \frac{p^r}{(1-q)^r} = 1.$

(f) $\sum_{k=0}^{\infty} \frac{\lambda^k e^{-\lambda}}{k!} = e^{-\lambda} \sum_{k=0}^{\infty} \frac{\lambda^k}{k!} = e^{-\lambda} e^{\lambda} = 1.$

In all cases the assertion follows from the fact that

$$\sum_i p_i = \sum_i P\{\omega : \xi(\omega) = x_i\} = P(\Omega) = 1.$$

9.2. We find $c = \frac{2}{3}$. (a) $P\{\xi \geqslant 10\} = 1/3^{10}$; (b) $P\{\xi \in A\} = \frac{1}{4}$; (c) $P\{\xi \in B\} = \frac{79}{351}$.

9.3. Here ξ assumes the values 1, 2, ..., 6 each with probability $\frac{1}{6}$ and $\eta = 7 - \xi$. Then for $\zeta = \xi\eta$ one easily gets $P\{\zeta = 6\} = P\{\xi = 10\} = P\{\xi = 12\} = \frac{1}{3}$.

9.4. It easily follows that ξ assumes the values 2, 3, 4, 6 and 8 respectively with probability 4/64, 8/64, 20/64, 16/64 and 16/64.

9.5. Here η assumes the values $n, n+1, \ldots, n+k, \ldots, N$ respectively with probability

$$p_k = P\{\eta = n + k\} = \binom{n+k-1}{n-1} / \binom{N}{n}, \quad k = 0, 1, \ldots, N-n.$$

9.6. Let $\zeta = \xi + \eta$. Then for $k = 0, 1, \ldots, 2n$ one has

$$P\{\zeta = k\} = \sum_{i=0}^{k} P\{\xi = i\} P\{\eta = k - i\} =$$

$$= \sum_{i=0}^{k} p_i r_{k-i} = p_0 \cdot \frac{1}{2} + p_k \cdot \frac{1}{2}.$$

For $k > 0$ the last expression equals $0 \cdot \frac{1}{2} + \frac{1}{n} \cdot \frac{1}{2} = \frac{1}{2n}$ and for $k = 0$ it equals $0 \cdot \frac{1}{2} = 0$

9.7. $P\{\xi_1 = i, \xi_2 = i\} = \frac{i}{36}$ for $i = 1, 2, \ldots, 6$.

$$P\{\xi_1 = i, \xi_2 = j\} = \begin{cases} \frac{1}{36}, & \text{if } i < j, \\ 0, & \text{if } i > j. \end{cases}$$

9.8. (a) $P\{\xi = x_i\} = \frac{1}{n}$, $i = 1, \ldots, n$, $E(\xi) = \frac{1}{n}\sum_{i=1}^{n} x_i$. (b) $E(\xi) = \frac{1}{2}$, $V(\xi) = (n + 1)/[12(n - 1)]$. (c) $E(\xi) = (a + b)/2$, $V(\xi) = (n + 1)(b - a)^2/[12(n - 1)]$. (d) $\lim_{n\to\infty} 3\sigma = \sqrt{3}/2 \approx 0.86$ in (b) and $\lim_{n\to\infty} 3\sigma = \sqrt{3}(b - a)/2$ in (c).

9.9. (a) Let $\xi = \xi_1 + \ldots + \xi_n$, where ξ_k, for $k = 1, \ldots, n$, equals 1 or 0 with probability p and q respectively, and $\xi_1, \ldots, \xi_n$ are independent. We have $E(\xi_k) = 1 \cdot p + 0 \cdot q = p$, $E(\xi_k^2) = 1^2 \cdot p + 0^2 \cdot q = p$, $V(\xi_k) = p - p^2 = pq$. Then $E(\xi) = \sum_{k=1}^{n} E(\xi_k) = np$, $V(\xi) = \sum_{k=1}^{n} V(\xi_k) = npq$.
(b) Consider Example 3.2. Lable the drawing of a black ball a success and let ξ_k be the number of successes obtained at the kth drawing for $k = 1, \ldots, n$. Then ξ_k is 1 or 0 and $P\{\xi_k = 1\} = M/N = 1 - P\{\xi_k = 0\}$. Hence $E(\xi_k) = M/N$ and $V(\xi_k) = M(N - M)/N^2$. Put $\xi = \xi_1 + \ldots + \xi_n$. Then $E(\xi) = nM/N$. Since all ξ_k, $k = 1, \ldots, n$, are not independent, one can use the generalization of (9.11) to compute $V(\xi)$. For $j \neq k$ we have $\xi_j\xi_k = 1$ if a black ball is drawn at both the kth and jth trial; $\xi_j\xi_k = 0$ otherwise. Hence one has $P\{\xi_j\xi_k = 1\} = M(M - 1)/[N(N - 1)]$ and $E\{\xi_j\xi_k\} = M(M - 1)/[N(N - 1)]$. According to (9.9)

$$\text{cov}(\xi_j, \xi_k) = E\{\xi_k\xi_j\} - E\{\xi_k\}E\{\xi_j\} = \frac{M(N - M)}{N^2(n - 1)}.$$

Then

$$V(\xi) = \sum_{k=1}^{n} V(\xi_k) + 2\sum_{i<j} \text{cov}(\xi_i, \xi_j) = \frac{nM(N - M)(N - n)}{N^2(N - 1)}.$$

9.10. (a) $g(s) = p(1 - sq)^{-1}$, $E(\xi) = q/p$, $V(\xi) = q/p^2$. (b) $g(s) = p^r(1 - sq)^{-r}$, $E(\xi) = rq/p$, $V(\xi) = rq/p^2$. (c) $g(s) = e^{\lambda(s-1)}$, $E(\xi) = \lambda$, $V(\xi) = \lambda$.

9.11. (a) $g_\eta(s) = sg(s)$. (b) $g_\eta(s) = s^k g(s)$. (c) $g_\eta(s) = g(s^2)$.

9.12. Let $p_n = P\{\xi = n\}$, $g(s) = \sum_{n=0}^{\infty} p_n s^n$. (a) $h(s) = s^0 p_0 + s^1(p_0 + p_1) + s^2(p_0 + p_1 + p_2) + \ldots + s^n(p_0 + p_1 + \ldots + p_n) + \ldots = g(s)(1 + s + s^2 + \ldots) = g(s)/(1 - s)$. (b) $sg(s)/(1 - s)$. (c) $(1 - sg(s)/(1 - s)$.
(d) $p_0/s + (1 - s^{-1}g(s))/(1 - s)$. (e) $\frac{1}{2}[g(\sqrt{s}) + g(-\sqrt{s})]$.

9.13. Here ξ assumes the values 0, 1, ..., 27 and for $k = 0, 1, \ldots, 27$ the probability $p_k = P\{\xi = k\}$ is exactly the coefficient at s^k in the quantity $g(s) = 10^{-3}(1 - s^{10})^3(1 - s)^{-3}$ (see (8.7) and (8.9)). Thus we have

ξ	0	1	2	3	4	5	6	7	8	9	10	11	12	13
$10^3 p_\xi$	1	3	6	10	15	21	28	36	45	55	63	69	73	75

ξ	14	15	16	17	18	19	20	21	22	23	24	25	26	27
$10^3 p_\xi$	75	73	69	63	55	45	36	28	21	15	10	6	3	1

.

9.14. $P\{\xi = 0\} = q^{n-1}$, $P\{\xi = k\} = pq^{n-1-k}$, $k = 1, \ldots, n - 1$, $q = 1 - p$; $P\{\eta = k\} = pq^{k-1}$, $k = 1, \ldots, n - 1$, $P\{\eta = n\} = q^{n-1}$; $g_\xi(s) = (ps^n + q^n s - q^n)/(s - q)$; $E(\xi) = (q^n + np - q)/p$, $V(\xi) = (q + (1 - 2n)pq^n - q^{2n})/p^2$, $\lim_{n\to\infty} E(\xi) = \infty$, $\lim_{n\to\infty} V(\xi) = q/p^2$; $g_\eta(s) = (q^n s^n - q^n s^{n+1} + ps)/(1 - qs)$; $E\{\eta\} = E\{n - \xi\} = n - E(\xi) = (1 - q^n)/p$; $V(\eta) = V\{n - \xi\} = V(n) + V(\xi) - 2\,\mathrm{cov}(n, \xi) = V(\xi)$, since n is a constant; $\lim_{n\to\infty} E(\eta) = 1/p$; $\lim_{n\to\infty} V(\eta) = q/p^2$.

9.15. The idea is to decompose g(s) into fractions which can be expanded as power series.

(a) $$\frac{1 - 4\alpha^2}{(1 - 4\alpha^2)s^2 - 4s + 4} = \frac{1 - 4\alpha^2}{8\alpha}\left(\frac{\frac{1}{2} + \alpha}{1 - (\frac{1}{2} + \alpha)s} - \frac{\frac{1}{2} - \alpha}{1 - (\frac{1}{2} - \alpha)s}\right).$$ Since g(s) admits a power series expansion for $|s| \leqslant 1$, we have the restrictions $|\frac{1}{2} + \alpha| < 1$ and $|\frac{1}{2} - \alpha| < 1$. Thus $|\alpha| < \frac{1}{2}$. The probability $P\{\xi = n\}$ coincides with the coefficient p_n of s^n in the expansion of g(s). Then we find that

$$p_n = \frac{1 - 4\alpha^2}{8\alpha}\,[(\tfrac{1}{2} + \alpha)^{n+1} - (\tfrac{1}{2} - \alpha)^{n+1}],\ n = 0, 1, 2, \ldots$$

Since $p_n > 0$, $n = 0, 1, \ldots$ and $g(1) = 1$, we conclude that $p_0 + p_1 + \ldots = 1$. Thus $\{p_n : n = 0, 1, \ldots\}$ is a probability distribution.

In (b) we obtain similarly $P\{\xi = 0\}$, $P\{\xi = n\} = 2 \cdot 3^{-n-1} + 2 \cdot (3/2)^{-n-1} - 2^{1-n}$, $n = 1, 2, \ldots$

In (c) we have $P\{\xi = n\} = \lambda^{2n}/((2n)!\,\mathrm{ch}\lambda)$, $n = 0, 1, \ldots$, $\lambda \in \mathbb{R}_1$.

9.16. Let $E(\xi) = \lambda$ and let us assume that there exist two integers $k_1 \geqslant 0$ and $k_2 \geqslant 0$ for which $P\{\xi = k_1\} = P\{\xi = k_2\}$; i.e., $\lambda^{k_1}e^{-\lambda}/k_1! = \lambda^{k_2}e^{-\lambda}/k_2!$ Assume also that $k_1 = k_2 + k$, $k > 0$. Then one can easily obtain that $\lambda^k = (k_2 + 1)(k_2 + 2) \ldots (k_2 + k)$. The last relation is possible only if

$k = 1$. (Why?) Hence $\lambda = k_2 + 1$; i.e., $k_1 = \lambda$ and $k_2 = \lambda - 1$.

9.17. If $p_k = P\{\xi = k\} = \lambda^k e^{-\lambda}/k!$, then $p_{k+1}/p_k = \lambda/(k + 1)$. If $\lambda = m$ is an integer then two values of ξ are most likely to occur, namely, $\xi = m - 1$ and $\xi = m$. If λ is not an integer, then the most likely value of ξ is $m = [\lambda]$.

9.18. (a) $P\{\xi = k\} = (1 - p)^{k-1}p$, $k = 0, 1, \ldots$ (b) It is most likely that one trial will precede the first success, since $p_k > p_{k+1}$ for any k.

9.19. (a) Obviously $p_r(k) \geqslant p_r(k - 1)$, if $k \leqslant q(r - 1)/p$ and $p_r(k) \leqslant p_r(k - 1)$, if $k \geqslant q(r - 1)/p$. (b) $m = [q(r - 1)/p]$.

9.20. We have $p_k = \binom{M}{k}\binom{N - M}{n - k}/\binom{N}{n} = \frac{M!}{k!(M - k)!} \cdot \frac{(N - M)!}{(n - k)!(N - M - n + k)!}$

$$\cdot \frac{n!(N - n)!}{N!} = \binom{n}{k} \frac{M}{N} \frac{\frac{M}{N} - \frac{1}{N}}{1 - \frac{1}{N}} \frac{\frac{M}{N} - \frac{2}{N}}{1 - \frac{2}{N}} \cdots \frac{\frac{M}{N} - \frac{k - 1}{N}}{1 - \frac{k - 1}{N}} \times$$

$$\times \frac{1 - \frac{M}{N}}{1 - \frac{k}{N}} \frac{1 - \frac{M}{N} - \frac{1}{N}}{1 - \frac{k + 1}{N}} \cdots \frac{1 - \frac{M}{N} - \frac{n - k - 1}{N}}{1 - \frac{n - 1}{N}}. \text{ Hence } \lim_{N\to\infty} p_k = \binom{n}{k}p^k q^{n-k}.$$

9.21. Since $p_n = \lambda/n$, then we have

$$\binom{n}{k}p_n^k(1 - p_n)^{n-k} = \frac{n(n - 1) \ldots (n - k + 1)}{k!} p_n^k(1 - p_n)^{n-k} =$$

$$= \frac{\lambda^k}{k!} \frac{n(n - 1) \ldots (n - k + 1)}{n^k} (1 - \frac{\lambda}{n})^{n-k} =$$

$$= \frac{\lambda^k}{k!} (1 - \frac{1}{n})(1 - \frac{2}{n}) \ldots (1 - \frac{k - 1}{n})(1 - \frac{\lambda}{n})^{n-k}.$$

Therefore $\lim_{n\to\infty} \binom{n}{k}p_n^k(1 - p_n)^{n-k} = \lambda^k e^{-\lambda}/k!$.

9.22. Let ζ be the total number of throwings of both players, including the final throwing. Put A_k = {A scores in the kth trial} and B_i = {B scores in the ith trial}. The list of the possible outcomes is the following: A_1, $\bar{A}_1B_1$, $\bar{A}_1\bar{B}_1A_2$, $\bar{A}_1\bar{B}_1\bar{A}_2B_2$, $\bar{A}_1\bar{B}_1\bar{A}_2\bar{B}_2A_3$, $\bar{A}_1\bar{B}_1\bar{A}_2\bar{B}_2\bar{A}_3B_3$, ... Thus ζ possesses the following distribution

ζ	1	2	3	4	5	6	...
p_ζ	p_1	q_1p_2	$q_1q_2p_1$	$q_1^2q_2p_2$	$q_1^2q_2^2p$	$q_1^3q_2^2p_2$	

The sum of all probabilities $P\{\zeta = 2k - 1\}$, $k = 1, 2, \ldots$ equals $p_1(1 + q_1q_2 + q_1^2q_2^2 + \ldots) = \frac{p_1}{1 - q_1q_2}$, which is exactly the probability that A wins. The sum of the remaining probabilities, i.e., the probability that B wins, then equals $q_1p_2(1 + q_1q_2 + q_1^2q_2^2 + \ldots) = \frac{q_1p_2}{1 - q_1q_2}$. The distributions of ξ and η can be easily found from the above reasoning.

9.23. (a) Let $m = 2$; $\eta = \xi_1 + \xi_2$ assumes the values $0, 1, \ldots, n_1 + n_2$ respectively with probability $p_k = P\{\eta = k\} = \sum_{j=1}^{k} \binom{n_1}{j} p^j q^{n_1-j} \times \binom{n_2}{k-j} p^{k-j} q^{n_2-k+j} = \binom{n_1+n_2}{k} p^k q^{n_1+n_2-k}$, $k = 0, 1, \ldots, n_1 + n_2$, which means that $\eta \in \mathcal{B}(n_1 + n_2, p)$. For an arbitrary $m \geqslant 2$ the statement follows by induction.

(b) For $m = 2$, one has

$$P\{\xi_1 + \xi_2 = k\} = \sum_{i=0}^{k} \frac{\lambda_1^i}{i!} e^{-\lambda_1} \frac{\lambda_2^{k-i}}{(k-i)!} e^{-\lambda_2} =$$

$$= \frac{(\lambda_1 + \lambda_2)^k}{k!} e^{-(\lambda_1+\lambda_2)}, \quad k = 0, 1, \ldots;$$

i.e., $\xi_1 + \xi_2 \in \mathcal{P}(\lambda_1 + \lambda_2)$. The general case again is proved by induction.

9.24. (a) Let k and n be natural numbers, $0 \leqslant k \leqslant n$. Then

$$P\{\xi = k | \xi + \eta = n\} = \frac{P\{\xi = k, \xi + \eta = n\}}{P\{\xi + \eta = n\}} =$$

$$= \frac{P\{\xi = k\}P\{\xi + \eta = n | \xi = k\}}{P\{\xi + \eta = n\}} =$$

$$= \frac{(\lambda^k e^{-\lambda}/k!)P\{\eta = n - k\}}{\sum_{j=0}^{n} P\{\xi = j\}P\{\eta = n - j\}} =$$

$$= \binom{n}{k}\left(\frac{\lambda}{\lambda + \mu}\right)^k \left(\frac{\mu}{\lambda + \mu}\right)^{n-k}.$$

For $k > n$ one has $P\{\xi = k | \xi + \eta = n\} = 0$. Thus the conditional distribution of ξ under the condition $\xi + \eta = n$ is binomial with parameters n and $p = \lambda/(\lambda + \mu)$.

(b) $P\{\xi - \eta = k\} = e^{-2\lambda} \sum_{i=0}^{\infty} [\lambda^{2i+|k|}(i!\Gamma(i + |k + 1|))^{-1}]$, $k = 0, \pm 1, \pm 2, \ldots$

9.25. Let η be the number of trials in which a success occurs. According to the theorem for total probability we have $P\{\eta = m\} = (\lambda p)^m e^{-\lambda p}/m!$, which means that $\eta \in \mathcal{P}(\lambda p)$.

9.26. By trivial calculations, based on the theorem of total probabilities, one easily gets $P\{\xi = k\} = \binom{M}{k}(pq)^k(1 - pq)^{M-k}$.

9.27. (a) 0. (b) 7/2. (c) 27/4.

9.28. (a) $\xi = \begin{pmatrix} 0.01 & 0.02 & 0.03 & 0.04 \\ 0.1 & 0.4 & 0.3 & 0.2 \end{pmatrix}$, $\eta = \begin{pmatrix} 0.002 & 0.004 & 0.006 & 0.008 \\ 0.11 & 0.48 & 0.30 & 0.11 \end{pmatrix}$.

(b) $E(\xi) = 0.026$, $E(\eta) = 0.00482$, $V(\xi) = 86 \cdot 10^{-6}$, $V(\eta) = 25281 \cdot 10^{-6}$.

(c) $\text{cov}(\xi, \eta) = 208.5 \cdot 10^{-6}$, $\rho(\xi, \eta) = 0.141$.

9.29. (a) Since η_1 and η_2 are also independent r.v.'s, $\rho(\eta_1, \eta_2) = 0$.
(b) Put $E(\xi_i) = a_i$ and $V(\xi_i) = \sigma^2$, $i = 1, \ldots, 5$. Then one easily gets $\rho(\zeta_1, \zeta_2) = \frac{1}{3}$.

9.30. It can easily be seen that $E(\xi_k) = np_k$, $V(\xi_k) = np_k(1 - p_k)$, $E\{\xi_i\xi_j\} = n(n - 1)p_ip_j$; hence we obtain the expression for $\rho(\xi_i, \xi_j)$.

9.31. We have $V\{\xi\eta\} - V(\xi)V(\eta) = E\{(\xi\eta)^2\} - (E\{\xi\eta\})^2 - [E\{\xi^2\} - (E[\xi])^2][E\{\eta^2\} - (E[\eta])^2] = E\{\xi^2\}E\{\eta^2\} - (E[\xi])^2(E[\eta])^2 - E\{\xi^2\}E\{\eta^2\} + (E[\xi])^2E\{\eta^2\} + (E[\eta])^2E\{\xi^2\} - (E[\xi])^2(E[\eta])^2 = (E[\xi])^2V(\eta) + (E[\eta])^2V(\xi) = 0$. The last equality holds only if $E(\xi) = E(\eta) = 0$.

9.32. Let A = {the first player wins}, B = {the second player wins}, H_i = {the first player wins i coins in the first tossing}, $i = -3, 0, 2$. We have $P(H_{-3}) = \frac{3}{16}$, $P(H_0) = \frac{5}{16}$, $P(H_2) = \frac{8}{16}$, $P(A|H_{-3}) = 0$, $P(A|H_0) = P(A)$, $P(A|H_2) = 1$. Then $P(A) = \frac{5}{16}P(A) + \frac{8}{16}$ and we obtain from the last relation: $P(A) = P(\bar{B}) = \frac{8}{11}$, $P(\bar{A}) = P(B) = \frac{3}{11}$.

Let ξ_1 and ξ_2 be the amounts of coins, won respectively by the first and the second player. Then ξ_1 and ξ_2 are distributed as follows:

ξ_1	-3	2
P_{ξ_1}	3/11	8/11

ξ_2	-2	3
P_{ξ_2}	8/11	3/11

Thus $E(\xi_1) = 7/11$ and $E(\xi_2) = -7/11$, which means that the game is not fair, since the first player has a distinct advantage.

9.33. The desired price is $c = \frac{2}{N} \cdot \sum_{i=1}^{n} m_ik_i$. Consider the r.v. ξ, which assumes the values $0, k_1, \ldots, k_n$, respectively with probability $p_0 = \frac{N - (m_1 + \ldots + m_n)}{N}$ and $p_i = \frac{m_i}{N}$ for $i = 1, \ldots, n$.

9.34. We have $1 = \sum_{k=1}^{\infty} p_k = b \sum_{k=1}^{\infty} (a^k/k) = bh(a)$, where $h(a) = \int\left(\sum_{m=0}^{\infty} a^m\right)da = -\ln(1 - a)$. Therefore $\{p_k, k \geq 1\}$ is a probability distribution for $b = -1/\ln(b - a)$. The p.g.f. $g(z)$ is given by $g(z) =$

$\sum_{k=1}^{\infty} p_k z^k = \ln(1 - az)/\ln(1 - a)$. Then $E(\xi) = \dfrac{a}{(1 - a)\ln(1 - a)}$ and

$V(\xi) = \dfrac{a^2 + a\ln(1 - a)}{(1 - a)^2 \ln^2(1 - a)}$.

9.35. According to Exercise 1.15 (e), we have $p_m = P\{\xi_r = m\} = \binom{m-1}{r-1}\binom{n-m}{k-r}/\binom{n}{r}$ for $m = r, r + 1, \ldots, n - k + r$. To find $E(\xi_r)$ and $\vec{V}(\xi_r)$, the following sums are useful:

$$S_1 = \sum_{m=r}^{n-k+r} m\binom{m-1}{r-1}\binom{n-m}{k-r}, \qquad S_2 = \sum_{m=r}^{n-k+r} m^2\binom{m-1}{r-1}\binom{n-m}{k-r}.$$

These sums can easily be calculated using the following identity (see Exercise 1.15 (e)) for $s = 1$ and $s = 2$:

$$\sum_{m=r}^{n-k+r} \binom{m+s-1}{r+s-1}\binom{n-m}{k-r} = \binom{n+s}{k+s}.$$

After some obvious transformations, we get

$$E(\xi_r) = \frac{r(n+1)}{k+1} \quad \text{and} \quad V(\xi_r) = \frac{r(n+1)(n-k)(k+1-r)}{(k+1)^2(k+2)}.$$

9.36. Put $\eta_i = N + 1 - \xi_i$, $i = 1, \ldots, n$. Obviously $\eta_1, \ldots, \eta_n$ are also independent r.v.'s, which are uniformly distributed over the set $\{1, 2, \ldots, N\}$. Hence

$$p_k = P\{\max[\xi_1, \ldots, \xi_n] = k\} = P\{\max[\eta_1, \ldots, \eta_n] = k\} =$$
$$= P\{N + 1 - \min[\xi_1, \ldots, \xi_n] = k\} = P\{\min[\xi_1, \ldots, \xi_n] =$$
$$= N + 1 - k\}.$$

Then

$$E\{\max[\xi_1, \ldots, \xi_n]\} + E\{\min[\xi_1, \ldots, \xi_n]\} =$$
$$= \sum_{k=1}^{N} kp_k + \sum_{k=1}^{N} kP\{\min[\xi_1, \ldots, \xi_n] = k\} =$$
$$= \sum_{k=1}^{N} kp_k + \sum_{k=1}^{N} (N + 1 - k)P\{\min[\xi_1, \ldots, \xi_n] = N + 1 - k\} =$$
$$= \sum_{k=1}^{N} kp_k + \sum_{k=1}^{N} (N + 1 - k)p_k = (N + 1)\sum_{k=1}^{N} p_k = N + 1$$

(see also Exercise 1.35).

9.37. Let A_n = {at least k 'heads' occur in the first n tossings and a 'tail' occurs at the nth tossing}, B_n = {at least k 'heads' occur in the first n tossings and a 'head' occurs at the nth tossing}. It is easily seen that $\{\tau_k = n\} = A_n \cup B_n$. Obviously $P\{\tau_k = n\} = 0$ if $n < 2k$. For $n \geqslant 2k$ we have

$$P(A_n) = P(B_n) = 2^{-n}\binom{n-1}{k-1}.$$

On the other hand $A_n B_n = \emptyset$ and $P\{\tau_k = n\} = 2^{-(n-1)}\binom{n-1}{k-1}$ for $n = 2k$, $2k+1, \ldots$ The events $\{\tau_k = n\}$ for $n = 2k, 2k+1, \ldots$ are mutually disjoint and therefore $\sum_{n=2k}^{\infty} 2^{-(n-1)}\binom{n-1}{k-1} = 1$. Using the last relation, one can easily calculate that

$$E(\tau_k) = \sum_{n=2k}^{\infty} nP\{\tau_k = n\} = 2k[2^{-k}\binom{2k}{k} + 1].$$

9.38. Let $\xi_1, \ldots, \xi_n$ be independent r.v.'s, for which $P\{\xi_k = 1\} = \frac{1}{k} = 1 - P\{\xi_k = 0\}$, $k = 1, \ldots, n$. The p.g.f. $g_n(z)$ of the r.v. $\mu_n = \xi_1 + \ldots + \xi_n$, which is exactly the number of successes, is given by

$$g_n(z) = \prod_{k=1}^{n} \frac{z+k-1}{k} = \frac{1}{n!} z(z+1) \ldots (z+n-1).$$

The last quantity is exactly the given function $p_n(z)$ and we can find

$$a_1(n) = E(\mu_n) = \sum_{k=1}^{n} \frac{1}{k}.$$

On the other hand $\sum_{k=1}^{n} \frac{1}{k} = \ln n + C + \varepsilon_n$, where C is the Euler constant and $\{\varepsilon_n\}$ is a decreasing sequence with $\lim_{n\to\infty} \varepsilon_n = 0$. Hence $\lim_{n\to\infty}[a_1(n)/\ln n] = 1$.

10. Normal and Poisson Approximation for the Binomial Distribution

10.1. We have $P\{1465 \leqslant \mu \leqslant 1535\} = \sum_{k=1465}^{1535} \binom{2000}{k} (\frac{3}{4})^k (\frac{1}{4})^{2000-k}$. We shall compute this probability approximately, using (10.5) and Table 1. Since $np = 1500$, $npq = 375$, we get

$$P\{1465 \leqslant \mu \leqslant 1535\} \approx 2\Phi(35/\sqrt{375}) - 1 \approx 0.9282.$$

10.2. Consider the Bernoulli scheme (n, p) with arbitrary $p \in (0, 1)$. Put $\mu_n^* = (\mu_n - np)/\sqrt{npq}$. A number $a = a(n, p)$ is to be found, such that $P\{|\mu_n^*| > a\}$ is close to 0.5 when n is large; i.e., $P\{|\mu_n^*| > a\} \approx 2\Phi(a) - 1 \approx 0.5$. The last relation gives $\Phi(a) = 0.75$. Unfortunately, the exact value of a cannot be found in Table 1. Approximately one can take $a = 0.67$ (in some more accurate tables of the normal distribution the value $a = 0.6745$ can be found). Thus for large n the events $\{|\mu_n - np| < 0.67\sqrt{npq}\}$ and $\{|\mu_n - np| > 0.67\sqrt{npq}\}$ are approximately equally likely to occur; i.e., each has probability close to 0.5. In case (a) $p = \frac{1}{2}$, $np = \frac{1}{2}n$, $0.67\sqrt{pq} \approx 0.335$; therefore, with probability close to 0.5 the number of heads obtained in n tossings will lie in the interval $(\frac{1}{2}n - 0.335\sqrt{n}, \frac{1}{2}n + 0.335\sqrt{n})$. In case (b) $p = \frac{1}{6}$, $np = \frac{n}{6}$, $0.67\sqrt{pq} \approx 0.251$; therefore, with probability close to 0.5 the number of times a six appears when a die is rolled n times will be contained in the interval $(\frac{1}{6}n - 0.251\sqrt{n}, \frac{1}{6}n + 0.251\sqrt{n})$.

10.3. (a) We have $P\{0.40 < \tilde{p}_n < 0.44\} = P\{0.40 < \frac{\mu_n}{n} < 0.44\} = P\{0 < \frac{\mu_n - np}{\sqrt{npq}} < \frac{60}{\sqrt{1500 \times 0.4 \times 0.6}}\} \approx \Phi(\frac{60}{360}) - \Phi(0) \approx 0.4992$.

(b) Since $\sqrt{pq} = 0.484$, we have $P\{|\tilde{p}_n - p| \leqslant 0.01\} \approx 2\Phi(0.0207\sqrt{n})$. Now the value of n, for which $2\Phi(0.0207\sqrt{n}) - 1 \geqslant 0.995$, is to be found. We have $\Phi(0.0207\sqrt{n}) \geqslant 0.9975$. From Table 1 we get $0.0207\sqrt{n} \geqslant 3$ and therefore $n \geqslant 21025$.

(c) Here we look for ε such that $P\{|\tilde{p}_n - p| < \varepsilon\} \geqslant 0.985$. We easily find that $P\{|\tilde{p}_n - p| < \varepsilon\} \approx 2\Phi(60\varepsilon) - 1)$ and the relation $2\Phi(60\varepsilon) - 1 \geqslant 0.985$ and Table 1 allow us to conclude that $\varepsilon \geqslant \varepsilon_{min} \approx 0.04$.

(d) Analogously to cases (a) - (c) we find $P\{|\tilde{p}_n - p| \leqslant 0.05\} \approx 2\Phi(0.6/\sqrt{p(1 - p)}) - 1 \geqslant 0.88$, or $\Phi(0.6/\sqrt{p(1 - p)}) \geqslant 0.94$. Since $\Phi(1.60) \approx 0.94$, the desired values of p must satisfy the inequality $0.6/\sqrt{p(1 - p)} \geqslant 1.60$. It follows that either $0 < p < 0.169$ or $0.830 < p < 1$.

10.4. We have $P\{|\mu_n - np| \geqslant 228\} = 1 - P\{|\mu_n - np| < 228\} = 1 - P\left\{\frac{|\mu_n - np|}{\sqrt{npq}} < \frac{228}{60}\right\} \approx 2 - 2\Phi(3.8) \approx 0.0002$.

10.5. $n \geqslant 230$.

10.6. Denote by ξ_i the number of damaged components of the ith type, $i = 1, 2, 3$. It follows from Poisson's theorem that when n is large in a Bernoulli scheme (n, p), the probability $P\{\mu_n = k\}$ is approximately equal to $\lambda^k e^{-\lambda}/k!$ with $\lambda = np$ for $k = 0, 1, \ldots$ This means that, approximately, the distribution of each of the r.v.'s ξ_1, ξ_2 and ξ_3 can be considered as a Poisson. The corresponding parameters are $\lambda_1 = n_1p_1 = 0.7$, $\lambda_2 = n_2p_2 = 0.6$, $\lambda_3 = n_3p_3 = 0.7$. The total number of damaged components is $\xi = \xi_1 + \xi_2 + \xi_3$, which has a Poisson distribution with parameter $\lambda = \lambda_1 + \lambda_2 + \lambda_3 = 2$ (see Exercise 9.44). Then $P\{\xi \geqslant 2\} = 1 - P\{\xi = 0\} - P\{\xi = 1\} \approx 1 - e^{-\lambda} - \lambda e^{-\lambda}$. For $\lambda = 2$, we obtain from Table 2 that $P\{\xi \geqslant 2\} \approx 0.5940$.

10.7. Consider an arbitrarily chosen pack. One fixed coffee bean can fall into this pack with probability $p = 0.0001$. We have 5000 unroasted coffee beans, which means that the number of trials (assumed independent) is $n = 5000$. Let μ be the number of successes in the Bernoulli scheme (n, p) with $n = 5000$ and $p = 0.0001$; i.e., μ is the number of unroasted coffee beans in the chosen pack. According to Poisson's theorem we find the desired probability:

$$P\{\mu \geqslant 1\} = 1 - P\{\mu = 0\} \approx 1 - e^{-0.5}.$$

For $\lambda = 0.5$ and $k = 0$, we obtain from Table 2 that $P\{\mu \geqslant 1\} \approx 0.3935$.

10.8. The reasoning is the same as in Exercise 10.7. The approximate value of the probability, which is to be found, is 0.0002.

10.9. $n \geqslant 1682$.

10.10. The birth of a girl will be referred to as a 'success'. Then we have $n = 10000$ independent trials and the probability for a success at each trial is given by $p = 0.488$; hence $np = 4880$ and $\sqrt{npq} = 50$. Denote by μ the number of successes in 10000 trials. Then:

(a) $P\{\mu \geqslant 10000 - \mu\} = P\{\mu \geqslant 5000\} \approx 1 - \Phi(2.4) \approx 0.0082$.

(b) $P\{10000 - \mu \geqslant \mu + 200\} = P\{\mu \leqslant 5100\} \approx \Phi(22/5) \approx 1$.

10.11. For an arbitrary $\varepsilon > 0$, we have

$$P\{|\tilde{p}_n - p| < \varepsilon\} = P\left\{\left|\frac{\mu_n}{n} - p\right| < \varepsilon\right\} =$$

$$= P\left\{\frac{|\mu_n - np|}{\sqrt{npq}} < \frac{\varepsilon\sqrt{n}}{\sqrt{pq}}\right\} \approx 2\Phi\left(\frac{\varepsilon\sqrt{n}}{\sqrt{pq}}\right) - 1,$$

which converges to 1 as $n \to \infty$, since $\Phi(x) \to 1$ as $x \to \infty$.

10.12. $m = 508$.

10.13. $P\{|\mu_n - np| < 1.96\sqrt{npq}\} \approx \Phi(1.96) - \Phi(-1.96) \approx 0.95$, $P\{|\mu_n - np| < 3\sqrt{npq}\} \approx \Phi(3) - \Phi(-3) \approx 0.9975$, $P\{|\mu_n - np| \geqslant 3\sqrt{npq}\} \approx 0.0025$.

10.14. Assume first that the coin is fair; i.e., $p = \frac{1}{2}$. Then $np = 10000$, $\sqrt{npq} \approx 70.9$ and the difference (in absolute value) between the number of the obtained heads and the expected number of heads np is 800; i.e., the event $\{|\mu_n - np| = 800\}$ has occurred. But $\{|\mu_n - np| = 800\} \subset \{|\mu_n - np| \geqslant 3 \times 70.9\} = \{|\mu_n - np| \geqslant 3\sqrt{npq}\}$. According to Exercise 10.13, $P\{|\mu_n - np| \geqslant 3\sqrt{npq}\} \approx 0.0025$. Thus $P\{|\mu_n - np| = 800\} < 0.0025$. Since this probability is so small, it is reasonable to think that the coin may be biased.

10.15. The local theorem of de Moivre-Laplace is to be used.

10.16. (a) Since $\mu_n \in \mathcal{B}(n, p)$, ξ_n assumes the values: $-n, 2 - n, 4 - n, \ldots, n - 2, n$. When k is one of these numbers, then

$$P\{\xi_n = k\} = P\{2\mu_n - n = k\} = P\{\mu_n = \frac{k+n}{2}\} = P_n(\frac{k+n}{2})$$

where $P_n(m)$, $m = 0, 1, \ldots, n$, is the binomial probability.

(b) Use the fact that for $np = \frac{1}{2}n$, we have $\sqrt{npq} = \frac{1}{2}n$ and hence

$$\lim_{n\to\infty} P\left\{\frac{\mu_n - n/2}{\sqrt{n/2}} < x\right\} = \Phi(x), \quad x \in \mathbb{R}_1.$$

10.17. The probability of a winning ticket is $p = M/N$. Let μ_n be the number of winning tickets among some n arbitrary chosen tickets. Then μ_n can be considered as the number of successes in a Bernoulli scheme (n, p), which can further be approximated by the Poisson distribution.

The probability of at least one winning ticket (among those n) is $1 - P\{\mu_n = 0\}$. If one assumes that p is small, which is quite realistic, and that r is close to 1, then it is clear that a great number of tickets must be bought in order to have at least one winning ticket with probability greater than r. This means that n is to be taken sufficiently large. According to the Poisson theorem, the probability to have k winning tickets among n bought tickets is approximately equal to $\lambda_n^k e^{-\lambda_n}/k!$, $k = 0, 1, \ldots$, where $\lambda_n = np = nM/N$. For $k = 0$ we have $P\{\mu_n = 0\} \approx e^{-\lambda_n}$ and $P\{\mu_n \geqslant 1\} \approx 1 - e^{-\lambda_n}$. Now we have to find the smallest n, for which $P\{\mu_n \geqslant 1\} > r$; i.e., $1 - e^{-\lambda_n} > r$. It is easy to show that $n \geqslant -(N/M)\ln(1 - r)$.

10.18. According to the integral theorem of de Moivre-Laplace we have $\lim_{n\to\infty} P(A_n) = \Phi(1) - \Phi(-1) = 2\Phi(1) - 1 \approx 0.6826$, $\lim_{n\to\infty} P(B_n) = \lim_{n\to\infty} (1 - P(A_n)) \approx 0.3174$; hence it is obvious that $\frac{0.6826}{0.3174} > 2$.

10.19. This problem can be solved algebraically, but here we give the following *probabilistic solution*: Consider a Bernoulli scheme $(2n + 1, p)$ and denote by μ the number of successes. Obviously $a_n(p) = P\{\mu \leqslant n\}$. However, $E(\mu) = (2n + 1)p$ and $V(\mu) = (2n + 1)p(1 - p)$. Therefore

$$a_n(p) = P\left\{\frac{\mu - (2n+1)p}{\sqrt{(2n+1)p(1-p)}} \leqslant \frac{n(1-2p) - p}{\sqrt{(2n+1)p(1-p)}}\right\}$$

According to the theorem of de Moivre-Laplace $\lim_{n\to\infty} a_n(p) = \Phi(c_p)$, where $c_p = \lim_{n\to\infty} \frac{n(1-2p) - p}{\sqrt{(2n+1)p(1-p)}}$; i.e., $a(p) = \Phi(c_p)$. It follows that

$$c_p = \begin{cases} -\infty, & \text{if } \frac{1}{2} < p < 1 \\ 0, & \text{if } p = \frac{1}{2} \\ \infty, & \text{if } 0 < p < \frac{1}{2} \end{cases} \quad \text{and} \quad a(p) = \begin{cases} 1, & \text{if } 0 < p < \frac{1}{2} \\ \frac{1}{2}, & \text{if } p = \frac{1}{2} \\ 0, & \text{if } \frac{1}{2} < p < 1. \end{cases}$$

11. General Definition of Probability and σ-Algebra of Events

11.1. (a) For any A. (b) For any A. (c) For $B \subset A$. (d) For $A \subset B$. (e) Only for $A = \emptyset$ and $B = \Omega$. (f) Only for $A = \Omega$ and $B = \emptyset$. (g) For $A = B$.

11.2. (a) $A \setminus B = A\bar{B} = A\bar{B} + A\bar{A} = A(\bar{B} \cup \bar{A}) = A(\overline{AB}) = A \setminus (AB)$, $A \setminus B = A\bar{B} = A\bar{B} + B\bar{B} = (A \cup B)\bar{B} = (A \cup B) \setminus B$.

(b) $A \Delta B = A\bar{B} + \bar{A}B = A\bar{B} + A\bar{A} + \bar{A}B + B\bar{B} = A(\overline{AB}) + B(\overline{AB}) = (A \cup B)\overline{AB} = (A \cup B) \setminus (AB)$.

The remaining equalities are proved analogously.

11.3. Only (c), (e) and (f) are true for arbitrary events.

11.4. No. For instance, if $A_i = A \neq \Omega$ and $B = \Omega$.

11.5. If $AB = \emptyset$, then any X such that $B \subset X \subset \bar{A}$, is a solution to the exercise. If $AB \neq \emptyset$, the exercise has no solution.

11.6. If $0 < i < j$, then $B_i \subset A_i$, and $B_j \subset \bar{A}_i$ and hence $B_iB_j = \emptyset$. If $j > 0$, $B_0 \subset \bar{A}_j$, $B_j \subset A_j$ and then $B_0B_j = \emptyset$. The equality $B_0 + B_1 + \dots + B_n = \Omega$ is easy to verify.

11.7. The method of mathematical induction can be used.

11.8. Let $\omega \in \bigcup_{i=1}^{\infty} A_i$ and let j be the least index such that $\omega \in A_j$. Then $\omega \in \bar{A}_1\bar{A}_2 \dots A_{j-1}A_j$ and $\bigcup_{i=1}^{\infty} A_i \subset A_1 + \bar{A}_1A_2 + \bar{A}_1\bar{A}_2A_3 + \dots$ The inverse inclusion is obvious.

11.9. Consider the events $A_1, A_2, \dots, A_n$ as elements of $\mathcal{P}(\Omega)$ and give a set-theoretic proof.

11.10. The equalities (a) and (c) are obtained directly from the definition of symmetric difference.

(b) If $A = B$, then $A \Delta B = A \Delta A = A\bar{A} + \bar{A}A = \emptyset$. If $A \Delta B = \emptyset$, then $A\bar{B} = \emptyset$ and $\bar{A}B = \emptyset$. Then $A \subset B$ and $B \subset A$. Therefore, $A = B$. (d) $(A \Delta B = M \Delta N) \Leftrightarrow ((A \Delta B) \Delta (M \Delta N) = \emptyset) \Leftrightarrow (A \Delta B \Delta M \Delta N = \emptyset) \Leftrightarrow (A \Delta M \Delta B \Delta N = \emptyset) \Leftrightarrow ((A \Delta M) \Delta (B \Delta N) = \emptyset) \Leftrightarrow (A \Delta M = B \Delta N)$. The results of (a), (b) and (c) are used here. The remaining assertions are easily proved.

11.11. Depending on the number of the elements, the subsets of Ω can be grouped in the following way (we shall be indicating the subsets themselves and the number of their elements):

$\{\emptyset\}$	$1 = \binom{n}{0}$
$\{\omega_1\}, \{\omega_2\}, \dots \{\omega_n\}$	$n = \binom{n}{1}$
$\{\omega_1, \omega_2\}, \{\omega_1, \omega_3\}, \dots, \{\omega_{n-1}, \omega_n\}$	$\binom{n}{2}$
$\vdots$	$\vdots$
$\{\omega_1, \dots, \omega_k\}, \dots, \{\omega_{n-k+1}, \dots, \omega_n\}$	$\binom{n}{k}$
$\vdots$	$\vdots$
$\{\omega_1, \dots, \omega_n\}$	$1 = \binom{n}{n}$

Hence the algebra of the subsets of Ω with $\nu(\Omega) = n$ has 2^n elements (see also Exercise 1.6).

11.12. (a) If $\omega \in A_k$ for infinitely many k, then $\omega \in \bigcup_{k=n}^{\infty} A_k$ for each n and therefore $\omega \in \bigcap_{n=1}^{\infty} \bigcup_{k=n}^{\infty} A_k$. If ω is not an element of infinite number of A_k, then there exists an n_0 such that $\omega \,\overline{\in}\, A_k$ when $k \geqslant n_0$; but then $\omega \,\overline{\in} \bigcup_{k=n_0}^{\infty} A_k$ and hence $\omega \,\overline{\in} \bigcap_{n=1}^{\infty} \bigcup_{k=n}^{\infty} A_k$. (b) The assertion is proved analogously or is deduced as a corollary of (a) using De Morgan's laws. (c) Follows from (a) and (b).

11.13. Use Exercise 11.12.

11.14. Let $\mathfrak{A}$ be a finite algebra and $\mathfrak{C}$ be the set of its atoms. (Show that $\mathfrak{C}$ is a non-empty set). To each $A \in \mathfrak{A}$ the set $\phi(A) \subset \mathfrak{C}$ of all the atoms which A contains is juxtaposed. It can be shown that ϕ is a one-to-one correspondence between $\mathfrak{A}$ and $\mathfrak{P}(\mathfrak{C})$, and moreover, that $\phi(\overline{A}) = \overline{\phi(A)}$, $\phi(A \cup B) = \phi(A) \cup \phi(B)$ and $\phi(AB) = \phi(A)\phi(B)$.

11.15. Follows from Exercise 11.14 and 11.11. If $\mathfrak{C}$ is the set of the atoms of $\mathfrak{A}$ and $\nu(\mathfrak{C}) = n$, then $\nu(\mathfrak{A}) = 2^n$.

11.16. Since Ω is finite, any algebra is a σ-algebra as well. An arbitrary algebra, containing A and B, must contain also $AB = \{1\}$, $A\overline{B} = \{2, 3\}$, $\overline{A}B = \{4\}$ and $\overline{A}\overline{B} = \{5, 6\}$. The sets {1}, {2, 3}, {4} and {5, 6} form a partition of Ω, and the minimal algebra $\mathfrak{A}$, containing A and B, has the four events pointed out as atoms. It is easy to see that there are three more algebras, containing A and B:

(a) with atoms {1}, {2}, {3}, {4} and {5, 6};

(b) with atoms {1},{2, 3}, {4}, {5} and {6};

(c) with atoms {1}, {2}, {3}, {4}, {5} and {6}.

Obviously, the algebra obtained in case (c) coincides with $\mathfrak{P}(\Omega)$.

11.17. (a) Let $a > 0$. Then $A = [a, b) \in \mathfrak{C}_1$, but $\overline{A} = [0, a) \cup [b, \infty) \,\overline{\in}\, \mathfrak{C}_1$ and hence $\mathfrak{C}_1$ is not an algebra. (b) A finite sum of finite sums of intervals of $\mathfrak{C}_1$ is also a finite sum of intervals. The complement to a finite sum of intervals can also be represented as a finite sum of intervals. Therefore $\mathfrak{C}_2$ is an algebra. $\mathfrak{C}_2$ is not a σ-algebra, because for instance $A_n = [0, 1/n) \in \mathfrak{C}_2$, $n = 1, 2, \ldots$, but $\bigcap_{n=1}^{\infty} A_n = \{0\}$ is not an element of $\mathfrak{C}_2$. (c) Obviously $\mathfrak{C}_1 \subset \mathfrak{C}_2$ and each algebra, containing $\mathfrak{C}_1$, should contain $\mathfrak{C}_2$ as well. (d) It follows from the representation $\{a\} = \bigcap_{n=1}^{\infty} [a, a + 1/n)$, $[a, b] = [a, b) + \{b\}$, $(a, b) = [a, b) \setminus \{a\}$ and $(a, b] = (a, b) + \{b\}$ that each σ-algebra, containing $\mathfrak{C}_2$ should also contain all the intervals in Ω.

11.18. (a) $P(\emptyset) = P(\emptyset + \emptyset) = P(\emptyset) + P(\emptyset) = 2P(\emptyset)$; hence, $P(\emptyset) = 0$. (b) $B = AB + \overline{A}B = A + \overline{A}B$ and therefore $P(B) = P(A) + P(\overline{A}B) \geqslant P(A)$. (c) $A \cup B = A + (B \setminus A)$ and $B = (B \setminus A) + AB$. Then $P(A \cup B) = P(A) + P(B \setminus A)$ and $P(B) = P(B \setminus A) + P(AB)$. After eliminating $P(B \setminus A)$ we obtain the desired probability.

The assertions in (d), (e) and (f) are obtained analogously.

11.19. Indeed in Ω there is no more than one simple event ω such that $\frac{1}{2} < P(\omega) \leqslant 1$, because, if $P(\omega_1) > \frac{1}{2}$ and $P(\omega_2) > \frac{1}{2}$, then $P(\{\omega_1, \omega_2\}) > 1$, while $P(A) \leqslant 1$ for any event A. Similarly there are no more than two simple events ω such that $\frac{1}{3} < P(\omega) \leqslant \frac{1}{2}$, ..., there are no more than n simple events ω with $1/(n + 1) < P(\omega) \leqslant 1/n$, etc. This reasoning shows that the set of the simple events ω, such that $P(\omega) > 0$, is a countable one.

11.20. Use De Morgan's laws: $\overline{\bigcup_i A_i} = \bigcap_i \overline{A}_i$; $\overline{\bigcap_i A_i} = \bigcup_i \overline{A}_i$.

11.21. Use Exercise 11.20.

11.22. The proof is obtained from the properties of the probability P and from the following relations:

(a) if $A_n \uparrow A$, then $A = (A \setminus A_n) + A_n$ and $(A \setminus A_n) \downarrow \emptyset$;

(b) if $A_n \downarrow A$, then $A_n = (A_n \setminus A) + A$ and $(A_n \setminus A) \downarrow \emptyset$.

11.23. Let P be a probability, satisfying the axioms (2) and (3), and let $\{A_n\}$ be a sequence of pairwise mutually exclusive events. Since $\sum_{i=1}^{n} A_i \uparrow \sum_{i=1}^{\infty} A_i$, from Exercise 11.22 and from axiom (2), it follows that

$$P\left(\sum_{i=1}^{\infty} A_i\right) = \lim_{n\to\infty} P\left(\sum_{i=1}^{n} A_i\right) = \lim_{n\to\infty} \sum_{i=1}^{n} P(A_i) = \sum_{i=1}^{\infty} P(A_i).$$

Let P satisfy the axiom (2'). Then obviously P satisfies the axiom (2) as well. If $\{B_n\}$ is a monotonically decreasing sequence and $B_n \downarrow \emptyset$, then $B_n = \sum_{i=n}^{\infty} (B_i \setminus B_{i+1})$ and hence

$$P(B_n) = P\left(\sum_{i=n}^{\infty} (B_i \setminus B_{i+1})\right) = \sum_{i=n}^{\infty} P(B_i \setminus B_{i+1}).$$

Obviously $\sum_{i=n}^{\infty} P(B_i \setminus B_{i+1}) \downarrow 0$ when $n \to \infty$.

11.24. See Exercise 11.17. Define $P\{(a, b)\} = F(b) - F(a + 0)$, $P\{[a, b]\} = F(b + 0) - F(a)$, where $F(x_0 + 0) = \lim_{x\downarrow x_0} F(x)$.

11.25. Since $A_* = \lim_{n\to\infty} \bigcap_{k=n}^{\infty} A_k$, then $P(A_*) = \lim_{n\to\infty} P\left(\bigcap_{k=n}^{\infty} A_k\right)$. We have $\bigcap_{k=n_0}^{\infty} A_k \subset A_n$ for an arbitrary n_0 and $n \geqslant n_0$; hence, $P\left(\bigcap_{k=n_0}^{\infty} A_k\right) \leqslant P(A_n)$. Therefore $P\left(\bigcap_{k=n_0}^{\infty} A_k\right) \leqslant \liminf_n P(A_n)$ for an arbitrary n_0, and when

$n_0 \to \infty$ we obtain $P(A_*) \leqslant \liminf_n P(A_n)$. The inequality for $P(A^*)$ is proved analogously; $\liminf_n P(A_n) \leqslant \limsup_n P(A_n)$ is trivial. The corollary is obtained directly from the above inequalities.

11.26. We have $P(A^*) = \lim_{k\to\infty} P\left(\bigcup_{n=k}^{\infty} A_n\right) \leqslant \lim_{k\to\infty} \sum_{n=k}^{\infty} P(A_n)$. There follows from the convergence of the series $\sum_{n=1}^{\infty} P(A_n)$ that $\sum_{n=k}^{\infty} P(A_n) \to 0$ when $k \to \infty$, whence $P(A^*) = 0$.

11.27. Use Exercise 11.25.

11.28. If $N_i \subset A_i$ and $P(A_i) = 0$, $i = 1, 2, \ldots$, then $\bigcup_{i=1}^{\infty} N_i \subset \bigcup_{i=1}^{\infty} A_i$ and $P\left(\bigcup_{i=1}^{\infty} A_i\right) \leqslant \sum_{i=1}^{\infty} P(A_i) = 0$. On the other hand, $\bigcap_{i=1}^{\infty} N_i \subset N_1 \subset A_1$ and $P(A_1) = 0$.

11.29. The assertions (a), (b) and (c) are proved directly by using the corresponding definitions. (d) If Q is a probability such that $Q(A) = P(A)$, where $A \in \mathfrak{A}$ and $N \subset B$ with $P(B) = 0$, then $P^*(A \cup N) = P(A)$ and $P(A) = Q(A) \leqslant Q(A \cup N) \leqslant Q(A \cup B) = P(A \cup B) \leqslant P(A) + P(B) = P(A)$; hence, $P^*(A \cup N) = Q(A \cup N)$.

11.30. Let M be a P^*-negligible set; i.e., $M \subset A \cup N$, $A \in \mathfrak{A}$, $N \in \mathfrak{N}$ and $P^*(A \cup N) = 0$. But then $P(A) = 0$ and $N \subset B$ with $P(B) = 0$. Therefore, $M \subset A \cup B$ and $P(A \cup B) = 0$, which means that $M \in \mathfrak{N}$. Obviously $M = \emptyset + M$ and hence $M \in \mathfrak{A}^*$.

11.31. Let $\mathfrak{G}$ be the class of the finite sums of non-intersecting elements from $\mathfrak{C}$. Obviously $\emptyset \in \mathfrak{G}$ and $\Omega \in \mathfrak{G}$. The class $\mathfrak{G}$ is closed with respect to the operation intersection, because if $A_i \in \mathfrak{G}$, $i = 1, \ldots, m$, and $B_j \in \mathfrak{G}$, $j = 1, \ldots, n$, then $\left(\sum_{i=1}^{m} A_i\right) \cap \left(\sum_{j=1}^{n} B_j\right) = \sum_{i,j} A_i B_j \in \mathfrak{G}$. The class $\mathfrak{G}$ is closed with respect to the operation complementation, because if $S \in \mathfrak{G}$ and $S = \sum_i A_i$, where $A_i \in \mathfrak{G}$, then $\bar{S} = \bigcap_i \bar{A}_i$ and $\bar{A}_i \in \mathfrak{G}$. Thus it follows that $\mathfrak{G}$ is an algebra. Obviously any algebra, containing $\mathfrak{C}$, will also contain the finite sums of elements of $\mathfrak{C}$; i.e., will contain $\mathfrak{G}$. Hence $\mathfrak{G} = \mathfrak{A}$.

11.32. According to the formula for total probability

$$P(A \Delta B) = P(\bar{A}B) + P(A\bar{B}) = P(\bar{A}BC) + P(A\bar{B}\bar{C}) + P(A\bar{B}C) + P(\overline{A}B\bar{C}).$$

We can write $P(A \Delta C)$ and $P(B \Delta C)$ analogously. We find that $P(A \Delta C) + P(B \Delta C) - P(A \Delta B) = 2P(AB\bar{C}) + 2P(\bar{A}\bar{B}C) \geqslant 0$. Equality is attained when $AB \subset C \subset A \cup B$.

11.33. Proved analogously to Exercise 11.32.

11.34. Let B and B' be non-intersecting sets from $\mathfrak{B}$, $B = \sum_{i=1}^{m} A_i$, $B' = \sum_{j=1}^{n} A'_j$; A_i, A'_j are sets from $\mathfrak{F}$. Obviously $A_i A'_j = \emptyset$ for arbitrary i and j. Then $P(B + B') = P\left(\sum_{i=1}^{m} A_i + \sum_{j=1}^{n} A'_j\right) = \sum_{i=1}^{m} P(A_i) + \sum_{j=1}^{n} P(A'_j) =$

$P(B) + P(B')$, and hence P is additive.

Obviously any one-point set $\{r\} \in \mathcal{B}$ and $P(\{r\}) = 0$. Likewise Ω is countable and $\Omega = \sum_{i=1}^{\infty} \{r_i\}$. We have $1 = P(\Omega) \neq \sum_{i=1}^{\infty} P(\{r_i\}) = 0$, which shows that P is not σ-additive.

11.35. (a) Let $x_n \to x \neq 0$ and $x_n = P(A_n) = \sum_{i:\omega_i \in A_n} p_i$. We are going to show that x can be represented in this way. We can choose such a subsequence $\{x_{n_k}\}$ from the sequence $\{x_n\}$, that in the representation of each x_{n_k} the greatest addend should be the same; let this be p_{i_1}. After that we are going to choose a subsequence of $\{x_{n_k}\}$ such that the second addends in the representation of each of its terms should coincide, and let this be the number p_{i_2} and so on. In this way we obtain the sequence $p_{i_1}, p_{i_2}, \ldots$, for which it is easy to see that $\sum_{k=1}^{\infty} p_{i_k} = x$. Now let $x = \sum_{k=1}^{\infty} p_{s_k}$. Obviously $x = \lim_{n\to\infty} x_n$ and $x \neq x_n$, where $x_n = \sum_{k=1} p_{s_k}$. If $x = \sum_{k=1}^{N} p_{i_k}$ (i.e., x is a finite sum), then $x_n = \sum_{k=1}^{N} p_{i_k} + p_{m+n}$ ($m = \max[i_1, \ldots, i_N]$) form a sequence such that $\lim_{n\to\infty} x_n = x$ and $x_n \neq x$.

(b) Let $p_n \leqslant \sum_{i=n+1}^{\infty} p_i$, $n = 1, 2, \ldots$, and let x be an arbitrary number from the interval $[0, 1]$. We are going to show that there exists an event A with $P(A) = x$, and at the same time we are going to find a representation of x and of $1 - x$ by means of the probabilities p_n. Let $x_1' = \max[x, 1 - x]$, $x_1'' = \min[x, 1 - x]$. Then obviously $x_1' \geqslant \frac{1}{2}$, and $p_1 \leqslant \frac{1}{2}$; let us include p_1 in the sum representing x_1'. Now let $x_2' = \max[x_1' - p_1, x_1'']$, $x_2'' = \min[x_1' - p_1, x_1'']$. We shall have $x_2' \geqslant (1 - p_1)/2$, and $p_2 \leqslant (1 - p_1)/2$ and let us include p_2 in the sum representing x_2'. which will be one of the sums, representing x or $1 - x$. Continuing this process infinitely, we shall obtain from the sequence $\{p_n\}$ two subsequences - one with a sum x and the other with a sum $1 - x$. Let us denote by A the set of the simple events ω_{i_k} corresponding to the first subsequence. Obviously A is an event with $P(A) = x$. The necessity of the condition $p_n \leqslant \sum_{i=n+1}^{\infty} p_i$ is easy to prove. If we assume that there exists n_0 such that $p_{n_0} > \sum_{i=n_0+1}^{\infty} p_i$ and if we choose the number x in such a way that $p_{n_0} > x > \sum_{i=n_0+1}^{\infty} p_i$, then it is easy to see that x is a probability of no event.

(c) The reasoning is analogous to that in (b).

11.36. Let $A \in \mathcal{A}$ and $P(A) > 0$. Then there exists B, $B \subset A$, with $P(B) \leqslant \varepsilon$. If $P(A) \leqslant \varepsilon$, we can take $B = A$. If $P(A) > \varepsilon$, there exists B, $B \subsetneq A$, $0 < P(B) < P(A)$, and then at least one of the numbers $P(B)$ and $P(A \setminus B)$ does not exceed $P(A)/2$. If $P(A)/2 < \varepsilon$, then an event with probability, not exceeding ε, has been found. If $P(A)/2 > \varepsilon$, we continue the process: when r is sufficiently large, then $2^{-r}P(A) \leqslant \varepsilon$, so that after a finite number of steps we can find an event B such that $0 < P(B) \leqslant \varepsilon$. Let us put $\mu_\varepsilon(A) = \sup_B P(B)$, $B \subset A$, $P(B) \leqslant \varepsilon$.

Let $A_1 \in \mathcal{A}$ be such that $0 < P(A_1) \leqslant \varepsilon$. Let A_2, $A_2 \subset \overline{A}_1$, be such that $\varepsilon \geqslant P(A_2) \geqslant \mu_\varepsilon(\overline{A}_1)/2$. In general, after we have chosen $A_1, \ldots, A_n$, we choose $A_{n+1} \in \mathcal{A}$ in such a way that $A_{n+1} \subset \bigcap_{i=1}^{n} \overline{A}_i$ and $\varepsilon \geqslant P(A_{n+1}) \geqslant \mu_\varepsilon(\overline{A_1 + A_2 + \ldots + A_n})/2$. Thus we obtain the sequence $\{A_n\}$ of mutually exclusive events for which $\sum_{n=1}^{\infty} P(A_n) \leqslant 1$. Therefore $P(A_n) \to 0$, whence

$$\lim_{n\to\infty} \mu_\varepsilon(\overline{A_1 + \ldots + A_n}) = 0.$$

Since $\mu_\varepsilon(A)$ is a monotone function, if we let $\sum_{i=1}^{\infty} A_i = B$, then $\mu_\varepsilon(\overline{B}) = 0$; hence it follows that $P(\overline{B}) = 0$ as well. We let $A_1' = A_1 + \overline{B}$. Then $A_1' + A_2 + \ldots + A_n + \ldots = \Omega$, $0 < P(A_1') \leqslant \varepsilon$, $0 < P(A_i) \leqslant \varepsilon$, $i = 2, 3, \ldots$ Now let N be so large that $\sum_{i=N}^{\infty} P(A_i) \leqslant \varepsilon$. We denote $A_N' = \sum_{i=N}^{\infty} A_i$. Then the union of $A_1', A_2, \ldots, A_{n-1}, A_N'$ is equal to Ω and the probability of any one of them does not exceed ε.

It follows that an arbitrary event $C \in \mathcal{A}$ can be represented as a union of a finite number of non-intersecting events such that the probability of any of them does not exceed ε.

It is sufficient to put $C_i = CA_i$, $i = 2, \ldots, N - 1$, $C_1 = CA_1'$, $C_N = CA_N'$. Obviously $C_iC_j = \emptyset$ for $i \neq j$. Also $C_1 + C_2 + \ldots + C_N = C$, and for any i, $i = 1, \ldots, N$, we have $P(C_i) \leqslant \varepsilon$.

11.37. We are going to show that for any x, $0 < x < 1$, there exists an event A with $P(A) = x$. Let Ω be partitioned to N_1 mutually exclusive events $A_{1,1}, A_{1,2}, \ldots, A_{1,N_1}$ in such a way that $P(A_{1,j}) \leqslant x/2$, $j = 1, 2, \ldots, N_1$ (see Exercise 11.36). Take $x_{1,r} = \sum_{j=1}^{r} P(A_{1,j})$ for $j = 1, 2, \ldots, N_1$. Then x lies in one of the intervals $[x_{1,r}, x_{1,r+1})$ and let this be the interval $[x_{1,r_1}, x_{1,r_1+1})$. If $x = x_{1,r_1}$, the proof is completed.

If $x > x_{1,r_1}$, then we divide A_{1,r_1+1} to mutually exclusive events $A_{2,1}$, $A_{2,2}$, ..., A_{2,N_2} so that $0 \leqslant P(A_{2,j}) \leqslant (x - x_{1,r_1})/2$, $j = 1, 2, \ldots, N_2$.

Let $x_{2,r} = P\left(\sum_{j=1}^{r_1} A_{1,j} + \sum_{j=1}^{r} A_{2,j}\right)$, $r = 1, \ldots, N_2$. Then x lies in one of the intervals $[x_{2,r}, x_{2,r+1})$ and let this be the interval $[x_{2,r_2}, x_{2,r_2+1})$. Continuing this process, we obtain the event

$$A = \sum_{j=1}^{r_1} A_{1,j} + \sum_{j=1}^{r_2} A_{2,j} + \ldots + \sum_{j=1}^{r_s} A_{s,j} + \ldots,$$

for which $P(A) = x$.

11.38. If A_1 and A_2 are two atoms of the probability P, then either $P(A_1A_2) = 0$ or $P(A_1 \Delta A_2) = 0$; i.e., with the exception of a set with a probability zero they either coincide or are mutually exclusive. Then there can be found a finite or an infinite sequence of atoms $A_1, A_2, \ldots$ such that $\Omega \setminus \Sigma A_i$ should not contain atoms. Let $\Sigma A_i = B$. We put $\mu_1(A) = P(AB)$, $\mu_2(A) = P(A\bar{B})$. Then $P(A) = \mu_1(A) + \mu_2(A)$, with μ_1 as a measure defined on $\mathfrak{P}(\{A_i\})$ and μ_2 a measure without atoms. The assertion of the exercise is obtained using Exercises 11.35 and 11.37.

11.39. We shall regard Ω, $\mathfrak{A}$ and $\mathfrak{C}$ as fixed, and we are going to consider different pairs P and Q, defined on $\mathfrak{A}$, such that $P = Q$ on $\mathfrak{C}$. We put $\mathfrak{A}_{P,Q} = \{A : A \in \mathfrak{A}$, such that $P(A) = Q(A)\}$. Obviously $\mathfrak{C} \subset \mathfrak{A}_{P,Q}$, $\emptyset \in \mathfrak{A}_{P,Q}$. Besides, if $A \in \mathfrak{A}_{P,Q}$, then $P(\bar{A}) = 1 - P(A) = 1 - Q(A) = Q(\bar{A})$ and $\bar{A} \in \mathfrak{A}_{P,Q}$. If $A_1, A_2, \ldots$ are pairwise mutually exclusive events from $\mathfrak{A}_{P,Q}$, then

$$P\left(\sum_i A_i\right) = \sum_i P(A_i) = \sum_i Q(A_i) = Q\left(\sum_i A_i\right).$$

Therefore $\mathfrak{A}_{P,Q}$ is closed with respect to the operations of complementation and countable union of mutually exclusive events.

Let us denote by $\mathfrak{B}$ the intersection of all $\mathfrak{A}_{P,Q}$; i.e., $\mathfrak{B}$ is the set such that $B \in \mathfrak{A}$ and $P(B) = Q(B)$ for arbitrary P and Q. Obviously $\emptyset \in \mathfrak{B}$ and $\mathfrak{C} \subset \mathfrak{B}$. Also $\mathfrak{B}$ is closed with respect to complementation and countable union of mutually exclusive events, because this holds true for any $\mathfrak{A}_{P,Q}$. We are going to prove that $\mathfrak{B}$ is closed with respect to finite intersections as well. First we are going to show that $BC \in \mathfrak{B}$, when $B \in \mathfrak{B}$, $C \in \mathfrak{C}$. Let P and Q be two probabilities, coinciding on $\mathfrak{C}$, and let $C \in \mathfrak{C}$. Then $P(BC) = Q(BC)$ for any $B \in \mathfrak{B}$. Indeed this holds true, if $P(C) = 0$. If $P(C) = Q(C) > 0$, let us consider the conditional probabilities $P(\cdot|C)$ and $Q(\cdot|C)$. We have $P(A|C) = P(AC)/P(C) = Q(AC)/Q(C) = Q(A|C)$ for any $A \in \mathfrak{C}$, because $\mathfrak{C}$ is closed with respect to intersection and therefore $P(\cdot|C)$ and $Q(\cdot|C)$ will coincide on $\mathfrak{B}$. Thus $P(BC) = Q(BC)$ for any $B \in \mathfrak{B}$.

Now we are going to show that $BB' \in \mathfrak{B}$ when $B \in \mathfrak{B}$ and $B' \in \mathfrak{B}$. Let P and Q be arbitrary and let $B \in \mathfrak{B}$. We have to show that $P(BB') = Q(BB')$ for any $B' \in \mathfrak{B}$. This holds true if $P(B) = 0$. If $P(B) = Q(B) > 0$, we consider the conditional probabilities $P(\cdot|B)$ and $Q(\cdot|B)$. Then using the above results,

$$P(A|B) = P(AB)/P(B) = Q(AB)/Q(B) = Q(A|B)$$

for any $A \in \mathfrak{C}$. Since $P(\cdot|B)$ and $Q(\cdot|B)$ coincide in $\mathfrak{C}$, they will coincide in $\mathfrak{B}$ as well; therefore we have $P(BB') = Q(BB')$ for any $B' \in \mathfrak{B}$. Thus we have shown that the class $\mathfrak{B}$, on which all the probabilities coinciding on $\mathfrak{C}$ coincide, is a σ-algebra. Since $\sigma(\mathfrak{C}) \subset \mathfrak{B}$, then each two probabilities P and Q, coinciding on $\mathfrak{C}$, will also coincide on $\sigma(\mathfrak{C})$.

11.40. Let us define the function

$$Q(B) = \sum_{\omega \in B} \frac{P(\omega)}{\nu(\omega)}$$

for an arbitrary $B \subset \Omega$, where $\nu(\omega)$ is the size of the sample $\omega \in \Omega$. Let $I^* \subset M$ be such that

$$Q(A(I^*, \overline{I}^*)) = \max_{I \subset M} Q(A(I, \overline{I})).$$

Then the following inequalities are obviously fulfilled:

$$0 \leqslant Q(A(I^*, \overline{I}^*)) - Q(A(I^* \cup \{k\}, \overline{I}^* \setminus \{k\})) \quad \text{for } k \in \overline{I}^*,$$

$$0 \leqslant Q(A(I^*, \overline{I}^*)) - Q(A(I^* \setminus \{k\}, \overline{I}^* \cup \{k\})) \quad \text{for } k \in I^*.$$

Let us denote the set of those samples from Ω which are composed only of elements of I by $A(I)$ for an arbitrary $I \subset M$; similarly let us denote the set of the samples which contain the element k for $k \in M$ by A_k. We put $B_k = (A(I^*) \cup A(\overline{I}^*)) \cap A_k$ and $C_k = A(I^*, \overline{I}_*) \cap A_k$. It is easily seen that for any $k \in M$ the following inequality holds:

$$Q(A(I^*, \overline{I}^*)) - Q(A(I^* \cup \{k\}, \overline{I}^* \setminus \{k\})) \leqslant$$
$$\leqslant Q(C_k) - Q(B_k) \quad \text{for } k \in \overline{I}^*,$$

or

$$Q(A(I^*, \overline{I}^*)) - Q(A(I^* \setminus \{k\}, \overline{I}^* \cup \{k\})) \leqslant$$
$$\leqslant Q(C_k) - Q(B_k) \quad \text{for } k \in I^*.$$

It follows from these and from the above inequalities that for any $k \in M$

we have $0 \leqslant Q(C_k) - Q(B_k)$; hence, $0 \leqslant \sum_{k=1}^{n} (Q(C_k) - Q(B_k))$. The relations

$$\sum_{k=1}^{n} Q(C_k) = \sum_{k=1}^{n} Q(A(I^*, \overline{I}^*) \cap A_k) = P(A(I^*, \overline{I}^*),$$

$$\sum_{k=1}^{n} Q(B_k) = P(A(I^*)) + P(A(\overline{I}^*))$$

are valid for the defined function Q. After substitution we obtain $0 \leqslant P(A(I^*, \overline{I}^*)) - P(A(I^*)) - P(A(\overline{I}^*))$, which together with the equality $P(A(I^*, \overline{I}^*)) + P(A(I^*)) + P(A(\overline{I}^*)) = 1$ yields the desired inequality $P(A(I^*, \overline{I}^*)) \geqslant \frac{1}{2}$.

Note. Additional details concerning the statement discussed in Exercise 11.40 can be found in the following paper: Vandev, D.L. (1975): A generalization of an urn scheme in probability theory. *Phys.-Math. J.* (Bulg. Acad. Sci.) 18, p. 27-29.

12. Random Variables and Integration

12.1. (a) Let $\{A_i\}$ be subsets of Ω_2. If $\omega \in \xi^{-1}\left(\bigcap_i A_i\right)$, then $\xi(\omega) \in \bigcap_i A_i$ and therefore $\xi(\omega) \in A_i$ for any i. Thus $\omega \in \xi^{-1}(A_i)$ for any i; i.e., $\omega \in \bigcap_i \xi^{-1}(A_i)$. If $\omega \in \xi^{-1}(A_i)$ for any i, then $\xi(\omega) \in A_i$ and therefore $\xi(\omega) \in \bigcap_i A_i$. Thus $\omega \in \xi^{-1}\left(\bigcap_i A_i\right)$. It can be shown analogously that ξ^{-1} preserves the operations of union and complementation.

(b) It follows from (a) that $\xi^{-1}(\mathfrak{A}_2)$ is isomorphic to $\mathfrak{A}_2$ and hence $\xi^{-1}(\mathfrak{A}_2)$ is a σ-algebra.

(c) The necessity is obvious. The sufficiency follows from the fact that the class of the sets, whose pre-images are measurable because of (a), is a σ-algebra, which contains $\mathfrak{C}$ and therefore also contains $\sigma(\mathfrak{C})$.

12.2. Show that the intervals of the kind $(-\infty, x)$ generate the σ-algebra $\mathfrak{B}_1$ and apply Exercise 12.1.

12.3. Since ξ is measurable, $\xi^{-1}(\mathfrak{A}_2) \subset \mathfrak{A}_1$; since η is measurable, it follows that $\eta^{-1}(\mathfrak{A}_3) \subset \mathfrak{A}_2$. Then $\xi^{-1}[\eta^{-1}(\mathfrak{A}_3)] \subset \mathfrak{A}_1$; i.e., $(\eta(\xi))^{-1}(\mathfrak{A}_3) \subset \mathfrak{A}_1$.

12.4. Obviously $P^*(A^*) = P(\xi^{-1}(A)) \leqslant 1$, $P^*(\Omega^*) = 1$. If A^* and B^* are mutually exclusive, $\xi^{-1}(A^*)$ and $\xi^{-1}(B^*)$ will also be mutually exclusive. Then $P^*(A^* + B^*) = P(\xi^{-1}(A^*) + \xi^{-1}(B^*)) = P(\xi^{-1}(A^*)) + P(\xi^{-1}(B^*)) = P^*(A^*) + P^*(B^*)$. Also if $A_n^* \downarrow \emptyset$, then $\xi^{-1}(A_n) \downarrow \emptyset$; hence, $P^*(A_n^*) = P(\xi^{-1}(A_n^*)) \downarrow 0$.

12.5. Let $c \in \mathbb{R}_1$. Then $\{\omega : \xi(\omega) < c\} = \Omega \setminus \bigcap_{m=1}^{\infty} \bigcap_{k=1}^{\infty} \bigcap_{n=k}^{\infty} \{\omega : \xi_n(\omega) \geqslant c - 1/m\}$. It then follows that $\{\omega : \xi(\omega) < c\} \in \mathcal{A}$ and according to Exercise 12.2 $\xi(\omega)$ is a r.v.

12.6. Let $\omega_0 \in \Omega$ be a fixed simple event such that $\lim_{n\to\infty} \xi_n(\omega_0)$ exists. Since $\{\xi_n(\omega_0)\}$ is a numerical sequence, according to Cauchy's test for $\forall k \exists n : \forall m_1, m_2 > n$ the inequality $|\xi_{m_1}(\omega_0) - \xi_{m_2}(\omega_0)| < 1/k$ is fulfilled.

To write A it is sufficient to note that the operation $\cap$ corresponds to the quantifier $\forall$, and the operation $\cup$ to the quantifier $\exists$. Therefore $A = \bigcap_{k=1}^{\infty} \bigcup_{n=1}^{\infty} \bigcap_{m_1,m_2>n}^{\infty} \{\omega : |\xi_{m_1}(\omega) - \xi_{m_2}(\omega)| < 1/k\}$. Obviously $B = \{\omega : \lim_{n\to\infty} \xi_n(\omega)$ does not exist$\} = \overline{A}$; hence,

$$B = \bigcup_{k=1}^{\infty} \bigcap_{n=1}^{\infty} \bigcup_{m_1,m_2>n}^{\infty} \{\omega : |\xi_{m_1}(\omega) - \xi_{m_2}(\omega)| \geqslant 1/k\}.$$

Since a countable number of operations $\cup$ and $\cap$ are used in the representations of A and B, then $A \in \mathcal{F}$ and $B \in \mathcal{F}$.

12.7. The assertions follow from the definition of an indicator.

12.8. (a) Obviously $\{c < t\}$ coincides either with $\emptyset$ or with Ω depending on whether $t \leqslant c$ or $t > c$. Therefore $\{c < t\} \in \mathcal{A}$ and c is measurable.

(b) If $c > 0$, then $\{c\xi < t\} = \{\xi < t/c\} \in \mathcal{A}$. If $c < 0$, then $\{c\xi < t\} = \{\xi > t/c\} \in \mathcal{A}$. If $c = 0$, then $c\xi = 0$ and the measurability of $c\xi$ follows from (a).

From the assertions (c) - (f) we shall prove (d) as an illustration. The remaining ones are proved analogously. First let ξ and η be simple r.v.'s and $\xi = \sum_i x_i I_{A_i}$, $\eta = \sum_j y_j I_{B_j}$. Then $\xi\eta = (\sum_i x_i I_{A_i})(\sum_j y_j I_{B_j}) = \sum_{i,j} x_i y_j I_{A_i B_j}$. Hence $\xi\eta$ is a simple r.v. as well. Now let ξ and η be arbitrary non-negative r.v.'s and let $\{\xi_k\}$ and $\{\eta_k\}$ be monotonically increasing sequences of such simple r.v.'s that $\xi_k \uparrow \xi$ and $\eta_k \uparrow \eta$. Then it is clear that $\xi_k\eta_k \uparrow \xi\eta$; therefore, $\xi\eta$ is a r.v. And last, if ξ and η are arbitrary r.v.'s, it is sufficient to make use of the representation

$$\xi\eta = (\xi^+\eta^+ + \xi^-\eta^-) - (\xi^+\eta^- + \xi^-\eta^+).$$

12.9. If we put $F_k = F_k(x_1, \ldots, x_n)$, then $\{F_k < t\} = \{F_k < t, F_{k+1} \geqslant t\} + \{F_{k+1} < t, F_{k+2} \geqslant t\} + \ldots + \{F_{n-1} < t, F_n \geqslant t\} + \{F_n < t\}$ for any $t \in \mathbb{R}_1$.

It is easy to see that each set from this representation is $\mathcal{B}_n$-measurable, since e.g. $\{F_j < t, F_{j+1} \geqslant t\} = \Sigma\{x_{i_1} < t, \ldots, x_{i_j} < t,$

$x_{i_{j+1}} \geqslant t, \ldots, x_{i_n} \geqslant t\}$, where the summing is over all possible permutations $\{i_1, \ldots, i_n\}$ of the indices 1, 2, ..., n.

12.10. Let $x \in \mathbb{R}_1$. Then $A = \{\xi < x\} = [a, b] \in \mathcal{A}$ when $x > b$ and $P(A) = 1$. But $P(A)$ is proportional to the length of the interval $[a, b]$; i.e., $P(A) = c \cdot (b - a) = 1$; hence, $c = 1/(b - a)$ and $F(x) = 1$ when $x > b$. When $x \in (a, b]$, we have $F(x) = P\{\xi < x\} = c(x - a) = (x - a)/(b - a)$, and when $x \leqslant a$, $F(x) = P\{\xi < x = P(\emptyset) = 0$.

12.11. The σ-additivity of the function $Q(A)$, $A \in \mathcal{A}$ follows from the properties of the Lebesgue integral. $Q(A)$ would be a probability if $\xi \geqslant 0$ (P-a.s.) and $E(\xi) = 1$.

12.12. Since $P(A \,\Delta\, B) = 0$, then $P(A\bar{B}) = P(\bar{A}B) = 0$; hence,

$$\int_{AB} \xi dP = \int_{\overline{AB}} \xi dP = 0$$

because of the continuity of the integral. Therefore $\int_A \xi dP = \int_A \xi dP + \int_{\overline{AB}} \xi dP = \int_{A \cup B} \xi dP = \int_B \xi dP + \int_{\overline{AB}} \xi dP = \int_B \xi dP.$

12.13. Let $A = \{\omega : \xi(\omega) \neq \eta(\omega)\}$. Then $\vec{P}(A) = 0$ and $\int_A \xi dP = \int_A \eta dP$. But $\vec{P}(\bar{A}) = 1$; i.e., $\vec{P}\{\omega : \xi(\omega) = \eta(\omega)\} = 1$; hence,

$$\int_\Omega \xi dP = \int_A \xi dP + \int_{\bar{A}} \xi dP = \int_{\bar{A}} \xi dP = \int_{\bar{A}} \eta dP =$$

$$= \int_{\bar{A}} \eta dP + \int_A \eta dP = \int_\Omega \eta dP.$$

12.14. Let $\xi = \xi(\omega_2) = I_B(\omega_2)$, $B \in \mathcal{A}_2$. Then

$$\int_{\Omega_2} \xi(\omega_2) dP_\alpha(\omega_2) = \int_{\Omega_2} I_B(\omega_2) dP_\alpha(\omega_2) = P_\alpha(B) =$$

$$= P\{\omega_1 : \alpha(\omega_1) \in B\} = \int_{\Omega_1} I_{\{\alpha \in B\}} dP(\omega_1) =$$

$$= \int_{\Omega_1} I_B(\alpha(\omega_1)) dP(\omega_1) = \int_{\Omega_1} \xi(\alpha(\omega_1)) dP(\omega_1).$$

The correctness of the equality under consideration for simple r.v.'s follows from the properties linearity and additivity of the integral. If ξ is a non-negative r.v., and if $\{\xi_n\}$ is such a sequence of simple r.v.'s such that $\xi_n \uparrow \xi$, then we have

$$\int_{\Omega_2} \xi_n(\omega_2)\,dP_\alpha(\omega_2) = \int_{\Omega_1} \xi_n(\alpha(\omega_1))\,dP(\omega_1).$$

After passing to the limit, we obtain the desired equality. If ξ is an arbitrary r.v., then $\xi = \xi^+ - \xi^-$ and

$$\int_{\Omega_2} \xi(\omega_2)\,dP_\alpha(\omega_2) = \int_{\Omega_2} \xi^+ dP_\alpha - \int_{\Omega_2} \xi^- dP_\alpha =$$

$$= \int_{\Omega_1} \xi^+(\alpha)\,dP - \int_{\Omega_1} \xi^-(\alpha)\,dP =$$

$$= \int_{\Omega_1} \xi(\alpha(\omega_1))\,dP(\omega_1).$$

12.15. First let $g(x) = I_B(x)$, where $B \in \mathcal{B}_1$. Then $\eta = g(\xi) = I_{\{\xi\in B\}}(\omega)$ and $E(\eta) = P\{\xi \in B\} = P_\xi(B)$. For that reason $\int_{\mathbb{R}_1} g(x)\,dP_\xi(x) = \int_{\mathbb{R}_1} I_B(x)\,dP_\xi(x) = P_\xi(B) = P\{\xi \in B\} = E(\eta)$. Further reasoning is analogous to that in Exercise 12.14.

12.16. From the representation $\{\xi > 0\} = \{\xi > 1\} + \sum_{n=1}^{\infty} \{\frac{1}{n+1} < \xi \leqslant \frac{1}{n}\}$ and the σ-additivity of P, it follows that

$$P\{\xi > 0\} = P\{\xi > 1\} + \sum_{n=1}^{\infty} P\{\frac{1}{n+1} < \xi \leqslant \frac{1}{n}\}.$$

Let us note that for an arbitrary r.v. ξ and a number $a > 0$, we have

$$P\{\xi > a\} = \int_\Omega I_{\{\xi>a\}}\,dP \leqslant \frac{1}{a}\int_{\{\xi>a\}} \xi\,dP \leqslant \frac{1}{a}\int_\Omega \xi\,dP = \frac{1}{a}E(\xi).$$

(As a matter of fact, this is the well-known Markov's inequality, see Section 21.)

We apply this inequality when $a = 1, 2, \ldots$ Taking into consideration that $E(\xi) = 0$, we find $0 \leqslant P\{\xi > 1\} \leqslant E(\xi) = 0$, or that $0 \leqslant P\{\frac{1}{n+1} < \xi \leqslant \frac{1}{n}\} \leqslant P\{\xi > \frac{1}{n+1}\} \leqslant (n+1)E(\xi) = 0$, $n = 1, 2, \ldots$ Therefore $P\{\xi > 0\} = 0$, which implies that $P\{\xi = 0\} = 1$.

12.17. For arbitrary $\xi, \eta \in \mathbf{L}_r$ the equality $d_r(\xi, \eta) = d_r(\eta, \xi)$ is obvious. It is also clear that $d_r(\xi, \eta) = 0$ if and only if ξ and η are equivalent; i.e., when $\xi = \eta$ a.s. The triangle inequality remains to be verified: $d_r(\xi_1, \xi_2) \leqslant d_r(\xi_1, \xi_3) + d_r(\xi_2, \xi_3)$, $\xi_1, \xi_2, \xi_3 \in \mathbf{L}_r$.

Use the following elementary inequality:

$$\frac{|x-y|^r}{1+|x-y|^r} \leq \frac{|x-z|^r}{1+|x-z|^r} + \frac{|y-z|^r}{1+|y-z|^r},$$

which holds true for arbitrary $x, y, z \in \mathbb{R}_1$, $r \in (0, 1]$.

Note. The space $\mathbb{L}_r$ is complete with respect to the metric d_r. Convergence in probability follows from convergence in this space (see Section 22).

12.18. Let $\varepsilon > 0$ be arbitrary and let $\alpha = M/\varepsilon$, where $M = \sup_n E(g(|\xi_n|))$. Let us choose the number c_0 so large that $g(x)/x > \alpha$ for any $x > c_0$. Then we shall have $|\xi_n| \leq g(|\xi_n|)/\alpha$ on the set $\{\omega : |\xi_n| \geq c_0\}$; hence,

$$\int_{\{\omega:|\xi_n|>c_0\}} \xi_n \, dP \leq (1/\alpha) \int_{\{\omega:|\xi_n|>c_0\}} g(|\xi_n|) \, dP \leq M/\alpha = \varepsilon$$

for any n; i.e., $\{\xi_n\}$ is uniformly integrable.

12.19. *Necessity*. Let us put $\xi_n^c = 0$, if $|\xi_n| \leq c$ and $\xi_n^c = \xi_n$, if $|\xi_n| > c$. Since ξ_n is integrable, for each $A \in \mathfrak{A}$ we have $\int_A |\xi_n| \, dP \leq cP(A) + E(|\xi_n^c|)$, where $E(|\xi_n^c|) = \int_{D_n} |\xi_n| \, dP$, $D_n = \{\omega : |\xi_n| > c\}$. If we take c large enough, from the uniform integrability of $\{\xi_n\}$ it follows that $\sup_n E(|\xi_n^c|) < \varepsilon/2$. Then for c large enough and for $A = \Omega$, we obtain the condition (1): $\sup_n E(|\xi_n|) \leq c + \varepsilon/2$. Then, if we put $\delta = \varepsilon/(2c)$, we obtain also the condition (2): $\sup_n \int_A |\xi_n| \, dP \leq \varepsilon/2 + \varepsilon/2 = \varepsilon$.

Sufficiency. For some $\varepsilon > 0$ and $\delta > 0$, let conditions (1) and (2) be fulfilled. We put $c = (1/\delta) \sup_n E(|\xi_n|)$. Obviously $0 < c < \infty$. It follows from the inequality obtained in the solution of Exercise 12.16 that

$$P\{|\xi_n| > c\} \leq \frac{1}{c} E(|\xi_n|) \leq \frac{1}{c} \sup_n E(|\xi_n|) = \delta.$$

From this inequality and from condition (2) we find that

$$\sup_n \int_{D_n} |\xi_n| \, dP \leq \varepsilon, \qquad \text{where } D_n = \{\omega : |\xi_n| > c\}.$$

13. Conditional Probability, Independence and Martingales

13.1. The assertion follows from the equalities

$$Q_B(A|C) = \frac{Q_B(AC)}{Q_B(C)} = \frac{P(AC|B)}{P(C|B)} = \frac{P(ABC)P(B)}{P(B)P(BC)} = P(A|BC).$$

13.2. Let us denote $Q_i(\cdot) = P(\cdot|A_i)$. We have the following from Exercise 13.1 and the formula for total probability (5.4):

$$\sum_{j=1}^{m} P(C|A_iB_j)P(B_j|A_i) = \sum_{j=1}^{m} Q_i(C|B_j)Q_i(B_j) = Q_i(C) = P(C|A_i).$$

The formula sought for P(C) is obtained from here after a second applying of the formula for total probability.

13.3. (a) Let $p = P(A_i)$. Then $P(A_{i_1} \dots A_{i_r}) = P(A_{i_1}) \dots P(A_{i_r}) = p^r$ and hence the events are mutually exchangeable.

(b) Let for instance $A = \{\text{tail}\}$, $B = \bar{A} = \{\text{head}\}$ when tossing a balanced coin. Obviously the events A and B are symmetrically dependent, but are not independent.

13.4. Use Exercise 6.11 and 6.12; the latter should be generalised for more than two trials.

13.5. (a) It is sufficient to calculate the corresponding probabilities.

(b) For instance, if each of two urns has a different composition and $p_1 = p_2 = \frac{1}{2}$, then the events A_1 and A_2 are not independent.

13.6. According to the problem's condition for any $t \in \mathbb{R}_1$ the event $A_t = \{f(\xi_1, \xi_2, \dots) < t\} \in \mathcal{A}_n$, n = 1, 2, ... Therefore $A_t \in \mathcal{A} = \bigcap_{n=1}^{\infty} \mathcal{A}_n$ and from 0-1 law it follows that $P(A_t) = 0$ for 1. This means that the d.f. of the r.v. $\xi = f(\xi_1, \xi_2, \dots)$ takes only the values 0 and 1; hence we conclude that ξ = constant, (P-a.s.).

13.7. Let $A \in \mathcal{F}$. According to the condition, the r.v.'s ξ and I_A are independent. Then

$$\int_A (E(\xi))dP = (E(\xi))P(A) = E(\xi)E(I_A) = E(\xi I_A) =$$

$$= \int_\Omega \xi I_A dP = \int_A \xi dP$$

and the desired equality follows from the definition of conditional mean.

13.8. Use Exercise 13.7.

13.9. Let $\xi_1, \dots, \xi_k$ be independent and non-degenerate r.v.'s; i.e., for any i = 1, ..., n there exists an event A_i, $A_i \in \mathcal{F}_{\xi_i}$, with

$0 < P(A_i) < 1$. Obviously A_i are independent. Let us consider the event B of the type $B = \bigcap_{i=1}^{k} C_i$, where $C_i = A_i$ or $\bar{A}_i$. The number of the different events of this kind is 2^k and they are mutually exclusive. From the mutual independence of $C_1, \ldots, C_k$ (why?) it follows that $\vec{P}(B) = \prod_{i=1}^{k} P(C_i) > 0$; i.e., any event B contains at least one element ω, $\omega \in \Omega$. Therefore $2^k \leqslant n$; hence, $k \leqslant [\ln n/\ln 2]$.

13.10. (a) If $A_i \in \mathfrak{C}_i$ and N, S are events from $\mathbf{A}$ with $P(N) = 0$ and $P(S) = 1$, then obviously $P(A_1 \ldots A_{i-1}NA_{i+1} \ldots A_n) = 0 = P(A_i) \ldots P(A_{i-1})P(N)P(A_{i+1}) \ldots P(A_n)$ and $P(A_1 \ldots A_{i-1}SA_{i+1} \ldots A_n) = P(A_1 \ldots A_{i-1}A_{i+1} \ldots A_n) - P(A_1 \ldots A_{i-1}\bar{S}A_{i+1} \ldots A_n) = \vec{P}(A_1) \ldots P(A_{i-1})P(A_{i+1}) \ldots P(A_n) - 0 = P(A_1) \ldots P(A_{i-1})P(S)P(A_{i+1}) \ldots P(A_n)$.

(b) If $A_i, B_i, C_i \in \mathfrak{C}_i$, $B_i \subset A_i$, then $P(C_1 \ldots C_{i-1}(A_i \setminus B_i)C_{i+1} \ldots C_n) = P(C_1 \ldots A_i \ldots C_n) - P(C_1 \ldots B_1 \ldots C_n) = P(C_1) \ldots P(A_i) \ldots P(C_n) - P(C_1) \ldots P(B_i) \ldots P(C_n) = P(C_1) \ldots P(C_{i-1})P(A_i \setminus B_i)P(C_{i+1}) \ldots P(C_n)$.

Using the properties of the probability P, the assertions (c) and (d) are proved easily.

Note. The assertions (a) - (d) are also true when the classes $\mathfrak{C}_i$ are infinite but countable.

13.11. In cases (a), (b) and (c) the events are independent, while in (d) they are not independent. See Exercises 13.10 and 4.16.

13.12. It can be verified directly.

13.13. It can be verified directly.

13.14. Let us assume that C is an atom (see Section 11) and $P(C) = p > 0$. Then for any $n = 1, 2, \ldots$ either $C \subset A_n$ or $C \subset \bar{A}_n$. Let B_n be that event from A_n and $\bar{A}_n$, which contains C. Then $P(B_n) \leqslant \max_n[p_n, 1 - p_n] = 1 - \alpha_n$ and for any n we have $p = P(C) \leqslant P(B_1 \ldots B_n) \leqslant \prod_{i=1}^{n} P(B_i) \leqslant \prod_{i=1}^{n} (1 - \alpha_i) \leqslant \exp\left(-\sum_{i=1}^{n} \alpha_i\right) \to 0$ as $n \to \infty$; hence, we come to a contradiction.

13.15. Use Exercise 13.10 and the following facts: (a) If $\emptyset$, Ω and the complements of the events from $\mathfrak{C}_i$ are supplemented to $\mathfrak{C}_i$, then the semi-algebra $\mathfrak{C}_i'$ (see Exercise 11.31) is obtained; (b) If the finite sums of non-intersecting events from $\mathfrak{C}_i'$ are supplemented to $\mathfrak{C}_i'$, then the algebra $\mathfrak{C}_i''$ is obtained; (c) If the limits of sequences of events from $\mathfrak{C}_i''$ are supplemented to $\mathfrak{C}_i''$, then the σ-algebra $\sigma(\mathfrak{C}_i)$ is obtained.

Note. The assertion holds true for an infinite number of independent calsses $\mathfrak{C}_i$ as well.

13.16. Use Exercise 13.15.

13.17. (a) Use Exercise 11.17. (b) Make use of the fact that if t is an irrational number and $r_n \uparrow t$, where r_n are rational numbers, then

$\{\xi < t\} = \bigcup_{n=1}^{\infty} \{\xi < r_n\}.$

13.18. The assertion follows from the definition of independence of r.v.'s, Exercise 11.17 and 13.15 (see also Exercise 13.17).

13.19. Let $A \in \mathfrak{A}$ and $B \in \mathcal{F}$. Then there exists an index m such that $A \in \mathfrak{A}_n$, $B \in \mathcal{F}_n$ for $n > m$ and according to the given condition $|P(AB) - P(A)P(B)| \to 0$ as $n \to \infty$; hence, $P(AB) = P(A)P(B)$. Thus A and B are independent and since they are arbitrary, the σ-algebras $\mathfrak{A}$ and $\mathcal{F}$ are independent.

13.20. We are to prove that $\int_B \xi d P = \int_B E(\xi|\mathfrak{D})dP$, (P-a.s.) for any $B \in \mathcal{F}$. Denote by $\mathfrak{B}$ the class of those sets for which the above equality holds. Let $D \in \mathfrak{D}$, $G \in \mathfrak{G}$. Then

$$\int_{DG} \xi dP = \int_\Omega \xi I_D I_G dP = \int_\Omega \xi I_D dP \cdot \int_\Omega I_G dP =$$

$$= \int_D \xi dP \int_\Omega I_G dP = \int_D E(\xi|\mathfrak{D})dP \int_\Omega I_G dP =$$

$$= \int_D E(\xi|\mathfrak{D}) I_G dP = \int_{DG} E(\xi|\mathfrak{D})dP, \quad \text{(P-a.s.)}.$$

We have used the independence of $\sigma(\mathfrak{D}, \mathfrak{C})$ and $\mathfrak{G}$, the definition of the conditional mean and the independence of $\mathfrak{D}$ and $\mathfrak{G}$. Therefore, $\mathfrak{D} \subset \mathfrak{B}$ and $\mathfrak{G} \subset \mathfrak{B}$. Let $B \in \mathfrak{B}$. Then

$$\int_{\bar{B}} \xi dP = \int_\Omega \xi dP - \int_B \xi dP = \int_\Omega E(\xi|\mathfrak{D})dP - \int_B E(\xi|\mathfrak{D})dP =$$

$$= \int_{\bar{B}} E(\xi|\mathfrak{D})dP, \quad \text{(P-a.s.)}.$$

It then follows that $\bar{B} \in \mathfrak{B}$. From the σ-additivity of the integral, it follows that $\mathfrak{B}$ is closed with respect to countable unions. Hence $\mathfrak{B}$ is a σ-algebra and $\mathcal{F} = \sigma(\mathfrak{D}, \mathfrak{C}) \subset \mathfrak{B}$.

13.21. Use Exercise 12.6 and the Kolmogorov 0-1 law.

13.22. $$E(\xi|\mathcal{F}) = \begin{cases} \dfrac{1}{P(A)} \displaystyle\int_A \xi dP, & \text{if } \omega \in A, \\ \dfrac{1}{P(A)} \displaystyle\int_{\bar{A}} \xi dP, & \text{if } \omega \in \bar{A}; \end{cases}$$

$$E(I_B|\mathcal{F}) = \begin{cases} P(B|A), & \text{if } \omega \in A, \\ P(B|\bar{A}), & \text{if } \omega \in \bar{A}. \end{cases}$$

13.23. Since f is a measurable function, we have $\mathcal{F}_\eta \subset \mathcal{F}_\xi$. But $\mathcal{F}_\xi$ and $\mathcal{F}_\eta$ are independent according to condition and therefore $\mathcal{F}_\eta$ is independent with itself. Thus it follows that $\vec{P}(A) = 0$ or 1 for any $A \in \mathcal{F}_\eta$; i.e.,

$\eta = f(\xi) =$ constant is fulfilled with probability 1.

13.24. Let ξ take the values a_1 and a_2 respectively when $\omega \in A$ and $\omega \in \bar{A}$, and let η take the values b_1 and b_2 when $\omega \in B$ and $\omega \in \bar{B}$. Then $\xi = a_1 I_A + a_2 I_{\bar{A}}$, $\eta = b_1 I_B + b_2 I_{\bar{B}}$ and we easily find that $I_A = (\xi - a_2)/(a_1 - a_2)$, $I_B = (\eta - b_2)/(b_1 - b_2)$. We have $P(AB) = E(I_A I_B) =$
$(a_1 - a_2)^{-1}(b_1 - b_2)^{-1}E(\xi\eta - a_2\eta - b_2\xi + a_2 b_2) =$
$(a_1 - a_2)^{-1}(b_1 - b_2)^{-1}(E(\xi)E(\eta) - a_2 E(\eta) - b_2 E(\xi) + a_2 b_2) =$
$E\{(\xi - a_2)/(a_1 - a_2)\} \times E\{(\eta - b_2)/(b_1 - b_2)\} = E(I_A)E(I_B) = P(A)P(B)$.
Hence it follows that the events A and B are independent. Therefore, the algebras $\mathcal{F}_\xi = \{\emptyset, A, \bar{A}, \Omega\}$ and $\mathcal{F}_\eta = \{\emptyset, B, \bar{B}, \Omega\}$ are also independent (see Exercise 13.16).

13.25. From the properties of the conditional means it follows that η_n is $\mathcal{F}_n$-measurable. Moreover, $E(|\eta_n|) = E\{|E(\xi|\mathcal{F}_n)|\} \leqslant E\{E(|\xi||\mathcal{F}_n)\} = E\xi < \infty$ and $E(\eta_{n+1}|\mathcal{F}_n) = E\{E(\xi|\mathcal{F}_{n+1})\mathcal{F}_n\} = E(\xi|\mathcal{F}_n) = \eta_n$, (P-a.s.).

13.26. (a) Justify and make use of the inequalities $E(\max[\xi_{n+1}, \eta_{n+1}]|\mathcal{F}_n) \geqslant \max[E(\xi_{n+1}|\mathcal{F}_n), E(\eta_{n+1}|\mathcal{F}_n) \geqslant \max[\xi_n, \eta_n]$, (P-a.s.).

(b) Can be verified directly.

13.27. Use *Jensen's inequality*: If ξ is a r.v., and if $f(x)$, $x \in \mathbb{R}_1$, is a continuous and convex function such that $E(f(|\xi|))$ exists, then $f(E(\xi|\mathcal{F})) \leqslant E(f(\xi)|\mathcal{F})$, (P-a.s.). See also Section 21.

13.28. Obviously η_n is $\mathcal{F}_n$-measurable and since η_n is a sum of a finite number of integrable r.v.'s, then $E|\eta_n| < \infty$. Besides, $E(\eta_{n+1}|\mathcal{F}_n) = E(\eta_n + \varepsilon_n(\xi_{n+1} - \xi_n)|\mathcal{F}_n) = \eta_n + \varepsilon_n E(\xi_{n+1} - \xi_n|\mathcal{F}_n) \geqslant \eta_n$, (P-a.s.). Therefore $\{\xi_n, \mathcal{F}_n\}$ is a submartingale. When $n = 1$ we have $E(\eta_1) \leqslant E(\xi_1)$. Let us suppose that $E(\eta_n) \leqslant E(\xi_n)$ for some n. Using the submartingale property of $\{\xi_n, \mathcal{F}_n\}$, we obtain

$$E(\xi_{n+1} - \eta_{n+1}) = E\{\xi_{n+1} - [\eta_n + \varepsilon_n(\xi_{n+1} - \xi_n)]\} =$$
$$= E\{(1 - \varepsilon_n) \cdot (\xi_{n+1} - \xi_n)\} + E(\xi_n - \eta_n) \geqslant$$
$$\geqslant \int_{\{\varepsilon_n = 0\}} (\xi_{n+1} - \xi_n)dP =$$
$$= \int_{\{\varepsilon_n = 0\}} E(\xi_{n+1} - \xi_n|\mathcal{F}_n)dP \geqslant 0.$$

If $\{\xi_n, \mathcal{F}_n\}$ is a martingale, the proofs of both assertions are analogous.

13.29. Let us put $m_0 = \xi_0$, $a_0 = 0$, $m_{n+1} - m_n = \xi_{n+1} - E(\xi_{n+1}|\mathcal{F}_n)$ and

$a_{n+1} - a_n = E(\xi_{n+1}|\mathcal{F}_n) - \xi_n$. Then $m_{n+1} = \sum_{k=0}^{n+1} \xi_k - \sum_{k=0}^{n} E(\xi_{k+1}|\mathcal{F}_k)$, $a_{n+1} = \sum_{k=0}^{n} E(\xi_{k+1}|\mathcal{F}_k) - \sum_{k=0}^{n} \xi_k$. Thus it is easily verified that $\{m_n, \mathcal{F}_n\}$ is a martingale, that the sequence $\{a_n\}$ increases and that $\xi_n = m_n + a_n$, (P-a.s.), $n = 0, 1, \ldots$ We are going to show that this decomposition is unique. Let $\xi_n = m_n + a_n = m'_n + a'_n$, where $\{m'_n, \mathcal{F}_n\}$ and a'_n are as given by $\{m_n, \mathcal{F}_n\}$ and $\{a_n\}$. Then $m_{n+1} - m_n + a_{n+1} - a_n = m_{n+1} - m'_n + a'_n - a'_n$.

After taking the conditional mean $E(\cdot|\mathcal{F}_n)$ of each side of this equality we obtain $a_{n+1} - a_n = a'_{n+1} - a'_n$. Since $0 = a_0 = a'_0$, by induction we find easily that $a'_n = a_n$. Analogously from $\xi_0 = m_0 = m'_0$, it follows that $m'_n = m_n$.

13.30. (a) Can be verified directly.

(b) The sequence $\{a_n\}$, where $a_0 = 0$, $a_n = \sigma_1^2 + \ldots + \sigma_n^2$, is an increasing one, and since a_n = constant, then a_{n+1} is $\mathcal{F}_n$-measurable for each n. We put $m_n = \xi_n^2 - a_n$ and prove that $\{m_n, \mathcal{F}_n\}$ is a martingale. It is clear that m_n is an integrable $\mathcal{F}_n$-measurable r.v. and $E(m_{n+1}|\mathcal{F}_n) = E\{(\eta_1 + \ldots + \eta_{n+1})^2|\mathcal{F}_n\} - a_{n+1} = E\{(\eta_1 + \ldots + \eta_n)^2 + 2(\eta_1 + \ldots + \eta_n)\eta_{n+1} + \eta_{n+1}^2|\mathcal{F}_n\} - a_{n+1} = (\eta_1 + \ldots + \eta_n)^2 + 0 + \sigma_{n+1}^2 - a_{n+1} = \xi_n^2 - a_n = m_n$ with probability 1. Hence $\xi_n^2 = m_n + a_n$, (P-a.s.). According to Doob's theorem this representation is unique.

13.31. Since the variables ξ_i are identically distributed, then P-a.s.

$$E\{\xi_1|S_{n+1}\} = E\{\xi_2|S_{n+1}\} = \ldots = E\{\xi_{n+1}|S_{n+1}\}.$$

Therefore

$$E\{\xi_i|S_{n+1}\} = \frac{1}{n+1} E\left\{\sum_{j=1}^{n+1} \xi_j|S_{n+1}\right\} = \frac{S_{n+1}}{n+1};$$

hence, we find

$$E\left\{\frac{S_n}{n}|S_{n+1}\right\} = \frac{1}{n}\sum_{i=1}^{n} E\{\xi_i|S_{n+1}\} = \frac{S_{n+1}}{n+1}.$$

13.32. From the existence of $E(\xi)$ it follows that either $E\{\xi^+\} < \infty$ or $E\{\xi^-\} < \infty$. Let $E\{\xi^+\} < \infty$. Then η is $\mathcal{G}$-measurable and $E(\xi I_A) = E(\eta I_B)$ for any $A \in \mathcal{G}$. For instance $E[\xi I_{\{\eta>0\}}] = E[\eta I_{\{\eta>0\}}] = E(\eta^+)$.

We shall now show the validity of the equality

$$E(\xi^+) = E(\eta^+) + E(|\xi| I_{\{\xi\eta<0\}}) + E(\xi^+ I_{\{\eta=0\}}). \tag{13.5}$$

We know that

$$E(\xi^+) = E(\xi^+ I_{\{\eta>0\}}) + E(\xi^+ I_{\{\eta<0\}}) + E(\xi^+ I_{\{\eta=0\}}),$$

$$E(\xi^+ I_{\{\eta>0\}}) = E(\xi I_{\{\eta>0\}}) + E(\xi^- I_{\{\eta>0\}}) = E(\eta^+) + E(\xi^- I_{\{\eta>0\}}),$$

$$E(\xi^- I_{\{\eta>0\}}) + E(\xi^+ I_{\{\eta<0\}}) = E(|\xi| I_{\{\xi\eta<0\}});$$

hence, (13.5) follows.

Since ξ and η have one and the same distribution, then $E(\eta^+) = E(\xi^+) < \infty$ and from (13.5) we obtain

$$E(|\xi| I_{\{\xi\eta<0\}}) + E(\xi^+ I_{\{\eta=0\}}) = 0.$$

But $|\xi| > 0$ on the set $\{\xi\eta < 0\}$; therefore,

$$P\{\xi\eta < 0\} = 0.$$

Let r be an arbitrary real number. Instead of the variables ξ and η we can consider $\xi - r$ and $\eta - r$, where $\eta - r = E\{(\xi - r)|\mathfrak{G}\}$. Then from the above reasoning it follows that

$$P\{(\xi - r)(\eta - r) < 0\} = 0.$$

Since the event $\{\xi \neq \eta\}$ is a countable union of events of the kind $\{(\xi - r)(\eta - r) < 0\}$, where r is a rational number, we conclude that $\xi = \eta$, (P-a.s.), and since $\mathfrak{G}$ is a complete σ-algebra, then ξ is also $\mathfrak{G}$-measurable.

14. Products of Measurable Spaces and Probabilities on Them

14.1. Show that if a point $\omega = (\omega_1, \omega_2)$ belongs to the set from the one side of the equality, it also belongs to the set from the other side of the equality, and vice versa.

14.2. Use Exercise 14.1. The assertion also holds true when $\mathfrak{C}$ and $\mathfrak{C}'$ are semi-algebras.

14.3. Use Exercise 14.2 and 11.31.

14.4. If we denote by $\mathfrak{C}_{\omega_1}$ the class of subsets $A \subset \Omega_1 \times \Omega_2$ such that $A_{\omega_1} \in \mathfrak{A}_2$, then:

(1) $\mathfrak{C}_{\omega_1}$ contains all the measurable rectangles;

(2) $\mathfrak{C}_{\omega_1}$ is closed with respect to the operations of complementation and countable intersection;

therefore, $\mathfrak{C}_{\omega_1}$ is a σ-algebra and $\mathbf{A}_1 \times \mathbf{A}_2 \subset \mathfrak{C}_{\omega_1}$.

14.5. We denote $P = P_1 \times P_2$ and $\Omega = \Omega_1 \times \Omega_2$. From Fubini's theorem

$$P(A) = \iint_{\Omega} I_A dP = \int_{\Omega_1} dP_1 \int_{\Omega_2} I_A dP_2 = \int_{\Omega_1} P_2(A_{\omega_1}) dP_1.$$

If $P(A) = 0$, then $P_2(A_{\omega_1}) = 0$ a.s., because $P_2(A_{\omega_1}) \geqslant 0$. On the contrary, if $P_2(A_{\omega_1}) = 0$ a.s., then $P(A) = 0$.

14.6. Let $\xi'(\omega_1, \omega_2) = \xi(\omega_1)$, $\eta'(\omega_1, \omega_2) = \eta(\omega_2)$. Obviously $\zeta(\omega_1, \omega_2) = \xi'(\omega_1, \omega_2) \cdot \eta'(\omega_1, \omega_2)$ with the r.v.'s ξ' and η' being defined in $\Omega_1 \times \Omega_2$. It is easy to see that ξ' and η' are independent (with respect to the probability P); hence,

$$\iint_{\Omega_1 \times \Omega_2} \xi'(\omega_1, \omega_2) dP \cdot \iint_{\Omega_1 \times \Omega_2} \eta'(\omega_1, \omega_2) dP = \iint_{\Omega_1 \times \Omega_2} \zeta(\omega_1, \omega_2) dP.$$

Applying Fubini's theorem, we obtain the desired equality.

14.7. (a) It is easy to see that $Q(\cdot) \geqslant 0$ and $Q(\Omega_1 \times \Omega_2) = 1$; the σ-additivity of Q follows from the properties of the integral.

(b) Making use of Fubini's theorem, we find

$$Q(A_1 \times \Omega_2) = 2\int_{\Omega} I_{A_1} I_B dP = 2 \int_{\Omega_1} dP_1 \int_{\Omega_2} I_{A_1} I_B dP_2 =$$

$$= \int_{\Omega_1} I_{A_1} dP_1 = P_1(A_1) = P(A_1 \times \Omega_2).$$

Ananlogously we obtain the equality $Q(\Omega_1 \times A_2) = P(\Omega_1 \times A_2)$.

(c) Since $P(B) = 1/2$, then $P(\overline{B}) = 1/2$. But

$$Q(\overline{B}) = 2\int_{\overline{B}} I_{\overline{B}} I_B dP = 2 \int_{\Omega} 0 \cdot dP = 0$$

Therefore $Q(\overline{B}) \neq P(\overline{B})$. (To find $P(B)$ see Exercise 11.18 (f)).

14.8. After applying Fubini's theorem, we obtain

$$P(A) = \iint_{\Omega_1 \times \Omega_2} I_A(\omega_1, \omega_2) dP = \int_{\Omega_1} dP_1 \int_{\Omega_2} I_A(\omega_1, \omega_2) dP_2 =$$

$$= \int_{\Omega_1} P_2(A_{\omega_1}) dP_1 = \int_{\Omega_1} P_2(B_{\omega_1}) dP_1 =$$

$$= \int_{\Omega_1} dP_1 \int_{\Omega_2} I_B(\ _1,\ _2) dP_2 = \iint_{\Omega_1 \times \Omega_2} I_B(\omega_1, \omega_2) dP = P(B).$$

14.9. The following relations are proved easily: (1) $\mathfrak{C}_1 \subset \mathfrak{C}_3$; (2) $\{(x_1, \ldots, x_n) : a_i < x_i < b_i,\ i = 1, \ldots, n\} = \bigcup_{k=1}^{\infty} \{(x_1, \ldots, x_n) : a_i + 1/k \leqslant x_i \leqslant b_i - 1/k,\ i = 1, \ldots, n\}$; (3) the complement of each open set is closed and vice versa; (4) $\mathfrak{C}_5 \subset \mathfrak{C}_3$; (5) any open set is a countable union of the open rectangles with rational ends a_i, b_i, $i = 1, \ldots, n$, which are contained in it; (6) $\{(x_1, \ldots, x_n) : a_i < x_i < b_i\} = \bigcup_{k=1}^{\infty} [\{(x_1, \ldots, x_n) : x_i < b_i\} \setminus \{(x_1, \ldots, x_n) : x_i < a_i + 1/k\}]$. From (1) - (6) it follows that

$$\sigma(\mathfrak{C}_1) = \sigma(\mathfrak{C}_2) = \sigma(\mathfrak{C}_3) = \sigma(\mathfrak{C}_4) = \sigma(\mathfrak{C}_5).$$

14.10. Using the notation set out in Exercise 14.9, we easily see that $\mathfrak{C}_1 \subset \mathcal{B}_1^{\times n}$, and therefore $\mathcal{B}_n \subset \mathcal{B}_1^{\times n}$.

Let us denote by $\mathfrak{C}$ the class of the closed intervals in $\mathbb{R}_1$. We know that $\mathcal{B}_1^{\times n}$ is generated by the rectangles of the kind $\{B_1 \times B_2 \times \ldots \times B_n\}$, $B_i \in \mathcal{B}_1$, $i = 1, \ldots, n$. Any such rectangle is an intersection of cylinders with bases in $\mathcal{B}_1$: $\{B_1 \times \ldots \times B_n\} = \bigcap_{i=1}^{n} \{A_1^i \times A_2^i \times \ldots \times A_n^i\}$, where $A_i^i = B_i$ and $A_j^i = \mathbb{R}_1$, $j \neq i$. For any i the cylinders $\{\mathbb{R}_1 \times \mathbb{R}_1 \times \ldots \times B_i \times \ldots \times \mathbb{R}_1\}$, $B_i \in \mathcal{B}_1$, form a σ-algebra, which we denote by $\mathcal{F}_i$. It is clear that $\mathcal{F}_i$ is generated by the sets of the kind $A_i = \{\mathbb{R}_1 \times \ldots \times C_i \times \ldots \times \mathbb{R}_1\}$, where C_i are closed intervals and therefore $A_i \in \mathfrak{C}_3$. Since $\mathfrak{C}_3 \subset \mathcal{B}_n$, then $\mathcal{F}_i \subset \mathcal{B}_n$, $i = 1, \ldots, n$. Then $\{B_1 \times \ldots \times B_n\} \in \mathcal{B}_n$; hence, $\mathcal{B}_1^{\times n} \subset \mathcal{B}_n$, and then $\mathcal{B}_1^{\times n} = \mathcal{B}_n$.

15. Distribution Function

15.1. $F_\xi(x) = P\{\xi < x\} = 0$ for $x \leqslant 0$ and $F_\xi(x) = P(\Omega) = 1$ for $x > n$. It is interesting to find $F_\xi(x)$ for $m - 1 < x \leqslant m$, where m is a natural number not exceeding n. We have

$$F_\xi(x) = P\{\xi < x\} = \sum_{k=0}^{m-1} P\{\xi = k\} = 1 - \sum_{k=m}^{n} \binom{n}{k} p^k (1 - p)^{n-k}.$$

Now we have only to find $\psi(p) = \sum_{k=m}^{n} \binom{n}{k} p^k (1 - p)^{n-k}$. First we calculate its derivative: $\psi'(p) = n\binom{n-1}{m-1} p^{m-1}(1 - p)^{n-m}$. Since $\psi(0) = 0$, hence $\psi(p) = \int_0^p \psi'(x)d\,x$ and therefore

$$\psi(p) = \frac{1}{B(m,\, n - m + 1)} \int_0^p x^{m-1}(1 - x)^{n-m}dx.$$

15.2. We have $P\{\xi \geqslant \lambda\} = \frac{1}{\Gamma(\alpha)} \int_\lambda^\infty u^{\alpha-1}e^{-u}du$. Apply integration by parts α times.

15.3. (a) $P\{0 \leqslant \xi < 3\} = (2\sqrt{2\pi})^{-1}\int_0^3 \exp\left[-\frac{(x - 2)^2}{2 \cdot 4}\right]d\,x =$ $\Phi(\frac{3-2}{2}) - \Phi(\frac{0-2}{2}) = \Phi(0.5) - \Phi(-1) = \Phi(0.5) + \Phi(1) - 1$. From Table 1 we find $\Phi(0.5) \approx 0.6915$ and $\Phi(1) \approx 0.8413$. Finally $P\{0 \leqslant \xi < 3\} \approx 0.5328$.
(b) $P\{|\xi| < 1\} = \Phi(-0.5) - \Phi(-1.5) = \Phi(1.5) - \Phi(0.5) \approx 0.2417$.
(c) $P\{-1 \leqslant \xi < 1 | 0 \leqslant \xi < 3\} = \frac{P\{-1 \leqslant \xi < 1,\ 0 \leqslant \xi < 3\}}{P\{0 \leqslant \xi < 3\}} =$ $\frac{P\{0 \leqslant \xi < 1\}}{P\{0 \leqslant \xi < 3\}} \approx \frac{\Phi(-0.5) - \Phi(-1)}{0.5328} \approx 0.281.$

15.4. According to (15.4) we have

$$1 = \int_{\mathbb{R}_1} f(x)dx = a\int_{-s}^{s} (s^2 - x^2)^{-1/2}dx = a\pi;$$

therefore, $a = 1/\pi$. So for the d.f. $F(x) = \int_{-\infty}^{x} f(u)du$ we have: $F(x) = 0$ for $x \leqslant -s$, $F(x) = 1$ for $x > s$, $F(x) = \frac{1}{\pi}\arcsin(\frac{x}{s}) + \frac{1}{2}$ for $-s < x \leqslant s$. It remains only to find $P\{0 \leqslant \xi < s\}$. We have $P\{0 \leqslant \xi < s\} =$ $F(s) - F(0) = \frac{1}{2}$.

15.5. (a) $f_\eta(y) = \lambda e^{\lambda y}$ for $y < 0$ and $f_\eta(y) = 0$ for $y > 0$. (b) $f_\eta(y) =$ $\frac{\lambda}{2}\exp(-\frac{\lambda}{2}(y + 1))$ for $y > -1$ and $f_\eta(y) = 0$ for $y < -1$. (c) $f_\eta(y) =$ $2\lambda y\exp(-\lambda y^2)$ for $y \geqslant 0$ and $f_\eta(y) = 0$ for $y < 0$. (d) $f_\eta(y) = (\frac{\lambda}{\alpha})y^{(1-\alpha)/\alpha}$ $\times \exp(-\lambda y^{1/\alpha})$ for $y \geqslant 0$ and $f_\eta(y) = 0$ for $y < 0$. (e) $f_\eta(y) = (\frac{\lambda}{c})\exp(-\frac{\lambda}{c}y)$ for $y \geqslant 0$ and $f_\eta(y) = 0$ for $y < 0$.

In case (e) the variable $\eta = c\xi \in E(\frac{\lambda}{c})$.

15.6. The events $A_k = \{[\xi] = 2k\} = \{2k \leqslant \xi < 2k + 1\}$, $k = 1, 2, \ldots$ are disjoint and their union gives A and $P\{\xi < x\} = 1 - e^{-x}$. So $P(A) =$ $\sum_k P(A_k) = e/(e + 1)$.

15.7. From the definition of conditional probability, it follows that

$$P\{\xi < x + y | \xi \geqslant y\} = \frac{P\{y \leqslant \xi < y + x\}}{P\{\xi \geqslant y\}} = \frac{F(x + y) - F(y)}{1 - F(y)}$$

and from the condition of the problem we find that

$$\frac{F(x + y) - F(y)}{1 - F(y)} = F(x).$$

If we set $G(x) = 1 - F(x)$, then from the last two equalities we obtain $G(x + y) = G(x) \cdot G(y)$, $x \geqslant 0$, $y \geqslant 0$. It is known that the solution of this functional equation, assuming continuity of $G(x)$, is $G(x) = e^{-\lambda x}$, $\lambda > 0$, $x \geqslant 0$. Hence $F(x) = 1 - e^{-\lambda x}$ for $x \geqslant 0$ and $F(x) = 0$ for $x < 0$.

15.8. From the condition of the problem we can easily show that $P\{\xi \geqslant t + \Delta t\} = P\{\xi \geqslant t\}(1 - (\Delta t)/\lambda - o(\Delta t))$. Then the function $\phi(t) = P\{\xi \geqslant t\}$ satisfies the equation $\phi(t + \Delta t) = \phi(t)(1 - (\Delta t)/\lambda - o(\Delta t))$. Therefore for any $t > 0$ $\phi'(t)$ exists and $\phi'(t) = -(\frac{1}{\lambda})\phi(t)$. All solutions of this linear differential equation are of the type $\phi(t) = c \cdot \exp(-t/\lambda)$, where c is an arbitrary constant. Hence $F_\xi(t) = 1 - \phi(t) = 1 - c \cdot \exp(-t/\lambda)$, $t \geqslant 0$. Since $\xi \geqslant 0$, then $F_\xi(0) = 0$; therefore, $c = 1$. Finally, $f_\xi(t) = \frac{1}{\lambda}\exp(-\frac{t}{\lambda})$ for $t \geqslant 0$ and $f_\xi(t) = 0$ for $t < 0$; i.e., $\xi \in E(\lambda^{-1})$.

15.9. Applying (15.6) we can show that

$$f_\eta(y) = (|c|\sqrt{2\pi})^{-1}\exp\{-[y - (ca + d)]^2/(2c^2\sigma^2)\}, \quad y \in \mathbb{R}_1,$$

which implies that $\eta \in N(ca + d, c^2\sigma^2)$.

15.10. Let ξ and η be defined on the probability space $(\Omega, \mathcal{F}, P)$. Denote $\mathcal{F}' = \{\omega : \xi(\omega) > \eta(\omega)\}$. From the inclusion

$$\{\omega : \xi(\omega) < x\} \cap \mathcal{F}' \subset \{\omega : \eta(\omega) < x\} \cap \mathcal{F}'$$

it follows that $F(x) \leqslant G(x)$ for $x \in \mathbb{R}_1$. The converse, however, is not true. For example, if $\Omega = (0, 1)$, $\mathcal{F} = \{B : B \in \mathcal{B}_1, B \subset (0, 1)\}$ and $\xi(\omega) = \omega$, $\eta(\omega) = q(1 - \omega)$, $\omega \in (0, 1)$, $0 < q < 1$, then $F(x) < G(x)$, $x \in (0, 1)$; but for $\omega \in (0, q/(1 + q))$ obviously we have $\xi(\omega) < \eta(\omega)$.

15.11. It is easy to verify that G is a d.f. and the equality $F(x) = G(x)$ for every $x \in \mathbb{R}_1$ is possible only if $F(x) = 0$ for $x \leqslant 0$ and $F(x) = 1$ for $x > 0$.

15.12. Take into account that the functions $g_1(x) = 1/x$, $g_2(x) = 2x/(1 - x^2)$ and $g_3(x) = \frac{3x - x^3}{1 - 3x^2}$ are monotonic and use (15.6).

15.13. The r.v. $|\xi - b| \in E(a)$.

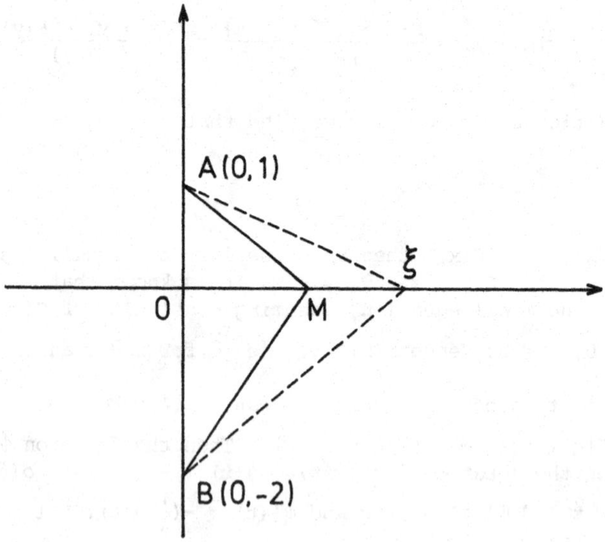

Figure 15.3

15.14. Let AMB be a rectangular triangle (see Figure 15.3). Then $\sphericalangle A|B < \frac{\pi}{2}$ only if $|\xi| > OM$. But $OM^2 = OA \times OB$ and therefore $OM = \sqrt{2}$. Hence $P\{\sphericalangle A\xi B < \frac{\pi}{2}\} = P\{|\xi| > \sqrt{2}\} = 1 - P\{|\xi| \leqslant \sqrt{2}\} = 1 - (\Phi(\sqrt{2}) - \Phi(-\sqrt{2})) = 1 - (2\Phi(\sqrt{2}) - 1) = 2 - 2\Phi(\sqrt{2}) \approx 0.16$.

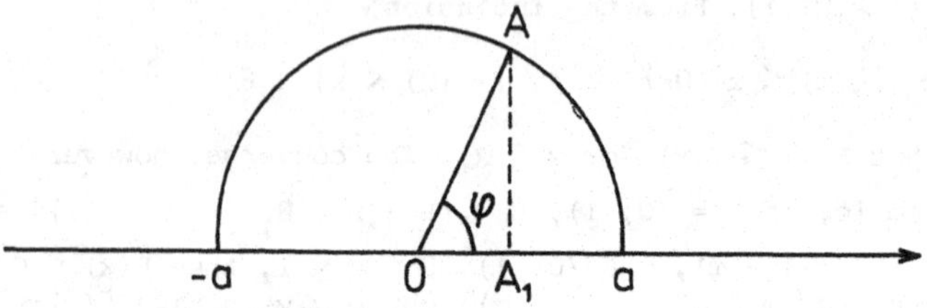

Figure 15.4

15.15. According to the notations in Figure 15.4, the r.v. $\xi = |AA_1| = a \sin \phi$. Since A is chosen at random on the arc with length $a\pi$, the angle $\phi = \sphericalangle A_1OA$ has a uniform distribution over the interval $[0, \pi]$. For the d.f. $F_\xi(t) = P\{\xi < t\}$, we find: $F_\xi(t) = 0$ if $t \leqslant 0$, $F_\xi(t) = 1$ if $t \geqslant a$. For $t \in (0, a)$ we have $F_\xi(t) = P\{0 \leqslant a \sin \phi < t\} =$

$P\{0 \leqslant \phi < \arcsin \frac{t}{a}\} + P\{\pi - \arcsin \frac{t}{a} < \phi \leqslant \pi\} = \frac{2}{\pi} \arcsin \frac{t}{a}$. Therefore after differentiation, we can easily determine the density f_ξ as well.

15.16. Let $A = \{|\xi - \eta| \leqslant \delta\}$, $B = \{\eta \leqslant x\}$. From the condition given for ξ and η and the equality $P(AB) + P(A \cup B) = P(A) + P(B)$, we find that $P\{\eta \leqslant x\} - \varepsilon \leqslant P\{|\xi - \eta| \leqslant \delta, \eta \leqslant x\}$. However, the joint occurrence of the events A and B implies the occurrence of the event $\{\xi \leqslant x + \delta\}$, and therefore $P\{\xi \leqslant x + \delta\} \geqslant P\{\eta \leqslant x\} - \varepsilon$. Similarly if we consider the pair of events A and $\bar{B} = \{\eta > x\}$, we arrive at the inequality $P\{\xi > x - \delta\} \geqslant P\{\eta > x\} - \varepsilon$. We can express the two inequalities obtained through the d.f.'s F and G: $F(x - \delta) - \varepsilon \leqslant G(x) \leqslant F(x + \delta) + \varepsilon$. Since $F(x - \delta) \leqslant F(x) \leqslant F(x + \delta)$, we see that $F(x - \delta) - F(x + \delta) - \varepsilon \leqslant F(x) - G(x) \leqslant F(x + \delta) - F(x - \delta) + \varepsilon$, and this is the desired result.

15.17. Denote $a_k = \arccos y + 2k\pi$, $b_k = 2(k + 1)\pi - \arccos y$. Then for $y \in (0, 1)$ we find:

$$F_\eta(y) = P\{\eta < y\} = P\{\cos \xi < y\} = P\{\xi \in \bigcup_{k=-\infty}^{\infty} (a_k, b_k)\} =$$

$$= \sum_{k=-\infty}^{\infty} \int_{a_k}^{b_k} f_\xi(t)\,d\,t = \int_{-\pi/2}^{a_0} \frac{1}{\pi}\,d\,t + \int_{a_0}^{\pi/2} \frac{1}{\pi}\,d\,t =$$

$$= \frac{2}{\pi}\left(\frac{\pi}{2} - \arccos y\right).$$

If $y \leqslant 0$, then $F_\eta(y) = 0$ and for $y > 1$ we have $F_\eta(y) = 1$. Hence, $(F_\eta(y))' = 2(\pi\sqrt{1 - y^2})^{-1}$ for $0 < y < 1$ and $(F_\eta(y))' = 0$ for $y < 0$ and $y > 1$ (at the points 0 and 1 the derivative $(F_\eta(y))'$ does not exist). Since $\int_{-\infty}^{y} F_\eta'(t)\,d\,t = F_\eta(y)$, then $F_\eta(y)$, $y \in \mathbb{R}_1$, is absolutely continuous and the r.v. η has density $f_\eta(y) = (F_\eta(y))'$, $y \in \mathbb{R}_1$.

15.18. (a) We have $Q_\xi(\alpha x) \leqslant Q_\xi(([\alpha] + 1)x) = \sup_{b \in \mathbb{R}_1} P\{b \leqslant \xi \leqslant b + x([\alpha] + 1)\} \leqslant ([\alpha] + 1) \sup_{b \in \mathbb{R}_1} P\{b \leqslant \xi \leqslant b + x\} = ([\alpha] + 1)Q_\xi(x)$.

(b) Show that for every r.v. ξ and any $x \in \mathbb{R}_1$ there exists $b \in \mathbb{R}_1$ such that $Q_\xi(x) = P\{b \leqslant \xi \leqslant b + x\}$. Use the fact that if x_0 ($x_0 \neq 0$) is a jump point of the function $Q_\xi(x)$ and if b is such that $Q_\xi(x_0) = P\{b \leqslant \xi \leqslant b + x_0\}$, then the points b and $b + x_0$ are jump points of the d.f. $F_\xi(x)$, $x \in \mathbb{R}_1$.

Note that m can be less than or greater than n as well. Give an example in which the equality $m = \binom{n}{2} + 1$ holds.

15.19. It is known that if ξ is a r.v., then $P\{\xi \leqslant x\}$ considered as a function of x possesses the properties of a distribution function with the only difference that instead of left continuity we have right continuity for every $x \in \mathbb{R}_1$.

We will show that the given function possesses these properties. It is directly verified that $\lim_{x\to-\infty} F(x) = 0$, $\lim_{x\to\infty} F(x) = 1$ and if $x_1 < x_2$, then $F(x_1) \leqslant F(x_2)$. It remains to show the right continuity of F(x). For the points $x \overline{\in} [0, 1)$ this is obvious. Let $0 \leqslant x_0 < 1$ and $\varepsilon > 0$. For every natural number N and for $0 \leqslant x \leqslant 1$ we have

$$\sum_{n=N}^{\infty} \frac{[2^{n-1}x + 2^{-1}]}{2^{2n-1}} \leqslant \sum_{n=N}^{\infty} \frac{2^{n-1}}{2^{2n-1}} = \frac{1}{2^{N-1}} .$$

It is clear that for sufficiently large N_0 and for every $0 \leqslant x \leqslant 1$,

$$\sum_{n=N_0}^{\infty} \frac{[2^{n-1}x + 2^{-1}]}{2^{2n-1}} < \frac{\varepsilon}{4} .$$

Since the function [y] is right-continuous, for fixed n we can choose δ_n, $0 < \delta_n < 1 - x_0$, in order to have

$$2^{-(2n-1)}|[2^{n-1}x_0 + 2^{-1}] - [2^{n-1}x + 2^{-1}]| \leqslant \frac{\varepsilon}{2(N_0 - 1)}$$

for every x, $x_0 \leqslant x \leqslant x_0 + \delta_n$.

If $\delta_\varepsilon = \min[\delta_1, \ldots, \delta_{N_0-1}]$, then for x, $x_0 \leqslant x \leqslant x_0 + \delta_\varepsilon$,

$$|F(x_0) - F(x)| \leqslant \sum_{n=1}^{N_0-1} \frac{|[2^{n-1}x_0 + 2^{-1}] - [2^{n-1}x + 2^{-1}]|}{2^{2n-1}} +$$

$$+ \sum_{n=N_0}^{\infty} \frac{[2^{n-1}x_0 + 2^{-1}]}{2^{2n-1}} + \sum_{n=N_0}^{\infty} \frac{[2^{n-1}x + 2^{-1}]}{2^{2n-1}} \leqslant$$

$$\leqslant \sum_{n=1}^{N_0-1} \frac{\varepsilon}{2(N_0 - 1)} + \frac{\varepsilon}{4} + \frac{\varepsilon}{4} = \varepsilon.$$

This implies that F(x), $x \in \mathbb{R}_1$, is a right-continuous function. Similarly, as the function [y] is continuous for every non-integer y, it is possible to show that for any x_0, $0 < x_0 < 1$, for which $2^{n-1}x_0 + 2^{-1}$ is not integer for n = 1, 2, ..., the function F(x) is continuous at the point

x_0. The remaining points x of the interval (0, 1) for which we do not know whether F(x) is continuous are $x = a_1 2^{-1} + a_2 2^{-2} + \ldots + a_{s-1} 2^{-(s-1)} + 1 \cdot 2^{-s}$, where $a_i = 0$ or 1, i = 1, 2, ..., s - 1 and s is a natural number (these points are called binary rational points). Let x be a fixed binary rational number. Then for the points $x_i = a_1 2^{-1} + a_2 2^{-2} + \ldots + a_{s-1} 2^{-(s-1)} + 2^{-(s+1)} + \ldots + 2^{-(s+j)}$, j = 1, 2, ..., we shall have $x_j < x$ and $\lim_{j\to\infty} x_j = x$.

We shall compute the values of F(x) for the binary rational points x and x_j we considered above. We have

$$F(x) = \sum_{n=1}^{\infty} \frac{[2^{n-1}x + 2^{-1}]}{2^{2n-1}} = \sum_{n=1}^{\infty} 2^{-(2n-1)}$$

$$= \sum_{n=1}^{\infty} \left\{2^{-(2n-1)}\left[2^{n-1}\left(\sum_{k=1}^{s-1} a_k 2^{-k} + 2^{-s}\right) + 2^{-1}\right]\right\} =$$

$$= \sum_{n=1}^{s} \{\cdot\} + \sum_{n=s+1}^{\infty} \{\cdot\} = \frac{a_1}{2^1} + \frac{a_1 + a_2}{2^3} + \frac{a_1 2^1 + a_2 2^0 + a_3}{2^5} +$$

$$+ \ldots + \frac{a_1 2^{s-2} + a_2 2^{s-3} + \ldots + a_{s-1} 2^0 + 1}{2^{2s-1}} +$$

$$+ \sum_{n=s+1}^{\infty} \left\{2^{-(2n-1)}\left(\sum_{k=1}^{s-1} a_k 2^{n-1-k} + 2^{n-1-2}\right)\right\} =$$

$$= a_1(2^{-1} + 2^{-3} + 2^{-4} + 2^{-5} + \ldots) + a_2(2^{-3} + 2^{-5} +$$

$$+ 2^{-6} + 2^{-7} + \ldots) + \ldots + a_{s-1}(2^{-2s+3} + 2^{-2s+1} + 2^{-2s} +$$

$$+ 2^{-2s-1} + \ldots) + 1 \cdot (2^{-2s+1} + 2^{-2s-1} + 2^{-2s-2} +$$

$$+ 2^{-2s-3} + \ldots) = a_1(2^{-1} + 2^{-2}) + a_2(2^{-3} + 2^{-4}) + \ldots +$$

$$+ a_{s-1}(2^{-2s+3} + 2^{-2s+2}) + 1 \cdot (2^{2s+1} + 2^{-2s}).$$

Similarly, we get $F(x_j) = a_1(2^{-1} + 2^{-2}) + \ldots + a_{s-1}(2^{-2s+3} + 2^{-2s+2}) + 0 \cdot (2^{-2s+1} + 2^{-2s}) + 1 \cdot (2^{-2s-1} + 2^{-2s-2}) + \ldots +$

$+ 1 \cdot (2^{-2(s+j)+1} + 2^{-2(s+j)})$. Now it is easily seen that $F(x) - \lim_{j\to\infty} F(x_j) = 2^{-2s+1}$; i.e., in every binary rational point given as $x = a_1 2^{-1} + \ldots + a_{s-1}2^{-(s-1)} + 1 \cdot 2^{-s}$, the function $F(x)$ has a jump of size 2^{-2s+1}. Since the number of these points for fixed s is equal to 2^{s-1}, for the sum of all jumps at all binary rational points in the interval (0, 1), we obtain

$$\sum_{s=1}^{\infty} 2^{s-1} \cdot 2^{-2s+1} = 1.$$

Hence, if ξ is a discrete r.v. with

$$P\{\xi = a_1 2^{-1} + a_2 2^{-2} + \ldots + a_{s-1}2^{-(s-1)} + 2^{-s}\} = 2^{-2s+1},$$

where $a_i = 0$ or 1, $i = 1, \ldots, s - 1$, $s = 1, 2, \ldots$, then $F(x) = P\{\xi \leqslant x\}$ for every $x \in \mathbb{R}_1$.

15.20. From the definition of symmetry, it follows that

$$F(x) = P\{\xi < x\} = P\{-\xi > -x\} = 1 - P\{-\xi \leqslant -x\} = 1 - F(-x + 0).$$

If F is a continuous function, then $F(x) = 1 - F(-x)$, $x \in \mathbb{R}_1$.

If F is absolutely continuous and $F' = f$, then $f(-x) = f(x)$, $x \geqslant 0$. Now it is trivial to verify the equality for F (or for f) for any of the distributions (a) - (e).

15.21. The determination of the maximum of f is an easy exercise. The density f does not have inflection points if $0 < p < 1$ and $0 < q < 1$.

15.22. Show that $f(x) = \exp(-\pi x^2)$, $x \in \mathbb{R}_1$. This means that $\xi \in N(0, 1/(2\pi))$ and if we put $\eta = \xi\sqrt{2\pi}$, then $\eta \in N(0, 1)$. From Table 1 for the 0.95-quantile of the d.f. of the r.v. η we find $y_{0.95} \approx 1.65$.

It is easily observed that p-quantiles x_p and y_p of the r.v.'s ξ and η satisfy the equality $x_p = y_p\sqrt{2\pi}$. Hence $x_{0.95} \approx 1.65/\sqrt{2\pi}$.

15.23. It can be directly seen that $\phi(x) \geqslant 0$, $x \in R_1$, and

$$\int_0^{\infty} \phi(x)dx = \sum_{k=0}^{\infty} \int_k^{k+1} \phi(x)d\,x = \sum_{k=0}^{\infty} \frac{\pi}{4} 2^{-k} \int_0^1 \sin\frac{\pi t}{2}\,dt =$$

$$= \sum_{k=0}^{\infty} \frac{\pi}{4} 2^{-k} \frac{2}{\pi} = 1;$$

i.e., ϕ is a density. Obviously for every point $x_k = k$, $k = 1, 2, \ldots$, the function ϕ has a local maximum and therefore it has infinitely many modes.

15.24. Both assertions can be directly verified.

15.25. It is easy to check that ϕ is a density and the median of F is $x_m = b \cdot 2^{1/\alpha}$.

15.26. (a) According to the definition of the median, we are looking for a number $x_{0.5}$ such that $F(x_{0.5}) = 0.5$; i.e.,

$$x_{0.5} \geqslant 0, \quad 1 - \exp\left\{-\frac{x_{0.5}^2}{2\sigma^2}\right\} = 0.5.$$

Hence we easily find that $x_{0.5} = \sigma\sqrt{2 \ln 2}$. (b) Differentiating F(x), we obtain that the density is $f(x) = (x/\sigma^2)\exp(-x^2/2\sigma^2)$ for $x \geqslant 0$ and $f(x) = 0$ for $x < 0$. (c) It is trivial to show that every point $x < 0$ is a local maximum of f; i.e., it is a mode (these points are not interesting). The only solution of the equation $f'(x) = 0$ for $x \geqslant 0$ is $x = \sigma$. Since $f'(x) > 0$ for $x < \sigma$ and $f'(x) < 0$ for $x > \sigma$, f(x) has a local maximum at $x = \sigma$. Hence the point $x = \sigma$ is (the only non-trivial) mode of the density f.

15.27. (a) The p-quantile x_p is given by $x_p = (-a \ln(1 - p))^{1/m}$. (b) $f(x) = 0$ for $x < 0$ and $f(x) = \frac{m}{a} x^{m-1} \exp(-x^m/a)$ for $x \geqslant 0$. (c) If $0 < m < 1$, then modes are the points $x \leqslant 0$ and if $m \geqslant 1$, then modes are the points $x < 0$ and $x = (a(m - 1)/m)^{1/m}$.

16. Multivariate Distributions. Functions of Random Variables

16.1. Obviously $p(i, j) \geqslant 0$ for $i, j = 1, \ldots, n$ and

$$\sum_{i=1}^{n} \sum_{j=1}^{n} p(i, j) = \left(\frac{2}{n(n + 1)}\right)^2 \sum_{i=1}^{n} i \sum_{j=1}^{n} j =$$

$$= \left(\frac{2}{n(n + 1)}\right)^2 \frac{n(n + 1)}{2} \frac{n(n + 1)}{2} = 1.$$

16.2. From the equality

$$\sum_{k=0}^{\infty} c \frac{\lambda^j \mu^k \nu^{jk}}{j!k!} = 1$$

we obtain

$$c^{-1} = \sum_{j=0}^{\infty} \sum_{k=0}^{\infty} \frac{\lambda^j \mu^k \nu^{jk}}{j!k!}.$$

We can verify that for $\lambda > 0$, $\mu > 0$ and $0 < \nu \leqslant 1$ this series converges. For the distribution of the r.v. ξ, we find

$$P\{\xi = j\} = \sum_{k=0}^{\infty} P\{\xi = j, \eta = k\} =$$

$$= c\lambda^j (j!)^{-1} \exp(\mu\nu^j), \quad j = 0, 1, 2, \ldots$$

Similarly we find $P\{\eta = k\} = c\mu^k (k!)^{-1} \exp(\lambda\nu^k)$. For $\nu = 1$ and $c = e^{-\lambda-\mu}$, we have $P\{\xi = j\} = \lambda^j e^{-\lambda}/j!$, $P\{\eta = k\} = \mu^k e^{-\mu}/k!$ and $P\{\xi = j, \eta = k\} = \lambda^j \mu^k (j!k!)^{-1} e^{-\lambda-\mu}$. Hence $P\{\xi = j, \eta = k\} = P\{\xi = j\} \cdot P\{\eta = k\}$ for $j, k = 0, 1, \ldots$; i.e., the r.v. ξ and η are independent. Vice versa, let ξ and η be independent. Then

$$c\lambda^j \mu^k \nu^{jk} (j!k!)^{-1} = c\lambda^j (j!)^{-1} \exp(\mu\nu^j) \cdot c\mu^k (k!)^{-1} \exp(\lambda\nu^k),$$

$j, k = 0, 1, \ldots$ In particular, for $j = 0$ and $k = 0$ we have $1 = ce^{\mu}e^{\lambda}$, i.e., $c^{-1} = e^{\lambda+\mu}$ and from the expression for c^{-1} we get

$$e^{\lambda+\mu} = \sum_{j=0}^{\infty} \sum_{k=0}^{\infty} \frac{\lambda^j \mu^k \nu^{jk}}{j!k!} \tag{16.12}$$

The equality (16.12) is fulfilled for $\nu = 1$ and since the right-hand side of (16.12) considered as a function of ν, is monotonically increasing and the left-hand side does not depend on ν, $\nu = 1$ is the only value satisfying (16.12). For the conditional probability $P\{\xi = j | \eta = k\}$ we find

$$P\{\xi = j | \eta = k\} = \frac{P\{\xi = j, \eta = k\}}{P\{\eta = k\}} = (\lambda\nu^k)^j (j!)^{-1} \exp(-\lambda\nu^k);$$

i.e., the r.v. ξ given the event $\{\eta = k\}$ has a Poisson distribution with parameter $\lambda\nu^k$.

16.3. (a) $P\{\zeta = 0\} = p^2$, $P\{\zeta = k\} = P\{\max[\xi, \eta] = k\} =$
$\sum_{j=0}^{k} P\{\xi = k, \eta = j\} + \sum_{j=0}^{k-1} P\{\xi = j, \eta = k\} = 2pq^k - pq^2k - pq^{2k+1}$;

(b) $P\{\zeta = i, \xi = j\} = \begin{cases} 0, & \text{if } i < j \\ pq^i(1 - q^{i+1}), & \text{if } i = j \\ p^2 q^{i+j}, & \text{if } i > j. \end{cases}$

16.4. Parts (a), (b) and (c) are easily verified. For (d) we have

$$F(2, 2) - F(-1, 2) - F(2, -1) + F(-1, -1) = -1 < 0.$$

16.5. According to (16.7) we have:

$$f_\xi(x) = g(x) \int_{-\infty}^{\infty} h(y)dy,$$

$$f_\eta(y) = h(y)\int_{-\infty}^{\infty} g(x)dx, \qquad \text{for } x, y \in \mathbb{R}_1,$$

Further,

$$\int_{-\infty}^{\infty}\int_{-\infty}^{\infty} f\xi(x, y)dx\,dy = 1;$$

therefore,

$$\int_{-\infty}^{\infty} h(y)dy = 1/\int_{-\infty}^{\infty} g(x)dx, \qquad \int_{-\infty}^{\infty} g(x)dx = 1/\int_{-\infty}^{\infty} h(y)dy.$$

Substituting these normalizing constants in the expressions for $f_\xi(x)$ and $f_\eta(y)$, it is easily seen that the r.v.'s ξ and η are independent. This problem can be extended to the case of more than two random variables.

16.6. Use the solution of Exercise 16.5. (a) $c = 1/5$, the r.v.'s ξ_1, ξ_2, ξ_3 are independent; (b) $c = 3/(4\pi)$, the r.v.'s ξ_1, ξ_2, ξ_3 are independent; (c) $c = \dfrac{\Gamma(m + n + k + 1)}{\Gamma(m)\Gamma(n)\Gamma(k)}$, the three r.v.'s are independent.

16.7. Let $g(x)$, $x \in \mathbb{R}_1$, be a strictly increasing function. Then

$$P\{\xi < x, \eta < y\} = P\{g(\xi) < g(x), \eta < y\} = P\{\eta < g(x), \eta < y\},$$

since the events $\{\xi < x\}$ and $\{g(\xi) < g(x)\}$ coincide. But $P\{\eta < g(x), \eta < y\} = P\{\eta < y\}$ when $y \leqslant g(x)$ and $P\{\eta < g(x), \eta < y\} = P\{\eta < g(x)\} = P\{\xi < x\}$, when $y > g(x)$. Hence $F_{\vec\zeta}(x, y) = F_\eta(y)$ for $y \leqslant g(x)$ and $F_{\vec\zeta}(x, y) = F_\xi(x)$ for $y > g(x)$. If g is a strictly decreasing function, then $F_\zeta(x, y) = 0$ for $y \leqslant g(x)$ and $F_{\vec\zeta}(x, y) = F_\eta(y) - F_\xi(x + y)$ for $y > g(x)$.

16.8. The density (if it exists) of the random vector $\vec\zeta$ should coincide, according to (16.4), with $\dfrac{\partial^2 F_{\vec\zeta}(x, y)}{\partial x \partial y}$ almost everywhere. For the d.f. $F_{\vec\zeta}(x, y)$ we have:

$$\begin{aligned}
F_{\vec\zeta}(x, y) &= P\{\eta_1 < x, \eta_2 < y\} = \\
&= P\{(\xi_1 < x) \cap (\xi_2 < x) \cap ((\xi_1 < y) \cap (\xi_2 < y))\} = \\
&= P\{\xi_1 < x, \xi_2 < x, \xi_1 < y\} + \\
&+ P\{\xi_1 < x, \xi_2 < x, \xi_2 < y\} - \\
&- P\{\xi_1 < x, \xi_2 < x, \xi_1 < y, \xi_2 < y\}.
\end{aligned}$$

If we denote $E(x) = 1 - \exp(-\lambda x)$ and use the independence of ξ_1 and ξ_2, we find

$$F_{\vec{\zeta}}(x, y) = \begin{cases} 2E(x)E(y) - E^2(y), & \text{if } x \geqslant y \geqslant 0 \\ E^2(x), & \text{if } 0 \leqslant x < y \\ 0, & \text{otherwise.} \end{cases}$$

After some manipulations using the density g of $\vec{\zeta}$, we get

$$g(x, y) = \begin{cases} 2\lambda^2 \exp(-\lambda(x + y)), & \text{if } x \geqslant y \geqslant 0 \\ 0, & \text{otherwise.} \end{cases}$$

16.9. Using reasoning similar to Exercise 16.8, we get that

$$G(x, y) = \begin{cases} F(x, x), & \text{if } x < y \\ F(x, y) + F(y, x) - F(y, y), & \text{if } x \geqslant y. \end{cases}$$

16.10. From (16.9) it follows that the random vector $\vec{\zeta} = (\xi, \eta)$ has a density $f_{\vec{\zeta}}(x, y) = (2\pi\sigma_1\sigma_2)^{-1} \exp\left(-\frac{(x - a_1)^2}{2\sigma_1^2} - \frac{(y - a_2)^2}{2\sigma_2^2}\right)$, $(x, y) \in \mathbb{R}_2$.

If we introduce the r.v. $\Theta = \xi + \eta$ and consider the transformation $u = x + y$, $v = y$, from (16.11) we can find the densities $f_{\Theta,\eta}(u, v)$ and $f_\Theta(u) = (2\pi(\sigma_1^2 + \sigma_2^2))^{-1} \exp(-(u - a_1 - a_2)^2/(2(\sigma_1^2 + \sigma_2^2)))$, $u \in \mathbb{R}_1$. This implies that $\xi + \eta \in N(a_1 + a_2, \sigma_1^2 + \sigma_2^2)$. Another proof of this fact will be given in Section 18.

16.11. The given function g(z) is even which means that each of the events $\{\Theta > 0\}$ and $\{\Theta < 0\}$ has probability equal to 0.5. However, if ξ and η have mean values correspondingly $a_1 > 0$ and $a_2 > 0$, it is easily observed that $P\{\Theta > 0\} > \frac{1}{2}$. Therefore the true density f_Θ could not be of the type of the given function g.

16.12. The ratio ξ_1/ξ_2 is Cauchy-distributed.

16.13. The density of $\xi + \eta$ is $f_{\xi+\eta}(x) = 0$ for $x < 0$ and $f_{\xi+\eta}(x) = (\lambda^k \exp(-\lambda))/k!$ for $k \leqslant x < k + 1$, where $k = 0, 1, \ldots$

16.14. Obviously $\{\measuredangle\, 0\xi\eta \leqslant 45°\} = \{|\xi| \geqslant |\eta|\}$. We easily find that $P\{\measuredangle\, 0\xi\eta \leqslant 45°\} = P\{|\xi| \geqslant |\eta|\} = \frac{1}{2}$.

16.15. The events A_c and B_c are independent if $c \leqslant 0$, $c = \frac{1}{3}$ or $c \geqslant \frac{2}{3}$.

16.16. Use the transformation $u = x + y$, $v = x/(x + y)$. (a) $f_\eta(u) = u \cdot \exp(-u)$ for $u > 0$ and $f_\eta(u) = 0$ for $u \leqslant 0$. (b) $f_\zeta(v) = 1$ for $0 < v < 1$ and $f_\zeta(v) = 0$ for $v \,\bar{\in}\, (0, 1)$. (c) The r.v.'s η and ζ are independent since $f_{\eta,\zeta}(u, v) = u \cdot \exp(-u)$ if $u > 0$, $0 < v < 1$ and $f_{\eta,\zeta}(u, v) = 0$, otherwise; i.e., $f_{\eta,\zeta}(u, v) = f_\eta(u) \cdot f_\zeta(v)$.

16.17. The r.v.'s ξ and η are independent.

16.18. Using the transformation $u = x - b\cos y$, $v = y$ we can use (16.11). We have $x = u + b\cos v$, $y = v$, $J = 1$, $U = \mathbb{R}_2$. We find $f_{\xi,\eta}(x, y) = 2/(a\pi)$ for $0 < u + b\cos v < a$, $0 < v < \frac{\pi}{2}$ and $f_{\xi,\eta}(u, v) = 0$, otherwise. Using (16.11) after some calculations we obtain $f_{\zeta,\eta}(u, v) = 2/(a\pi)$ for $0 < u + b\cos v < a$, $0 < v < \pi/2$ and $f_{\zeta,\eta}(u, v) = 0$, otherwise with $\zeta = \xi - b\cos\eta$. Then

$$P\{\xi < b\cos\eta\} = P\{\xi - b\cos\eta < 0\} = P\{\zeta < 0\} = (2b)/(a\pi).$$

16.19. The density of $\vec{\zeta}$ is $f_{\vec{\zeta}}(u, v) = (2\pi)^{-1}\exp(-(u^2 + v^2)/2)$, $(u, v) \in \mathbb{R}_2$; hence η_1 and η_2 are independent.

16.20. The density of the random vector (a, b, c) is $f(x, y, z) = 1$ for $0 < x < 1$, $0 < y < 1$, $0 < z < 1$ and $f(x, y, z) = 0$, otherwise. The probability that this equation has real roots is $P\{b^2 - 4ac \geqslant 0\}$. Consider the transformation $u = y^2 - 4xz$, $v = y$, $w = z$ and put $\xi_1 = u(a, b, c) = b^2 - 4ac$, $\xi_2 = v(a, b, c) = b$, $\xi_3 = w(a, b, c) = c$. Then from (16.11) after easy calculations for the density of the random vector $\vec{\eta} = (\xi_1, \xi_2, \xi_3)$ we obtain $f_{\vec{\eta}}(u, v, w) = (4w)^{-1}$ for $0 < \frac{v^2 - u}{4w} < 1$, $0 < v < 1$, $0 < w < 1$ and $f_{\vec{\eta}}(u, v, w) = 0$, otherwise. Hence we find

$$P\{b^2 - 4ac \geqslant 0\} = P\{\xi_1 \geqslant 0\} = \int_0^{\infty}\int_{-\infty}^{\infty}\int_{-\infty}^{\infty} f_{\vec{\eta}}\, du\, dw\, dv = \frac{5 + 6\ln 2}{36}.$$

16.21. The roots of the given equation are real and different only when $b^2 - c > 0$. Denote $p = P\{b^2 - c > 0\}$ and $D = \{(x, y) : x > 0, y > 0, x^2 - y > 0\}$. Since the joint density of b and c is $\lambda^2 e^{-\lambda(x+y)}$, then $p = \iint_D \lambda^2 e^{-\lambda(x+y)}\, dx\, dy$; hence we find

$$p = 1 - \sqrt{\pi\lambda}\, e^{\lambda/4}(1 - \Phi(\sqrt{\lambda/2})).$$

16.22. (a) Check conditions (a), (b), (c) and (d) from the Introductory Notes of Section 16 for the functions H_1, H_2 and H. For example, condition (d) for H_2 is:

$$H_2(b_1, b_2) - H_2(a_1, b_2) - H_2(b_1, a_2) + H_2(a_1, a_2) \geqslant 0$$

for every $a_1 \leqslant b_1$ and $a_2 \leqslant b_2$.

Let us consider the special case when $F(a_1) \leqslant G(a_2) \leqslant F(b_1) \leqslant G(b_2)$.

We have

$$H_2(b_1, b_2) - H_2(a_1, b_2) - H_2(b_1, a_2) + H_2(a_1, a_2) =$$

$$= \min[F(b_1), G(b_2)] - \min[F(a_1), G(b_2)] -$$

$$- \min[F(b_1), G(a_2)] + \min[F(a_1), G(a_2)] =$$

$$= F(b_1) - F(a_1) - G(a_2) + F(a_1) = F(b_1) - G(a_2) \geqslant 0.$$

Similarly one can check condition (d) for the rest of the cases as well. For the marginal distributions of $H_2(x, y)$, we have

$$H_2(x, \infty) = \min[F(x), G(\infty)] = \min[F(x), 1] = F(x);$$

similarly, $H_2(\infty, y) = G(y)$.

(b) Let the random vector (ξ, η) have d.f. $W(x, y)$, $(x, y) \in \mathbb{R}_2$, with marginal d.f.'s $F(x) = W(x, \infty)$ and $G(y) = W(\infty, y)$. We have $W(x, y) = P\{(\xi < x) \cap (\eta < y)\} \leqslant P\{\xi < x\} = F(x)$ and also $W(x, y) = P\{(\xi < x) \cap (\eta < y)\} \leqslant P\{\eta < y\} = G(y)$. Hence $W(x, y) \leqslant \min[F(x), G(y)] = H_2(x, y)$.

From $1 - W(x, y) = P\{(\xi \geqslant x) \cup (\eta \geqslant y)\} \leqslant P(\xi \geqslant x) + P(\eta \geqslant y) = 1 - F(x) + 1 - F(y)$ we obtain $W(x, y) \geqslant F(x) + F(y) - 1$, and combining the latter with the fact that $W(x, y) \geqslant 0$, we get

$$W(x, y) \geqslant \max[0, F(x) + F(y) - 1] = H_1(x, y).$$

Let us note that Frechet's minimal and maximal d.f.'s for the n-dimensional random vector $(\xi_1, \ldots, \xi_n)$ with d.f. $W(x_1, \ldots, x_n)$ and marginals $F_i(x_i)$, $i = 1, \ldots, n$, are

$$H_1(x_1, \ldots, x_n) = \max\left[0, \sum_{i=1}^{n} F_i(x_i) - n + 1\right]$$

and

$$H_2(x_1, \ldots, x_n) = \min[F_1(x_1), \ldots, F_n(x_n)];$$

i.e.,

$$H_1(x_1, \ldots, x_n) \leqslant W(x_1, \ldots, x_n) \leqslant$$

$$\leqslant H_2(x_1, \ldots, x_n), \quad (x_1, \ldots, x_n) \in \mathbb{R}_n.$$

The functions H_1 and H_2 are d.f.'s and if $F_i(x_i)$, $i = 1, \ldots, n$, are continuous, then H_1 and H_2 are singular d.f.'s with respect to the

Lebesgue measure in $\mathbb{R}_n$.

16.23. $f_{\xi+\eta}(y) = (f_\xi * f_\eta)(y) = (b_1 + b_2)(\pi((b_1 + b_2)^2 + (y - a_1 - a_2)^2))^{-1}$, $y \in \mathbb{R}_1$; i.e., the Cauchy-distributions form a closed class with respect to the operation "convolution" (see Example 16.1).

16.24. $f_\Theta(x) = 2\pi^{-2}(x^2 - 1)^{-1} \ln|x|$, $x \in \mathbb{R}_1$.

16.25. Using the independence of the ξ_i and the transformation $u_1 = x_1 + \ldots + x_n$, $u_2 = x_2$, $u_3 = x_3$, ..., $u_n = x_n$, we find: (a) $f_\eta(x) = \lambda^n(x - n\beta)^{n-1}((n - 1)!)^{-1} \exp(-\lambda(x - n\beta))$ for $x > n\beta$ and $f_\eta(x) = 0$ for $x \leqslant n\beta$; (b) $f_\eta(x) = 0$ for $x \leqslant 0$, $f_\eta(x) = (-1)^{n-1}\lambda_1 \ldots \lambda_n \sum_{k=1}^{n} \exp(-\lambda_k x)/\prod_{i\neq k}(\lambda_k - \lambda_i)$ for $x > 0$.

16.26. Using the transformation $u_i = \alpha_i x_i + \beta_i$, $i = 1, \ldots, n$, we find

$$f_{\vec\eta}(u_1, \ldots, u_n)|\alpha_1\alpha_2\ldots\alpha_n|^{-1} f_{\vec\eta}\left(\frac{u_1 - \beta_1}{\alpha_1}, \frac{u_2 - \beta_2}{\alpha_2}, \ldots, \frac{u_n - \beta_n}{\alpha_n}\right).$$

16.27. Let x and y satisfy the inequalities $\alpha \leqslant x \leqslant y \leqslant \beta$. The d.f. of the random vector $\vec\zeta$ is given by

$$F_{\vec\zeta}(x, y) = P\{\eta_1 < x, \eta_2 < y\} =$$

$$= P\left\{\left(\bigcup_{i=1}^{n} (\xi_i < x)\right) \cap \left(\bigcap_{i=1}^{n} (\xi_i < y)\right)\right\} =$$

$$= P\left\{\bigcap_{i=1}^{n} ((\xi_1 < y) \cap \ldots \cap (\xi_{i-1} < y) \cap (\xi_i < x) \cap (\xi_{i+1} < y) \cap \ldots \cap (\xi_n < y))\right\}.$$

Set $A_i = \{(\xi_1 < y) \cap \ldots \cap (\xi_{i-1} < y) \cap (\xi_i < x) \cap (\xi_{i+1} < y) \cap \ldots \cap (\xi_n < y)\}$ for $i = 1, \ldots, n$. Since $F_{\vec\zeta}(x, y) = P\left(\bigcup_{i=1}^{n} A_i\right)$, hence we can use (5.2). Because of the independence of the r.v.'s $\xi_1, \ldots, \xi_n$, we have $P(A_i) = (y - \alpha)^{n-1}(x - \alpha)/(\beta - \alpha)^n$, $i = 1, \ldots, n$; $P(A_iA_j) = (y - \alpha)^{n-2}(x - \alpha)^2/(\beta - \alpha)^n$ for $1 \leqslant i < j \leqslant n$; $P(A_iA_jA_k) = \dfrac{(y - \alpha)^{n-3}(x - \alpha)^3}{(\beta - \alpha)^n}$ for $1 \leqslant i < j < k \leqslant n$; ..., $P(A_1A_2 \ldots A_n) = (x - \alpha)^n/(\beta - \alpha)^n$.

Simplifying $F_{\vec\zeta}(x, y)$, we obtain $F_{\vec\zeta}(x, y) = \left(\dfrac{y - \alpha}{\beta - x}\right)^n - \left(\dfrac{y - x}{\beta - \alpha}\right)^n$ for

for $\alpha \leqslant x \leqslant y \leqslant \beta$. It is easily seen that

$$F_{\vec{\zeta}}(x, y) = \begin{cases} F_{\vec{\zeta}}(y, y), & \text{if } \alpha \leqslant y \leqslant x \leqslant \beta \\ 1 - ((\beta - x)(\beta - \alpha)^{-1})^n, & \text{if } \alpha \leqslant x \leqslant \beta < y \\ 1, & \text{if } \beta < x \leqslant y \\ 0, & \text{otherwise.} \end{cases}$$

Thus for the density of $\vec{\zeta}$, we find

$$f_{\vec{\zeta}}(x, y) = \begin{cases} n(n - 1)(y - x)^{n-2}/(\beta - \alpha)^n, & \text{if } \alpha < x \leqslant y < \beta \\ 0, & \text{otherwise.} \end{cases}$$

16.28. The function $f(x_1, \ldots, x_n) = \pi^{-n/2} \exp(-x_1^2 - \ldots - x_n^2)$ is the density of the n-dimensional random vector $(\xi_1, \ldots, \xi_n)$, where ξ_i, $i = 1, \ldots, n$, are independent and each has distribution $N(0, \frac{1}{2})$. Let $\tilde{\xi}_k = \xi_k/\sqrt{k}$, $\tilde{S}_n = \tilde{\xi}_1 + \ldots + \tilde{\xi}_n$; then $\tilde{\xi}_k \in N(0, \frac{1}{2k})$, $S_n \in N(0, B_n^2)$, where $B_n^2 = \frac{1}{2}(1 + \frac{1}{2} + \ldots + \frac{1}{n})$. It is easily seen that $I_n(x) = P\{|\tilde{S}_n| \leqslant x\}$. If $\Theta \in N(0, 1)$, the r.v.'s $\tilde{S}_n$ and $B_n \cdot \Theta$ have the same distribution; since $B_n \to 0$ when $n \to \infty$, we see that $I_n(x) \to 0$ for every $x > 0$.

16.29. (a) We have

$$P\{\xi_i < x\} = P\{\xi_1 < \infty, \ldots, \xi_{i-1} < \infty, \xi_i < x, \xi_{i+1} < \infty, \ldots, \xi_n < \infty\} =$$

$$= \frac{1}{n!} F(x)(F(\infty) + 1) \ldots (F(\infty) + n - 1) = F(x).$$

(b) Show that

$$P\{\xi_1 < x_1, \ldots, \xi_n < x_n\} = \frac{1}{n!} F(x_1)F(x_2) \ldots F(x_n) +$$

$$+ (1 - \frac{1}{n!})G(x_1, \ldots, x_n),$$

where the d.f. $G(x_1, \ldots, x_n)$, $(x_1, \ldots, x_n) \in \mathbb{R}_n$, is such that $\int \ldots \int_{\mathbb{R}_n} 1_B d\, G = 0$ for $B = \{(x_1, \ldots, x_n) : x_1 < x_2 < \ldots < x_n\}$. Further we find

$$P(\xi_1 < \xi_2 < \ldots < \xi_n) = \int \ldots \int_{\mathbb{R}_n} 1_A d\, P\{\xi_1 < x_1, \ldots, \xi_n < x_n\} =$$

$$= \frac{1}{n!} \int \ldots \int_{\mathbb{R}_n} 1_A d(F(x_1) \ldots F(x_n)) +$$

$$+ (1 - \frac{1}{n!}) \int_{\mathbb{R}_n} \cdots \int 1_A d\, G(x_1, \ldots, x_n) =$$

$$= \frac{1}{n!} \cdot \frac{1}{n!} + (1 - \frac{1}{n!}) \cdot 0 = (\frac{1}{n!})^2.$$

This distribution arises in the construction of a negative binomial point process considered in the following paper:

Z. Ignatov and G. Chobanov (1980): Construction of a negative binomial process. In: *Mathematics and education in mathematics*. Vol. 9. (Proc. Spring Conf. Union Bulg. Math.), pp. 138-141. Academia, Sofia.

16.30. From the solution of Example 16.2(a) and the fact that the r.v.'s ξ and η are independent, it follows that

$$f_{\xi,\eta}(x, y) = \begin{cases} \dfrac{\exp(-\frac{1}{2}(x + y))}{4\Gamma(\frac{1}{2} m)\Gamma(\frac{1}{2} n)} (\frac{1}{2} x)^{(m-2)/2} (\frac{1}{2} y)^{(n-2)/2}, & \text{if } \min[x, y] > 0 \\ 0, & \text{if } \min[x, y] \leqslant 0. \end{cases}$$

Consider the transformation $u = bx/(my)$, $v = y$. The Jacobian $J = mv/n$ and although it vanishes over the set $\{(u, v) : v = 0\}$, formula (16.11) is still valid. From the latter formula we obtain

$$f_{\xi,\eta}(u, v) = \begin{cases} \dfrac{(muv)^{(m-2)/2} \exp(-\frac{1}{2} v(1 + mu/n))}{2^{(m+1)/2}\Gamma(\frac{1}{2} m)\Gamma(\frac{1}{2} n)} v^{(n-2)/2} (mv/n), & \text{if } \min[u, v] > 0 \\ 0, & \text{if } \min[u, v] \leqslant 0. \end{cases}$$

Further from (16.7) we find

$$f_\zeta(u) = \begin{cases} \dfrac{(mu)^{m/2}}{u(2n)^{m/2} 2^{n/2}\Gamma(\frac{1}{2} m)\Gamma(\frac{1}{2} n)} \displaystyle\int_0^\infty v^{(m+n-2)/2} \exp(-\frac{v}{2}(1 + \frac{mu}{n}))d\, v, & \text{if } u > 0. \\ 0, & \text{if } u \leqslant 0. \end{cases}$$

Substituting $w = v(mu + n)/(2n)$ under the integral, we obtain

$$f_\zeta(u) = \begin{cases} \dfrac{m^{m/2} n^{n/2}}{\Gamma((m + n)/2)\Gamma(m/2)\Gamma(n/2)} u^{m/2-1} (mu + n)^{-(m+n)/2}, & \text{if } u > 0, \\ 0, & \text{otherwise.} \end{cases}$$

16.31. Put $\theta = \sqrt{\eta/n}$. From the independence of ξ and η it follows that ξ and Θ are independent; i.e., $f_{\xi,\Theta}(x, y) = f_\xi(x) f_\Theta(y)$. From the solution of Example 16.2 (b) and formula (16.11) with the transformation $u = x/y$, $v = y$, we get

$$f_{\zeta,\Theta}(u, v) = \begin{cases} 2(n/2)^{n/2}(\sqrt{2\pi}\Gamma(n/2))^{-1}\exp(-\frac{1}{2}v^2(u^2 + n^2)), & \text{if } v > 0,\ u \in \mathbb{R}_1 \\ 0, & \text{otherwise.} \end{cases}$$

From (16.17) we find

$$f_\zeta(u) = \Gamma((n+1)/2)(\sqrt{\pi n}\Gamma(n/2))^{-1}(1 + u^2/n)^{-(n+1)/2}, \quad u \in \mathbb{R}_1.$$

16.32. First method of solution: Use formula (16.13) from the solution of Example 16.2.

Second method of solution: Let the r.v.'s $\eta_1, \eta_2, \ldots, \eta_n, \eta_{n+1}$ be independent and $\eta_i \in N(0, 1)$, $i = 1, \ldots, n+1$. Then, according to Example 16.2 and Exercise 16.31, the r.v. $\zeta = \eta_1/\sqrt{(\eta_2^2 + \ldots + \eta_{n+1}^2)/n}$ has a t-distribution with n degrees of freedom as does the r.v. ξ.

Hence ζ^2 and ξ^2 are also identically distributed. But $\zeta^2 = \eta_1^2/[(\eta_2^2 + \ldots + \eta_{n+1}^2)/n]$; i.e., (see Exercise 16.30) ζ^2 is F-distributed with (1, n) degrees of freedom.

16.33. Use (16.13) from the solution of Example 16.2. The r.v. ξ^2 has a F-distribution with parameters (1, 1).

16.34. (a) The statement follows from the equality $F_{\xi,\eta}(x, y) = F_\xi(x)F_\eta(y)$ and from the formula (16.2).

(b) Use the fact that $f_{\xi,\eta}(x, y) = f_\xi(x)f_\eta(y)$ and formula (16.6).

(c) $$(f_\xi * f_\eta)(y) = \begin{cases} \alpha\beta(\beta - \alpha)^{-1}(\exp(-\alpha y) - \exp(-\beta y)), & \text{if } y > 0 \\ 0, & \text{if } y \leq 0. \end{cases}$$

(d) We find that

$$(f_\xi^{*n})(y) = \begin{cases} \lambda^n y^{n-1}((n-1)!)^{-1}\exp(-\lambda y), & \text{if } y > 0 \\ 0, & \text{if } y \leq 0; \end{cases}$$

i.e., (see (15.10)) the sum of n independent and identically exponentially distributed r.v.'s has gamma distribution.

(e) We have

$$f_\xi(x) = \begin{cases} 1, & \text{if } 0 < x < 1 \\ 0, & \text{if } x \,\overline{\in}\, (0, 1). \end{cases}$$

Let $\xi_1, \xi_2, \ldots, \xi_n$ be independent and identically distributed realizations of the r.v. ξ. Then, according to Exercise 16.34 (b), $f_{\xi_1+\ldots+\xi_n}(y) = (f_\xi^{*n})(y)$. Since $P\{0 \leq \xi_1 + \ldots + \xi_n \leq n\} = 1$, then $f_{\xi_1+\ldots+\xi_n}(y) = 0$, if $y \,\overline{\in}\, (0, n)$; hence, $(f_\xi^{*n})(y) = 0$, if $y \,\overline{\in}\, (0, n)$. On the other hand, $(f_\xi^{*n})(y) = (f_\xi^{*(n-1)} * f_\xi)(y) =$

$$= \int_{-\infty}^{+\infty} f_\xi(y - x)(f_\xi^{*(n-1)})(x)dx = \int_{y-1}^{y} (f_\xi^{*(n-1)})(x)dx.$$

For n = 2 we have

$$(f_\xi^{*2})(y) = \begin{cases} y, & \text{if } 0 < y < 1 \\ y - 2(y - 1), & \text{if } 1 \leqslant y < 2 \\ 0, & \text{if } y \,\overline{\in}\, (0, 2). \end{cases}$$

For n = 3 we find

$$(f_\xi^{*3})(y) = \begin{cases} \frac{y^2}{2}, & \text{if } 0 < y < 1 \\ \frac{y^2 - 3(y - 1)^2}{2}, & \text{if } 1 \leqslant y < 2 \\ \frac{y^2 - 3(y - 1)^2 + 3(y - 2)^2}{2}, & \text{if } 2 \leqslant y < 3 \\ 0, & \text{if } y \,\overline{\in}\, (0, 3) \end{cases}$$

By induction we find in the general case that

$$(f_\xi^{*n})(y) = \begin{cases} \frac{1}{(n - 1)!} y^{n-1} - \binom{n}{1}(y - 1)^{n-1} + \\ \qquad + \binom{n}{2}(y - 2)^{n-1} - \ldots, & \text{if } y \in (0, n) \\ 0, & \text{if } y \,\overline{\in}\, (0, n). \end{cases}$$

where the summation is taken as long as the quantities y, y - 1, y - 2, ... remain positive. The graph of the density $(f_\xi^{*2})(y)$ is given in Figure 16.1. Such a density is called a triangular distribution.

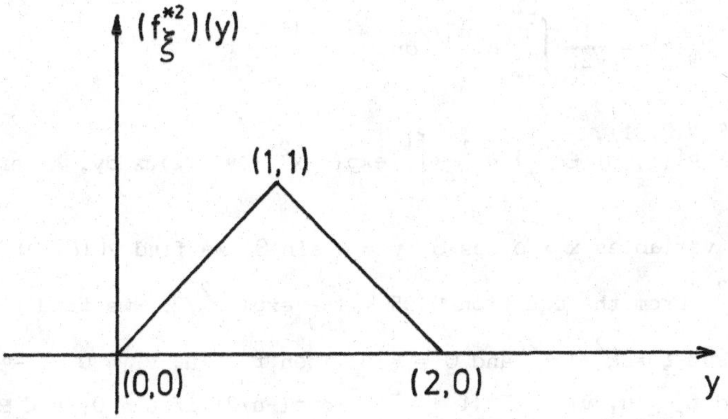

Fig. 16.1.

16.35. From Exercise 16.34 (a), we find that

$$F_\zeta(y) = F_{\xi+\eta}(y) = \int_{-\infty}^{\infty} F_\xi(y - x)\,dF_\eta(x) =$$

$$= \int_{-\infty}^{\infty}\left(\int_{-\infty}^{y-x} f_\xi(u)\,du\; dF_\eta(x) = \int_{-\infty}^{y}\left(\int_{\infty}^{\infty} f_\xi(u - x)\,dF_\eta(x)\right)du,\right.$$

where we have used Fubini's theorem (see Section 12). Hence the r.v. ζ is absolutely continuous with density

$$f_\zeta(u) = \int_{-\infty}^{\infty} f_\xi(u - x)\,dF_\eta(x), \qquad u \in \mathbb{R}_1.$$

From the properties of Lebesgue-Stieltjes integral it can be shown that if only the d.f. $F_\xi(x)$ is continuous and $F_\eta(x)$ is discrete, then $(F_\xi * F_\eta)(y)$ is a continuous function.

16.36. The density of the random vector (ξ, η, ζ) is

$$f(x, y, z) = (\sqrt{2\pi})^{-3}\exp(-(x^2 + y^2 + z^2)/2), \qquad (x, y, z) \in \mathbb{R}_3.$$

We define the set $D_t = \left\{(x, y, z) : \dfrac{x + yz}{\sqrt{1 + z^2}} < t\right\}$ in $\mathbb{R}_3$. Then

$$F_\Theta(t) = P\left\{\frac{\xi + \eta\zeta}{\sqrt{1 + \zeta^2}} < t\right\} = \iiint_{D_t} f(x, y, z)\,dx\,dy\,dz.$$

Changing variables we find

$$F_\Theta(t) = \frac{1}{\sqrt{2\pi}}\int_{-\infty}^{t} e^{-u^2/2}\,du;$$

hence, $\Theta \in N(0, 1)$.

16.37. (a) $P\{(\xi, \eta) \in C_r\} = \dfrac{1}{2\pi}\displaystyle\iint_{C_r} \exp(-(x^2 + y^2)/2)\,dx\,dy$. After the change of variables $x = \rho\cos\Theta$, $y = \rho\sin\Theta$, we find $P\{(\xi, \eta) \in C_r\} = 1 - e^{-r^2/2}$. From the equation $0.95 = 1 - \exp(-r^2/2)$, we find $r = \sqrt{2\ln 20} \approx 2.45$.

(b) Put $\zeta = \xi^2 + \eta^2$ and $\Theta = \xi/\eta$. Then $f_{\zeta,\Theta}(u, v) = 0$ if $-\infty < u < 0$, $v \in \mathbb{R}_1$ and $f_{\zeta,\Theta}(u, v) = (2\pi(1 + v^2))^{-1}\exp(-u/2)$ if $u \geqslant 0$, $v \in \mathbb{R}_1$; hence it follows that ζ and Θ are independent.

(c) $f_{\vec{\delta}}(u, v, w) = (\frac{1}{\sqrt{2\pi}})^3\exp(-\frac{1}{2}(u^2 + v^2 + w^2))$, $u, v, w \in \mathbb{R}_1$.

It follows that the r.v.'s α, β and γ are independent since

$$f_{\vec{\delta}}(u, v, w) = \frac{1}{\sqrt{2\pi}} \exp(-\frac{1}{2} u^2) \frac{1}{\sqrt{2\pi}} \exp(-\frac{1}{2} v^2) \frac{1}{\sqrt{2\pi}} \exp(-\frac{1}{2} w^2),$$

$u, v, w \in \mathbb{R}_1$.

(d) $$f_{\vec{\Theta}}(u, v, w) = \begin{cases} (2\pi)^{-3/2} u^2 \cos v \exp(-\frac{1}{2} u^2), & \text{if } u > 0,\ v \in (-\frac{\pi}{2}, \frac{\pi}{2}),\ w \in (0, 2\pi) \\ 0, & \text{otherwise.} \end{cases}$$

Thus the r.v.'s ρ, ϕ and ψ are independent since

$$f_{\vec{\Theta}}(u, v, w) = f_\rho(u) f_\phi(v) f_\psi(w), \quad u, v, w \in \mathbb{R}_1,$$

where

$$f_\rho(u) = \begin{cases} (\frac{2}{\pi})^{1/2} u^2 \exp(-\frac{1}{2} u^2), & \text{if } u > 0 \\ 0, & \text{if } u \leqslant 0, \end{cases}$$

$$f_\phi(v) = \begin{cases} \frac{1}{2} \cos v, & \text{if } v \in (-\frac{\pi}{2}, \frac{\pi}{2}) \\ 0, & \text{if } v \overline{\in} (-\frac{\pi}{2}, \frac{\pi}{2}), \end{cases}$$

$$f_\psi(w) = \begin{cases} \frac{1}{2\pi}, & \text{if } w \in (0, 2\pi) \\ 0, & \text{if } w \overline{\in} (0, 2\pi). \end{cases}$$

16.38. (a) Since $g(x, y)$ is a positive definite, then $A > 0$, $C > 0$, $AC - B^2 > 0$ and $D \exp(-g(x, y))$ is a density of a two-dimensional normal distribution. Comparing with (16.12), we get $A_{11} = 2A$, $A_{12} = A_{21} = 2B$, $A_{22} = 2C$. Hence,

$$\begin{aligned} D &= (2\pi)^{-2/2} (\det \Sigma)^{-1/2} = (2\pi)^{-1}\left((\det \Sigma)^{-1}\right)^{1/2} = \\ &= (2\pi)^{-1}\left(\det(\Sigma^{-1})\right)^{1/2} = \left(2\pi)^{-1} (2A \cdot 2C - (2B)^2\right)^{1/2} = \\ &= \pi^{-1} (AC - B^2)^{1/2}. \end{aligned}$$

(b) $$f_\xi(x) = \left((AC - B^2)/(\pi C)\right)^{1/2} \exp(-(AC - B^2)(x - a)^2/C),\ x \in \mathbb{R}_1,$$

$$f_\eta(y) = \left((AC - B^2)/(\pi A)\right)^{1/2} \exp(-(AC - B^2)(y - b)^2/A),\ y \in \mathbb{R}_1.$$

Obviously $\xi \in N(a, \sigma_1^2)$ and $\eta \in N(b, \sigma_2^2)$, where $\sigma_1^2 = \dfrac{C}{2(AC - B^2)}$ and

$\sigma_2^2 = \dfrac{A}{2(AC - B^2)}$. Put $r = (1 - \sqrt{1 + 16B^2\sigma_1^2\sigma_2^2})/(4B\sigma_1\sigma_2)$ for $B \neq 0$ and $r = 0$ for $B = 0$ and express A, B and C in terms of σ_1, σ_2 and r. Then for $-1 < r < 1$ we find

$$f_{\xi,\eta}(x, y) = \frac{1}{2\pi\sigma_1\sigma_2\sqrt{1 - r^2}} \exp\left\{- \frac{1}{2(1 - r^2)} \left(\frac{(x - a)^2}{\sigma_1^2} - 2r\frac{(x - a)(y - b)}{\sigma_1\sigma_2} + \frac{(y - b)^2}{\sigma_2^2}\right)\right\}.$$

(c) $P\{\zeta \in E_k\} = 1 - \exp(-k^2)$.

16.39. Obviously $f_{\vec{\zeta}}(x, y)$ is not a two-dimensional normal density since it vanishes in the second and the fourth quadrant and $D \exp(-g(x, y)) > 0$ for $(x, y) \in \mathbb{R}_2$. It is directly calculated that $\xi \in N(0, 1)$ and $\eta \in N(0, 1)$.

16.40. From Example 16.1 one can easily obtain that for every $(u_1, \ldots, u_n) \in \mathbb{R}_n$ we have

$$f_{\vec{\eta}}(u_1, u_2, \ldots, u_n) = (2\pi)^{-n/2}\exp\left(- \frac{1}{2} u_1^2 - \frac{1}{2}(u_2 - u_1)^2 - \ldots - \frac{1}{2}(u_n - u_{n-1})^2\right)$$

16.41. Show that for arbitrary x and y the equation $g(x^2 + y^2) = Kg(x^2)g(y^2)$ holds, where K is a positive constant. Hence derive that the unique continuous solution of this functional equation is given by $g(t) = Ae^{Bt}$, where A and B are suitable constants.

16.42. (a) The event $\{\xi_{(k)} < x\}$ takes place if at least k of the r.v.'s $\xi_1, \ldots, \xi_n$ are less than x. Hence

$$P\{\xi_{(k)} < x\} = \begin{cases} 0, & \text{if } x \leqslant 0 \\ \sum_{j=k}^{n} \binom{n}{j} x^j (1 - x)^{n-j}, & \text{if } 0 < x < 1 \\ 1, & \text{if } x \geqslant 1. \end{cases}$$

Differentiating $P\{\xi_{(k)} < x\}$ we find

$$f_{\xi_{(k)}}(x) = \begin{cases} n\binom{n - 1}{k - 1} x^{k-1} (1 - x)^{n-k}, & \text{if } 0 \leqslant x \leqslant 1 \\ 0, & \text{otherwise.} \end{cases}$$

(b) Since $\eta = \xi_{(1)} + \dots + \xi_{(n)} = \xi_1 + \dots + \xi_n$, then the solution is given by Exercise 16.34 (e). If we let $(x - k)^+ = \max[x - k, 0]$, these answers could be written in the form:

$$F_\eta(x) = \frac{1}{n!} \sum_{k=0}^{n} (-1)^k \binom{n}{k} ((x - k)^+)^n, \quad x \in \mathbb{R}_1;$$

$$f_\eta(x) = \frac{1}{(n - 1)!} \sum_{k=0}^{n} (-1)^k \binom{n}{k} ((x - k)^+)^n, \quad x \in \mathbb{R}_1.$$

For these expressions, however, it is not evident that $f_\eta(x) = 0$ for $x \overline{\in} (0, n)$; $F_\eta(x) = 0$ for $x < 0$ and $F_\eta(x) = 1$ for $x > n$. The verification is left to the reader.

16.43. (a) Show that $P\{U_1 > t_1, \dots, U_{n+1} > t_{n+1}\} = ((1 - t_1 - t_2 - \dots - t_{n+1})^+)^n$ for $t_i \geqslant 0$, $i = 1, \dots, n + 1$. Therefore, the density

$$f_{(U_1,\dots,U_n)}(t_1, \dots, t_n) = (-1)^n \frac{\partial^n P\{U_1 > t_1, \dots, U_n > t_n\}}{\partial t_1 \dots \partial t_n} =$$

$$= (-1)^n \frac{\partial^n P\{U_1 > t_1, \dots, U_n > t_n, U_{n+1} > 0\}}{\partial t_1 \dots \partial t_n} =$$

$$= (-1)^n \frac{\partial^n ((1 - t_1 - \dots - t_n)^+)^n}{\partial t_1 \dots \partial t_n} =$$

$$= \begin{cases} n!, & \text{if } \sum_{i=1}^{n} t_i < 1,\ t_i > 0,\ i = 1, \dots, n \\ 0, & \text{otherwise.} \end{cases}$$

Let us note that the random vector $(U_1, \dots, U_{n+1})$ has a singular d.f. because of the fact that $(-1)^{n+1} \frac{\partial^{n+1}}{\partial t_1 \dots \partial t_{n+1}} ((1 - t_1 - \dots - t_n)^+)^n = 0$ almost everywhere. (b) If $f(x_1, \dots, x_n)$ is the joint density of the order statistics $\xi_{(1)}, \dots, \xi_{(n)}$, then $f(x_1, \dots, x_n) = n!$ for $0 < x_1 < \dots < x_n < 1$ and $f(x_1, \dots, x_n) = 0$, otherwise. We make a change of variables: $v_1 = x_1/x_2$, $v_2 = x_2/x_3$, $\dots$, $v_{n-1} = x_{n-1}/x_n$, $v_n = x_n$. The Jacobian of this transformation is $(v_2 v_3^2 \dots v_n^{n-1})^{-1}$. Then the density $g(v_1, \dots, v_n)$ of $(V_1, \dots, V_n)$ is given by $g(v_1, \dots, v_n) =$

$n!v_2v_3^2 \ldots v_n^{n-1}$, $0 < v_i < 1$, $i = 1, \ldots, n$. Thus it is easily seen that the r.v.'s $V_1, \ldots, V_n$ are independent and V_k has density kv_k^{k-1}, $0 < v_k < 1$, $k = 1, \ldots, n$.

16.44. Let $D(\nu_1, \ldots, \nu_n; \nu_{n+1})$ denote the Dirichlet distribution with parameters $\nu_1 > 0, \ldots, \nu_{n+1} > 0$. This is an n-dimensional distribution with density

$$f(x_1, \ldots, x_n) = \frac{\Gamma(\nu_1 + \ldots + \nu_{n+1})}{\Gamma(\nu_1) \ldots \Gamma(\nu_{n+1})} x_1^{\nu_1 - 1} \ldots x_n^{\nu_n - 1}(1 - x_1 - \ldots - x_n)^{\nu_{n+1}-1}$$

over the n-simplex $\{x_i \geq 0, i = 1, \ldots, n, x_1 + \ldots + x_n \leq 1\}$ and $f(x_1, \ldots, x_n) = 0$ otherwise. It can be shown that each of the random vectors $(\eta_1, \ldots, \eta_n)$ and $(\zeta_1, \ldots, \zeta_n)$ has distribution $D(1, 1, \ldots, 1; 1)$. (a) Use also the equality

$$P\left\{\sum_{i=1}^{n+1} \eta_i = 1\right\} = P\left\{\sum_{i=1}^{n+1} \zeta_i = 1\right\} = 1.$$

16.45. Consider the r.v. $\eta_i = F(\xi_i)$, $i = 1, \ldots, n$ and use the equalities $\eta_{(i)} = F(\xi_{(i)})$, $i = 1, \ldots, n$, in Exercise 15.18 and 16.42 (a). The answer is:

$$F_{\xi_{(k)}}(x) = \frac{n!}{(k-1)!(n-k)!}\int_0^{F(x)} y^{k-1}(1-y)^{n-k}dy, \quad x \in \mathbb{R}_1.$$

16.46. (a) From Exercise 16.45 we find that

$$F_{\xi_{(1)}}(x) = 1 - (1 - F(x))^n,$$

$$f_{\xi_{(1)}}(x) = n(1 - F(x))^{n-1}f(x), \quad x \in \mathbb{R}_1.$$

(b) $F_{\xi_{(n)}}(x) = (F(x))^n$, $\quad f_{\xi_{(n)}}(x) = n(F(x))^{n-1}f(x)$, $\quad x \in \mathbb{R}_1$.

(c) Reasoning analogous to that in Exercise 16.27 shows that:

$$F_{\vec{\eta}}(x, y) = \begin{cases} (F(y))^n - (F(y) - F(x))^n, & \text{if } -\infty < x \leqslant y < \infty \\ (F(y))^n, & \text{if } -\infty < y < x < \infty; \end{cases}$$

$$f_{\vec{\eta}}(x, y) = \begin{cases} n(n-1)(F(y) - F(x))^{n-2} f(x) f(y), & \text{if } -\infty < x \leqslant y < \infty \\ 0, & \text{if } y < x. \end{cases}$$

(d) $$F_\Theta(u) = \begin{cases} \int_{-\infty}^{\infty} n(F(u+x) - F(x))^{n-1} f(x)\,d\,x, & \text{if } u > 0 \\ 0, & \text{if } u \leqslant 0; \end{cases}$$

$$f_\Theta(u) = \begin{cases} \int_{-\infty}^{\infty} n(n-1)(F(u+x) - \\ \quad - F(x))^{n-2} f(u+x) f(x)\,d\,x, & \text{if } u > 0 \\ 0, & \text{if } u \leqslant 0. \end{cases}$$

16.47. (a) $(f^{*2})(y) = (1/4)(1 + |y|)\exp(-|y|)$, $y \in \mathbb{R}_1$; $(f^{*3})(y) = (1/16)(3 + 3|y| + y^2)\exp(-|y|)$, $y \in \mathbb{R}_1$.

(b) $(f^{*k})(y) = kb(\pi(k^2b^2 + y^2))^{-1}$, $y \in \mathbb{R}_1$, $k = 2, 3, 4, \ldots$

16.48. $P\{\eta = k\} = P\{\xi_1 \geqslant \xi_2 \geqslant \ldots \geqslant \xi_{k-1} < \xi_k\} = \dfrac{k-1}{k!}$, $k = 2, 3, \ldots$

16.49. The scheme of the proof is as follows: (a) According to Exercise 15.18 the r.v. $\eta_i = F_i(\xi_i)$; hence, $1 - \eta_i$, $i = 1, \ldots, n$, are independent and uniformly distributed over the interval (0, 1). (b) If Θ is uniformly distributed over (0, 1), then $\xi = -\ln \Theta \in E(1)$. (c) The sum of independent and exponentially distributed r.v.'s has a gamma distribution (see Exercise 16.25 for $\beta = 0$, $\lambda = 1$). (d) Use the relation between the gamma and the chi-square distribution.

16.50. $$f_{\eta_n}(x) = \frac{1}{\sqrt{n\pi}} \frac{\Gamma(\frac{n}{2})}{\Gamma(\frac{n-1}{2})} \left(1 - \frac{x^2}{n}\right)^{(n-3)/2}, \quad x \in \mathbb{R}_1.$$

16.51. (a) We use induction. If $n = 1$, obviously $f_1(x) = k\lambda \exp(-\lambda kx)$, $x > 0$. Assume $f_n(x)$ has the desired form. Since $S_{n+1} = S_n + \xi_{n+1}$ and S_n and ξ_{n+1} are independent, $f_{S_{n+1}}$ is a convolution of f_n and the density of ξ_{n+1}. After some transformations we find that $f_{S_{n+1}}$ is equal to f_{n+1}.

(b) Assuming that $S_n \in (u, u + d\,u)$ where $u < x$, then the event $\{S_{n+1} > x\}$ will occur only if $\{\xi_{n+1} > x - u\}$. This event has probability $\exp(-\lambda(n + k)(x - u))$. Hence,

$$P\{S_n \leqslant x < S_{n+1}\} = \int_0^x (\exp(-\lambda(n+k)(x-u))) f_n(u)\,du.$$

Denoting this integral by $I_n(x)$, we find that

$$I_n(x) = [\lambda(n + k)]^{-1} f_{n+1}(x) = \binom{n + k - 1}{n} e^{-\lambda kx}(1 - e^{-\lambda x})^n.$$

16.52. $P\{\eta = \xi_i\} = P\left\{\bigcap_{j=1}^{n} (\xi_i \geqslant \xi_j)\right\} =$

$$\int_0^{a_i} d\,x_i \int_0^{a_1\wedge x_i} \cdots \int_0^{a_n\wedge x_i} \frac{dx_1 \ldots dx_{i-1}\, dx_{i+1} \ldots dx_n}{a_1 \ldots a_n} =$$

$$\frac{1}{a_1 \ldots a_n}\int_0^{a_i} \prod_{j\neq i} \min[a_j, x_i]dx_i. \text{ (Here } a \wedge x = \min[a, x].)$$

Putting $a_0 = 0$ and integrating by parts we get

$$P\{\eta = \xi_i\} = \sum_{k=1}^{i} \int_{a_{k-1}}^{a_k} \frac{x_i^{n-k}\, dx_i}{a_k a_{k+1} \ldots a_n} =$$

$$= \sum_{k=1}^{i} \frac{a_k^{n-k+1} - a_{k-1}^{n-k+1}}{(n - k + 1)a_k a_{k+1} \ldots a_n}.$$

16.53. (a) Using Exercise 16.43 (d), we find that

$$P\{\Theta_t = n\} = P\{S_n \leqslant t, S_{n+1} > t\} = P\{S_n \leqslant t\} - P\{S_{n+1} \leqslant t\} =$$

$$= \frac{(\lambda t)^n}{n!} e^{-\lambda t}, \quad t > 0,\ n = 0, 1, \ldots$$

Since $\sum_{n=0}^{\infty} (\lambda t)^n e^{-\lambda t}/n! = 1$, $P\{\Theta_t = +\infty\} = 0$; i.e., Θ_t is a non-negative integer-valued r.v.

(b) For $x \leqslant 0$ it is obvious that $P\{\eta_t < x\} = 0$. For $x > 0$ we have

$$P\{\eta_t < x\} = P\left\{[\eta_t < x]\left(\bigcup_{k=0}^{\infty} [\Theta_k = k]\right)\right\} =$$

$$= \sum_{k=0}^{\infty} P\{\xi_{k+1} < x,\ S_k \leqslant t,\ S_{k+1} > t\} =$$

$$= \sum_{k=0}^{\infty} \int_{A_{k+1}}\!\!\cdots\!\int \lambda^{k+1}\exp[-\lambda(x_1 +..+ x_{k+1})]\,dx_1 \ldots dx_{k+1},$$

where $A_1 = \{x_1 : t < x_1 < x\}$ and for $k = 1, 2, \ldots$ we set $A_{k+1} = \{(x_1, \ldots, x_{k+1}) : x_1 + \ldots + x_k \leq t,\ x_1 + \ldots + x_{k+1} > t,\ x_{k+1} < x,\ x_1 > 0, \ldots, x_{k+1} > 0\}$.

If $x \leq t$, the set A_1 is empty. Therefore, the first integral vanishes, and after calculating the rest of the integrals and summing we get

$$P\{\eta_t \leq x\} = 1 - \exp(-\lambda x) - \lambda x \cdot \exp(-\lambda x).$$

Hence, we find the density of η_t:

$$f_{\eta_t}(x) = \begin{cases} \lambda^2 x \exp(-\lambda x), & \text{if } 0 < x \leq t \\ (1 + \lambda t)\,\lambda\exp(-\lambda x), & \text{if } x > t \\ 0, & \text{if } x \leq 0. \end{cases}$$

It is easy to see that $\lim_{t\to\infty} f_{\eta_t}(x) = \lambda^2 x \exp(-\lambda x)$, if $x > 0$ and $\lim_{t\to\infty} f_{\eta_t}(x) = 0$ if $x \leq 0$.

(c) $f_{\zeta_t}(x) = \lambda \exp(-\lambda x)$, if $x > 0$ and $f_{\zeta_t}(x) = 0$ if $x \leq 0$.

(d) $F_{\nu_t}(x) = 1$ if $x \geq t$, $F_{\nu_t}(x) = 0$ if $x \leq 0$ and $F_{\nu_t}(x) = 1 - \exp(-\lambda x)$ if $0 < x < t$. Let us note that the d.f. $F_{\nu_t}(x)$ has a jump of size $\exp(-\lambda t)$ at the point $x = t$; i.e., the r.v. ν_t is not absolutely continuous and, therefore, has no density. Finally we find that $\lim_{t\to\infty} F_{\nu_t}(x) = 1 - \exp(-\lambda x)$, if $x > 0$ and $\lim_{t\to\infty} F_{\nu_t}(x) = 0$, if $x \leq 0$.

17. Expectation, Variance and Moments of Higher Order

17.1. (a) $E\{|\xi - np|\} = \sum_{k=0}^{n} |k - np| \binom{n}{k} p^k q^{n-k}$, $q = 1 - p$. Put $m = [np + 1]$. Then $E\{|\xi - np|\} = -\sum_{k=0}^{m-1} (k - np)\binom{n}{p} p^k q^{n-k} + \sum_{k=m}^{n} (k - np)\binom{n}{k} p^k q^{n-k} = -\Sigma_1 + \Sigma_2$. For the first sum we have

$$-\Sigma_1 = \sum_{k=0}^{m-1} [(n - k)p - k(1 - p)]\binom{n}{k} p^k q^{n-k} =$$

$$= \sum_{k=0}^{m-1} n\binom{n-1}{k} p^{k+1} q^{n-k} - \sum_{k=0}^{m-1} n\binom{n-1}{k-1} p^k q^{n-k+1} =$$

$$= \sum_{k=0}^{m-1} n\binom{n-1}{k} p^{k+1} q^{n-k} - \sum_{j=0}^{m-2} n\binom{n-1}{j} p^{j+1} q^{n-j} =$$

$$= n\binom{n-1}{m-1} p^m q^{n-m+1} .$$

Similary for the second sum we find:

$$\Sigma_2 = n\binom{n-1}{m-1} p^m q^{n-m+1} .$$

Finally we get $E\{|\xi - np|\} = 2n\binom{n-1}{m-1} p^m (1-p)^{n-m+1}$.

(b) From (17.7) and the fact that ξ has variance $V(\xi) = npq$, we obtain

$$V\{|\xi - np|\} = E\{|\xi - np|^2\} - (E\{|\xi - np|\})^2 =$$

$$= np(1-p) - 4n^2\binom{n-1}{m-1}^2 p^{2m}(1-p)^{2(n-m+1)},$$

where $m = [np + 1]$.

17.2. Use the equality

$$\sum_{n=1}^{m} P\{\xi \geqslant n\} = \sum_{n=1}^{m} nP\{\xi = n\} + mP\{\xi \geqslant m + 1\}$$

and show that $E(\xi)$ exists only if $\lim_{m\to\infty}[mP\{\xi \geqslant m + 1\}] = 0$.

17.3. (a) The variable τ_c takes some of the values 1, 2, ..., and the events $\{\tau_c \geqslant n\}$ and $\{S_{n-1} \leqslant c\}$ coincide for $n = 2, 3, \ldots$ Then

$$P\{\tau_c = \infty\} = P\{\tau_c \geqslant n,\ n = 1, 2, \ldots\} = \lim_{n\to\infty} P\{\tau_c \geqslant n\} =$$

$$= \lim_{n\to\infty} P\{S_{n-1} \leqslant c\} =$$

$$= \lim_{n\to\infty}\left[\frac{1}{(n-1)!}\sum_{k=0}^{n-1} (-1)^k \binom{n-1}{k} ((c-k)^+)^n\right] = 0.$$

Hence it follows that τ_c is a r.v. with $P\{\tau_c < \infty\}$. According to Exercise 17.2 we have

$$E(\tau_c) = \sum_{n=1}^{\infty} P\{\tau_c \geqslant n\} = P\{\tau_c \geqslant 1\} + \sum_{n=2}^{\infty} P\{\tau_c \geqslant n\} =$$

$$= 1 + \sum_{n=2}^{\infty} P\{S_{n-1} \leqslant c\} = 1 + \sum_{j=1}^{\infty} P\{S_j \leqslant c\} =$$

$$= 1 + \sum_{j=1}^{\infty}\left[(j!)^{-1} \sum_{k=0}^{j} (-1)^k \binom{j}{k}((c-k)^+)^j\right] =$$

$$= 1 + \sum_{j=1}^{\infty} (c^j/j!) + \sum_{k=1}^{\infty} \sum_{j=k}^{\infty} \frac{(-1)^k}{j!} \binom{j}{k}((c-k)^+)^j =$$

$$= e^c + \sum_{k=1}^{\infty}\left[\frac{(-1)^k}{k!}((c-k)^+)^k \cdot \sum_{j=k}^{\infty} \frac{((c-k)^+)^{j-k}}{(j-k)!}\right] =$$

$$= e^c + \sum_{k=1}^{[c]} \frac{(-1)^k (c-k)^k}{k!} e^{c-k} = e^c \sum_{k=0}^{[c]} \frac{1}{k!} \left(\frac{k-c}{e}\right)^k.$$

(b) Proceed as in case (a).

17.4. According to the assumptions of this exercise, $p_0 + \sum_{k=1}^{\infty} p_k = 1$, $\sum_{k=1}^{\infty} kp_k = a$, $\sum_{k=1}^{\infty} k^2 p_k = b$. Since p_k, $k = 1, 2, \ldots$ form a geometric progression, $p_k = p_1 q^{k-1}$, $k = 1, 2, \ldots$, where q is a suitable constant, $0 < q < 1$. Substituting p_k and using the equalities

$$\sum_{k=1}^{\infty} q^{k-1} = \frac{1}{1-q}, \quad \sum_{k=1}^{\infty} kq^{k-1} = \frac{1}{(1-q)^2},$$

$$\sum_{k=1}^{\infty} k^2 q^{k-1} = \frac{2q}{(1-q)^3} + \frac{1}{(1-q)^2},$$

we find that $p_0 + p_1/(1-q) = 1$, $p_1(1-q)^2 = a$, $p_1[2q/(1-q)^3 + 1/(1-q)^2] = b$; hence $q = (b-a)/(b+a)$, $p_0 = 1 - 2a^2/(a+b)$, $p_1 = 4a^3/(a+b)^2$. Thus $p_0 = 1 - 2a^2/(a+b)$, $p_k = 4a^3(b-a)^k/[(b^2-a^2)(a+b)^k]$, $k = 1, 2, \ldots$ The conditions $p_k > 0$, $k = 0, 1, 2, \ldots$, are equivalent to the following ones: $a + b \geqslant 2a^2$, $b \geqslant a \geqslant 0$. Hence we see that it is not possible to have $a = 4$, $b = 25$.

17.5. Let n be fixed and let $U_1, U_2, \ldots, U_{n+1}$ be the spacings introduced in Exercise 16.43 for the order statistics $\xi_{1,n} \leqslant \ldots \leqslant \xi_{n,n}$. Put $x_0 = 0$, $x_i = \sum_{j=1}^{i} a_j$. Prove that η_n is distributed as $\sum_{i=0}^{n} x_i U_i$. The joint distribution of $U_1, \ldots, U_{n+1}$ is given in Exercise 16.43. The next step is to find the density function f_{η_n}:

$$f_{\eta_n}(x) = n \sum_{j=0}^{n} \frac{((x_j - x)^+)^{n-1}}{h'(x_j)},$$

where $h(x) = (x - x_0)(x - x_1) \ldots (x - x_n)$ and $(z)^+ = \max[0, z]$.

Note that the density f_{η_n} coincides with the so-called B-spline of order n with knots $x_0 < x_1 < \ldots < x_n$. For a precise definition and properties of B-splines we refer the reader to the following book:

Shumaker, L.L. (1981): *Spline functions. Basic theory*. John Wiley & Sons, New York.

Further, the probabilistic interpretation of B-splines is given in the paper:

Ignatov, Z and V. Kaishev (1985): *B-splines and linear combinations of uniform order statistics*. MRC Techn. Summary Report No. 2817, Univ. of Wisconsin, Madison.

(b) This relation combines the interpretation of f_{η_n} as a B-spline and the corresponding relation for the moments of B-splines.

(c) The recurrence relation follows directly if we take into account the so-called de Boor-Cox recurrence formula for B-splines.

(d) In the case of arbitrary $a_1, \ldots, a_n$, we have

$$\mu_{1,n} = \frac{1}{n+1} \sum_{j=1}^{n} x_j,$$

$$\mu_{2,n} = \frac{2}{(n+1)(n+2)}\left[\left(\sum_{i=1}^{n} x_i\right)^2 - \sum_{1 \leqslant i < j \leqslant n} x_i x_j\right].$$

If $a_1 = \ldots = a_n = 1$, $\mu_{1,n} = \frac{n}{2}$ and $\mu_{2,n} = \frac{n(3n+1)}{12}$. Obviously $\mu_{1,n}$ and $\mu_{2,n}$ can be calculated directly.

Note that the conclusions in case (b) and case (c) are based essentially on the paper of Ignatov and Kaishev quoted above.

17.6. From the equality $1 = \sum_{k=0}^{\infty} p_k = \frac{c}{2d(d+1)}$, we find $c = 2d(d+1)$. The second equation for c and d is $\frac{3}{2+d} = E(\xi) = \frac{c}{2(d+1)}$; hence $d = 1$, $c = 4$ and $d = -3$, $c = 12$. However, the second solution is not possible for our problem since p_1, p_2 and p_3 are not defined for these d and c. For the desired expectation we have $E(\xi) = 1$; however, the series $\sum_{k=1}^{\infty} k^2 p_k$ is divergent and therefore $E\{\xi^2\}$ does not exist.

17.7. $E\{S_n^3\} = na_3$, $E\{S_n^4\} = na_4 + 3n(n-1)a_2^2$.

17.8. If we denote η_n/n by $\tilde{\eta}_n$, then $\tilde{\eta}_n = \max[p_n, 1 - p_n]$, where $p_n = \mu_n/n$ is the relative frequency of success. Let $\delta_n = p_n - \frac{1}{2}$. Obviously $\tilde{\eta}_n = p_n$ if $\delta_n > 0$ and $\tilde{\eta}_n = 1 - p_n$ if $\delta_n < 0$. Then we find that $n^{-1}E\{\eta_n\} = \frac{1}{2} + E\{|\delta_n|\}$. If $p = \frac{1}{2}$, we have $E\{|\delta_n|\} = (E\{\delta_n^2\})^{1/2} = \sqrt{1/2n} \to 0$ for $n \to \infty$; hence for $p = \frac{1}{2}$ we have $\lim_{n\to\infty} n^{-1}E\{\eta_n\} = \frac{1}{2}$. It remains to estimate $E\{|\delta_n|\}$. Since $\delta_n = \frac{\mu_n}{n} - \frac{1}{2}$, then $E\{|\delta_n|\} = E\left\{\left|\frac{\mu_n}{n} - p + (p - \frac{1}{2})\right|\right\} \leqslant E\left\{\left|\frac{\mu_n}{n} - p\right|\right\} + \left|p - \frac{1}{2}\right|$. However, μ_n has $E(\mu_n) = np$, $V(\mu_n) = npq$ and from the inequality $E\{|\xi|\} \leqslant (E\{\xi^2\})^{1/2}$ it is easily seen that $E\left\{\left|\frac{\mu_n}{n} - p\right|\right\} \to 0$. Then if $p > \frac{1}{2}$, we have $E\{|\delta_n|\} \to p - \frac{1}{2}$. If $p < \frac{1}{2}$, $E\{|\delta_n|\} \to \frac{1}{2} - p$. Since $n^{-1}E\{\eta_n\} = \frac{1}{2} + E\{|\delta_n|\}$, the solution can be given as $\lim_{n\to\infty} n^{-1}E\{\eta_n\} = \max[p, 1 - p]$.

17.9. Let $u > 0$. We integrate the Stieltjes integral $\int_0^u x^\alpha d\,F(x)$ by parts to get:

$$\int_0^u x^\alpha dF(x) = -u^\alpha[1 - F(u)] + \alpha\int_0^u x^{\alpha-1}[1 - F(x)]dx. \tag{17.11}$$

The point 0 is included in the integral in the left-hand side of (17.11), while u is excluded. Recall that the d.f. F(x) is left-continuous for every $x \in \mathbb{R}_1$. Let $I_1(\alpha) = \int_0^\infty x^\alpha dF(x) < \infty$. Then

$$I_1(\alpha) = \int_0^u x^\alpha dF(x) + \int_{u^+}^\infty x^\alpha dF(x) \geqslant$$

$$\geqslant -u^\alpha[1 - F(u)] + \alpha\int_0^u x^{\alpha-1}[1 - F(x)]dx + u^\alpha[1 - F(u)] =$$

$$= \alpha\int_0^u x^{\alpha-1}[1 - F(x)]dx.$$

Passing to the limit in u, as $u \to \infty$, we obtain

$$I_2(\alpha) = \alpha\int_0^\infty x^{\alpha-1}[1 - F(x)]dx \leqslant I_1(\alpha). \tag{17.12}$$

Again from (17.11), $\int_0^u x^\alpha dF(x) \leqslant \alpha \int_0^u x^{\alpha-1}[1 - F(x)]dx \leqslant I_2(\alpha)$; hence,

$$I_1(\alpha) \leqslant I_2(\alpha). \tag{17.13}$$

From (17.12) and (17.13) it follows that $I_1(\alpha) = I_2(\alpha)$ for every $\alpha > 0$.

17.10. Use Exercise 17.9.

17.11. Since $\xi^2 + \eta^2 \leqslant (\xi + \eta)^2 + 2|\xi\eta|$ it suffices to prove that $|\xi\eta|$ is integrable. However ξ and η are independent and according to Fubini's theorem applied to the joint distribution of ξ and η, it is enough to show that $|\xi|$ and $|\eta|$ are integrable. From the assumption $E(|\xi|) = \infty$ and the inequality $|\xi| \leqslant |x| + |x + \xi|$, it follows that $E(|x + \xi|) = \infty$ for every $x \in \mathbb{R}_1$. Again by Fubini's theorem we obtain $E(|\eta + \xi|) = \infty$, which contradicts the assumption $E\{(\xi + \eta)^2\} < \infty$ in view of the inequality $E\{|\xi + \eta|\} \leqslant (E\{(\xi + \eta)^2\})^{1/2}$. Therefore $E(|\xi|) < \infty$ and analogously $E(|\eta|) < \infty$.

17.12. We set $a_{n+1} = \int_n^{n+1} xdF(x) + \int_{-(n+1)}^{-n} |x|dF(x)$ (the bounds of integration n and $-(n + 1)$ are included). Then, since $E(\xi)$ exists, $a_{n+1} \geqslant 0$ for $n = 0, 1, 2, \ldots$ and $\sum_{n=0}^{\infty} a_{n+1} = \int_{-\infty}^{\infty} |x|dF(x) < \infty$. Also

$$\int_n^{n+1} x^2 dF(x) + \int_{-(n+1)}^{-n} x^2 dF(x) \leqslant (n + 1) \cdot a_{n+1}$$

and hence

$$\sum_{n=1}^{N} (1/n^2)\int_{-n}^{n} x^2 d\,F(x) \leqslant \sum_{n=1}^{N} \left\{\frac{1}{n^2} \sum_{k=1}^{n} ka_k\right\} = \sum_{k=1}^{N} \left\{ka_k \sum_{n=k}^{N} \frac{1}{n^2}\right\} \leqslant$$

$$\leqslant \sum_{k=1}^{N} \left\{ka_k \sum_{n=k}^{\infty} \frac{1}{n^2}\right\} \leqslant$$

$$\leqslant \sum_{k=2}^{\infty} \left\{ka_k \sum_{n=k}^{\infty} \frac{1}{n(n - 1)}\right\} + a_1 \sum_{n=1}^{N} \frac{1}{n^2} =$$

$$= \sum_{k=2}^{N} \frac{k}{k - 1} a_k + a_1 \sum_{n=1}^{\infty} \frac{1}{n^2} \leqslant$$

$$\leqslant 2 \sum_{k=2}^{\infty} a_k + 2a_1 \leqslant 2 \sum_{k=1}^{\infty} a_k \leqslant \infty.$$

Thus the desired inequality easily follows.

17.13. Let $E(\xi)$ exist. According to Exercise 17.9, for $\alpha = 1$ we have

$$\int_0^\infty x dF(x) = \int_0^\infty [1 - F(x)]dx < \infty.$$

We shall assume that $F(x) < 1$ for every $x \in \mathbb{R}_1$, since in the opposite case the condition $\lim_{n\to\infty} x(1 - F(x)) = 0$ is directly verified. Using the equality

$$\int_0^\infty (1 - F(x))dx = \int_0^1 F^{-1}(y)dy,$$

where $F^{-1}(y) = \sup_x \{x : F(x) < y\}$ we obtain that

$$\infty > \int_0^1 F^{-1}(y)dy = \int_0^{1-\varepsilon} F^{-1}(y)dy + \int_{1-\varepsilon}^1 F^{-1}(y)dy$$

for every $0 < \varepsilon < 1$; hence,

$$\lim_{\varepsilon\downarrow 0} \int_{1-\varepsilon}^1 F^{-1}(y)dy = 0.$$

Put $x(\varepsilon) = F^{-1}(1 - \varepsilon)$. Taking into account that $\lim_{\varepsilon\downarrow 0} x(\varepsilon) = \infty$, $F(F^{-1}(1 - \varepsilon)) \leqslant 1 - \varepsilon$ and the inequalities

$$\int_{1-\varepsilon}^1 F^{-1}(y)dy \geqslant \varepsilon \cdot F^{-1}(1 - \varepsilon) \geqslant (1 - F(x(\varepsilon)))x(\varepsilon)$$

we easily find that

$$\lim_{x\to\infty} x(1 - F(x)) = 0.$$

To show that the condition is not sufficient we consider the d.f. $F(x) = 0$ for $x \leqslant 1$, $F(x) = 1 - \frac{1}{kx}$ for $e^{k-1} < x \leqslant e^k$, $k = 1, 2, \ldots$ It is directly verified that $\lim_{x\to\infty} x[1 - F(x)] = 0$, while $\int_0^\infty [1 - F(x)]d\,x = \infty$.

17.14. From formula (17.9) for $k = 1$ and from formula (17.10) we find: (a) $E(\xi) = (a + b)/2$, $V(\xi) = (b - a)^2/12$; (b) $E(\xi) = a$, $V(\xi) = \sigma^2$; (c) $E(\xi) = 1/\lambda$, $V(\xi) = 1/\lambda^2$; (d) $E(\xi) = \frac{\Gamma(\alpha + 1)}{\beta\Gamma(\alpha)} = \frac{\alpha}{\beta}$, $E\{\xi^2\} = \frac{(\alpha + 1)\alpha}{\beta^2}$, $V(\xi) = \frac{\alpha}{\beta^2}$; (e) $E(\xi) = \frac{p}{p + q}$, $V(\xi) = \frac{pq}{(p + q)^2(p + q + 1)}$; (f) $E(\xi)$ and

$V(\xi)$ do not exist; (g) Since χ^2-distribution with n degrees of freedom is the distribution of the r.v. $\eta_1^2 + \ldots + \eta_n^2$, where η_i are independent and $\eta_i \in N(0, 1)$, $i = 1, \ldots, n$, then $E(\xi) = n$. Similarly $V(\xi) = 2n$, since $V\{\eta_i^2\} = E\{\eta_i^4\} - (E\{\eta_i^2\})^2 = 3 - 1 = 2$, $i = 1, \ldots, n$. (h) $E(\xi) = \frac{n}{n-2}$ for $n > 2$ and $E(\xi)$ does not exist for $n \leqslant 2$, $V(\xi) = \frac{2n^2(m+n-2)}{m(n-2)^2(n-4)}$ for $n > 4$ and $V(\xi)$ does not exist for $n \leqslant 4$; (i) $E(\xi) = 0$ for arbitrary n, $V(\xi) = \frac{n}{n-2}$ for $n > 2$ and $V(\xi)$ does not exist for $n \leqslant 2$.

17.15. $E\{\zeta^n\} = n![\lambda^{-n} + \mu^{-n} - (\lambda + \mu)^{-n}]$, $n = 0, 1, 2, \ldots$

17.16. Let

$$\zeta = \frac{n(\xi_{n+1}^2 + \xi_{n+2}^2 + \ldots + \xi_{n+m}^2)}{m(\xi_1^2 + \xi_2^2 + \ldots + \xi_n^2)} .$$

To find $E(\eta)$ and $V(\eta)$ use the equality $\eta = 1 + \frac{m}{n}\zeta$ and Exercises 16.29, 16.30 and 17.11.

(a) $E(\eta) = \frac{n + m - 2}{n - 2}$ for $n > 2$ and $E(\eta)$ does not exist for $n \leqslant 2$; $V(\eta)$ does not exist for $n \leqslant 4$ and $V(\eta) = \frac{2m(m + n - 2)}{(n-2)^2(n-4)}$ for $n > 4$.

(b) Show that the r.v. η has beta distribution with parameters $p = \frac{m}{2}$ and $q = \frac{n}{2}$; hence, $E(\eta) = \frac{m}{m+n}$, $V(\eta) = \frac{2nm}{(n+m)^2(n+m+2)}$.

17.17. We have $E\{\max[\xi_1, \xi_2]\} = E\{\max[\xi_1, \xi_2] - a\} + a =$

$$a + \int_{-\infty}^{\infty}\int_{-\infty}^{\infty} (\max[x_1, x_2] - a)(2\pi\sigma^2)^{-1}\exp\left[-\frac{(x_1 - a)^2 + (x_2 - a)^2}{2\sigma^2}\right] d\,x_1 d\,x_2 .$$

After representing this integral as a sum of two integrals over the regions $\{x_1 > x_2\}$ and $\{x_2 \geqslant x_1\}$, simple calculations yield $E\{\max[\xi_1, \xi_2]\} = a + \sigma/\sqrt{\pi}$.

17.18. $E\{|\xi - a|\} = \sigma\sqrt{2/\pi}$, $V\{|\xi - a|\} = \sigma^2(1 - 2/\pi)$.

17.19. (a) $E(\eta) = 1/4$; (b) $E\{\frac{\eta}{1-\eta}\} = \ln\frac{4}{e}$; (c) $E\{\frac{1-\eta}{\eta}\}$ does not exist.

17.20. We find $E(\xi_1) = 0$, $V(\xi_1) = 1$, $E(\xi_2) = 0$, $V(\xi_2) = E\{(\xi_1^2 - 1)^2\} = E\{\xi_1^4 - 2\xi_1^2 + 1\} = 3 - 2 + 1 = 2$. Then $\mathrm{cov}(\xi_1, \xi_2) = E\{\xi_1\xi_2\} = E\{\xi_1(\xi_1^2 - 1)\} = E\{\xi_1^3\} - E\{\xi_1\} = 0$ and $\rho(\xi_1, \xi_2) = \frac{1}{\sqrt{2}}\mathrm{cov}(\xi_1, \xi_2) = 0$. Here the interesting point is that ξ_1 has a normal distribution, $\rho(\xi_1, \xi_2) = 0$, but ξ_1 and ξ_2 are functionally dependent.

17.21. (a) $\rho = 0$; i.e., the r.v.'s η and ζ are non-correlated but they are dependent. (b) $\rho = 0$, but η and ζ are dependent.

17.22. The quantity $R(y_1, \ldots, y_n) = \max[y_1, \ldots, y_n] - \min[y_1, \ldots, y_n]$

is called the *range* of the set of the real numbers $\{y_1, \ldots, y_n\}$. For $n = 3$ it is easy to verify that the equality

$$R(y_1, y_2) + R(y_2, y_3) + R(y_3, y_1) = 2R(y_1, y_2, y_3) \qquad (17.14)$$

holds.

Now, we shall replace the real numbers y_i in (17.14) with the r.v.'s ξ_1, ξ_2, ξ_3. Then

$$R(\xi_1, \xi_2) + R(\xi_2, \xi_3) + R(\xi_3, \xi_1) = 2R(\xi_1, \xi_2, \xi_3),$$

with probability 1. From the assumption of the problem, each of the pairs (ξ_1, ξ_2), (ξ_2, ξ_3) and (ξ_3, ξ_1) has the same two-dimensional distribution. From this fact and from (17.14), we easily deduce that $3E\{R_2\} = 2E\{R_3\}$.

17.23. Use (17.4) and (17.5) to show that $E\{\zeta^4\} = 3(E\{\zeta^2\})^2$. From the conditions imposed on ξ two solutions are obtained, namely $p = (3 + \sqrt{3})/6$, $q = (3 - \sqrt{3})/6$ and $p = (3 - \sqrt{3})/6$, $q = (3 + \sqrt{3})/6$.

17.24. $E(\eta) = 1/a$ given $a \neq 0$; $V(\eta)$ does not exist.

17.25. (a) According to the notation introduced in the solution of Exercise 16.38, we have $E(\xi) = a$, $V(\xi) = \sigma_1^2$. (b) $E(\eta) = b$, $V(\eta) = \sigma_2^2$. (c) $\mathrm{cov}(\xi, \eta) = r\sigma_1\sigma_2$. (d) $\rho = r$. (e) $\rho = \cos \pi q$. (f) $\rho = r\sqrt{3/\pi}$.

17.26. (a) $E(\xi) = 0$, $V(\xi) = 1$. (b) $E(\eta) = 0$, $V(\eta) = 1$. (c) $\rho = 0$. (d) $\rho = 0$. (e) The r.v.'s ξ and η are dependent.

17.27. $\mathrm{cov}(\overline{\xi}_n, s_n^2) = 0$.

17.28. Show that for every $c \in \mathbb{R}_1$, we have

$$E\{|\xi - c|\} = E\{|\xi - m|\} + 2\int_c^m (x - c)f(x)\,dx.$$

17.29. Use the representation $E\{(\xi - c)^2\} = E\{(\xi - E(\xi))^2\} + (E(\xi) - c)^2$.

17.30. If f_ξ and f_η are the densities of the r.v.'s ξ and η, it is sufficient to establish the equality

$$\int_{-\infty}^{\infty} F_\xi(x)f_\eta(x)\,dx + \int_{-\infty}^{\infty} F_\eta(x)f_\xi(x)\,dx = 1,$$

which can be illustrated integrating by parts.

17.31. Show that the random vector $\vec{\zeta} = (\eta_1, \ldots, \eta_k)$ has density $f_{\vec{\zeta}}(x_1, \ldots, x_k) = k!$ for $x_1 > 0, \ldots, x_k > 0$, $x_1 + \ldots + x_k < 1$ and $f_{\vec{\zeta}}(x_1, \ldots, x_k) = 0$, otherwise. (a) For the covariance matrix we get $b_{ii} = \dfrac{k}{(k + 1)^2(k + 2)}$ for $i = 1, \ldots, k$, and $b_{ij} = \dfrac{1}{(k + 1)^2(k + 2)}$

for $i \neq j$, $i, j = 1, \ldots, k$. (b) $a_{r_1 \ldots r_k} = \dfrac{r_1!r_2! \ldots r_k!k!}{(r_1 + \ldots + r_k + k)!}$, where $r_1, \ldots, r_k$ are non-negative integers.

17.32. $f_\Theta(z) = (1 - r^2)^{1/2}\pi^{-1}/(1 - 2rz + z^2)$, $z \in \mathbb{R}_1$.

17.33. (a) $E\{\xi_1\xi_2\xi_3\xi_4\} = b_{12}b_{34} + b_{13}b_{24} + b_{14}b_{23}$. (b) The r.v. $\eta \in N\left(\sum_{i=1}^{n} c_i a_i, \sum_{i,j=1}^{n} b_{ij}c_ic_j\right)$ and according to Exercise 17.14 (b) we have $E(\eta) = \sum_{i=1}^{n} c_i a_i$, $V(\eta) = \sum_{i,j=1}^{n} b_{ij}c_ic_j$. (c) The r.v. Θ has gamma distribution with parameters $\alpha = n/2$ and $\beta = 1$ (see (15.10)) and according to Exercise 17.14, $E(\Theta) = \frac{1}{2}n$, $V(\Theta) = \frac{1}{2}n$.

17.34. (a) If r is odd, then $E\{\xi^r\} = 0$. If $r = 2k$ we easily show that $E\{\xi^{2k}\} = (2k - 1)E\{\xi^{2k-2}\}$; hence, $E\{\xi^{2k}\} = 1 \cdot 3 \cdot \ldots \cdot (2k - 1) = (2k - 1)!! = (2k)!/(2^k k!)$. Similarly it can be shown that

$$E\{|\xi^r|\} = \begin{cases} \sqrt{2/\pi}\,2^{(r-1)/2}\left(\frac{r-1}{2}\right)!, & \text{if } r \text{ is odd} \\ r!/(2^{r/2}(r/2)!), & \text{if } r \text{ is even.} \end{cases}$$

(b) $$E\{\xi^r\} = \frac{\Gamma(\alpha + r)}{\beta^r\Gamma(\alpha)}, \quad r = 1, 2, \ldots$$

(c) $$E\{\xi^r\} = \frac{p(p + 1) \ldots (p + r - 1)}{(p + q)(p + q + 1) \ldots (p + q + r - 1)}, \quad r = 1, 2, \ldots$$

(d) $$E\{\xi^r\} = \frac{\Gamma((n + 1)/2}{\sqrt{\pi n}\Gamma(n/1)} \int_{-\infty}^{\infty} x^r\left(1 + \frac{x^2}{n}\right)^{-(n+1)/2} dx.$$

The latter integral is absolutely convergent for $r < n$, and therefore only the moments $E\{\xi^r\}$ for $r = 1, 2, \ldots, n - 1$ exist. If r is odd and $r < n$, obviously $E\{\xi^r\} = 0$. For r even, $r = 2m < n$,

$$E\{\xi^{2m}\} = \frac{1 \cdot 3 \cdot 5 \cdot \ldots \cdot (2m - 1)n^m}{(n - 2)(n - 4) \ldots (n - 2m)}.$$

(e) $$a_{k_1,\ldots,k_n} = \frac{\Gamma(\nu_1 + k_1) \ldots \Gamma(\nu_n + k_n)\Gamma(\nu_1 + \ldots + \nu_{n+1})}{\Gamma(\nu_1) \ldots \Gamma(\nu_n)\Gamma(\nu_1 + \ldots + \nu_{n+1} + k_1 + \ldots + k_n)}.$$

17.35. The area of the triangle $\Delta A_1A_2A_3$ is equal to 1/2 (see Figure 17.1). Let $M = (p, q)$. For the coordinates of the points B_1, B_2 and B_3 we find $B_1 = \left(\frac{p}{p + q}, \frac{q}{p + q}\right)$, $B_2 = \left(0, \frac{q}{1 - p}\right)$, $B_3 = \left(\frac{p}{1 - q}, 0\right)$. Then for the mean value of the area of $\Delta A_1B_2B_3$ we obtain

$$S_1 = \int_{\substack{p>0,q>0\\p+q<1}} \frac{pq}{(1-p)(1-q)}\,dp\,dq =$$

$$= \int_0^1 \left(\frac{p}{1-p}\int_0^{1-p}\frac{q\,dq}{1-q}\right)dp = \frac{\pi^2}{6} - \frac{3}{2}\,.$$

Similarly, for the mean values S_2 and S_3 of the areas of $\Delta A_2B_1B_3$ and $\Delta A_3B_1B_2$, respectively, we get $S_2 = S_3 = \frac{\pi^2}{6} - \frac{3}{2}$. For the mean value S of the area of $\Delta B_1B_2B_3$, we have

$$S = \frac{1}{2} - 3\left(\frac{\pi^2}{6} - \frac{3}{2}\right) = \frac{10 - \pi^2}{2}\,.$$

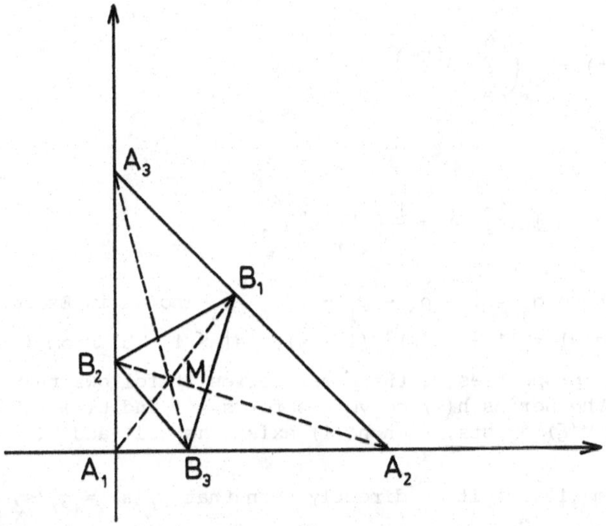

Figure 17.1

<u>17.36.</u> $m = \frac{9 - 5\beta}{2(\beta - 3)}$, $a = \sqrt{2\sigma^2\beta/(3 - \beta)}$, $b = 1/(aB(\frac{1}{2}, m + 1))$.

<u>17.37.</u> (a) If we put $k = \left(\int_0^\infty f(x)dx\right)^{-1}$, obviously, f will be a density of a certain r.v. which we denote by ξ. Using equality

$$\int_0^\infty x^n \exp(-\sqrt[4]{x})\sin\sqrt[4]{x}\,dx = 0\,,$$

which is valid for every integer $n \geq 0$ and the fact that f is a density, we conclude that the function g is also a density of some r.v., say η. Note that the proof of the above equality is based on the properties of the gamma function.

(b) It is obvious that $f \neq g$. Nevertheless, for every natural number n we have $E\{\xi^n\} = E\{\eta^n\}$.

18. Generating Functions and Characteristic Functions

18.1. Denote by η the number of points on the face of the first die and by ζ the number of points on the second one. Then $\xi = \eta + \zeta$ and for the ch.f. $\phi_\xi(t)$ of the r.v. ξ, we have

$$\phi_\xi(t) = E\{e^{it\xi}\} = E\{e^{it(\eta+\zeta)}\} = E\{e^{it\eta}\}E\{e^{it\zeta}\} = \phi_\eta(t)\phi_\zeta(t).$$

According to (18.2), $\phi_\eta(t) = \phi_\zeta(t) = \frac{1}{6}\sum_{k=1}^{6} e^{itk}$, and hence

$$\phi_\xi(t) = \frac{1}{36}\left(\sum_{k=1}^{6} e^{itk}\right)^2.$$

Then

$$g_\xi(s) = \phi_\xi\left(\frac{\ln s}{i}\right) = \frac{1}{36}\left(\sum_{k=1}^{6} s^k\right)^2.$$

18.2. (a) We have $q_j = 1 - p_0 - p_1 - \dots - p_j$ and as in Exercise 9.12 we find that $h(s) = [1 - g_\xi(s)]/(1 - s)$, $|s| < 1$. (b) From (a), (9.12), (9.13) and the properties of the power series it follows that if $E(\xi)$ exists, then the series h(s) converges for s = 1 and $h(1) = E(\xi)$. Similarly, if $V(\xi)$ exists, then $h'(1)$ exists as well and $V(\xi) = 2h'(1) + h(1) - h^2(1)$.

18.3. (a) From (18.10) it is directly seen that $g_\xi(s) = g_\zeta(s, 1)$ and $g_\eta(s) = g_\zeta(1, s)$. (b) $g_\zeta(s, s) = g_{\xi+\eta}(s)$. (c) Put $p_j = P\{\xi = j\}$, $q_k = P\{\eta = k\}$, j, k = 0, 1, ... If the r.v.'s ξ and η are independent, then $p_{jk} = p_j q_k$, and hence

$$g_\zeta(s_1, s_2) = \sum_{j=0}^{\infty}\sum_{k=0}^{\infty} p_j q_k s_1^j s_2^k = g_\xi(s_1)g_\eta(s_2), \quad |s_i| < 1,\ i = 1, 2.$$

Vice versa, let

$$\sum_{j=0}^{\infty}\sum_{k=0}^{\infty} p_{jk} s_1^j s_2^k = \left(\sum_{j=0}^{\infty} p_j s_1^j\right)\left(\sum_{k=0}^{\infty} q_k s_2^k\right), \quad |s_i| < 1,\ i = 1, 2.$$

Then the coefficients of the similar terms of these power series coincide; i.e., $p_{jk} = p_j q_k$ and we conclude that the r.v.'s ξ and η are independent.

18.4. (a) According to formula (18.1), $g(s) = [(1 - s^m)(m(1 - s))^{-1}]^n$.
(b) Using Exercise 9.11, we find $h(s) = (1 - s^m)^n/[m^n(1 - s)^{n+1}]$.

18.5. (a) $g(t) = e^{\lambda(t-1)}$. (b) $h(t) = e^{\lambda(e^t-1)}$. (c) $m(t) = e^{\lambda(e^t-1-t)}$.
(d) $k(t) = \lambda(e^t - 1 - t) = \lambda \sum_{r=2}^{\infty} t^r/r!$. (e) $k_r = \lambda$, $r = 2, 3, \ldots$

18.6. $\phi(t) = \sum_{k=0}^{\infty} e^{itk} P\{\xi = k\} = (\frac{a}{1 + a})^{\lambda}(1 - \frac{e^{it}}{1 + a})^{-\lambda}$, $E(\xi) = \frac{\lambda}{a}$,
$V(\xi) = \frac{\lambda(a + \lambda)}{a^2}$.

18.7. In Exercise 18.6 we put $p = a/(1 + a)$, $q = 1 - p = 1/(1 + a)$, $\lambda = r$. Then $\eta = \xi + r$, where the r.v. ξ is given in Exercise 18.6 and we can easily find that the ch.f. is $\phi_\eta(t) = [p/(1 - qe^{it})]^r$. We can then compute the p.g.f. $g(t)$ and the moment g.f. $h(t)$:

$$g(t) = p^r(1 - qt), \quad h(t) = p^r/(1 - qe^t).$$

18.8. From (18.2) we find:
(a) $\phi(t) = t^{-2}\sin t$; (b) $\phi(t) = 4c^{-2}t^2e^{iat}\sin^2(ct/2)$; (c) $\phi(t) = (e^{ibt} - e^{iat})(it(b - a))^{-1}$; (d) $\phi(t) = e^{iat}(1 + \sigma^2t^2)^{-1}$; (e) $\phi(t) = \frac{\Gamma(p + q)}{\Gamma(p)} \sum_{j=0}^{\infty} \frac{(it)^j\Gamma(p + j)}{\Gamma(p + q + j)\Gamma(j + 1)}$.

18.9. (a) $V\{\sin \xi\} + V\{\cos \xi\} = 1 - |\phi(1)|^2$. (b) $\phi_\zeta(t) = e^{it\beta}\phi(\alpha t)$.

18.10. (a) $\phi'(t)$ exists for $t = 0$. (b) $E(\xi)$ does not exist.

18.11. Let $h_i(t)$ be the moment g.f. of ξ_i. Then

$$h_1(t) = (q + pe^t)^n,$$

$$h_2(t) = e^{Nt/2}\, \frac{\sin h[(N + 1)t/2]}{(N + 1)t/2} \left(\frac{\sin h(t/2)}{t/2}\right)^{-1},$$

$$h_3(t) = \exp[at + \frac{1}{2}\sigma^2t^2], \quad h_4(t) = (\frac{\alpha}{\alpha - t})^{\beta},$$

$$h_5(t) = \frac{e^{tb} - e^{ta}}{(b - a)t}.$$

18.12. The function f is a density of the r.v. $\eta = \xi_1\xi_2$, where ξ_1 and ξ_2 are independent with ξ_1 being uniformly distributed over the interval (0, 1) and $\xi_2 \in E(1)$. For the moments of η we find: $E\{\eta^r\} = E\{\xi_1^r\xi_2^r\} = E\{\xi_1^r\}E\{\xi_2^r\} = r!/(r + 1)$, $r = 0, 1, \ldots$ Since (18.8) is valid for every

$|c| < 1$, it follows from (18.9) that $\phi_\eta(t) = -(it)^{-1}\text{Ln}(1 - it)$.

18.13. $k_2 = E\{(\xi - E(\xi))^2\}$, $k_3 = E\{(\xi - E(\xi))^3\}$, $k_4 = E\{(\xi - E(\xi))^4 - 3(E\{(\xi - E(\xi))^2\})^2$.

18.14. The r.v. $\xi_1 \in E(\lambda)$ has ch.f. $\phi(t) = \dfrac{-\lambda}{it - \lambda}$. We have

$$\phi_{S_\tau}(t) = E\left\{e^{itS_\tau}\right\} = \sum_{n=0}^{\infty} E\left\{e^{itS_\tau}|\tau = n\right\}P\{\tau = n\} =$$

$$= \sum_{n=0}^{\infty} [\phi_1(t)]^m qp^{n+1} = \frac{-\lambda q}{it - \lambda q}.$$

hence, $S_\tau \in E(\lambda q)$.

18.15. $\phi_\eta(t) = 1/(1 + \sigma^4 t^2)$.

18.16. The continuous distribution with density $\dfrac{1}{\pi} \cdot \dfrac{1 - \cos x}{x^2}$, $x \in \mathbb{R}_1$, corresponds to the function ϕ_1. The function ϕ_2 is a ch.f. of the r.v. ξ whose values are integers with

$$P\{\xi = 0\} = \frac{1}{2}, \quad P\{\xi = 2n + 1\} = \frac{2}{\pi^2}\frac{1}{(2n + 1)^2}, \; n = \pm 1, \pm 2, \ldots$$

18.17. (a) $\phi(t) = \frac{1}{4}(1 + e^{it})^2 = \frac{1}{4} + \frac{1}{2}e^{it} + \frac{1}{4}e^{2it}$. Hence $\phi(t)$ is a ch.f. of the r.v. taking values 0, 1 or 2 with probabilities 1/4, 1/2 and 1/4. (b) $\phi(t) = (2 - e^{it})^{-1} = \frac{1}{2}(1 - \frac{1}{2}e^{it})^{-1} = \frac{1}{2} + (\frac{1}{2})^2 e^{it} + (\frac{1}{2})^3 e^{2it} + \ldots$ As in case (a), we find $P\{\xi = k\} = 2^{-(k+1)}$, $k = 0, 1, \ldots$ (c) $\phi(t) = \cos t = \frac{1}{2}(e^{it} + e^{-it}) = \frac{1}{2}\exp[it(-1)] + \frac{1}{2}\exp[it \cdot 1]$, and hence $P\{\xi = \pm 1\} = \frac{1}{2}$. (d) $P\{\xi = -2\} = \frac{1}{4}$, $P\{\xi = 0\} = \frac{1}{2}$ and $P\{\xi = 2\} = \frac{1}{4}$. (e) $\phi(t) = \sum_{k=0}^{\infty} a_k \cos kt = \sum_{k=0}^{\infty} [(a_k/2)e^{itk} + (a_k/2)e^{it(-k)}]$; hence, $P\{\xi = 0\} = a_0$, $P\{\xi = k\} = \frac{1}{2}a_k$, $P\{\xi = -k\} = \frac{1}{2}a_k$, $k = 1, 2, \ldots$ (f) $\phi(t) = e^{-\lambda}[\exp(\lambda e^{it})] = e^{-\lambda}\sum_{k=0}^{\infty} \lambda^k e^{i\lambda k}/k!$; hence, the r.v. $\xi \in P(\lambda)$.

18.18. Let $G(x) = (E\{\xi^2\})^{-1}\int_{-\infty}^{x} y^2\, dF(y)$. Obviously $\lim_{x\to\infty} G(x) = 1$, $\lim_{x\to-\infty} G(x) = 0$ and $G(x)$, $x \in \mathbb{R}_1$, is monotonically increasing. Therefore, G is a d.f. Since $E\{\xi^2\} < \infty$, $\phi''(t)$ exists for every $t \in \mathbb{R}_1$ and $\phi''(t) = -\int_{-\infty}^{\infty} x^2 e^{itx} dF(x)$. On the other hand, the Fourier transformation of G is

$$\int_{-\infty}^{\infty} e^{itx} dG(x) = (E\{\xi^2\}^{-1} \int_{-\infty}^{\infty} x^2 e^{itx} dF(x) =$$

$$= -(E\{\xi^2\})^{-1}\left(- \int_{-\infty}^{\infty} x^2 e^{itx}\, dF(x)\right) =$$

$$= -(E\{\xi^2\})^{-1}\phi''(t) = \psi(t);$$

hence, $\phi(t)$, $t \in \mathbb{R}_1$ is the ch.f. of the d.f. $G(x)$, $x \in \mathbb{R}_1$.

18.19. (a) From (18.6) we find

$$F(x) - F(0) = \frac{1}{\pi} \int_0^{\infty} \frac{\sin tx}{t} e^{-a^2t^2}\, dt.$$

Differentiating with respect to x, we get

$$F'(x) = -\frac{1}{\pi} \int_0^{\infty} te^{-a^2t^2} \sin tx\, dt,$$

and after integrating by parts, $F'(x) = (-2a^2/x)F(x)$; hence,

$$F(x) = c_1 \int_{-\infty}^{x} e^{-u^2/4a^2}\, du + c_2.$$

where the constants c_1 and c_2 are determined from the conditions $\lim_{x\to\infty} F(x) = 1$ and $\lim_{x\to-\infty} F(x) = 0$. Finally

$$F(x) = (2a\sqrt{\pi})^{-1} \int_{-\infty}^{x} e^{-u^2/4a^2}\, du,$$

and hence F is a d.f. of the r.v. $\xi \in N(0, 2a^2)$. (b) Since $(1 - t^2)e^{-t^2/2} = -\frac{d^2}{dt^2}\left(e^{-t^2/2}\right)$, according to Exercise 18.18 for $a = 1/\sqrt{2}$, we obtain $\psi(t) = -(E\{\xi^2\})^{-1}\phi''(t)$, where ϕ is a ch.f. of $N(0, 1)$. Hence the d.f. we are looking for is

$$G(x) = (2\pi)^{-1/2} \int_{-\infty}^{x} y^2 e^{-y^2/2}\, dy.$$

18.20. (a) Find $\phi_{\xi_i}(t) = t^{-1} \sin t$, $i = 1, 2, \ldots$ From (18.5) it follows that $\phi_\eta(t) = (t^{-1} \sin t)^n$. (b) If we proceed as in the solution of

Exercise 18.19, for $F'_\eta(x)$, the density of η, we find:

$$F'_\eta(x) = \frac{1}{2\pi}\int_{-\infty}^{\infty} (t^{-1}\sin t)^n \cos(tx)\, dt.$$

Then

$$P\{-a < \eta < a\} = \int_{-a}^{a} dF\ (x) = \frac{1}{\pi}\int_{-\infty}^{\infty} t^{-n-1}\sin^n t \sin at\, dt.$$

The computation of this integral is difficult; however, we can find its value using Exercises 16.34 and 16.26:

$$P\{-a < \eta < a\} = P\{\eta < a\} - P\{\eta \leqslant -a\} =$$

$$= \frac{1}{n!}\sum_{k=0}^{n} (-1)^k \binom{n}{k}\left[\left(\left(\frac{a+n}{2} - k\right)^+\right)^n - \left(\left(\frac{n-a}{2} - k\right)^+\right)^n\right].$$

18.21. From Exercise 18.19 it follows that $\phi_{\xi_k}(t) = \exp(ita - \frac{1}{2}\sigma^2 t^2)$, $k = 1, 2, \ldots$ From (18.11) and the independence of ξ_1 and ξ_2 we find the ch.f. $\phi_\eta(t, s)$ of the random vector $\eta = (\xi_1 + \xi_2, \xi_1 - \xi_2)$ is given by $\phi_\eta(t, s) = \eta_{\xi_1}(t + s)\eta_{\xi_2}(t - s)$; however, $\phi_{\xi_1+\xi_2}(t) = \exp(2ita - \sigma^2 t^2)$, $\phi_{\xi_1-\xi_2}(s) = \exp(-\sigma^2 s^2)$, $\phi_\eta(t, x) = \exp(2ita - \sigma^2 t^2 - \sigma^2 s^2)$. Now comparing $\phi_\eta(t, s)$, $\phi_{\xi_1+\xi_2}(t)$ and $\phi_{\xi_1-\xi_2}(s)$, we conclude that the r.v.'s $\xi_1 + \xi_2$ and $\xi_1 - \xi_2$ are independent.

18.22. (a) $\phi_{\xi\eta}(t) = \int_{-\infty}^{\infty} \psi(t/x)f(x)\, dx$. (b) $\phi_{\xi\eta}(t) = \int_{-\infty}^{\infty} \phi(t/x)g(x)\, dx$.

18.23. $\phi_a(t) = [2\phi(t) - \phi(t + a) - \phi(t - a)][2(1 - \phi(a))]^{-1}$. It is interesting to note that $\lim_{a\to\infty} \phi_a(t) = \phi(t)$ for every $t \in \mathbb{R}_1$, while $f_a(x)$ does not converge to $f(x)$ as $a \to \infty$.

18.24. We find that

$$f_\xi(x) = \begin{cases} \frac{1}{2}, & \text{if } |x| \leqslant 1 \\ 0, & \text{if } |x| > 1, \end{cases} \qquad \phi_\xi(t) = t^{-1}\sin t,$$

$$f_\eta(y) = \begin{cases} \frac{1}{2}, & \text{if } |y| \leqslant 1 \\ 0, & \text{if } |y| > 0, \end{cases} \qquad \phi_\eta(t) = t^{-1} \sin t,$$

$$f_{\xi+\eta}(u) = \begin{cases} \frac{1}{4}(2 + 4), & \text{if } 2 \leqslant u \leqslant 0 \\ \frac{1}{4}(2 - u), & \text{if } 0 < u \leqslant 2 \\ 0, & \text{if } |u| > 2, \end{cases} \qquad \phi_{\xi+\eta}(t) = t^{-2} \sin^2 t.$$

18.25. For the ch.f. of the random vector $\vec{\zeta} = (\eta_1, \eta_2)$ we have

$$\phi_{\vec{\zeta}}(t, s) = \phi_1(a_{11}t)\phi_2(a_{12}t)\phi_1(a_{21}s)\phi_2(a_{22}s).$$

On the other hand, $\phi_{\vec{\zeta}}(t, s) = \phi_1(a_{11}t + a_{21}s)\phi_2(a_{12}t + a_{22}s)$; hence, ϕ_1 and ϕ_2 satisfy the functional equation

$$\phi_1(a_{11}t + a_{21}s)\phi_2(a_{12}t + a_{22}s) =$$

$$= \phi_1(a_{11}t)\phi_2(a_{12}t)\phi_1(a_{21}s)\phi_2(a_{22}s). \qquad (18.12)$$

Taking the logarithm of both sides of (18.12) and differentiating with respect to t and s, it is easy to show that all solutions $\phi_i(t)$, $i = 1, 2$, of (18.12), for which $|\phi_i(0)| = 1$, $|\phi_i(t)| \leqslant 1$, $t \in \mathbb{R}_1$, and $\phi_i(t)$ is an even function, are of the type $\phi_i(t) = \exp(-c_i t^2)$, $c_i > 0$, $i = 1, 2$. According to Exercise 18.19, ξ_1 and ξ_2 are normally distributed.

18.26. (a) $\phi(t, s) = \exp(-|t| - |s|)$; (b) $\phi_\xi(t) = \phi(t, 0) = \exp(-|t|)$.

18.27. (a) The r.v. ξ from Exercise 18.26 (b) has the same distribution as the r.v. $(\xi_1 - a)/b$. Hence $\phi_{\xi_1}(t) = \exp(ita - b|t|)$. For the ch.f. of η we have $\phi_\eta(t) = \prod_{i=1}^n \phi_{\xi_1}(\frac{t}{n}) = \exp(ita - b|t|)$. Hence, the r.v. η coincides in distribution with the r.v. ξ_1. (b) Use Exercise 18.19 (a) for $a = 1/\sqrt{2}$ and the equality $\phi_\zeta(t) = (\phi_{\xi_1}(t/\sqrt{n}))^n$. (c) Use Exercise 18.17 (f) and the equality

$$\prod_{i=1}^n \exp[\lambda_i(e^{it} - 1)] = \exp[(\lambda_1 + \dots + \lambda_n)(e^{it} - 1)].$$

18.28. (a) From (18.8) and (18.9) we find the ch.f. $\phi_\xi(t)$, and from (18.7) it follows that $f(x) = \delta x^{\delta-1}$ for $0 < x < 1$ and $f(x) = 0$ for $x \overline{\in} (0, 1)$. (b) Considerations similar to case (a) show that $f(x) = \frac{x^n e^{-x}}{n!}$ for $x > 0$ and $f(x) = 0$ for $x \leqslant 0$.

18.29. For the ch.f. of ξ we find $\phi_\xi(t) = (1 - \frac{it}{\beta})^{-\alpha}$. Since the χ^2-distribution with n degrees of freedom is a gamma distribution with parameters $\alpha = \frac{1}{2}n$ and $\beta = \frac{1}{2}$ (see Exercise 16.29), then $\phi_\eta(t) = (1 - 2it)^{-n/2}$.

18.30. Let $\phi_1(t)$ and $\phi_2(t)$ be ch.f.'s of ξ_1 and ξ_1^2, respectively. Then the r.v. ξ_1^2 has χ^2-distribution with one degree of freedom and from Exercise 18.29 we find $\phi_2(t) = (1 - 2it)^{-1/2}$. From the definition of η and the properties of the ch.f. it follows that

$$\phi_\eta(t) = E\{e^{it\eta}\} = (E\{\exp(it\xi_1\xi_2)\})^2, \quad \phi_1(t) = \exp(-t^2/2).$$

From the properties of the conditional expectation we have

$$E\{\exp(it\xi_1\xi_2)\} = E\{E[\exp(it\xi_1\xi_2)\,|\,\xi_2]\} = E\{\exp(-\xi_2^2t^2/2)\}.$$

But ξ_2^2 has a ch.f. $\phi_2(t) = (1 - 2it)^{-1/2}$. Then

$$E\{\exp(-\xi_2^2t^2/2)\} = E\{\exp[i\xi_2^2(it^2/2)]\} =$$

$$= \phi_2(it^2/2) = (1 + t^2)^{-1/2}.$$

Hence, $\phi(t) = 1/(1 + t^2)$, $t \in \mathbb{R}_1$, which is the ch.f. of the Laplace distribution with parameters $a = 0$ and $\sigma = 1$ (see Exercise 18.8).

18.31. From the assumptions of Exercise 18.31 it follows that the series

$$\phi_{\xi_1}(t_1) = \sum_{k=0}^{\infty} \frac{(it_1)^k E\{\xi^k\}}{k!}, \quad \phi_\eta(t_2) = \sum_{j=0}^{\infty} \frac{(it_2)^j E\{\xi^j\}}{j!}$$

are absolutely convergent for arbitrary complex values of t_1 and t_2.

For the ch.f. of the random vector $\vec{\zeta} = (\xi, \eta)$ we find

$$\phi_{\vec{\zeta}}(t_1, t_2) = \sum_{k=0}^{\infty}\sum_{j=0}^{\infty} \frac{(it_1)^k(it_2)^j E\{\xi^k\}E\{\xi^j\}}{k!j!} = \phi_\xi(t_1)\phi_\eta(t_2);$$

hence, it follows that ξ and η are independent.

18.32. (a) Since $\phi(t) = (\text{ch}t)^{-1} = 2(e^t + e^{-t})^{-1}$ is continuous, bounded and integrable over $\mathbb{R}_1$, we use the Bochner criterion to verify that ϕ is ch.f. Obviously $\phi(0) = 1$ and for $f(x)$ we have

$$f(x) = \frac{1}{2\pi}\int_{-\infty}^{\infty} (\text{ch}t)^{-1}e^{-itx}dt = \frac{1}{\pi}\int_{0}^{\infty} (1 + u^2)^{-1}e^{ix\ln n}\,du.$$

We consider the integral $I = \int_C g(z)dz$, $g(z) = (1 + z^2)^{-1}e^{ix\text{Ln}z}$ ($\text{Ln}z$ is that branch of the logarithmic function for which $\text{Ln}1 = 0$) along the closed contour C defined by the segment $-R < u < R$ on the real axis and the semicircle $\Gamma : z = Re^{i\Theta}$, $0 \leqslant \Theta \leqslant \pi$ (see Figure 18.1).

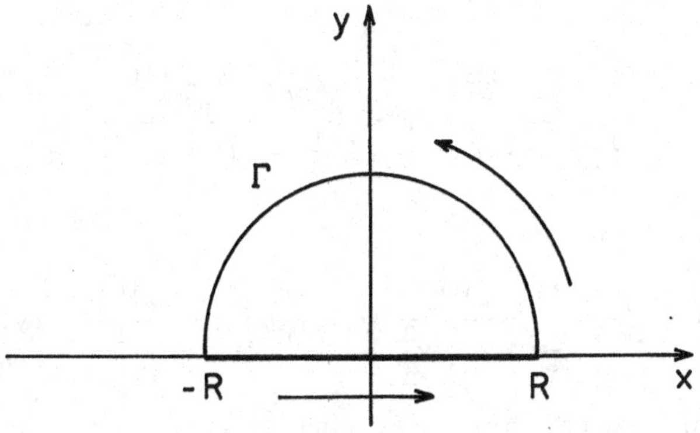

Fig. 18.1

Then

$$I = \int_{-R}^{0} + \int_{0}^{R} + \int_{\Gamma} = 2\pi i\Sigma\text{Res}g(z). \tag{18.13}$$

The function $g(z)$ has only one singular point for $R > 1$ in the region bounded by C. This is the point $z = i$ and we easily find that

$$\text{Res}g(z)\Big|_{z=i} = \lim_{z\to i}\,[(z - i)(1 + z^2)^{-1}\exp(-ix\text{Ln}z)] =$$

$$= \frac{1}{2i}\exp(\pi x/2).$$

Since $|g(z)| \leqslant e^{x\Theta}/R^2$ over the semicircle Γ, then $\lim_{R\to\infty}\int_{\Gamma} g(z)dz = 0$. If in (18.13) we pass to the limit in R as $R \to \infty$ and set $z = -u$ in the integral $\int_{-R}^{0}$, we obtain

$$(1 + e^{\pi x}) \int_0^\infty (1 + u^2)^{-1} e^{-ixLnu} du = 2\pi i \operatorname{Res} g(z)\big|_{z=i}.$$

Finally $f(x) = (1 + e^{\pi x})^{-1} e^{\pi x/2} = (2\operatorname{ch}\frac{\pi x}{2})^{-1}$, $x \in \mathbb{R}_1$, is a probability density. Hence, $\phi(t) = (\operatorname{ch}t)^{-1}$, $t \in \mathbb{R}_1$, is a ch.f. (b) From (a) it follows that $(\operatorname{ch}t)^{-2}$, $t \in \mathbb{R}_1$, is a ch.f. whose density is equal to the convolution $(f * f)(x)$, $x \in \mathbb{R}_1$. (c) The density is $\pi[4\operatorname{ch}^2 \frac{\pi x}{2}]^{-1}$, $x \in \mathbb{R}_1$.

18.33. Denote by Θ the r.v. whose density is $\frac{1}{2} f_{\xi_1\sqrt{3}}(y) + \frac{1}{2} f_{\xi_2}(y)$ and by ϕ_η and ϕ_Θ the ch.f.'s of η and Θ. Then

$$\phi_\eta(t) = \frac{1}{\pi} \int_{-\infty}^{\infty} \frac{e^{\frac{1}{2}\sqrt{3}xit}}{1 + x^2} dx \cdot \frac{2}{\sqrt{3}\pi} \int_{-\infty}^{\infty} \frac{e^{\frac{1}{2}xit}}{(1 + \frac{1}{3}x^2)^2} dx.$$

and

$$\phi_\Theta(t) = \frac{1}{2\pi\sqrt{3}} \int_{-\infty}^{\infty} \frac{e^{ity}}{1 + \frac{1}{3}y^2} dy + \frac{1}{\sqrt{3}\pi} \int_{-\infty}^{\infty} \frac{e^{ity}}{(1 + \frac{1}{3}y^2)} dy.$$

Using the Cauchy integral theorem, we find

$$\phi_\eta(t) = \phi_\Theta(t) = e^{-|t|\sqrt{3}}(\tfrac{1}{2}|t|\sqrt{3} + 1), \quad t \in \mathbb{R}_1.$$

The ch.f., however, defines uniquely the corresponding distribution. Hence η and Θ are identically distributed.

18.34. For $0 < \alpha \leqslant 2$ use Bochner's criterion. Obviously for $\alpha \leqslant 0$ the function $\phi(t)$ does not satisfy the conditions $\phi(0) = 1$ and $|\phi(t)| \leqslant 1$. For $\alpha > 2$ we have $\phi''(0) = 0$ and hence $\phi(t)$ cannot be a ch.f. as well. We note that $\phi(t)$ is a ch.f. for $0 < \alpha \leqslant 1$ and this fact follows directly from Polya's criterion.

18.35. (a) Let $G(x, y)$, $(x, y) \in \mathbb{R}_2$, be the d.f. of the random vector (ξ, η). Then

$$G(x, y) = \begin{cases} F(x)(1 - e^{-\lambda y}), & \text{if } x \geqslant 0 \text{ and } y \geqslant 0 \\ 0, & \text{otherwise.} \end{cases}$$

Thus we easily find that

$$P\{\eta > \xi\} = P\{\xi - \eta > 0\} = \iint_{\{y>x\}} dG(x, y) = \int_0^\infty e^{-\lambda x} dF(x) =$$

$$= \psi_\xi(\lambda).$$

(b) *First solution*: According to (17.5) we have

$$\psi_{\xi_1+\xi_2}(\lambda) = E\left\{e^{-\lambda(\xi_1+\xi_2)}\right\} = E\left\{e^{-\lambda\xi_1}\right\}E\left\{e^{-\lambda\xi_2}\right\} = \psi_{\xi_1}(\lambda)\psi_{\xi_2}(\lambda).$$

Second solution: Let $\zeta \in E(\lambda)$ and ξ_1, ξ_2 and ζ be independent r.v.'s. Show that $P\{\zeta > \xi_1 + \xi_2\} = P\{\zeta > \xi_1\}P\{\zeta > \xi_2\}$ and use the conclusion in case (a) of this exercise.
(c) If $F_1(x)$ and $F_2(y)$ are d.f.'s of ξ_1 and ξ_2, respectively, then

$$\psi_\eta(\lambda) = \int_0^\infty \left(\int_0^\infty \exp(-\lambda x^{1/\alpha} y)\, d\, F_2(y)\right) dF_1(x) =$$

$$= \int_0^\infty \exp(-\ (\lambda x^{1/\alpha})^\alpha)\, dF_1(x) =$$

$$= \int_0^\infty \exp(-\lambda^\alpha x)\, dF_1(x) = \psi_{\xi_1}(\lambda^\alpha).$$

18.36. (a) Show that the function $u(s) = \psi(\lambda + c - cs) - s$, $0 \leqslant s \leqslant 1$, for fixed $\lambda > 0$, is a convex function. Since $u(0) > 0$ and $u(1) < 0$, then the equation $u(s) = 0$ has a unique root in the interval $(0, 1)$. We denote this root by $\beta(\lambda)$. (b) Put $\beta_0(\lambda) = 0$, $\beta_{n+1}(\lambda) = \psi(\lambda + c - c\beta_n(\lambda))$ and show that: $\beta_n(\lambda) \leqslant \beta_{n+1}(\lambda) \leqslant 1$, $n = 0, 1, \ldots$; $\lim_{n\ \infty} \beta_n(\lambda) = \beta(\lambda)$; the functions $\beta_n(\lambda)$, $n = 0, 1, \ldots$ and $\beta(\lambda)$ are absolutely monotone functions; $\beta(0) = 1$. Further use the Bernstein theorem (see the Introdcutory Notes to this section).
18.37. From $|\phi(t_0)| = 1$ it follows that $\phi(t_0) = \exp(it_0 a)$ for a certain real a. Then

$$e^{it_0 a} = \int_{-\infty}^\infty e^{it_0 x}\, dF(x);$$

hence,

$$\int_{-\infty}^\infty e^{it_0(x-a)}\, dF(x) = 1, \qquad \int_{-\infty}^\infty \cos[t_0(x-a)]\, dF(x) = 1,$$

$$\int_{-\infty}^\infty [1 - \cos(t_0(x-a))]\, dF(x) = 0.$$

Since the function $1 - \cos(t_0(x - a)) \geqslant 0$, the d.f. F is non-negative and the integral vanishes; then $\cos(t_0(\xi - a)) = 1$ with probability one.

Thus it follows that $t_0(\xi - a) = 2\pi n$, $n = 0, \pm 1, \pm 2, \ldots$, and therefore the possible values of ξ are $a + \frac{2\pi}{t_0} \cdot n$, $n = 0, \pm 1, \pm 2, \ldots$

19. Infinitely Divisible and Stable Distributions

19.1. Let $\xi \in P(\lambda)$. Then $\phi(t) = \exp(\lambda(e^{it} - 1)) = (\exp[\frac{\lambda}{n}(e^{it} - 1)])^n$. Since $\exp[\frac{\lambda}{n}(e^{it} - 1)]$ is the ch.f. of the Poisson distribution with parameter $\frac{\lambda}{n}$, it follows that ξ is infinitely divisible. By reasoning similar to those in Example 19.1, we obtain that $\gamma = \lambda$, $G(x) = 0$ for $x \leqslant 1$ and $G(x) = \lambda$ for $x \geqslant 1$. If $\eta = c_1 + c_2\xi$, then $\phi_\eta(t) = \exp[\lambda(e^{itc_2} - 1) + itc_1]$. In this case we have that $\gamma = c_1 + c_2\lambda$, $G(x) = 0$ for $x \leqslant c_2$ and $G(x) = c_2^2\lambda$ for $x > c_2$.

19.2. (a) Show first that the ch.f. of the Cauchy distribution can be written as $\phi(t) = \exp(iat - b|t|)$. Therefore $\phi(t) = [\phi_n(t)]^n$, where $\phi_n(t) = \exp(i\frac{a}{n}t - \frac{b}{n}|t|)$ is obviously the ch.f. of the Cauchy distribution with parameters $\frac{a}{n}$ and $\frac{b}{n}$.

To prove (b) and (c) use Exercise 18.29.

(d) See Exercise 18.8 (d).

(e) This is a particular case of (b).

(f) See Exercise 18.7.

(g) For the corresponding ch.f. we have $\phi(t) = e^{iat} = (e^{iat/n})^n$.

19.3. The statement follows from the definition of infinitely divisible ch.f. and from the relation between ch.f. and p.g.f.: for any integer n there exists a function $g_n(s)$ which is a p.g.f. and which is such that $g(s) = [g_n(s)]^n$.

19.4. For every n we have $\xi = \xi_{n1} + \ldots + \xi_{nn}$. Assume that with probability 1, it holds that $|\xi| \leqslant c$ for some finite constant c. Since the ξ_{ni} are identically distributed, then $|\xi_{ni}| \leqslant \frac{c}{n}$, $V(\xi_{ni}) \leqslant \frac{c^2}{n^2}$, $V(\xi) \leqslant \frac{c^2}{n}$. Since n was arbitrary chosen, it follows that $V(\xi) = 0$; i.e., ξ turns out to be a degenerated r.v. This contradiction shows that ξ takes its values on the entire real line $\mathbb{R}_1$.

19.5. The statement follows from the well-known properties of the ch.f. and the explicit form of the ch.f.'s of the normal distribution and the Poisson distribution. The given function turns out to be a ch.f. of the r.v. $\xi + c_1\xi_1 + \ldots + c_n\xi_n$, where $\xi \in N(\gamma, \sigma^2)$, $\xi_k \in P(\lambda_k)$, $k = 1, \ldots, n$, c_i are constants and $\xi, \xi_1, \ldots, \xi_n$ are independent r.v.'s.

19.6. Let $h(t)$ and $\phi(t)$ be the ch.f.'s corresponding to $H(x)$ and $F(x)$. Then

$$\phi(t) = e^{-p} \sum_{k=0}^{\infty} \frac{p^k}{k!} h^k(t) = \exp[ph(t) - 1],$$

$$\ln \phi(t) = i\gamma t + \int_{-\infty}^{\infty} (e^{itx} - 1 - \frac{itx}{1 + x^2}) d\, L(x),$$

where

$$d\, L(x) = p\, d\, H(x), \qquad \gamma = p \int_{-\infty}^{\infty} \frac{x}{1 + x^2}\, d\, H(x).$$

19.7. $\phi_\eta(t) = \exp(\lambda\phi(t) - \lambda) = [\exp(\frac{\lambda}{n}\, \phi(t) - \frac{\lambda}{n})]^n$; hence, η is infinitely divisible.

19.8. The statement follows from the definition of infinite divisibility and also from the fact that any product of finitely many ch.f.'s which are infinitely divisible, is again an infinitely divisible ch.f. The converse statement is trivial to prove.

19.9. Let $\phi(t)$ be an infinitely divisible ch.f. Then for every n $|\phi_n(t)|^2 = |\phi(t)|^{2/n}$ is again a ch.f. We have that $\phi_0(t) = \lim_{n\to\infty} |\phi_n(t)|^2$ is 0 or 1, depending on whether $\phi(t) = 0$ or not. But $\phi(t)$ is continuous and $\phi(0) = 1$. Therefore $\phi(t) \neq 0$ in some neighbourhood of 0 and thus everywhere in this neighbourhood $\phi_0(t) = 1$. Then ϕ_0 is continuous at $t = 0$ and, being a limit of ch.f.'s, it is again a ch.f. Thus $\phi_0(t)$ turns out to be continuous at every t, which is possible only if $\phi_0(t)$ 1. Then it follows that $\phi(t) \neq 0$ for every t.

Since (see Exercise 18.8 (c)) the ch.f. of the uniform distribution is given by $\phi(t) = e^{iat}(bt)^{-1} \sin(bt)$ and since the last function does have real roots, we conclude from the above reasoning that the uniform distribution is not infinitely divisible.

19.10. According to Exercise 19.8, it suffices to show that $\phi^{1/n}$ is a ch.f. for every n. We then have

$$\phi_\lambda^{1/n}(t) = (\frac{\lambda}{\lambda + 1})^{1/n}(1 - \frac{1}{\lambda + 1}\gamma(t))^{-1/n} =$$

$$= (\frac{\lambda}{\lambda + 1})^{1/n} + \sum_{k=1}^{\infty} c_k^{(n)}\gamma^k(t),$$

where $c_k^{(n)} = \frac{1}{n}(1 + \frac{1}{n})(2 + \frac{1}{n}) \ldots (k - 1 + \frac{1}{n})\frac{1}{k!}\lambda^{1/n}(1 + \lambda)^{-k-1/n}$.

Obviously $c_0^{(n)} = (\frac{\lambda}{\lambda + 1})^{1/n} > 0$, $c_k^{(n)} > 0$, $\sum_{k=0}^{\infty} c_k^{(n)} = 1$ and $\phi_\lambda^{1/n}$ turns out to be a mixture of ch.f.'s. Then ϕ_λ is infinitely divisible.

19.11. The ch.f. of ξ is given by $\phi(t) = (1 - \frac{it}{\beta})^{-\alpha}$. Since ϕ is infinitely divisible, we have that $\phi^{1/n}(t)$ is a ch.f. Denote by F_n the corresponding d.f. Then $n(\phi^{1/n}(t) - 1) = \phi(t) + o(1)$ and we obtain that $\phi(t) = \lim_{n\to\infty} n(\phi^{1/n}(t) - 1)$. However,

$$n(\phi^{1/n}(t) - 1) = n\int_{-\infty}^{\infty} (e^{itx} - 1)\,dF_n(x) =$$

$$= ia_n t + \int_{-\infty}^{\infty}\left(e^{itx} - 1 - \frac{itx}{1 + x^2}\right)\frac{1 + x^2}{x^2}\,dN_n(x).$$

On the other hand, F_n is Γ-distributed and we easily find that

$$a = \lim_{n\to\infty} a_n = \alpha\int_0^{\infty} e^{-\beta u}(1 + u^2)^{-1}\,du,$$

and $N(x) = 0$, if $x < 0$, $N(x) = \lim_{n\to\infty} N_n(x) = \alpha\int_0^{x} u(1 + u^2)^{-1}e^{-\beta u}\,du$, if $x > 0$.

19.12. We have

$$\int_{-\infty}^{\infty}\left(e^{itx} - 1 - \frac{itx}{1 + x^2}\right)dL(x) = \int_{-\infty}^{\infty}(e^{itx} - 1 - itx)\,dL(x) +$$

$$+\int_{-\infty}^{\infty}\frac{itx^3}{1 + x^2}\,dL(x) =$$

$$= \int_{-\infty}^{\infty}(e^{itx} - 1 - itx)\,dL(x) + itc,$$

$$c = \int_{-\infty}^{\infty}\frac{x^3}{1 + x^2}\,dL(x).$$

From our assumptions and (19.3), it follows that c is finite. To obtain the desired representation one only needs to put $\tilde{\gamma} = \gamma + c$.

19.13. From (19.3) and from the condition $\phi(-t) = \phi(t)$, it follows that

$$i\gamma t - \frac{vt^2}{2} + \int_{-\infty}^{\infty}\left(e^{itx} - 1 - \frac{itx}{1 + x^2}\right)dL(x) =$$

$$= -i\gamma t - \frac{vt^2}{2} + \int_{-\infty}^{\infty}\left(e^{itx} - 1 - \frac{itx}{1 + x^2}\right)dL(-x).$$

Since the representation (19.3) is unique, we get $\gamma = 0$ and $d\,L(x) = d\,L(-x)$; i.e., $L(x)$ is symmetric. Hence,

$$\ln \phi(t) = -\frac{vt^2}{2} + \int_{-\infty}^{\infty} [\cos tx - 1]\, dL(x) =$$

$$= -\frac{vt^2}{2} + \int_{+0}^{\infty} [\cos tx - 1]\, dQ(x),$$

where $Q(x) = 2L(x)$, $x > 0$. Then the properties of Q can be easily derived from the properties of L in (19.3).

19.14. According to Exercise 18.34, ϕ is a ch.f. To verify its stability, it suffices to check validity of (19.5).

19.15. Use the explicit form of the ch.f. and the result obtained in Exercise 19.14.

19.16. See Theorem 5.8.4 in the book [20] cited at the end of the present Manual.

19.17. Let $\xi_1, \xi_2, \ldots$ be i.i.d. r.v.'s with d.f. F. From the assumptions it follows that for every n the r.v. $\frac{1}{\sqrt{n}}(\xi_1 + \ldots + \xi_n)$ has the same d.f. F, which is stable. On the other hand $\{\xi_n\}$ obeys the CLT (see Section 24); i.e., $\frac{1}{\sqrt{n}}(\xi_1 + \ldots + \xi_n) \xrightarrow{d} \Theta \in N(0, 1)$ as $n \to \infty$. Therefore F is standard normal distribution.

19.18. (a) Use the Bernstein theorem (see Introductory Notes to Section 18). (b) Since $(\psi_\alpha(\lambda))^n = \psi_\alpha(n^{1/\alpha}\lambda)$, using the result of Exercise 18.35, one gets $F_\alpha^{*n}(x) = F_\alpha(n^{-1/\alpha}x)$ and therefore $c_n = n^{-1/\alpha}$.

19.19. For the ch.f. $\phi_{\eta_n}(t)$ of η_n we obtain in limit

$$\phi_{\eta_n}(t) = \exp\left[\frac{c}{n} \sum_{k=-n^2}^{n^2} \frac{1}{(|k|/n)^{1+\alpha}} (e^{itk/n} - 1)\right] \to$$

$$\to \exp\left[-c|t|^\alpha \int_{-\infty}^{\infty} \frac{1 - \cos x}{|x|^{1+\alpha}}\, dx\right],$$

as $n \to \infty$. This completes the proof.

19.20. Let ϕ_n be the ch.f. of F. From the assumptions and from the continuity theorem (see Section 24), we have $\phi_n(t) \to \phi(t)$, $t \in \mathbb{R}_1$, where ϕ is the ch.f. corresponding to F. But ϕ_n is infinitely divisible, which means that $\phi_n(t) = [\phi_n^{(k)}(t)]^k$ for any integer k, where $\phi_n^{(k)}$ is a ch.f. for which $\phi_n^{(k)}(t) \neq 0$ for any $t \in \mathbb{R}_1$. Therefore, according to the continuity theorem (see Section 24), $\phi_n^{(k)}(t) \to \phi^{(k)}(t)$ as $n \to \infty$ for every fixed k, where $\phi^{(k)}(t)$, $t \in \mathbb{R}_1$, is again a ch.f. Since $\phi(t) = [\phi^{(k)}(t)]^k$ for every k, we find that F is infinitely divisible.

19.21. Obviously ϕ admits the following representation: $\phi(t) = \sum_{n=0}^{\infty} (1 - b)b^n e^{int}$. Now we conclude that ϕ is a ch.f. of a r.v. ξ, which assumes only non-negative integer values with probabilities $P\{\xi = n\} = (1 - b)b^n$, $n = 1, 2, \ldots$ (geometric distribution with parameter b). Since $\ln \phi(t) = \sum_{k=1}^{\infty} (e^{itk} - 1)b^k/k$, all the terms in the last sum are logarithms of ch.f.'s of Poisson distributions with some parameters. Since the Poisson distribution is infinitely divisible, we can conclude that ϕ is an infinitely divisible ch.f.

19.22. From the relation $\zeta(z) = \prod_p (1 - p^{-z})^{-1}$, we easily get

$$\ln \phi(t) = \sum_p [\ln(1 - p^{-s}) - \ln(1 - p^{-s-it})] =$$

$$= \sum_p \sum_{m=1}^{\infty} \frac{p^{-ms}(p^{imt} - 1)}{m} = \sum_p \sum_{m=1}^{\infty} \frac{p^{-ms}(e^{-imt\cdot\ln p} - 1)}{m},$$

where $\sum_p$ denotes summation over all prime numbers p. Now each term $p^{-ms}(e^{-imt\cdot\ln p} - 1)/m$ is a logarithm of the ch.f. of some Poisson distribution. Thus the solution follows from Exercise 19.21.

20. Conditional Distribution and Conditional Expectations

20.1. Let j and s be integers, $0 \leq j \leq n_1$, $0 \leq s \leq n_2$, $j + s = k$. Then

$$P\{\xi = j, \eta = s | \xi + \eta = k\} = \frac{P\{\xi = j, \eta = s, \xi + \eta = k\}}{P\{\xi + \eta = k\}} =$$

$$= \frac{P\{\xi = j, \eta = k - j\}}{P\{\xi + \eta = k\}} =$$

$$= \frac{\binom{n_1}{j} p^j q^{n_1 - j} \binom{n_2}{k-j} p^{k-j} q^{n_2 - k + j}}{\binom{n_1 + n_2}{k} p^k q^{n_1 + n_2 - k}} =$$

$$= \frac{\binom{n_1}{j}\binom{n_2}{k-1}}{\binom{n_1+n_2}{k}}.$$

If j and s do not satisfy the above conditions, then it is obvious that $P\{\xi = j,\ \eta = s \mid \xi + \eta = k\} = 0$. It is interesting to note that the conditional distribution of the random vector (ξ, η) given $\xi + \eta = k$ is the so-called *hypergeometric distribution* and does not depend on the parameter p (see Section 9).

20.2. If j and s are integer values $j \geqslant 0$, $s \geqslant 0$, $j + s = k$, then

$$P\{\xi = j,\ \eta = s \mid \xi + \eta = k\} = \binom{k}{j} \frac{B(\lambda_1 + j,\ \lambda_2 + s)}{(\lambda_1 + \lambda_2)}.$$

For the rest of the values of 1 and s we have $P\{\xi = j,\ \eta = s \mid \xi + \eta = k\} = 0$. This discrete distribution is called *negative hypergeometric distribution.*

20.3. Let m_j be such that $\sum_{j=0}^{\infty} jm_j = n$ and $\sum_{j=0}^{\infty} m_j = N$. Otherwise we put $P\{\xi_j = m_j,\ j = 0, 1, 2, \ldots \mid \eta_1 = n,\ \eta_2 = N\} = 0$. The joint distribution of the r.v.'s ξ_j, $j = 0, 1, 2, \ldots$ is

$$P\{\xi_j = m_j,\ j = 0, 1, 2, \ldots\} = \prod_{j=0}^{\infty} \frac{z^{jm_j} x^{m_j}}{(j!)^{m_j} (m_j)!} \exp\left(-\frac{xz^j}{j!}\right) \qquad (20.7)$$

Since $P\{\xi_j \neq 0\} = 1 - e^{\lambda_j} \leqslant \lambda_j = xz^j/j!$ and the series $\sum_{j=0}^{\infty} P\{\xi_j \neq 0\}$ converges, then according to the Borel-Cantelli lemma (see Exercise 11.26) only a finite number of r.v.'s ξ_j are non-zero. Hence η_1 and η_2 are finite with probability 1. Therefore (20.7) can be rewritten as

$$P\{\xi_j = m_j,\ j = 0, 1, \ldots\} = [z^n x^N \exp(-xe^z)] \cdot$$

$$\cdot \left[\prod_{j=0}^{\infty} (j!)^{m_j} m_j!\right]^{-1}. \qquad (20.8)$$

Let us find the distribution of the random vector (η_1, η_2), using its ch.f. $\phi(t_1, t_2)$. We have

$$\phi(t_1, t_2) = E\{\exp[i(t_1\eta_1 + t_2\eta_2)]\} =$$

$$= E\left\{\exp\left[i \sum_{j=0}^{\infty} (jt_1 + t_2)\xi_j\right]\right\} = \prod_{j=0}^{\infty} E\{\exp[i(jt_1 + t_2)\xi_j]\}.$$

Using Exercise 18.17 (f), we obtain

$$\phi(t_1, t_2) = \prod_{j=0}^{\infty} \exp\left\{\lambda_j [e^{i(jt_1+t_2)} - 1]\right\} =$$

$$= \exp\left\{\sum_{j=0}^{\infty} \left[\frac{xz^j}{j!} e^{i(jt_1+t_2)} - 1)\right]\right\} =$$

$$= \exp(-xe^z)\exp\left\{xe^{it_2}\left[\exp\left(ze^{it_1}\right)\right]\right\}.$$

Now expanding the function $\exp\left\{xe^{it_2}\left[\exp\left(ze^{it_1}\right)\right]\right\}$ in powers of x and z, we have

$$\phi(t_1, t_2) = \exp(-xe^z) \sum_{N=0}^{\infty} \sum_{n=0}^{\infty} \left\{\frac{N^n z^n x^N}{N!n!} \exp[i(t_1 n + t_2 N)]\right\}.$$

Thus

$$P\{\eta_1 = n, \eta_2 = N\} = \frac{N^n z^n x^N}{N!n!} \exp(-xe^z). \tag{20.9}$$

From (20.8) and (20.9) we find

$$P\{\xi_j = m_j, j = 0, 1, \ldots | \eta_1 = n, \eta_2 = N\} =$$

$$= \frac{P\{\xi_j = m_j, j = 0, 1, \ldots, \eta_1 = n, \eta_2 = N\}}{P\{\eta_1 = n, \eta_2 = N\}} =$$

$$= \frac{\left[\prod_{j=0}^{\infty} (j!)^{m_j} m_j!\right]^{-1} z^n x^N \exp(-xe^z)}{N^n z^n x^N \exp(-xe^z)[N!n!]^{-1}} =$$

$$= N!n!\left[N^n\left[\prod_{j=0}^{\infty} (j!)^{m_j} m_j!\right]\right]^{-1}.$$

Remark. This distribution is related to the following occupancy problem: n identical particles are allocated at random into N different cells, with equal probability for all the cells and independent of the number of particles already allocated. Let η_j denote the number of cells containing exactly j particles, j = 0, 1, 2, ... If the conditions $\sum_{j=0}^{\infty} jm_j = n$ and $\sum_{j=0}^{\infty} m_j = N$ are satisfied, then

$$\vec{P}\{\eta_j = m_j, j = 0, 1, 2, \ldots\} = N!n!\left[N^n \prod_{j=0}^{\infty} (j!)^{m_j} m_j!\right]^{-1}.$$

The probability $P\{\eta_j = m_j, j = 0, 1, 2, \ldots\} = 0$ otherwise.

20.4. (a) The random vector (ζ, ξ) has density $f_{\zeta,\xi}(x, y) = \frac{1}{4}$ for $0 < x + y < 2$, $0 < y < 2$ and $f_{\zeta,\xi}(x, y) = 0$ otherwise. From formula (20.1) we find that if $0 < y < 2$, then $f_{\zeta|\xi}(x|y) = \frac{1}{2}$ for $-y < x < 2 - y$ and $f_{\zeta|\xi}(x|y) = 0$ for $x \bar{\in} (-y, 2 - y)$. We note that $f_{\zeta|\xi}(x|y)$ is defined for almost all y since $P\{\eta \bar{\in} (0, 2)\} = 0$. From (20.3) it follows that $P\{|\zeta| \leqslant 1|\xi = 1\} = \int_{-1}^{1} (1/2)dx = 1$. Using formula (20.2), we find that if $0 < y < 2$, then $F_{\zeta|\xi}(x|y) = \int_{-y}^{x} (1/2)du = (x + y)/2$ for $-y < x < 2 - y$, $F_{\zeta|\xi}(x|y) = 0$ for $x \leqslant -y$ and $F_{\zeta|\xi}(x|y) = 1$ for $x \geqslant 2 - y$. The conditional d.f. $F_{\zeta|\xi}(x|y)$ is defined for almost all y as well. In the special case when x = 0 and y = 1, we obtain $F_{\zeta|\xi}(0|1) = \frac{1}{2}$, $f_{\zeta|\xi}(0|1) = \frac{1}{2}$. (b) $P\{|\zeta| \leqslant 1|\xi = 1\} = 0.865$, $F_{\zeta|\xi}(0|1) \approx 0.632$, $f_{\zeta|\xi}(0|1) \approx 0.368$.

20.5. (a) $P\{|\zeta| \leqslant 1|\eta = 1\} \approx 0.276$. (b) $F_{\zeta|\xi}(0|1) = \frac{1}{2}$. (c) $f_{\zeta|\xi}(0|1) = 1/\sqrt{8\pi}$. (d) Using (20.4) we obtain $P\{\eta \geqslant 0||\xi| \leqslant 1\} = \frac{1}{2}$.

20.6. $P\{\xi_1 + \ldots + \xi_r < r, r = 1, \ldots, n|\xi_1 + \ldots + \xi_n = k\} = 1 - k/n$.

20.7. Let $\tau, \xi_1, \xi_2, \ldots$ be independent r.v.'s, where $\tau \in P(\lambda)$ and ξ_i, i = 1, 2, ..., are identically distributed and have ch.f. $\phi(t)$, $t \in \mathbb{R}_1$. Consider the r.v. $\eta = \xi_1 + \ldots + \xi_\tau$ for $\tau \geqslant 1$ and $\eta = 0$ for $\tau = 0$. Using the properties of the conditional mean (see Section 13) for the ch.f. of the r.v. η, we find

$$\phi_\eta(t) = E\{\exp(it\eta)\} = E\{E\{\exp(it\eta)|\tau\}\} =$$

$$= \sum_{k=0}^{\infty} P\{\tau = k\} \times E\{\exp(it\eta)|\tau = k\} =$$

$$= \sum_{k=0}^{\infty} ((\lambda^k \exp(-\lambda))/k!)E\{\exp(it(\xi_1 + \ldots + \xi_k))\} =$$

$$= (\exp(-\lambda)) \sum_{k=0}^{\infty} \lambda^k (\phi(t))^k / k! = \exp(\lambda\phi(t) - \lambda).$$

20.8. (a) $f(x|y) = f_1(x) f_2(y - x) \Big/ \int_{-\infty}^{\infty} f_1(x) f_2(y - x) dx$. (b) $f(x|y) = 1/y$ for $0 < x < y$ and $f(x|y) = 0$ for $x \leqslant 0$ or $x \geqslant y$; i.e., the conditional distribution of ξ_1 given $\eta = y$ is uniform over the interval $(0, y)$.

20.9. Let us find the p.g.f. $g(s)$ of the r.v. ζ. We have

$$g(s) = E\{s^{\zeta}\} = E\{E\{s^{\zeta}|\eta\}\} = \sum_{k=0}^{\infty} P\{\eta = k\} E\{s^{\zeta}|\eta = k\} =$$

$$= \sum_{k=0}^{\infty} P\{\eta = k\} \left(E\left\{ s^{\xi_1} \right\} \right)^k .$$

Since $E\left\{s^{\xi_1}\right\} = (\ln(1 - q))^{-1} \ln(1 - qs)$, after transformation we find that $g(s) = (1 - q)^r (1 - qs)^{-r}$. Expanding $g(s)$ in power series, we find that the r.v. ζ has a negative binomial distribution (see Section 9) with parameters r and $p = 1 - q$.

20.10. Show that $\frac{1}{2} \cdot 1_A(z) + \frac{1}{2} + \cdot 1_A(-z)$ is a version of $P\{A|\xi\}$.

20.11. For the conditional density $f(x|x_1, \ldots, x_{n-1})$ of the r.v. ξ_n given $\xi_1 = x_1, \ldots, \xi_{n-1} = x_{n-1}$, we have

$$f(x|x_1, \ldots, x_{n-1}) =$$

$$f_{\xi}(x_1, x_2, \ldots, x_{n-1}, x) \Big/ \int_{-\infty}^{\infty} f_{\xi}(x_1, \ldots, x_{n-1}, u) du;$$

hence, $\int_{-\infty}^{\infty} (xf(x|x_1, \ldots, x_{n-1}) dx$ is a version of

$$E\{\xi_n | \xi_1 = x_1, \xi_2 = x_2, \ldots, \xi_{n-1} = x_{n-1}\}.$$

Finally,

$$E\{\xi_n | \xi_1, \ldots, \xi_{n-1}\} = \sum_{k=1}^{n-1} \alpha_k \xi_k \quad (P\text{-a.s.}),$$

where $\alpha_1, \ldots, \alpha_{n-1}$ is the unique solution of the system of equations

$$\sum_{k=1}^{n-1} b_{kj}\alpha_k = b_{nj}, \quad j = 1, 2, \ldots, n-1.$$

Compare the result obtained for $n = 2$ with that in Example 20.1.
20.12. $g^*(x) = E\{\eta | \xi = x\}$.
20.13. $\operatorname{cov}(\xi, \eta) = 0$.
20.14. (a) $E\{\xi_n | \xi_{n-1}\} = \frac{1}{2}(\xi_{n-1} + 1)$ (P-a.s.). (b) $E\{\xi_n\} = 1 - 2^{-n}$.
20.15. $a = (V(\xi))^{-1} \operatorname{cov}(\xi, \eta)$, $b = E(\eta) - (V(\xi))^{-1} E(\xi) \operatorname{cov}(\xi, \eta)$, $V(\zeta) = (1 - (V(\xi)V(\eta))^{-1}(\operatorname{cov}(\xi, \eta))^2)V(\eta)$.
20.16. If we denote the desired density by $f(x_1, \ldots, x_k | y)$, then

$$f(x_1, \ldots, x_k | y) = \begin{cases} k!/y^k, & \text{if } 0 < x_1 < x_2 < \ldots < x_k < y \\ 0, & \text{otherwise.} \end{cases}$$

20.17. We put $p_k = P\{\tau = k\}$, $k = 1, 2, \ldots$; $\eta_k = 0$ if $\tau < k$ and $\eta_k = 1$ if $\tau \geqslant k$; $S_\tau = \xi_1 + \ldots + \xi_\tau = \sum_{k\geqslant 1} \eta_k \xi_k$. Obviously η_k is a r.v. independent of ξ_k and $P\{\eta_k = 1\} = P\{\tau \geqslant k\}$. Besides,

$$|E\{\eta_k \xi_k\}| = |E(\eta_k)E(\xi_k)| \leqslant cP\{\tau \geqslant k\} = c \sum_{i\geqslant k} p_i;$$

hence,

$$\sum_{k\geqslant 1} |E\{\eta_k \xi_k\}| \leqslant c \sum_{k\geqslant 1} kp_k = cE(\tau) < \infty.$$

Thus $E(S_\tau)$ exists and

$$E(S_\tau) = E\{E(S_\tau | \tau)\} = \sum_{k\geqslant 1} E(S_\tau | \tau = k)P\{\tau \geqslant k\} =$$

$$= \sum_{k\geqslant 1} E(S_k)P\{\tau = k\} = \sum_{k\geqslant 1} akP\{\tau = k\} = aE(\tau).$$

20.18. The proof is similar to that given in Exercise 20.7.

21. Inequalities for Random Variables

21.1. Denote $A = \{\omega : |\xi(\omega)| \geqslant c\}$. Then

$$E\{f(\xi)\} = \int_\Omega f(\xi(\omega))dP = \int_A f(\xi(\omega))dP + \int_{\bar{A}} f(\xi(\omega))dP \geqslant$$

$$\geqslant \int_A f(\xi(\omega))dP \geqslant \int_A f(c)d\,P = f(c)P(A).$$

Thus the Markov inequality is proved. It is interesting to consider the particular case when $\xi = \eta - E\eta$ for some r.v. η and $f(x) = x^2$. Obviously, in this case we obtain Chebyshev's inequality.

21.2. Consider the r.v. ζ defined by

$$\zeta = \max\left[\left(\frac{\xi - E(\xi)}{\sqrt{V(\xi)}}\right)^2, \left(\frac{\eta - E(\eta)}{\sqrt{V(\eta)}}\right)^2\right].$$

Then using Markov's inequality (see Exercise 21.1), we obtain

$$P\{|\xi - E(\xi)| \geqslant \varepsilon\sqrt{V(\xi)} \quad \text{or} \quad |\eta - E(\eta)| \geqslant \varepsilon\sqrt{V(\eta)}\} =$$

$$= P\{\zeta \geqslant \varepsilon^2\} \leqslant \frac{1}{\varepsilon^2} E(\zeta).$$

It only remains to apply the inequality established in Example 21.2.

21.3. (a) Let $A = \{\omega : |\xi(\omega)| \geqslant c\}$. Then

$$\int_\Omega f(\xi(\omega))dP = \int_{\bar{A}} f(\xi(\omega))dP + \int_A f(\xi(\omega))dP \leqslant$$

$$\leqslant f(c)P(\bar{A}) + KP(A) \leqslant f(c) + KP(A).$$

(b) The proof is similar to that of (a).

21.4. Use Exercise 21.1 and Exercise 21.3 with $f(x) = |x|^r/(1 + |x|^r)$.

21.5. We find that $P\{|\xi| \geqslant \varepsilon\} = P\{\xi = -\varepsilon\} + P\{\xi = \varepsilon\} = \frac{\sigma^2}{\varepsilon^2}$. Since $E(\xi) = 0$ and $V(\xi) = \sigma^2$, according to Chebyshev's inequality we have

$$P\{|\xi| \geqslant \varepsilon\} \leqslant V(\xi)/\varepsilon^2 = \sigma^2/\varepsilon^2.$$

Then it is obvious that the exact value of $P\{|\xi| \geqslant \varepsilon\}$ coincides with its bound, given by Chebyshev's inequality.

21.6. Consider the r.v.'s $\eta_i = \log_{b_i} \xi_i$, $i = 1, \ldots, n$, and then apply Chebyshev's inequality to the r.v. $\frac{1}{n}(\eta_1 + \ldots + \eta_n)$.

21.7. (a) $P\{\xi \geqslant 1\} = 1 - P\{\xi = 0\} = 1 - e^{-\lambda} \leqslant \lambda$, since $e^{-\lambda} \geqslant 1 - \lambda$ for every λ. (b) $P\{\xi \geqslant 2\} = e^{-\lambda} \sum_{k=2}^{\infty} \frac{\lambda^k}{k!} = \frac{\lambda^2}{2} e^{-\lambda} \sum_{k=2}^{\infty} \frac{2\lambda^{k-2}}{k!} \leqslant \frac{\lambda^2}{2}$, since $\frac{2}{k!} - \frac{1}{(k-2)!} \leqslant 0$, $k \geqslant 2$.

21.8. Let Θ be another r.v., which is independent of ξ_1 and also such that $\Theta \in P(\lambda)$, where $\lambda = \lambda_2 - \lambda_1$. Then $\xi_1 + \Theta \in P(\lambda_1 + (\lambda_2 - \lambda_1))$; i.e., $\xi_1 + \Theta$ has the same distribution as ξ_2. Therefore, for every $k = 0, 1, \ldots$ we have

$$P\{\xi_2 \leqslant n\} = P\{\xi_1 + \Theta \leqslant n\} =$$

$$= \sum_{k=0}^{n} P\{\Theta = k\}P\{\xi_1 \leqslant n - k\} \leqslant$$

$$\leqslant P\{\Theta \leqslant n\}P\{\xi_1 \leqslant n\} < P\{\xi_1 \leqslant n\}.$$

21.9. Since $E(\xi) = m + 1$ and $V(\xi) = m + 1$, we have

$$P\{0 < \xi < 2m + 2\} = P\{-2m - 2 < \xi < 2m + 2\} =$$

$$= P\{|\xi - E\xi| < m + 1\}.$$

To finish the proof it suffices to apply Chebyshev's inequality.

21.10. Use Exercise 21.1 with $f(x) = |x|^r$ to prove (a), (b) and (c).

(d) Let $f(x) = (x + c)^2$, where $c > 0$, $x \in \mathbb{R}_1$. Then $f(x) \geqslant 0$ for every $x \in \mathbb{R}_1$, $f(x) \geqslant (\lambda + c)^2$ for $x \geqslant \lambda > 0$ and

$$P\{|\xi - a| \geqslant \lambda\} \leqslant \frac{E\{(\xi - a + c)^2\}}{(\lambda + c)^2} = \frac{\sigma^2 + c^2}{(\lambda + c)^2}.$$

But the minimum w.r.t. c of $(\sigma^2 + c^2)/(\lambda + c)^2$ is attained when $c = \sigma^2/\lambda$; therefore, $P\{|\xi - a| \geqslant \lambda\} \leqslant \dfrac{\sigma}{\lambda^2 + \sigma^2}$. Cantelli's inequality can easily be obtained from the last relation.

(e) Use Markov's inequality with an appropriately chosen function $f(x)$, $x \in \mathbb{R}_1$.

21.11. Use the relation $\{\xi + \eta < x + y\} \subset \{\xi < x\} \cup \{\eta < y\}$, as well as the properties of the probability P.

21.12. We have $F(x_1, \ldots, x_n) = \sqrt[n]{F(x_1, \ldots, x_n) \ldots F(x_1, \ldots, x_n)}$ (with n multipliers under the root), and $F_1(x_1) = F(x_1, \infty, \ldots, \infty)$, ..., $F_n(x_n) = F(\infty, \ldots, \infty, x_n)$. Since $F(x_1, \ldots, x_n)$ is non-decreasing function w.r.t. each of its arguments, the desired inequality follows from the above representation of F.

21.13. Consider the function $H(a, b) = \Phi(-a - b)/\Phi(-a)$, $a, b \geqslant 0$. By standard techniques we obtain $\partial H(a, b)/\partial a \leqslant 0$, which means that $H(a, b)$ is non-increasing w.r.t. a. Thus $H(a, b) \leqslant H(0, b) = \Phi(-b)/\Phi(0) = 2\Phi(-b)$, which relation proves the desired inequality.

21.14. Each of the three quantities is to be represented as an integral of the corresponding derivative and then the three integrands are to be compared.

21.15. Since $x > 0$ and $c > 0$, directly from the definition of conditional probability, we get

$$P\{\xi - x > \tfrac{c}{x} | \xi > x\} = P\{\xi > x + cx^{-1}\}/P\{\xi > x\}.$$

Let

$$\overline{\Phi}_\sigma(x) = P\{\xi > x\} = \frac{1}{\sqrt{2\pi}} \int_x^\infty e^{-u^2/e\sigma^2} du.$$

Then the problem is reduced to establishing the following inequality:

$$\overline{\Phi}_\sigma(x + cx^{-1}) < e^{-c/\sigma^2} \overline{\Phi}_\sigma(x), \quad x \in \mathbb{R}_1.$$

21.16. We have

$$|P\{\xi < x\} - P\{\xi < cx\}| = \frac{1}{\sqrt{2\pi}} \left| \int_{cx}^x e^{-u^2/2} du \right| \leqslant$$

$$\leqslant \frac{1}{\sqrt{2\pi}} |x - cx| \max\left[e^{-x^2/2}, e^{-c^2x^2/2}\right].$$

On the other hand, for any real x, $|x|e^{-x^2/2} \leqslant 1/\sqrt{e}$, and $|x|e^{-c^2x^2/2} \leqslant 1/(c\sqrt{e})$. When these quantities are replaced above, we obtain the desired inequality.

21.17. Put $\zeta_n = \max[|\xi_i| : 1 \leqslant i \leqslant n]$. Then

$$F_n(x) = P\{\zeta_n < x\} = P\{|\xi_1| < x, \ldots, |\xi_n| < x\} =$$

$$= \left(\frac{2}{\sqrt{2\pi}} \int_0^x e^{-u^2/2} du\right)^n.$$

Now the smallest n, for which $P\{\zeta_n \geqslant 2\} = 1 - F_n(2) \geqslant \frac{1}{2}$, is to be found. From Table 1 we get n = 15.

21.18. From the Cauchy-Bunyakovski-Schwarz inequality, it follows that

$$(g(x) - g(a))^2 = \left(\int_0^x g'(u) du\right)^2 \leqslant \int_a^x (g'(u))^2 du \int_a^x du;$$

therefore,

$$V\{g(\xi)\} \leqslant E\{(g(\xi) - g(a))^2\} \leqslant$$

$$\leqslant \int (x - a) f(x) \int_a^x (g'(u))^2 du\, dx =$$

$$= -\int H'(x) dx \int_a^x (g'(u))^2 du = \int H(x)(g'(x))^2 dx \leqslant$$

$$\leqslant c\int f(x)(g'(x))^2dx = cE(g'(\xi))^2.$$

Note that if $H(x) = cf(x)$, then $cf'(x) = -(x - a)f(x)$ and $f(x) = \frac{1}{\sqrt{2\pi c}}\exp\left(-\frac{(x-a)^2}{2c}\right)$; i.e., $\xi \in N(a, c)$.

21.19. Since $g(k) = \sum_{j=0}^{k-1} \Delta g(j) + g(0)$, then

$$V\{g(\xi)\} = V\left\{\sum_{j=0}^{\xi-1} \Delta g(j)\right\} \leqslant E\left\{\left(\sum_{j=0}^{\xi-1} \Delta g(j)\right)^2\right\} \leqslant$$

$$\leqslant E\left\{\sum_{j=0}^{\xi-1} 1^2 \sum_{j=0}^{\xi-1} (\Delta g(j))^2\right\} = E\left\{\xi \sum_{j=0}^{\xi-1} (\Delta g(j))^2\right\} =$$

$$= \sum_{j=1}^{\infty}\left[j\left\{\sum_{k=0}^{j-1} (\Delta g(k))^2 p_k\right\}\right] = \sum_{j=0}^{\infty}\left[(\Delta g(j))^2 \sum_{k=j+1}^{\infty} kp_k\right].$$

21.20. The statement follows from Exercise 21.28.

21.21. (a) If $E\{\xi^2\} = E\{\eta^2\} = 0$, then $\xi = \eta$ (P-a.s.) and the inequality is obvious. Without loose of generality, we may assume that $E\{\eta^2\} \neq 0$. Then for every $t \in \mathbb{R}_1$ we have $0 \leqslant E\{(|\xi| + t|\eta|)^2\} = E\{\xi^2\} + 2tE\{|\xi\eta|\} + t^2E\{\eta^2\}$, which is possible only if

$$(E\{|\xi\eta|\})^2 - E\{\xi^2\}E\{\eta^2\} \leqslant 0.$$

(b) For a fixed $y \in \mathbb{R}_1$ and arbitrary $x \in R_1$ we have $g(x) \geqslant g(y) + k_y(x - y)$, where k_y is the slope of the tangent $z - g(y) = k_y(x - y)$ to the graph of the convex functions g at the point y. For $x = \xi$ and $y = E(\xi)$, we obtain $g(\xi) \geqslant g(E(\xi)) + k_y(\xi - E(\xi))$. Taking expectations on the both sides, we get

$$E\{g(\xi)\} \geqslant g(E(\xi)).$$

(c) For $0 \leqslant h \leqslant r$ it follows from (a) that $(E\{|\xi|^r\})^2 \leqslant E\{|\xi|^{r+h}\} \times E\{|\xi|^{r-4}\}$. Taking logarithms we get

$$g\left(\frac{x_1 + x_2}{2}\right) \leqslant \frac{1}{2}\, g(x_1) + \frac{1}{2}\, g(x_2),$$

where $g(r) = \ln E\{|\xi|^r\}$, $r \geqslant 0$, $x_1 = r - h$, $x_2 = r + h$. Hence $g(r)$, $r > 0$, is a convex downwards function.

(d) Consider the function $\ln((E\{|\xi|^r\})^{1/r})$ and use (c).

(e) Show that for arbitrary $x, y \in \mathbb{R}_1$ the following inequality is valid:

$$|xy| \leqslant \frac{|x|^r}{r} + \frac{|y|^s}{s}, \tag{21.5}$$

where $r > 1$, $r^{-1} + s^{-1} = 1$. If $E\{|\xi|^r\} = 0$ or $E\{|\eta|^s\} = 0$, then the desired inequality is trivial. If $E\{|\xi|^r\} \neq 0$ and $E\{|\eta|^s\} \neq 0$, one can write (21.5) with $x = \xi/(E\{|\xi|^r\})^{1/r}$ and $y = \eta/(E\{|\eta|^s\})^{1/s}$. Then Hölder's inequality is obtained by taking expectations on both sides. Note that the Cauchy-Bunyakovski-Schwarz inequality is a particular case of Hölder's inequality with $r = s = 2$.

(f) We ignore the trivial case $r = 1$. Then the following inequality can be used:

$$E\{|\xi + \eta|^r\} \leqslant E\{|\xi||\xi + \eta|^{r-1}\} + E\{|\eta||\xi + \eta|^{r-1}\}.$$

The last step is to apply Hölder's inequality to $E\{|\xi||\xi + \eta|^{r-1}\}$ and to $E\{|\eta||\xi + \eta|^{r-1}\}$

(g) Use the following inequality, which is elementary to prove:

$$|x + y|^r \leqslant c_r|x|^r + c_r|y|^r, \quad r > 0,\ x, y \in \mathbb{R}_1,$$

where $c_r = 1$ for $r \leqslant 1$ and $c_r = 2^{r-1}$ for $r > 1$.

21.22. After some elementary calculations we get $E\{|\xi - \frac{1}{2}|\} = E\{|\eta - \frac{1}{2}|\} = \frac{1}{4}$. To finish the proof we only need to use the inequality: $|x - y| \leqslant |x + y|$.

21.23. We have

$$n + 2\binom{n}{2}c = E\{\xi_1^2 + \dots + {}^2_n + 2\sum_{i<j} \xi_i\xi_j\} =$$

$$= E\{(\xi_1 + \dots + \xi_n)^2\} = V\{\xi_1 + \dots + \xi_n\} \geqslant 0.$$

From these relations it follows that $c \geqslant -\frac{1}{n-1}$.

21.24. Since $\mathrm{Re}(1 - \phi(t)) = \int_{-\infty}^{\infty} (1 - \cos tx)dF(x) \leqslant |1 - \phi(t)|$, we have that $\int_{-c}^{c} (1 - \cos tx)dF(x) \leqslant |1 - \phi(t)|$. Now it only remains to use the following inequality: $x^2 \leqslant 3(1 - \cos x)$ for $|x| \leqslant 1$.

21.25. First prove that $|x + y|^r + |x - y|^r \leqslant 2(|x|^r + |y|^r)$, $x, y \in \mathbb{R}_1$, $r \in [1, 2]$. Because of the symmetry of η, the distribution of $\xi + \eta$ is the same as that of $\xi - \eta$. Therefore, $E\{|x + \eta|^r\} = \frac{1}{2}(E\{|x + \eta|^r\} +$

$E\{|\xi - \eta|^r\})$. Finally the above inequality is to be used.

21.26. Use Jensen's inequality and convexity of the function $f(x) = |x|^r$, $r \geq 1$.

21.27. Let $A = \{\max_{1\leq k\leq n} S_k \geq \varepsilon\}$, $A_k = \{|S_i| < \varepsilon,\ i \leq k - 1,\ |S_k| \geq \varepsilon\}$. The events $A_1, \ldots, A_n$ are disjoint, $A = \bigcup_k A_k$ and $E\{S_n^2\} \geq E\{S_n^2 I_A\} = \sum_{k=1}^n E\{S_n^2 I_{A_k}\}$. We find that

$$E\{S_n^2 I_{A_k}\} = E\{(S_k + (\xi_{k+1} + \ldots + \xi_n))^2 I_{A_k}\} =$$

$$= E\{S_k^2 I_{A_k}\} + 2E\{S_k(\xi_{k+1} + \ldots + \xi_n)I_{A_k}\} +$$

$$+ E\{(\xi_{k+1} + \ldots + \xi_n)^2 I_{A_k}\} \geq E\{S_k^2 I_{A_k}\}.$$

In the above expression, the independence of ξ_i and the condition $E(\xi_i) = 0$ were used to conclude that

$$E\{S_k(\xi_{k+1} + \ldots + \xi_n)I_{A_k}\} = E\{S_k I_{A_k}\}E\{\xi_{k+1} + \ldots + \xi_n\} = 0.$$

Then we have

$$E\{S_n^2\} \geq \sum_k E\{S_k^2 I_{A_k}\} \geq \varepsilon^2 \sum_k P(A_k) = \varepsilon^2 P(A),$$

which proves the first inequality. To prove (b), we observe that

$$E\{S_n^2 I_A\} = E\{S_n^2\} - E\{S_n^2 I_{\bar{A}}\} \geq E\{S_n^2\} - \varepsilon^2 P(\bar{A}) =$$

$$= E\{S_n^2\} - \varepsilon^2 + \varepsilon^2 P(A). \tag{21.5}$$

If $\omega \in A_k$, then $|S_{k-1}| < \varepsilon$ and therefore $|S_k| < \varepsilon + c$. Thus

$$E\{S_n^2 I_A\} = \sum_{k=1}^n E\{S_k^2 I_{A_k}\} + \sum_{k=1}^n E\{I_{A_k}(S_n - S_k)^2\} \leq$$

$$\leq (\varepsilon + c)^2 \sum_{k=1}^n P(A_k) + \sum_{k=1}^n \left(P(A_k) \sum_{j=k+1}^n \sigma_j^2\right) \leq$$

$$\leq \left((\varepsilon + c)^2 + \sum_{j=1}^n \sigma_j^2\right)P(A).$$

The validity of the second inequality follows from the last relations and from relations (21.5).

21.28. Denote $S_k = \xi_1 + \ldots + \xi_k$, $\eta = \sum_{k=m}^{n-1} S_k^2(c_k^2 - c_{k+1}^2) + c_n^2 S_n^2$. We easily find that

$$E(\eta) = c_m^2 \sum_{k=1}^{m} \sigma_k^2 + \sum_{k=m+1}^{n} c_k^2 \sigma_k^2.$$

The events $A_i = \{c_j|S_j| < \varepsilon,\ m \leqslant j < i,\ c_i|S_i| > \varepsilon\}$ are disjoint and

$$P\{\max_{m \leqslant k \leqslant n} c_k|S_k| \geqslant \varepsilon\} = \sum_{i=m}^{n} P(A).$$

Put $A_0 = \{c_j|S_j| < \varepsilon,\ m \leqslant j \leqslant n\}$. Then $A_0, A_1, \ldots, A_n$ are disjoint and with $\sum_{i=0}^{n} P(A_i) = 1$. Hence

$$E(\eta) = \sum_{i=0}^{n} E\{\eta|A_i\}P(A_i) \geqslant \sum_{i=1}^{n} E\{\eta|A_i\}P(A_i).$$

From the properties of the ξ_j's, we obtain that for $k \geqslant 1$,

$$E\{S_k^2|A_i\} = E\{S_i^2 + (\xi_{i+1} + \ldots + \xi_k)^2 + 2S_i(\xi_{i+1} + \ldots + \xi_k)|A_i\} \geqslant$$

$$\geqslant E\{S_i^2|A_i\} + 2E\{S_i(\xi_{i+1} + \ldots + \xi_k)|A_i\}.$$

Since $\vec{E}\{S_i\xi_j|A_i\} = 0$ when $j > i$, we have that $E\{S_k^2|A_i\} \geqslant E\{S_i^2|A_i\} \geqslant \varepsilon^2/c_i^2$. Thus

$$E\{\eta|A_i\} = \sum_{k=m}^{n-1} E\{S_k^2|A_i\}(c_k^2 - c_{k+1}^2) + c_n^2 E\{S_n^2|A_i\} \geqslant$$

$$\geqslant \sum_{k=i}^{n-1} E\{S_k^2|A_i\}(c_k^2 - c_{k+1}^2) + c_n^2 E\{S_n^2|A_i\} \geqslant$$

$$\geqslant \frac{\varepsilon^2}{c_i^2}\left[\sum_{k=i}^{n-1} (c_k^2 - c_{k+1}^2) + c_n^2\right] = \varepsilon^2.$$

Hence

$$E(\eta) \geqslant \varepsilon^2 \sum_{i=m}^{n} P(A_i),$$

which completes the proof of Hájek-Rényi's inequality.

21.29. Use DeMorgan's relation: $\overline{\bigcap_{i=1}^{n} A_i} = \bigcup_{i=1}^{n} \overline{A_i}$, and the semi-additivity of the probability P.

21.30. The first inequality follows from Exercise 21.29. To prove the second inequality, first we shall establish the following relation

$$P\left(\bigcup_{i=1}^{n} C_i\right) \geqslant \frac{Q_1^2}{Q_1 + 2Q_2}. \tag{21.6}$$

Consider the indicators $I_i = I_{C_i}$, $I_{ij} = I_{C_i C_j}$. We obviously have $2\sum_{i<j} I_{ij} = \left(\sum_{i=1}^{n} I_i\right)^2 - \sum_{i=1}^{n} I_i^2$. Taking expectations on both sides, we obtain

$$2Q_2 = E\left\{\left(\sum_{i=1}^{n} I_i\right)^2\right\} - \sum_{i=1}^{n} E\{I_i^2\}.$$

Consider now the r.v. $\eta = \sum_{i=1}^{n} I_i$ and let $\Theta = 1$ when $\eta > 0$ and $\Theta = 0$ when $\eta \leqslant 0$. Since $(E\{\Theta\eta\})^2 = (E(\eta))^2$, the Cauchy-Bunyakovski-Schwarz inequality yields $(E(\eta))^2 \leqslant P(\eta > 0)E\{\eta^2\}$. However, $EI_i = E\{I_i^2\} = P(C_i)$ and $P(\eta > 0) = P\left(\bigcup_{i=1}^{n} C_i\right)$. Then we get

$$2Q_2 \geqslant \frac{Q_1^2}{P\left(\bigcup_{i=1}^{n} C_i\right)} - Q_1.$$

This proves (21.6), which, as is easily seen, is equivalent to the second inequality.

21.31. For the complementary events $\overline{A}_1, \ldots, \overline{A}_n$, we find that

$$P\left(\bigcup_{i=1}^{n} \overline{A_i}\right) \leqslant \sum_{i=1}^{n} P(\overline{A_i}) = n - \sum_{i=1}^{n} P(A_i).$$

The equality above can be reached only if $\overline{A}_1, \ldots, \overline{A}_n$ are mutually disjoint. If $\sum_{i=1}^{n} P(A_i) > n - 1$, then

$$P(A_1 \ldots A_n) = P(\overline{\overline{A}_1 \cup \ldots \cup \overline{A}_n}) = 1 - P\left(\bigcup_{i=1}^{n} \overline{A_i}\right) \geqslant 1 - n + \sum_{i=1}^{n} P(A_i).$$

Since it is always true that $P(A_1 \dots A_n) \geqslant 0$, the desired inequality is obtained.

21.32. If $u(x)$ is a non-negative function of x and $\vec{E}\{u(\xi)\}$ exists, then according to Markov's inequality,

$$P\{u(\xi) < c\} \geqslant 1 - \frac{1}{c} E\{u(\xi)\}$$

for any number $c > 0$. If we take $u(x) = \left(x - \frac{x_1 + x_2}{2}\right)^2$ and $c = \left(\frac{x_2 - x_1}{2}\right)^2$, the above inequality yields

$$P\left\{\left(\xi - \frac{x_1 + x_2}{2}\right)^2 < \left(\frac{x_2 - x_1}{2}\right)^2\right\} \leqslant$$

$$\leqslant 1 - \left(\frac{x_2 - x_1}{2}\right)^{-2} E\left\{\left(\xi - \frac{x_1 + x_2}{2}\right)^2\right\}.$$

However,

$$E\left\{\left(\xi - \frac{x_1 + x_2}{2}\right)^2\right\} = \sigma^2 + \left(a - \frac{x_1 + x_2}{2}\right)^2,$$

and hence

$$P\left\{\left|\xi - \frac{x_1 + x_2}{2}\right| < \frac{x_2 - x_1}{2}\right\} \geqslant 1 - \left(\frac{x_2 - x_1}{2}\right)^{-2}\left[\sigma^2 + \right.$$

$$\left. + \ a - \frac{x_1 + x_2}{2}\right)^2\right].$$

After some standard transformations, the last relation yields the desired inequality.

21.33. Let Θ_1 and Θ_2 be independent r.v.'s with $\Theta_1, \Theta_2 \in N(0, 1)$. Put $\xi_1 = \frac{1}{\sqrt{2}}(\Theta_1 - \Theta_2)$, $\xi_2 = \frac{1}{\sqrt{2}}(\Theta_1 + \Theta_2)$, $\xi_3 = \Theta_1$. The probabilities on both sides of the inequality that we are going to prove, can be expressed in terms of Θ_1 and Θ_2. Then the difference Δ between the right-hand side and the left-hand side can be written as

$$\Delta = c[\Phi(-x\sqrt{2} - \varepsilon) - 2\Phi^2(-x)],$$

where $c > 0$ (the explicit form of c is not important). Using the well-known properties of the standard normal d.f. Φ, we conclude that, for every $x > 0$, there exists sufficiently small $\varepsilon > 0$, for which $\Delta > 0$.

21.34. For $\alpha \in (0, \frac{1}{2}]$ the desired inequality is trivial. Therefore we may assume that $\alpha \in (\frac{1}{2}, 1)$. Take $A_k = \{|s_1| \leqslant 2x, \dots, |s_{k-1}| \leqslant 2x, |s_k| > 2x\}$, $B_k = \{|s_n - s_k| \leqslant x\}$ $k = 1, \dots, n$. The events $A_1, \dots, A_n$

are disjoint and for every fixed k, A_k and B_k are independent. The following relation is also true

$$\{|S_n| > x\} \supset \bigcup_{k=1}^{n} (A_k B_k).$$

This yields

$$1 - \alpha > P\{|S_n| > x\} \geqslant P\left\{\bigcup_{k=1}^{n} (A_k B_k)\right\} =$$

$$= \sum_{k=1}^{n} P(A_k)P(B_k) \geqslant \alpha \sum_{k=1}^{n} P(A_k) = \alpha P\{\max_{1\leqslant k\leqslant n} |S_k| > 2x\},$$

which completes the proof.

21.35. According to Exercise 21.21 (d), $(E\{\theta^{1/r}\})^r$ is a monotonically non-increasing function of r. Hence $E\{\xi^k\} \geqslant (E\{\xi^{k/r}\})^r$. To prove the second inequality, we observe first that for arbitrary $a > b > 0$ and $r \geqslant 1$ we have $a^r - b^r \geqslant (a - b)^r$. Then using the equality $\frac{1}{a}[b + (a - b)] = 1$, we conclude that

$$\frac{1}{a^r}[b + (a - b)]^r \geqslant \frac{1}{a^r}[b^r + (a - b)^r], \quad r \geqslant 1.$$

This is obviously equivalent to $a^r \geqslant b^r + (a - b)^r$. The second inequality can easily be derived from this last relation.

21.36. Put $A_k = \{S_1 \leqslant x, \ldots, S_{k-1} \leqslant x, S_k > x\}$, $B_k = \{S_n - S_k \geqslant - 2\sqrt{n}\}$, $A = \{S_n \geqslant x - 2\sqrt{n}\}$. Clearly $A_k B_k \subset A$, $k = 1, \ldots, n$, $A_i A_j = \emptyset$ for $i \neq j$, A_k and B_k are independent for every fixed k. Here

$$\sum_{k=1}^{n} P(A_k)P(B_k) = \sum_{k=1}^{n} P(A_k B_k) = P\left(\sum_{i=1}^{n} A_k B_k\right) \leqslant P(A).$$

According to Chebyshev's inequality, for every $k = 1, \ldots, n$ we have

$$1 - P(B_k) \leqslant P\{ \xi_{k+1} + \ldots + \xi_n| > 2\sqrt{n}\} \leqslant \frac{n - k}{4n} \leqslant \frac{1}{4} ;$$

we then obtain $P(B_k) \geqslant \frac{3}{4}$. Thus

$$P\{S_n > x - 2\sqrt{n}\} = P(A) \geqslant \sum_{k=1}^{n} P(A_k B_k) \geqslant$$

$$\geq \frac{3}{4}\sum_{k=1}^{n} P(A_k) = \frac{3}{4} P\left(\sum_{k=1}^{n} A_k\right) = \frac{3}{4} P\{\max_{1\leq k\leq n} S_k > x\}.$$

21.37. Let $A_1 = \{\xi_1 \geq \varepsilon\}$, $A_k = \{\xi_1 < \varepsilon, \ldots, \xi_{k-1} < \varepsilon, \xi_k \geq \varepsilon\}$, $A = \{\max_{1\leq i\leq n} \xi_i \geq \varepsilon\}$. Then $A = \bigcup_{k=1}^{n} A_k$. Since $A_k \in \mathcal{F}_k$, we conclude from the definition of conditional expectation that

$$\int_A \xi_n dP = \sum_{k=1}^{n} \int_{A_k} \xi_n dP = \sum_{k=1}^{n} \int_{A_k} E(\xi_n | \mathcal{F}_k) dP \geq$$

$$\geq \sum_{k=1}^{n} \int_{A_k} \xi_k dP \geq \varepsilon \sum_{k=1}^{n} P(A_k) = \varepsilon P(A)..$$

Therefore

$$\varepsilon P\{\max_{1\leq i\leq n} \xi_i \geq \varepsilon\} \leq \int_{[\max_i \xi_i \geq \varepsilon]} \xi_n dP \leq E\{|\xi_n|\}.$$

22. Types of Convergence of Sequences of Random Variables

22.1. The statement follows from the assumptions that for a fixed k and for $\varepsilon = 2^{-k}$ there exists m_k, such that for arbitrary n and m with $n > m_k$ and $m > m_k$, we have $P\{\omega : f_n^m(\omega) < 2^{-k}\} < 2^{-k}$. This means that the probability of the event $A_k = \{\omega : f_n^m(\omega) < 2^{-k}\}$ is $P(A_k) < 2^{-k}$. For the sequence $\{A_k\}$, we obviously have $\sum_{k=1}^{\infty} P(A_k) = 1$. According to the Borel-Cantelli lemma, for a.a. $\omega \in \Omega$ and all sufficiently large k, it holds that $f_{m_{k+1}}^{m_k}(\omega) < 2^{-k}$. Taking into account that $f_n^m \leq f_s^m + f_n^s$, we find that for arbitrary k and s there exist m_k and m_s such that $f_{m_s}^{m_k}(\omega) \leq \varepsilon$ for almost all ω and for arbitrary $\varepsilon > 0$. On the other hand, $f_n^m(\omega)$ are non-negative r.v.'s and hence $f_{m_s}^{m_k} \to 0$ with probability 1, as $k \to \infty$ and $s \to \infty$.

22.2. The assertion follows from Exercise 22.1 with $f_n^m = |\xi_n - \xi_m|$.

22.3. If $\xi_n \to \xi$ in some definite sense, then

$$|\xi_n - \xi_k| \leq |\xi_n - \xi| + |\xi_k - \xi|,$$

$$\sup_{n\geqslant k} |\xi_n - \xi_k| \leqslant \sup_{n\geqslant k} |\xi_n - \xi| + |\xi_k - \xi|,$$

$$|\xi_n - \xi_k|^r \leqslant c_r|\xi_n - \xi|^r + c_r|\xi_k - \xi|^r, \quad c_r > 0.$$

This proves the necessity of (a), (b) and (c). The sufficiency of (a) and (b) follows from Exercise 22.2. The proof of (c) requires some analytical methods (see the books [19] and [37] cited at the end of this Manual).

22.4. According to the Cauchy-Bunyakovski-Schwarz inequality (see Exercise 21.21),

$$|E\{\xi_n - \xi\}| \leqslant E\{|\xi_n - \xi|\} \leqslant (E\{1^2\})^{1/2}(E\{|\xi_n - \xi|^2\})^{1/2} \to 0.$$

(b) The following representation is to be used:

$$\xi_n\eta_m - \xi\eta = (\xi - \xi_n)(\eta - \eta_m) - (\xi - \xi_n)\eta - \xi(\eta_m - \eta).$$

(c) For every $\varepsilon > 0$ and for all sufficiently large n, we have

$$\varepsilon \geqslant E\{(\xi_n - \xi)^2\} = V\{\xi_n - \xi\} + (E\{\xi_n - \xi\})^2 \geqslant (E\xi_n - E\xi)^2.$$

Therefore $|E\xi_n - E\xi| \to 0$ as $n \to \infty$.

(d) Since $E\{\xi_n^2\} = E\{(\xi_n - \xi + \xi)^2\} = E\{(\xi_n - \xi)^2 + E\{\xi^2\} + 2E\{\xi(\xi_n - \xi)\}$ and since $\xi_n \xrightarrow{L_2} \xi$, we easily obtain by the Cauchy-Bunyakovski-Schwarz inequality:

$$E\{\xi_n^2\} - E\{\xi^2\} \to 0.$$

22.5. Use Markov's inequality.

22.6. We find that $P\{|\xi_n - \xi_m| < \varepsilon\} = \frac{1}{\pi} \operatorname{arc}[\tan(\varepsilon\sigma_{mn}) - \operatorname{arc}\tan(-\varepsilon\sigma_{mn})]$, which converges to 1 as $m, n \to \infty$. From Exercise 22.3 it follows that there exists a r.v. ξ, such that $\xi_n \xrightarrow{P} \xi$.

22.7. In Exercise 21.4 take $g(x) = |x|^r/(1 + |x|^r)$. Then

$$E\{|\xi_n - \xi|^r/(1 + |\xi_n - \xi|^r)\} \leqslant P\{|\xi_n - \xi| \geqslant \varepsilon\} + \varepsilon^r/(1 + \varepsilon^r),$$

$$P\{|\xi_n - \xi| \geqslant \varepsilon\} \leqslant [(1 + \varepsilon^r)/\varepsilon^r]E\{|\xi_n - \xi|^r/(1 + |\xi_n - \xi|^r)\}.$$

Now taking the limit as $n \to \infty$ and $\varepsilon \to 0$, we obtain both assertions.

22.8. Let the sequence $\{\eta_n\}$ be monotonically decreasing and let $\eta_n \xrightarrow{P} 0$. Then for every $\varepsilon > 0$ we have $\{\eta_{n+1} \geqslant \varepsilon\} \subset \{\eta_n \geqslant \varepsilon\}$ and thus

$\{\lim_{n\to\infty} \eta_n \geq \varepsilon\} = \bigcap_n \{\eta_n \geq \varepsilon\}$ and $P\{\lim_{n\to\infty} \eta_n \geq \varepsilon\} = \lim_{n\to\infty} P\{\eta_n \geq \varepsilon\} = 0$. When $\{\eta_n\}$ is monotonically increasing, we come to the same conclusion. The statement follows from the above reasoning with $\eta_n = \xi_n - \xi$.

22.9. Let $P\{\xi = c\} = 1$ and let F and F_n be the d.f.'s of ξ and ξ_n. Obviously $F(x) = 0$ for $x \leq c$ and $F(x) = 1$ for $x > c$. The assumption says that $F_n(x) \to 0$ if $x < c$ and $F_n(x) \to 1$ if $x > c$. Therefore $P\{\xi_n < c - \varepsilon\}$ and $P\{\xi_n > c + \varepsilon\}$ both converge to 0 as $n \to \infty$. This means that $P\{|\xi_n - c| > \varepsilon\} \to 0$; i.e., $\xi_n \overset{P}{\to} = c = \xi$.

22.10. (a) Let F_n and G be the d.f.'s of ξ_n and η, respectively, let $\zeta_n = \eta_n - \xi_n$ and let x be a continuity point for G. Then

$$F_n(x) = P\{\xi_n < x\} = P\{\eta_n < x + \zeta_n\} =$$

$$= P\{\eta_n < x + \zeta_n, \zeta_n < \varepsilon\} + P\{\eta_n < x + \zeta_n, \zeta_n \geq \varepsilon\} \leq$$

$$\leq P\{\eta_n < x + \varepsilon\} + P\{\zeta_n \geq \varepsilon\}.$$

The last relation shows that $\limsup_n F_n(x) \leq G(x + \varepsilon)$. Similarly, we can show that $\liminf_n F_n(x) \geq G(x - \varepsilon)$. Therefore $\lim_{n\to\infty} F_n(x) = G(x)$; i.e., $\xi_n \overset{d}{\to} \eta$.

(b) We have $\xi_n + c \overset{d}{\to} \xi + c$, $(\xi_n + \xi_n) - (\xi_n + c) = \eta_n - c \overset{P}{\to} 0$. Combining the last relation with (a), we conclude that $\xi_n + \eta_n \overset{d}{\to} \xi + c$. The other assertions in (b) can be proved analogously.

(c) This assertion follows from (a) and Exercise 22.9. It can be proved directly as well.

22.11. From (22.5) we get

$$E\{\exp[itg(\xi_n)]\} = \int_{-\infty}^{\infty} e^{itg(x)} dF(x) =$$

$$= \int_{-\infty}^{\infty} [\cos(tg(x)) + i \sin(tg(x))] dF_n(x) \to$$

$$\to \int_{-\infty}^{\infty} [\cos(tg(x)) + i \sin(tg(x))] dF(x) =$$

$$= E\{\exp[itg(\xi)]\}.$$

To complete the proof, refer to the Continuity theorem, given in Section 24.

(b) Let δ, ε, $\varepsilon_1 > 0$ and let I_ε be a finite interval, for which

$P\{\xi \in I_\varepsilon\} = 1 - \varepsilon/2$ and $P\{|\xi_n - \xi| > \delta\} > 1 - \varepsilon/2$ for $n \geq n_0$.

It follows from the assumptions for $g(x)$ that $|g(\xi_n) - g(\xi)| < \varepsilon_1$ a.s., if $|\xi_n - \xi| < \delta$ for every $\xi \in I_\varepsilon$. Hence for $n \geq n_0$ we have

$$P\{|g(\xi_n) - g(\xi)| < \varepsilon_1\} \geq P\{|\xi_n - \xi| < \delta, \ \xi \in I_\varepsilon\} \geq$$

$$\geq P\{|\xi_n - \xi| < \delta\} - P\{\xi \overline{\in} I_\varepsilon\} \geq 1 - \varepsilon.$$

22.12. By Markov's inequality we have $P\{|\xi_n| \geq \varepsilon\} \leq \varepsilon^{-1}E\{|\xi_n|\}$. Thus if $E\{|\xi_n|\} \to 0$, then $P\{|\xi_n| \geq \varepsilon\} \to 0$, i.e., $\xi_n \xrightarrow{P} 0$. Now assume that $\xi_n \xrightarrow{P} 0$. If $A_n(\delta) = \{|\xi_n| > \delta\}$, then

$$E\{|\xi_n|\} = E\{|\xi_n||A_n(\delta)\}P(A_n(\delta)) + E\{|\xi_n||\overline{A_n(\delta)}\}P(\overline{A_n(\delta)}) \leq$$

$$\leq cP(A_n(\delta)) + \delta.$$

Since $P(A_n(\delta)) \to 0$ for every $\delta > 0$, it is easy to see that $E\{|\xi_n|\} \to 0$.

22.13. Let $\varepsilon > 0$ be arbitrarily chosen. One can find an integer n_ε such that for every $n \geq n_\varepsilon$, we have

$$P\{|\xi_n - 0| > 0\} = P\{|\xi_n| > 0, \ \xi_n = 0\} + P\{|\xi_n| > 0, \ \xi_n = n^{\alpha/r}\} =$$

$$= \frac{1}{n} \to 0;$$

i.e., $\xi_n \xrightarrow{P} 0$. Further $E\{|\xi_n|^r\} = n^{\alpha-1}$, which obviously does not converge to 0, because α is assumed to be greater than 1.

22.14. The first assertion can be proved directly. It also follows from Exercise 22.10 (a) with $\eta_n = \eta$, $n = 1, 2, \ldots$ Now let ξ be a r.v., which is 0 or 1 with probability $\frac{1}{2}$ and let $\Theta = 1 - \xi$. Obviously ξ and Θ are identically distributed and $|\xi - \Theta| = 1$. Consider the sequence $\{\xi_n\}$ with $\xi_n = \Theta$. The variables ξ_n have the same distribution as ξ and thus $\xi_n \xrightarrow{d} \xi$. At the same time $\{\xi_n\}$ does not converge to ξ in probability, since $|\xi_n - \xi| = |\Theta - \xi| = 1$.

22.15. It suffices to consider the case $c = 0$. The constant $e^{it0} = 1$ is the ch.f. of a r.v. which equals 0 with probability 1. On the other hand this r.v. has d.f. F given by $F(x) = 0$ for $x \leq 0$ and $F(x) = 1$ for $x > 0$. The equivalence of (b) and (c) follows from the relation between ch.f. and d.f. According to Exercise 22.14, (a) implies (b) and according to Exercise 22.9, (b) implies (a). Thus the three conditions are equivalent.

22.16. Obviously $E\{|\xi_n|^r\} = 1/n^r$ and we conclude that $\xi_n \xrightarrow{L_r} 0$. Put $A_{k,\delta} = \{|\xi_k| \leq \delta\}$ and note that $|\xi_j| > |\xi_k|$ when $j < k$. Hence $A_{j,\delta} \subset A_{k,\delta}$

and therefore $B_{n,\delta} = \bigcap_{j=n}^{\infty} A_{j,\delta} = A_{n,\delta}$. Let $\delta > 0$ be fixed. For $n > 1/\alpha$ we have $P(B_{n,\delta}) = P(A_{n,\delta}) = P(|\xi_n| \leqslant \delta) = 1$. Hence $\xi_n \xrightarrow{a.s.} 0$. Then we have (see Exercise 22.1) $\xi_n \xrightarrow{P} 0$ and we conclude that (see Exercise 22.14) $\xi_n \xrightarrow{d} 0$.

22.17. It is easy to show that $\xi_n \xrightarrow{P} 1$, $E(\xi_n) = 1 + 4/(n + 4)$. Then obviously $\lim_{n\to\infty} E(\xi_n) = 1 \neq E\{P\text{-}\lim_{n\to\infty} \xi_n\} = -1$.

22.18. We have

$$0 \leqslant E\left\{\left[\frac{1}{n} \sum_{i=1}^{n} (\xi_i - E\xi_i)\right]^2\right\} = \frac{1}{n^2} \sum_{i=1}^{n} E\{(\xi_i - E\xi_i)^2\} +$$

$$+ \frac{2}{n^2} \sum_{i<j} E\{(\xi_i - E\xi_i)(\xi_j - E\xi_j)\} \leqslant \frac{1}{n} \sigma^2$$

which converges to 0 as $n \to \infty$.

22.19. It is obvious that $F_{\eta_n}(y) = P\{\eta_n < y\} = 0$ for $y \leqslant 0$ and $F_{\eta_n}(y) = 1$ for $y > 1$. Since $0 < y \leqslant 1$, we have $F_{\eta_n}(y) = P\{\max[\xi_1, \ldots, \xi_n] < y\} = (P\{\xi_1 < y\})^n$. Now it is easy to see that ζ_n assumes its values in the interval $(0, n)$ and for $x \in (0, n)$ we get

$$F_{\zeta_n}(x) = P\{0 < \zeta_n < x\} = P\{1 - \frac{x}{n} < \eta_n < 1\} = 1 - (1 - \frac{x}{n})^n.$$

Obviously $F_{\zeta_n}(x) \to 1 - e^{-x}$. The last expression is the d.f. of a r.v. $\zeta \in \mathcal{E}(1)$, and hence $\zeta_n \xrightarrow{d} \zeta$ as $n \to \infty$.

22.20. We have $F_{\eta_n}(x) = [F(x)]^n$. Then $F_{\zeta_n}(x) = [F(x + \ln(nb))]^n$. It follows from the assumptions that $F(x) = 1 - be^{-x} + o(e^{-x})$, $x \to \infty$. Hence $F_{\zeta_n}(x) \to \exp(-e^{-x}) = G(x)$, $x \in \mathbb{R}_1$. The reader should check that G is a d.f. of some r.v. ξ. Thus $\zeta_n \xrightarrow{d} \zeta$.

22.21. Let $x > 0$ be fixed. For sufficiently large n, namely for $nx > 1$, we have

$$P\{n^{-1/\alpha}\eta_n < x\} = P\{\eta_n < xn^{1/\alpha}\} = (1 - \frac{x^{-\alpha}}{n})^n,$$

which converges to $\exp(-x^{-\alpha})$ as $n \to \infty$. For $x < 0$ it holds that $\lim_{n\to\infty} P\{n^{-1/\alpha}\eta_n < x\} = 0$. Finally we get $n^{-1/\alpha}\eta_n \xrightarrow{d} \zeta$, where

$$P\{\zeta < x\} = \begin{cases} \exp(-x^{-\alpha}), & \text{if } x > 0, \\ 0, & \text{if } x \leqslant 0. \end{cases}$$

Note that the above distribution is one of the three possible distributions of the maximum order statistics (see also Exercise 22.19 and Exercise 22.20).

22.22. One can use Exercise 22.3 and Exercise 21.21.

22.23. According to Exercise 16.42, for $x \in (0,1)$ we have

$$P\{n\xi_{(k)} < x\} = \sum_{j=0}^{k-1} \binom{n}{j}\left(\frac{x}{n}\right)^j\left(1 - \frac{x}{n}\right)^{n-j},$$

which yields

$$\lim_{n\to\infty} P\{n\xi_{(k)} < x\} = e^{-1}\left[1 + \frac{x}{1!} + \frac{x^2}{2!} + \dots + \frac{x^{k-1}}{(k-1)!}\right].$$

Denote by $F_k(x)$, $x \in R_1$, the right-hand side of the last relation. Thus we have shown that $n\xi_{(k)} \xrightarrow{d} \zeta_k$ as $n \to \infty$, where the d.f. of ζ_k is exactly F_k (*Erlang's distribution* with k degrees of freedom). It is also easily seen that F_k is the d.f. of a Γ-distributed r.v. with parameters k and 1 (see (15.10)).

22.24. Use Chebyshev's inequality and Exercise 22.15.

22.25. (a) Using Kolmogorov's inequality (Exercise 21.28) show that the sequence $\{S_n\}$, $S_n = \xi_1 + \dots + \xi_n$ is fundamental in the sense of convergence with probability 1. Then it only remains to use Exercise 22.3.

(b) Again Exercise 22.3 is to be used.

22.26. According to the Borel-Cantelli lemma, only finitely many events $\{|\xi_n| \geqslant c\}$ might occur. Since $c > 0$ is arbitrarily chosen, this is equivalent to the assertion that $\xi_n \to 0$ with probability 1.

22.27. From the convergence of the series $\sum_{n=1}^{\infty} V(\xi_n)$, from Kolmogorov's inequality (Exercise 21.28) and from the Borel-Cantelli lemma, it follows that $\sum_{n=1}^{\infty} (\xi_n - E(\xi_n))$ is convergent with probability 1. Since it is assumed that $\sum_{n=1}^{\infty} E(\xi_n) < \infty$, we conclude that $\sum_{n=1}^{\infty} \xi_n$ is also convergent with probability 1.

Let us now assume that $\xi_1, \xi_2, \dots$ are uniformly bounded and let $\sum_{n=1}^{\infty} \xi_n$ be convergent with probability 1. Consider a new sequence $\{\tilde{\xi}_n\}$ of independent r.v.'s which are also independent of the r.v.'s $\{\xi_n\}$, and which are chosen such that $\tilde{\xi}_n$ and ξ_n have the same distribution, n = 1,

2, ... Then $\sum_{n=1}^{\infty} \tilde{\xi}_n$ and $\sum_{n=1}^{\infty} (\xi_n - \tilde{\xi}_n)$ are both convergent with probality 1. But $E\{\xi_n - \tilde{\xi}_n\} = 0$ and it follows from the convergence of $\sum_{n=1}^{\infty} (\xi_n - \tilde{\xi}_n)$, and from Kolmogorov's inequality as well, that $\sum_{n=1}^{\infty} V\{\xi_n - \tilde{\xi}_n\} < \infty$. From the last relation we get $\sum_{n=1}^{\infty} V(\xi_n) = \frac{1}{2} \sum_{n=1}^{\infty} V\{\xi_n - \tilde{\xi}_n\} < \infty$. It follows from the above reasoning that $\sum_{n=1}^{\infty} (\xi_n - E(\xi_n))$ is convergent with probability 1, which together with the convergence of $\sum_{n=1}^{\infty} \xi_n$ yields the convergence of $\sum_{n=1}^{\infty} E(\xi_n)$.

22.28. Since $V(\xi_n) = p_n(1 - p_n)$, it follows from Exercise 22.27 that the desired condition is $\sum_{n=1}^{\infty} p_n(1 - p_n) < \infty$.

22.29. It is easy to see that $\lim_{n\to\infty} p_n = 0$ is necessary and sufficient condition for the convergence $\xi_n \xrightarrow{P} 0$ and also for the convergence $\xi_n \xrightarrow{L_r} 0$. On the other hand $\xi_n \xrightarrow{a.s.} 0$ only if $\sum_{n=1}^{\infty} p_n < \infty$. Hence for $p_n = \frac{1}{n}$ we obtain a sequence $\{\xi_n\}$, which is not convergent with probability 1, although it is convergent in L_r-sense (and therefore convergent also in probability).

22.30. We choose a point M at random in the unit cube in $\mathbb{R}_n$. Let $\xi_1, \ldots, \xi_n$ be the coordinates of M. Then $\xi_1, \ldots, \xi_n$ are i.i.d. r.v.'s, which are uniformly distributed over the interval (0, 1). Put

$$\eta_n = \frac{1}{n}[f(\xi_1) + \ldots + f(\xi_n)], \qquad \zeta_n = \frac{1}{n}[g(\xi_1) + \ldots + g(\xi_n)].$$

As it is easily seen that the following relations hold:

$$\eta_n \xrightarrow{P} a_1 = \int_0^1 f(x)dx, \qquad \zeta_n \xrightarrow{P} a_2 = \int_0^1 g(x)dx \qquad \text{as } n \to \infty.$$

However, $a_2 > 0$ and therefore $(\eta_n/\zeta_n) \xrightarrow{P} (a_1/a_2)$ as $n \to \infty$. Since $0 \leq \eta_n/\zeta_n < c$ then $\lim_{n\to\infty} E\{\eta_n/\zeta_n\} = a_1/a_2$.

Note. A more general result can be found in the following paper: Stoyanov, J. (1986): Probabilistic proof of the convergence of a class of n-fold integrals. *Glasnik Matem. (Zagreb)* 21, p. 101-114.

22.31. Show first that the density of Θ is given by $\frac{1}{2} e^{-|x|}$, $x \in \mathbb{R}_1$, and then use Exercise 22.27.

22.32. It is easy to see that $1 - \eta_n = 1 - \max[\xi_1, \ldots, \xi_n] = \min[1 - \xi_1, \ldots, 1 - \xi_n] = \min[\Psi_1, \ldots, \Psi_n]$, where $\Psi_1 = 1 - \xi_1, \ldots, \Psi_n = 1 - \xi_n$, are also i.i.d. r.v.'s, which are uniformly distributed over the interval (0, 1). For an arbitrary $x > 0$, we have

$$P\{n \cdot \min \Psi_1, \ldots, \Psi_n > x\} = P\{\min[\Psi_1, \ldots, \Psi_n] > \frac{x}{n}\} =$$

$$= (1 - \frac{x}{n})^n,$$

which converges to e^{-x} as $n \to \infty$. This means that the sequence $\{n(1 - \eta_n)\}_{n=1}^{\infty}$ converges weakly to the r.v. Ψ, where $\Psi \in E(1)$. Since g is continuous and since $P\{\Psi > 0\} = 1$, we have, according to Exercise 22.11, that $g(n(1 - \eta_n)) \xrightarrow{d} g(\Psi)$. Applying (16.11), one can easily find the density f_ζ of $\zeta = g(\Psi)$.

22.33. Put $A_{n,N} = \{\sup_{k,j \geq n} |S_k - S_j| > \frac{1}{N}\}$. Then the series $\sum_{n=1}^{\infty} \xi_n$ is divergent over the set $D = \bigcup_{N=1}^{\infty} \bigcap_{n=1}^{\infty} A_{n,N}$. Next we shall compute the probability of D.

Let $\varepsilon > 0$ and $\delta > 0$ be chosen arbitrarily. According to the imposed assumptions, $\sum_{n=1}^{n} \xi_n$ converges in probability. Then it follows that there exists $n_0 = n_0(\varepsilon, \delta)$, such that for arbitrary $k_1, j > n_0$ it holds that $P\{|S_k - S_j| > \varepsilon\} < \delta$. The inequality in Exercise 21.34 shows that

$$P\{\max_{n_0 < i < k} |S_i| > 2\varepsilon\} = P\{\max_{n_0 < i < k} |S_i - S_{n_0}| > 2\varepsilon\} \leq \frac{\delta}{1 - \delta}$$

for any $k > n_0$. Thus we get

$$P\{\sup_{i > n_0} |S_i - S_{n_0}| > 2\varepsilon\} \leq \frac{\delta}{1 - \delta},$$

where $\delta > 0$ can be chosen arbitrarily small. This obviously means that $P(A_{n,N}) \leq \frac{\delta}{1 - \delta}$ and $P\left(\bigcap_{n=1}^{\infty} A_{n,N}\right) = 0$ for any N. Thus $\sum_{n=1}^{\infty} \xi_n$ is divergent over the set D with $P(D) = 0$ and it is convergent over the set $\overline{D}$ with $P(\overline{D}) = 1$; i.e., $\sum_{n=1}^{\infty} \xi_n$ is convergent with probability 1.

22.34. Obviously $\zeta \geq \sup_n [\xi_n - nc]$. On the other hand $\eta_n \leq \max[\xi_k - kc, k = 1,\dots, n]$, which yields $\zeta \leq \sup_n[\xi_n - nc$, and therefore $\zeta = \sup_n[\xi_n - nc]$.

Now let F be the common d.f. of the given variables. Since it is assumed that $E\{|\xi_1|\} < \infty$, for any $\alpha > 0$, we have

$$\sum_{n=1}^{\infty} P\{\xi_1 > n\alpha\} = \sum_{n=1}^{\infty} (1 - F(n\alpha)) \leq \int_0^{\infty} (1 - F(\alpha x))dx =$$

$$= \frac{1}{\alpha}\int_0^{\infty} x\ dF(x) < \infty.$$

Taking into account that $\xi_1, \xi_2, \dots$ are identically distributed, for $\alpha = \frac{c}{2}$ we obtain $\sum_{n=1}^{\infty} P\{\xi_n > n\frac{c}{2}\} < \infty$. According to the Borel-Cantelli lemma, only for finitely many n we have $\xi_n > n\frac{c}{2}$. This means that $\lim_{n\to\infty} (\xi_n - nc) = -\infty$ with probability 1. On the other hand $\zeta = \sup_n[\xi_n - nc]$ and hence $P\{\zeta < \infty\} = 1$. Now we wish to prove that $\eta_n \xrightarrow{a.s.} -\infty$. Consider the r.v. $\Theta_n = \max[\xi_1, \dots, \xi_n] - \frac{1}{2}nc$, $n = 1, 2, \dots$ Put $\tau = \sup_n \Theta_n$. The above reasoning implies that $P\{\tau < \infty\} = 1$. Since $\eta_n = \Theta_n - \frac{1}{2}nc \leq \tau - \frac{1}{2}nc$, $n = 1, 2, \dots$, we conclude that $\eta_n \xrightarrow{a.s.} -\infty$ as $n \to \infty$.

23. Laws of Large Numbers

23.1. It is easy to check that each of the four sequences satisfies the Markov's condition (23.3) and thus each of these sequences obeys the WLLN. Even more, in the four cases Kolmogorov's condition (23.4) holds. Hence the given sequences obey the SLLN.

23.2. (a) If $-1 < \alpha < 1$, the sequence $\{\eta_n\}$ obeys both the WLLN and SLLN. (b) The sequence $\{\zeta_n\}$ obeys the WLLN and the SLLN for $\alpha < \frac{1}{2}$.

23.3. Using the explicit form of the distributions of η_j and ζ_j, we can easily check the validity of Markov's condition (23.3) and conclude that each of the sequences $\{\eta_n\}$ and $\{\zeta_n\}$ satisfies the WLLN. However, Markov's condition is not fulfilled for the sequence $\{\xi_n\}$. Nevertheless, $\{\xi_n\}$ also satisfies the WLLN. This follows from the general fact that $X_n \xrightarrow{P} c$, c is a constant, if $\phi_{X_n}(t) \to e^{itc}$ (see Exercise 22.15 (c)). Moreover, from the sequence $\{\xi_n\}$ we have shown that Markov's condition (23.3) is not necessary for the validity of the WLLN.

23.4. Let $S_n = \xi_1 + \dots + \xi_n$, $E(\xi_1) = a$, $V(\xi_1) = \sigma^2$. According to the CLT,

$P\{(S_n - A_n)/B_n < x\} \to \Phi(x)$ as $n \to \infty$, where Φ is the standard normal d.f., $A_n = E(S_n)$, $B_n^2 = V(S_n)$. In our case $A_n = na$, $B_n^2 = n\sigma^2$ and we easily find that for every fixed $\varepsilon > 0$

$$P\left\{\left|\frac{S_n}{a} - a\right| < \varepsilon\right\} = P\left\{\left|\frac{S_n - A_n}{B_n}\right| < \frac{n\varepsilon}{B_n}\right\} \approx 2\Phi\left(\frac{n\varepsilon}{B_n}\right) - 1.$$

The last quantity converges to 1 as $n \to \infty$ since $\frac{n\varepsilon}{B_n} \to \infty$ and $\Phi(x) \to 1$ as $x \to \infty$. This means that $\{\xi_n\}$ obeys the WLLN.

23.5. We have $V\{\xi_1 + \ldots + \xi_n\} = \sum_{k=1}^{\infty} V(\xi_k) + \sum_{i \neq j} a_{ij}$, where $a_{ij} = \rho_{ij}\sqrt{V(\xi_i)V(\xi_j)} < 0$ for arbitrary $i, j, i \neq j$. Then

$$\frac{1}{n^2} V\{\xi_1 + \ldots + \xi_n\} < \frac{1}{n^2} \cdot nc \to 0,$$

and hence $\{\xi_n\}$ obeys the WLLN.

23.6. $V\{\xi_1 + \ldots + \xi_n\} = \sum_{k=1}^{n} V(\xi_k) + 2(a_{12} + a_{13} + \ldots + a_{n-1,n})$, where a_{ij} is the covariance between ξ_i and ξ_j. The assumptions and the Cauchy-Bunyakovski-Schwarz inequality show that $|a_{i,i+1}| \leqslant \sqrt{a_{i,i}a_{i+1,i+1}}$. Also for $|i - j| > 1$, we have $a_{ij} = 0$. Hence

$$\frac{1}{n^2} V\{\xi_1 + \ldots + \xi_n\} < \frac{1}{n^2}(nc + 2nc) \to 0 \quad \text{as } n \to \infty,$$

and according to Markov's theorem $\{\xi_n\}$ obeys the WLLN.

23.7. We have $V\{\xi_1 + \ldots + \xi_n\} = \sum_{k=1}^{n} V(\xi_k) + 2\sum_{i<j} \rho_{ij}\sqrt{V(\xi_i)V(\xi_j)}$. Let $\varepsilon > 0$ be arbitrary chosen. Then one can find N_ε, such that for arbitrary i, j with $|i - j| > N_\varepsilon$ it holds that $|\rho_{ij}| < \varepsilon$. Consider the second sum $\sum_{i<j}$ above, assuming that $n > N_\varepsilon$. Then for every fixed $i = 1, \ldots, n$ the inequality $\varepsilon < |\rho_{ij}| < 1$ holds for no more than $2N$ values of j. Hence

$$\frac{1}{n^2} V\{\xi_1 + \ldots + \xi_n\} < \frac{1}{n^2}[nc + 4cnN + \varepsilon cn(n + 1)] =$$

$$= \frac{1}{n^2}(c_1 n + c_2 \varepsilon n^2).$$

The last quantity converges to 0 as $n \to \infty$ and $\varepsilon \to 0$. Thus $\{\xi_n\}$ obeys the WLLN.

23.8. Use Chebyshev's inequality.

23.9. (a) We have $E(\xi_k = 0$, $V(\xi_k) = k^{2\beta-\alpha}$. According to Exercise 23.8 it needs to determine α and β for which $\frac{1}{2} V(\xi_n) \to 0$ as $n \to \infty$. Obviously $\alpha > 2\beta - 1$. The same result can be obtained if condition (23.3) is checked.

(b) Since the series $\sum_{k=1}^{\infty} k^{-2} V(\xi_k) = \sum_{k=1}^{\infty} 1/k^{2-2\beta+\alpha}$ is convergent for $2 - 2\beta + \alpha > 1$, we can conclude that: If $\beta > \frac{1}{2}$ and $\alpha > 2\beta - 1$, then the sequence $\{\xi_n\}$ obeys the WLLN and the SLLN.

23.10. Consider the Bernoulli scheme (n, x) and let μ_n be the number of successes. It is easily seen that

$$B_n(f; x) = E\left\{f\left(\frac{\mu_n}{n}\right)\right\} = \sum_{k=0}^{n} f(\tfrac{k}{n}) P_n(k),$$

where $P_n(k) = \binom{n}{k} x^k (1 - x)^{n-k}$. Let $\varepsilon > 0$ be fixed and let A be the set of those k, k = 0, 1, ..., n, for which $|\frac{k}{n} - x| < \varepsilon$. Then

$$|B_n(f; x) - f(x)| = \left| \sum_{k=0}^{n} [f(\tfrac{k}{n}) - f(x)] P_n(k) \right| \leqslant$$

$$\leqslant \sum_{k=0}^{n} |f(\tfrac{k}{n}) - f(x)| P_n(k) \leqslant$$

$$\leqslant \sum_{k:k\in A} + \sum_{k:k\in \bar{A}} \leqslant \varepsilon + \frac{M}{n\varepsilon^2}.$$

When the first sum above was estimated, the uniform continuity of f was used. When the second sum was estimated, it was used that f is bounded and Chebyshev's inequality was applied. Hence $B_n(f; x) \to f(x)$ as $n \to \infty$ uniformely in $x \in [0, 1]$.

The above theorem of Bernstein is a stronger result than the classical theorem of K. Weierstrass, which only states that every function, which is continuous on a finite interval, can be uniformly approximated with polynomials.

23.11. Put $C_n(f; x) = \sum_{k=0}^{n} [f(\frac{k}{n}) \binom{n}{k}] x^k (1 - x)^{n-k}$, where $[\cdot]$ stands for the greatest integer not exceeding the quantity in the bracket. Obviously $[f(0)\binom{n}{0}] = f(0)$ and $[f(1)\binom{n}{n}] = f(1)$ are integers. Then we proceed as in Exercise 23.10.

23.12. From the assumptions it follows that for small t we have $\phi(t) = 1 + iat + o(t)$. If $\eta_n = \frac{1}{n}(\xi_1 + \ldots + \xi_n)$, then $\phi_{\eta_n}(t) = [\phi(\frac{t}{n})]^n =$

$[1 + \frac{iat}{n} + o(\frac{t}{n})]^n$. Clearly $\phi_{\eta_n}(t) \to e^{iat}$ as $n \to \infty$, for every $t \in \mathbb{R}_1$. The assertion follows from Exercise 22.15.

23.13. Denote the ch.f. of ξ_n by $\phi(t)$, $t \in \mathbb{R}_1$. Since $E(\xi_n)$ exists, for small t the function $\phi(t)$ can be written as $\phi(t) = 1 + iat + o(t)$, where $a = E(\xi_n)$. Then one can use the same reasoning as in Exercise 23.12.

23.14. Put $\eta_n = \frac{1}{n}(\xi_1 + \ldots + \xi_n)$. Then

$$\phi_n(t) = E\left\{e^{it\eta_n}\right\} = \left[\int_{-\infty}^{\infty} e^{itx/n} dF(x)\right]^n =$$

$$= \left[1 + \int_{-\infty}^{\infty} (e^{itx/n} - 1) dF(x)\right]^n =$$

$$= \left[1 + \frac{1}{n}\left(it \int_{-n}^{n} x dF(x) + \delta_n\right)\right]^n,$$

where

$$|\delta_n| \leqslant n[F(-n) + 1 - F(n)] + \frac{t^2}{n} \int_{-n}^{n} x^2 \, dF(x).$$

According to (b), $n[F(-n) + 1 - F(n)] \to 0$. Since $\frac{1}{n}\int_{-n}^{n} x^2 dF(x) \to 0$ (from (a) and (b)), we have that $\delta_n \to 0$ as $n \to \infty$ and therefore $\lim_{n\to\infty} \phi_n(t) = 1$. According to Exercise 22.5, the last relation means that $\eta_n \xrightarrow{P} 0$.

23.15. We shall use Exercise 22.15. We can easily see that the ch.f. of ξ_k is $\cos(t\sqrt{k})$. Then the ch.f. of the r.v. $\eta_n = \frac{1}{n}(\xi_1 + \ldots + \xi_n)$ is given by $\phi_n(t) = \prod_{k=1}^{n} \cos\frac{t\sqrt{k}}{n}$. For the convergence of $\{\eta_n\}$ in probability to 0, it is necessary that $\lim_{n\to\infty} \phi_n(t) = e^{ito} = 1$. But $\cos\frac{t\sqrt{k}}{n} = 1 - \frac{t^2 k}{2n^2} + o(\frac{1}{n^3})$ and we obtain that $\lim_{n\to\infty} \phi_n(t) = e^{-t^2/4}$. Hence $\{\eta_n\}$ does not converge in probability to 0, although $E(\eta_n) = 0$ for each n.

Remark. From the Continuity theorem (see Section 24) it follows that $\eta_n \xrightarrow{d} \eta$, where $\eta \in N(0, \frac{1}{2})$.

23.16. Let F_n be the d.f. of S_n. Then, for every sufficiently small $\varepsilon > 0$ and $A_n = E(S_n)$, we have:

$$V(S_n) = \int_{-cn}^{cn} (x - A_n)^2 dF_n(x) = \int_{|x-A_n|\leqslant n} (x - A_n)^2 dF_n(x) +$$

$$+ \int_{\varepsilon n\leqslant|x-A_n|\leqslant cn} (x - A_n)^2 dF_n(x) <$$

$$< \varepsilon^2 n^2 + c^2 n^2 \int_{\varepsilon n\leqslant|x-A_n|\leqslant c_n} dF_n(x) = \varepsilon^2 n^2 + c^2 n^2 I_n.$$

If the integral I_n converges to 0 as $n \to \infty$, then we must have $V(S_n) = \varepsilon^2 n^2 + o(n^2)$ and choosing ε sufficiently small, we come to a contradiction with the relation $V(S_n) > \alpha n^2$. Hence $I_n \geqslant c_1 > 0$, which means that $P\left\{\left|\frac{S_n - A_n}{n}\right| > \varepsilon\right\} > c_1 > 0$, and we conclude that $\{\xi_n\}$ does not obey the WLLN.

23.17. We have that $P\{\xi^{(n)} \in B_{\sqrt{n},\varepsilon}\} = P\left\{\left|\frac{\xi_1^2 + \ldots + \xi_n^2}{n} - 1\right| \leqslant \varepsilon\right\}$ and the WLLN, applied to the sequence $\{\xi_n^2\}$ (obviously these r.v.'s are i.i.d. and $E\{\xi_n^2\} = 1 < \infty$), yields the first assertion. The sequence $\{|\xi_n|\}$ also obeys the WLLN, since the variables $|\xi_n|$ are independent, $E\{|\xi_n|\} = \sqrt{2/\pi}$ and $V\{|\xi_n|\} = 1 - \frac{2}{\pi} < \infty$. Now we conclude that

$$P\{\xi^{(n)} \in C_{n\sqrt{2/\pi},\varepsilon}\} = P\left\{\left|\frac{|\xi_1| + \ldots + |\xi_n|}{n\sqrt{2/\pi}} - 1\right| \leqslant \varepsilon\right\} \to 1$$

as $n \to \infty$.

23.18. Use the fact that $\{\xi_n\}$ obeys the SLLN.

23.19. Let $V(\xi_1) = \sigma^2$. From the independence of the given r.v.'s we have

$$E\left\{\left[\sum_{k=1}^{n} (\xi_k - a)\right]^4\right\} = nb + 6\binom{n}{2}\sigma^2 \leqslant cn^2.$$

According to Markov's inequality (see Exercise 21.1), we have

$$P\left\{\left|\frac{1}{n}\sum_{k=1}^{n} \xi_k - a\right| > \varepsilon\right\} \leqslant \frac{cn^2}{(\varepsilon n)^4}.$$

Since $\sum_{n=1}^{\infty} \frac{cn^2}{(\varepsilon n)^4}$ is convergent, we can now apply the Borel- Cantelli lemma.

To obtain Borel's SLLN it is necessary to check that the sequence

$\{\xi_n\}$ of i.i.d. r.v.'s, where each ξ_n is 1 or 0 with probability p and $q = 1 - p$, respectively, obeys the following conditions: $E\{\xi_1\} < \infty$ and $E\{|\xi_1 - a|^4\} < \infty$. As it is easily seen the last two relations are satisfied.

23.20. Let $E(\xi_k) = 0$, $V(\xi_k) = \sigma^2$, $\zeta(n) = \frac{1}{n}(\xi_1 + \ldots + \xi_n)$. According to Chebyshev's inequality, $P\{|\zeta(n^2)| \geqslant \varepsilon\} \leqslant \frac{\sigma^2}{\varepsilon^2 n^2}$. Then $\sum_{n=1}^{\infty} P\{|\zeta(n^2)| \geqslant \varepsilon\}$ is convergent, and it follows from the Borel-Cantelli lemma that for all sufficiently large n,

$$|\zeta(n^2)| \leqslant \varepsilon \quad (\vec{P}\text{-a.s.}). \tag{23.5}$$

On the other hand,

$$P\left\{\max_{n^2 \leqslant N \leqslant (n+1)^2} \left|\sum_{k=n^2+1}^{N} \xi_k\right| \geqslant \varepsilon n^2\right\} \leqslant \sum_{N=n^2-1}^{(n+1)^2-1} P\left\{\left|\sum_{k=n^2+1}^{N} \xi_k\right| \geqslant \varepsilon n^2\right\}.$$

We again use Chebyshev's inequality and the fact that the variables ξ_n are pairwise uncorrelated to obtain

$$P\left\{\max_{n^2 \leqslant N \leqslant (n+1)^2} \left|\sum_{k=n^2+1}^{N} \xi_k\right| \geqslant \varepsilon n^2\right\} \leqslant \sum_{N=n^2-1}^{(n+1)^2-1} \frac{1}{\varepsilon^2 n^4} \sigma^2 (N - n^2) \leqslant$$

$$\leqslant \frac{4\sigma^2}{\varepsilon^2 n^2}.$$

From the Borel-Cantelli lemma we conclude that, when n is sufficiently large, then with probability 1 it holds that

$$\max_{n^2+1 \leqslant N \leqslant (n+1)^2} \left|\sum_{k=n^2+1}^{N} \xi_k\right| \leqslant \varepsilon n^2. \tag{23.6}$$

From (23.5) and (23.6) it follows that when n is sufficiently large then with probability 1 we have $|\zeta(n)| \leqslant 2\varepsilon$; i.e., $\frac{1}{n}(\xi_1 + \ldots \xi_n) \xrightarrow{\text{a.s.}} 0$. The last relation means that $\{\xi_n\}$ obeys the SLLN.

23.21. We have $E(\xi_n) = 0$, $V(\xi_n) = \sigma_n^2 = 1 - 2^{-n} + 2^n$ and $\sum_{n=1}^{\infty} \frac{\sigma_n^2}{n^2} = \infty$; i.e., (23.4) is not satisfied. We form a new sequence of r.v.'s $\{\Theta_n\}$: $P\{\Theta_n = 1\} = P\{\Theta_n = -1\} = \frac{1}{2}(1 - 2^{-n})$, $P\{\Theta_n = 0\} = 2^{-n}$. Then $P\{\Theta_n \neq \xi_n\} = P\{\Theta_n = 0\} = 2^{-n}$. Hence $\sum_{n=1}^{\infty} P\{\Theta_n \neq \xi_n\} = \sum_{n=1}^{\infty} 2^{-n}$ is convergent, and we conclude from the Borell-Cantelli lemma that $\{\xi_n\}$ and $\{\Theta_n\}$ are two equivalent sequences of r.v.'s. (This is an equivalence in the sense of A. Ya. Khintchine.) We have $E(\Theta_n) = 0$, $V(\Theta_n) = b_n^2 = 1 - 2^{-n}$ and $\sum_{n=1}^{\infty} b_n^2/n^2 < \sum_{n=1}^{\infty} n^{-2} < \infty$. Hence (23.4) is satisfied and thus $\{\Theta_n\}$ obeys the SLLN. The equivalence of $\{\xi_n\}$ and $\{\Theta_n\}$ implies that $\{\xi_n\}$ also obeys the SLLN, although it does not obey (23.4).

23.22. Let $c > 0$ be arbitrarily chosen. Consider the events $A_n = \{\omega : |\xi_n(\omega)| \geqslant cn\}$, $n = 1, 2, \ldots$ It is assumed that the A_n are independent and since $E\{|\xi_n|\} = \infty$, we have $\sum_{n=1}^{\infty} P(A_n) = \infty$. From the Borel-Cantelli lemma, we conclude that $P\{|\xi_n| \geqslant cn \text{ for sufficiently large } n\} = 1$.

Since c was arbitrarily chosen, the assertion thus follows.

23.23. We have $E\{|\xi_1|\} = \sum_{k=1}^{\infty} |(-1)^k K| \frac{6}{\pi^2 k^2} = \frac{6}{\pi^2} \sum_{k=1}^{\infty} \frac{1}{k} = \infty$ and thus $E\{\xi_1\}$ does not exist. According to Kolmogorov's theorem, $\{\xi_n\}$ does not obey the SLLN.

To prove that $\frac{1}{n} S_n \xrightarrow{P} \frac{6 \ln 2}{\pi^2}$, we shall use the statement of Exercise 23.12; i.e., we shall prove that the ch.f. ϕ of ξ_1 is differentiable at $t = 0$ and $\phi'(0) = i \frac{6 \ln 2}{\pi^2}$ with $i = \sqrt{-1}$. Indeed,

$$\phi(t) = \sum_{k=1}^{\infty} [\exp(it(-1)^{k-1}k)] \frac{6}{\pi^2 k^2} =$$

$$= \frac{6}{\pi^2} \sum_{j=1}^{\infty} \frac{(e^{it})^{2j-1}}{(2j-1)^2} + \frac{6}{\pi^2} \sum_{s=1}^{\infty} \frac{(e^{-it})^{2s}}{(2s)^2} .$$

Consider then the following two functions

$$h_1(u) = \sum_{j=1}^{\infty} \frac{u^{2j-1}}{(2j-1)^2}, \quad |u| \leqslant 1 ;$$

$$h_2(v) = \sum_{s=1}^{\infty} \frac{v^{2s}}{(2s)^2}, \quad |v| \leqslant 1.$$

It can be shown (we omit some of the details) that

$$h_1'(u) = \frac{1}{2u} \ln \frac{1+u}{1-u}, \quad h_2'(v) = -\frac{1}{2v} \ln(1 - v^2).$$

Then

$$h_1'(u) - h_2'(u) = \frac{1}{u}\ln(1+u), \qquad \phi'(0) = i\,\frac{6}{\pi^2}(h_1'(1) - h_2'(1)) =$$

$$= i\,\frac{6\ln 2}{\pi^2}\,.$$

According to Exercise 23.12, if $n \to \infty$, we have

$$\frac{1}{n}(\xi_1 + \ldots + \xi_n) \xrightarrow{P} \frac{6\ln 2}{\pi^2}\,.$$

23.24. Let $(\xi_1, \ldots, \xi_n)$ and $(\eta_1, \ldots, \eta_n)$ be the chosen points. Then $\xi_1, \ldots, \xi_n, \eta_1, \ldots, \eta_n$ are i.i.d. r.v.'s, which are uniformly distributed over the interval $(0, 1)$. The distance, we are interested in, is

$$\rho_n = \left(\sum_{k=1}^{n} (\xi_k - \eta_k)^2\right)^{1/2}.$$

Obviously $n^{-1/2}E(\rho_n) = E(\sqrt{\zeta_n})$, where $\zeta_n = \frac{1}{n}\sum_{k=1}^{n}(\xi_k - \eta_k)^2$. Let $\Theta_k = (\xi_k - \eta_k)^2$. It is easy to verify that $\{\Theta_k\}$ obeys the WLLN. Since $E(\Theta_k) = \frac{1}{6}$, then $E(\zeta_n) = \frac{1}{6}$; hence, $\zeta_n \xrightarrow{P} \frac{1}{6}$ as $n \to \infty$. Now we easily obtain

$$\lim_{n\to\infty} n^{-1/2}E(\rho_n) = \lim_{n\to\infty} E(\sqrt{\zeta_n}) = \sqrt{1/6}.$$

24. Central Limit Theorem and Related Topics

24.1. Lindeberg's condition (24.2) can be easily checked from the assumptions. Another solution, based on the Continuity theorem, is also possible.

24.2. Since the ch.f. of ξ_1 equals $\exp[\lambda(e^{it} - 1)]$, then

$$\phi_{\eta_n}(t) = [\exp(it\sqrt{n\lambda})[\,]\exp(n\lambda(e^{it/\sqrt{n\lambda}} - 1))]$$

and

$$\ln\phi_{\eta_n}(t) = -it\sqrt{n\lambda} + n\lambda(e^{it/\sqrt{n\lambda}} - 1) = -\frac{t^2}{2} - \frac{1}{6}\frac{it^3}{\sqrt{n\lambda}} + \ldots$$

Thus $\lim_{n\to\infty}\ln\phi_{\eta_n}(t) = -t^2/2$, $\lim_{n\to\infty}\phi_{\eta_n}(t) = e^{-t^2/2}$ for $t \in \mathbb{R}_1$, and the assertion follows from the Continuity theorem.

24.3. The reasoning is similar to that in Exercise 24.2.
24.4. Let μ be a binomially distributed r.v. with parameters n and p. Then its ch.f. is given by $\phi(t) = (pe^{it} + q)^n$.
24.5. The proof follows from the Continuity theorem and from the relation between ch.f. and p.g.f.
24.6. Calculate the p.g.f. of a binomially distributed r.v. and use Exercise 24.5. Passing to the limit as $n \to \infty$ and $np_n \to \lambda$, we obtain the desired result.
24.7. For the ch.f. of the r.v. ξ_n we obtain

$$\phi_n(t) = \sum_{k=1}^{n} \frac{1}{n} e^{itk/n} \quad \text{and} \quad \phi_n(t) = [n(e^{-it/n} - 1)]^{-1}(1 - e^{it}).$$

But $\lim_{n\to\infty} [n(e^{-it/n} - 1)] = -it$, and then $\lim_{n\to\infty} \phi_n(t) = (it)^{-1}(e^{it} - 1)$. The last expression coincides with the ch.f. of a r.v. ξ, which is uniformly distributed over the interval (0, 1). Hence $\xi_n \xrightarrow{d} \xi$ as $n \to \infty$.
24.8. Find the ch.f. ϕ_n of ξ_n and show that $\{\phi_n(t), t \in R_1\}$ does not converge when $n \to \infty$ to any function, which is a ch.f.
24.9. From the assumptions we have that $V(\xi_n) = \sigma_n^2$ and $\sigma_n^2 \to V(\xi)$ as $n \to \infty$. Let $V(\xi) = \sigma^2$. The existence of variances implies existence of means and $\xi_n \xrightarrow{L_2} \xi$ implies $E(\xi_n) \to E(\xi)$. But $E(\xi_n) = a_n$, and hence the following limit exists: $\lim_{n\to\infty} E(\xi_n) = \lim_{n\to\infty} a_n = E(\xi)$. Put $E(\xi) = a$ and consider the ch.f. ϕ_n of the r.v. ξ_n. As we know, $\phi_n(t) = \exp(ia_n t - \sigma_n^2 t^2/2)$. It follows from the above reasoning that $\lim_{n\to\infty} \phi_n(t) = \exp(iat - \sigma^2 t^2/2)$. From the Continuity theorem we have $\xi \in N(a, \sigma^2)$.

24.10. It follows from Exercise 22.25 that $\sum_{n=1}^{\infty} \xi_n$ is convergent with probability 1. Put $S_n = \xi_1 + \ldots + \xi_n$ and $B_n^2 = \sigma_1^2 + \ldots + \sigma_n^2$. Since ξ_k, $k = 1, 2, \ldots$ are independent r.v.'s, then $\phi_n(t) = E\{\exp(itS_n)\} = \exp(-t^2 B_n^2/2)$. The assertion follows from the Continuity theorem and from the assumption $\lim_{n\to\infty} B_n^2 = \sigma^2$.
24.11. The ch.f. $\phi(t)$, $t \in \mathbb{R}_1$, of ξ_k, $k = 1, \ldots, n$ is given by $\phi(t) = (e^{it} - 1)/(it)$. Then we proceed as in Exercise 23.2.
24.12. According to Exercise 18.29, $\phi_{\xi_k}(t) = (1 - it\beta)^{-\alpha}$, $k = 1, \ldots, n$. Then

$$\phi_{\eta_n}(t) = \frac{\exp(it\alpha n/\sqrt{\alpha n})}{(1 - it/\sqrt{\alpha n})^{\alpha n}}, \quad \ln \phi_{\eta_n}(t) = -\frac{t^2}{2} + \sum_{k=3}^{\infty} \frac{(it)^k}{k(\alpha n)^{(k-2)/2}},$$

and

$$\phi_{\eta_n}(t) \to e^{-t^2/2} \quad \text{as } n \to \infty.$$

It follows from the above relations that $\eta_n \xrightarrow{d} \eta \in N(0, 1)$.
24.13. Use Exercise 18.29.
24.14. For the sequence $\{\xi_n\}$, we have $E(\xi_k) = 0$, $E\{\xi_k^2\} = 2p/k^{2\alpha}$, $E\{|\xi_k|^3\} = 2p/k^{3\alpha}$. According to the assumption, $2\alpha \in (\frac{2}{3}, 1)$ and $3\alpha \in (1, \frac{3}{2})$. Then for $n \to \infty$, we obtain

$$B_n^2 = 2p \sum_{k=1}^{n} \frac{1}{k^{2\alpha}} \to \infty \quad \text{and} \quad \rho_n^3 = 2p \sum_{k=1}^{n} \frac{1}{k^{3\alpha}} \to \sum_{k=1}^{\infty} \frac{1}{k^{3\alpha}} < \infty;$$

hence the condition given in (24.4) is satisfied. Similar reasoning shows that $\{\eta_n\}$ also satisfies this condition. Hence both sequences obey the CLT.
24.16. The reasoning is similar to that in Exercise 24.15.
24.17. (a) It is easy to see that the CLT holds for the sequence $\{\xi_n\}$; hence $P\{|\eta_{4500} - a| \leqslant 0.04\} \approx 2\Phi(1.2) - 1 \approx 0.7698$. (b) We have to find n such that $P\{|\xi_n - a| \leqslant 0.2\} \geqslant 0.8$. Applying again the CLT we get $P\{|\eta_n - a| \leqslant 0.2\} \approx 2\Phi(0.089\sqrt{n}) - 1 \geqslant 0.8$, which implies that $n \geqslant 202$.
24.18. (a) We have: $E(\xi_1) = 0$, $\sigma^2 = E\{\xi_1^2\} = 0.6$, $\{\xi_n\}$ obeys the CLT and $P\{S_n < \sqrt{n}\} \approx \Phi(1/\sqrt{0.6}) \approx 0.9032$.

(b) $P\{S_n < \sqrt{n}\} \approx \Phi(1) \approx 0.8413$.

24.19. We find $V(\xi_n) = \frac{1}{2} + 5/(3 \cdot 2^{2n+7})$ and thus $\frac{1}{2} < V(\xi_n) < 1$. Lindeberg's condition is therefore satisfied and $\{\xi_n\}$ obeys the CLT.
24.20. We have $E(\xi_k) = 0$, $\sigma_k^2 = E\{\xi_k^2\} = k^{2\delta}$, $E\{|\xi_k|^3\} = k^{3\delta}$, $B_n^2 \approx (2\delta + 1)^{-1} n^{2\delta+1}$, $\rho_n^3 \approx (3\delta + 1)^{-1} n^{3\delta+1}$. Also it is assumed that $\delta > -\frac{1}{3}$; hence the validity of Lyapunov's condition (24.3) is easily verified.
24.21. Lyapunov's condition is not valid in (a) and (b), but it does hold in (c). Consider for example (b). We have $E(\xi_k) = 0$, $\sigma_k^2 = E\{\xi_k^2\} = 1$, $E\{|\xi_k|^{2+\delta}\} = 2^{k\delta}$. Then

$$\frac{1}{B_n^{2+\delta}} \sum_{k=1}^{n} E\{|\xi_k - a_k|^{2+\delta}\} = \frac{1}{n^{(2+\delta)/2}} \sum_{k=1}^{n} 2^{k\delta} = \frac{2^{\delta}(2^{n\delta} - 1)}{(2^{\delta} - 1)n^{1+\delta/2}}.$$

Clearly there is no $\delta > 0$ for which the last expression converges to 0 as $n \to \infty$.
24.22. Lindeberg's theorem can be applied, but it is not necessary to

check (24.2). It is sufficient to note that $\xi_1 + \ldots + \xi_n \in N(0, B_n^2)$, where $B_n^2 = V\{\xi_1 + \ldots + \xi_n\} = (c^{-(n+1)/2} - 1)/(c^{-1/2} - 1)$; therefore, $\eta_n = \frac{1}{B_n}(\xi_1 + \ldots + \xi_n) \in N(0, 1)$, $n \geq 1$; i.e., the distribution of η_n does not depend on n.

24.23. $P\{S_n < 13\} \approx 0.7734$.

24.24. It is easy to show that $E(\xi_k) = 0$, $V(\xi_k) = \frac{1}{2} k^2\alpha^2$, $B_n^2 = \frac{\alpha^2}{2}(1^2 + 2^2 + \ldots + n^2) = o(n^3)$. Hence, for any $\varepsilon > 0$ and for sufficiently large n, we have $\varepsilon B_n > \alpha n$, which implies that $\int_{|x|>\varepsilon B_n} x^2 dF_k(x) \to 0$ as $n \to \infty$.. Thus (24.2) holds.

24.25. Obviously F_n is a d.f. Further, for all $x \in \mathbb{R}_1$, we have $\lim_{n\to\infty} F_n(x) = \frac{1}{2}$, which, however, is not a d.f. Why has this occurred? The ch.f. ϕ_n of F_n is given by $\phi_n(t) = \frac{1}{nt}\sin(nt)$; then the $\lim_{n\to\infty} \phi_n(t) = \phi_0(t)$, where $\phi_0(t) = 1$ for $t = 0$ and $\phi_0(t) = 0$ for $t \neq 0$. Obviously the limit ϕ_0 is not continuous at $t = 0$.

24.26. Let $\xi_i \in P(\lambda)$, $i = 1, \ldots, n$, be i.i.d. r.v.'s and let $S_n = \xi_1 + \ldots + \xi_n$. Then $S_n \in P(n\lambda)$ and we can easily show that

$$G_n(\lambda) = \sum_{k=0}^{n} P\{S_n = k\} = P\{S_n \leq n\} = P\left\{\frac{S_n - n\lambda}{\sqrt{n\lambda}} \leq \frac{n(1 - \lambda)}{\sqrt{n\lambda}}\right\}.$$

Further we have $E(S_n) = n\lambda$, $V(S_n) = n\lambda$. Also, according to the CLT, we have $\lim_{n\to\infty} G_n(\lambda) = \Phi(s)$, where $s = \lim_{n\to\infty} \frac{n(1 - \lambda)}{\sqrt{n\lambda}}$. Obviously $s = -\infty$ for $\lambda > 1$; $s = 0$ for $\lambda = 1$; and $s = +\infty$ for $0 < \lambda < 1$. Since $\Phi(-\infty) = 0$, $\Phi(0) = \frac{1}{2}$ and $\Phi(\infty) = 1$, we obtain the three possible values of $\lim_{n\to\infty} G_n(\lambda)$.

24.27. It follows from Exercise 16.34 that $\lim_{n\to\infty} E\{\xi_n^k\} = 0$, when k is odd, and $\lim_{n\to\infty} E\{\xi_n^k\} = 1 \cdot 3 \cdot 5 \cdot \ldots \cdot (k - 1)$, when k is even. Now we shall use the following result (see the book [32] cited at the end of this Manual): For every $u \in \mathbb{R}_1$ and $k = 1, 2, \ldots$, it is true that

$$\left| e^{iu} - \sum_{m=0}^{k-1} \frac{(iu)^m}{m!} \right| \leq \frac{|u|^k}{k!} . \qquad \square$$

If we put $u = t\xi_N$ and $k = 2s$ in this result we find that

$$\left| E\left\{e^{it\xi_N}\right\} - \sum_{m=1}^{2s-1} \frac{(it)^m E\{\xi_N^m\}}{m!} \right| \leqslant \frac{t^{2s} E\{\xi_N^{2s}\}}{(21)!}.$$

Passing to the limit first as $N \to \infty$ and then as $1 \to \infty$, we get $E\left\{e^{it\xi_N}\right\} \to e^{-t^2/2}$. This means that $\xi_N \xrightarrow{d} \xi \in N(0, 1)$.

24.28. The ch.f. of each r.v. equals cos t. Then $\phi_{\eta_n}(t) = \prod_{k=1}^{n} \cos(t/2^k)$. It can be proved that $\phi_{\eta_n}(t) \to t^{-1} \cdot \sin t$ for every t. But $t^{-1} \cdot \sin t$ is a ch.f. of an uniformly distributed r.v. over the interval $(-1, 1)$. From the Continuity theorem we get $\eta_n \xrightarrow{d} \eta$ as $n \to \infty$.

24.29. We have $E(\xi_n) = 0$, $V(\xi_n) = \frac{1}{3} n^2$, $B_n^2 = V\{\xi_1 + \dots + \xi_n\} = n(n + 1)(2n + 1)/18$. We need to determine α so that Lindeberg's condition (24.2) will be satisfied. In our case this condition has the following form: $L(n, \varepsilon) = \frac{1}{B_n^2} \sum_{k=1}^{n} \int_{|x|>\varepsilon B_n} x^2 dP\{\xi_k < x\} \to 0$ as $n \to \infty$ for every $\varepsilon > 0$. Put $c(\varepsilon,n) = \frac{\varepsilon}{3} \sqrt{\frac{1}{6} n(n + 1)(2n + 1)}$ and let $\varepsilon > 0$ and n be fixed. If the integer k is chosen such that $k^\alpha > c(\varepsilon, n)$, then

$$\int_{|x|>\varepsilon B_n} x^2 dP\{\xi_k < x\} = V(\xi_k).$$

Let $k_0 = k_0(n, \varepsilon, \alpha)$ be the smallest integer for which $k_0 > c(n, \varepsilon)$. Then $L(n, \varepsilon) = \frac{1}{B_n^2} \sum_{k=1}^{n} V(\xi_k)$, if $k_0 \leqslant n$ and $L(n, \varepsilon) = 0$, if $k_0 > n$. When $n \to \infty$, we have $c(n, \varepsilon) \approx \frac{\varepsilon}{3} n^{3/2}$. Let $\alpha < \frac{3}{2}$ and let $\varepsilon > 0$ be fixed. Then if n is chosen sufficiently large, it holds that $k_0(n, \varepsilon, \alpha) > n$. This means that $L(n, \varepsilon) = 0$; therefore, $L(n, \varepsilon) \to 0$ as $n \to \infty$. For $\alpha > \frac{3}{2}$ it is easy to see that $L(n, \varepsilon) \to 1$ as $n \to \infty$. For $\alpha = \frac{3}{2}$ the quantity $L(n, \varepsilon)$ converges either to 0 or to 1, depending on the value of ε. Hence only for $\alpha < \frac{3}{2}$ is Lindeberg's condition satisfied.

24.30. It is easy to see that $\{\xi_n\}$ obeys the CLT; i.e., if $S_n = \xi_1 + \dots + \xi_n$, then $P\{\frac{1}{\sqrt{n}} S_n < x\} \to \Phi(x)$ or $\eta_n = \frac{1}{\sqrt{n}} S_n \xrightarrow{d} \eta \in N(0, 1)$. Since all r.v.'s in the sequence $\{\xi_n^2\}$ are independent and their mean is given by $E\{\xi_n^2\} = 1$, it follows that $\{\xi_n^2\}$ obeys the WLLN; i.e., $\frac{1}{n}(\xi_1^2 + \dots + \xi_n^2) \xrightarrow{P} 1$. Thus for $\zeta_n = n/(\xi_1^2 + \dots + \xi_n^2)$, we also have $\zeta_n \xrightarrow{P} 1$. Note first that $\gamma_n = \eta_n \zeta_n$, and then use Exercise 22.10 to complete the proof.

24.31. The reasoning is similar to that in Exercise 24.30.

24.32. Let $\phi(t)$ be the ch.f. of ξ_k and let $a = E(\xi_k)$. Then we get $\phi_\eta(t) = p/(1 - q\phi(t))$. According to the Wald identity (see Exercise 20.17), $E(\eta) = aE(\tau)$; on the other hand $E(\tau) = \frac{q}{p}$. Obviously $E(\tau) \to \infty$ if $p \to 0$. Since $\phi_\Theta(t) = p/(1 - q\phi(pt/aq)))$ and $\phi(t) = 1 + iat + o(t)$, we can replace ϕ in ϕ_Θ taking the limit as $p \to 0$, we obtain $\phi_\Theta(t) \to 1/(1 - it)$, which is exactly the ch.f. of a r.v. $\zeta \in E(1)$. Then the Continuity theorem yields the desired assertion.

24.33. We shall carry this proof in three steps:

(1) The assertion that $\eta_n \xrightarrow{L_2} \eta$ is equivalent to the following two assertions: $\eta_n \xrightarrow{P} \eta$ and $E\{\eta_n^2\} \to E\{\eta^2\} < \infty$ (see Exercise 22.5 and Exercise 22.22).

(2) Let $\eta_n \xrightarrow{P} \eta$. From Exercise 22.14, $\eta_n \xrightarrow{d} \eta \in N(0, 1)$. It can be checked that $E\{\xi_n^2\} = 1$. Thus $E\{\eta_n^2\} \to E\{\eta^2\} = 1$, and from Step (1) we get $\eta_n^2 \xrightarrow{L_2} \eta$.

(3). We now need to prove that $\{\eta_n\}$ cannot converge in mean square. Indeed, if $\eta_n \xrightarrow{L_2} \eta$, then for every $\varepsilon > 0$ there should exist $N = N_\varepsilon$ such that for arbitrary $m > n > N_\varepsilon$,

$$V\{\eta_m - \eta_n\} < \varepsilon. \tag{24.8}$$

Put $\zeta_k = (\xi_k - a)/\sigma$. It is easy to check that

$$\eta_m - \eta_n = \left(\frac{1}{\sqrt{m}} - \frac{1}{\sqrt{n}}\right)(\zeta_1 + \ldots + \zeta_n) + \frac{1}{\sqrt{m}}(\zeta_{n+1} + \ldots + \zeta_m).$$

Since $\zeta_1, \ldots, \zeta_m$ are independent, one has

$$V\{\eta_m - \eta_n\} = 2 - 2\sqrt{n/m}$$

and thus (24.8) will not hold if m is sufficiently large. This completes the proof.

24.34. We shall prove that

$$\lim_{n\to\infty} \int_{\mathbb{R}_1} |f_n(x) - f(x)|dx = 0.. \tag{24.9}$$

Put $\delta_n(x) = f_n(x) - f(x)$, $\delta_n^+ = \frac{1}{2}(\delta_n + |\delta_n|)$, $\delta_n^- = \frac{1}{2}(\delta_n - |\delta_n|)$. It is assumed that $\delta_n^+(x) \to 0$ and $\delta_n^-(x) \to 0$ as $n \to \infty$ for almost all $x \in \mathbb{R}_1$. Since f_n and f are density functions and since $\delta_n = \delta_n^+ + \delta_n^-$, then

$$\int_{\mathbb{R}_1} (f_n(x) - f(x))dx = \int_{\mathbb{R}_1} \delta_n^+(x)dx + \int_{\mathbb{R}_1} \delta_n^-(x)dx = 0. \tag{24.10}$$

Next we have $f_n(x) - f(x) = \delta_n(x) \geqslant -f(x)$ for $\delta_n(x)$ positive and for $\delta_n(x)$ negative. Then $0 \geqslant \delta_n^-(x) \geqslant -f(x)$; i.e., $|\delta_n^-| \leqslant f$, and according to the Lebesgue dominated-convergence theorem

$$\int_{\mathbb{R}_1} \delta_n^-(x)dx \to \int_{\mathbb{R}_1} \lim_{n\to\infty} \delta_n^-(x)dx = 0. \tag{24.11}$$

From (24.10) and (24.11) we get $\int_{\mathbb{R}_1} \delta_n^+(x)dx \to 0$; i.e., (24.9) holds.

Then for any Borel set $B \in \mathcal{B}_1$ we have $\int_B |f_n(x) - f(x)|dx \to 0$, which leads to the conclusion that $\xi_n \overset{d}{\to} \xi$ as $n \to \infty$.

24.35. The scheme of the proof is the following:

(1) For every fixed n the function $f_n(x) = \sqrt{n}p_n([c_n + x\sqrt{n}])$, $x \in \mathbb{R}_1$, is a density function. Then (24.6) yields the relation $\lim_{n\to\infty} f_n(x) = f(x)$ for almost all $x \in \mathbb{R}_1$.

(2) From Scheffe's theorem (see Exercise 24.34) we conclude that

$$\lim_{n\to\infty} \int_{-\infty}^{b} f_n(x)dx = \int_{-\infty}^{b} f(x)dx \tag{24.12}$$

for every $b \in \mathbb{R}_1$.

(3) For any $\varepsilon > 0$ and for all n such that $\varepsilon\sqrt{n} > 1$, the following inequalities hold:

$$\int_{-\infty}^{b-\varepsilon} f_n(x)dx \leqslant \sum_{r=-\infty}^{r_n-1} P_n(r) \leqslant$$

$$\leqslant \int_{-\infty}^{b} f_n(x)dx \quad \text{with } r_n = [c_n + b\sqrt{n}]. \tag{24.13}$$

(4) We have

$$\sum_{r=-\infty}^{r_n-1} p_n(r) = P\{\xi_n < [c_n + b\sqrt{n}]\} = P\{\xi_n < c_n + b\sqrt{n}\} =$$

$$= P\left\{\frac{\xi_n - c_n}{\sqrt{n}} < b\right\}. \tag{24.14}$$

(5) From (24.12), (24.13) and (24.14), we conclude that

$$\lim_{n\to\infty} P\left\{\frac{\xi_n - c_n}{n} < b\right\} = \int_{-\infty}^{b} f(x)dx,$$

for every $b \in \mathbb{R}_1$; i.e., $\frac{1}{\sqrt{n}}(\xi_n - c_n) \overset{d}{\to} \xi$ as $n \to \infty$.

24.36. Put $c_N = Np\tilde{p}$, $p_N(k) = P\{\xi_N = k\}$. Using Stirling's formula (see Exercise 24.39), it can be easily verified that (24.6) (see Exercise 24.35) holds, namely $\lim_{N\to\infty} \sqrt{N}p_N([c_N + x\sqrt{N}]) = \phi_0(x)$, where $\phi_0(x) = \frac{1}{\sqrt{2\pi}} e^{-x^2/2}$. The desired assertion follows from Okamoto's theorem (see Exercise 24.35).

24.37. It is easy to show that ζ_n possesses a Student's t-distribution with n - 1 degrees of freedom (see Exercise 16.39). The density f_n of ζ_n can be written explicitly and it is easy to see that $\lim_{n\to\infty} f_n(x) = \phi_0(x) = \frac{1}{\sqrt{2\pi}} \exp(-\frac{x^2}{2})$. The Scheffé theorem (see Exercise 24.34) can now be applied.

24.38. We have $E(\xi_j^{(n)}) = 0$ and $V(\xi_j^{(n)}) = 1$. Then the p.g.f. of η_n is given by

$$g_n(s) = \left(E\left\{s^{\xi_1^{(n)}/\sqrt{n}}\right\}\right)^n = \left(1 - \frac{s - 1}{2n} + \frac{s^{-1} - 1}{2n}\right)^n \to$$

$$\to \exp(\frac{s - 1}{2}) \exp\left(\frac{s^{-1} - 1}{2}\right).$$

However $\exp(\frac{s - 1}{2})$ is a p.g.f. of a r.v. $\zeta \in \mathcal{P}(\frac{1}{2})$ and $\exp\left(\frac{s^{-1} - 1}{2}\right) = E\{s^{-\zeta}\}$. Then we conclude that $\eta_n \overset{d}{\to} \zeta_1 - \zeta_2$ as $n \to \infty$, where ζ_1 and ζ_2 are independent r.v.'s with $\zeta_1, \zeta_2 \in \mathcal{P}(\frac{1}{2})$.

24.39. Let $\{\xi_n\}$ be a sequence of i.i.d. r.v.'s where $\xi_n \in \mathcal{E}(1)$, n = 1, 2, ... For every n we have $E(\xi_n) = 1$, $V(\xi_n) = 1$ and for $S_n = \xi_1 + \xi_2 + \ldots + \xi_n$ we have $E(S_n) = n$, $V(S_n) = n$. The CLT yields $n^{-1/2}(S_n - n) \overset{d}{\to} \Theta \in \mathcal{N}(0, 1)$ as $n \to \infty$. Since $E\{[n^{-1/2}(S_n - n)]^2\} < \infty$ for every n, then $E\{|n^{-1/2}(S_n - n)|\} \to E\{|\Theta|\}$. But $E\{|\Theta|\} = \sqrt{2/\pi}$ and hence $I_n = \sqrt{\pi/2}E\{|n^{-1/2}(S_n - n)|\} \to 1$. Now we shall use the fact that S_n has a Γ-distribution with parameters n and 1; i.e., S_n has density $f_n(x) = \frac{1}{\Gamma(n)} x^{n-1}e^{-x}$, $x > 0$. Then

$$I_n = \sqrt{\frac{\pi}{2}}\, E\left\{\frac{S_n - n}{\sqrt{n}}\right\} = \frac{\sqrt{\pi/2}}{\Gamma(n)} \int_0^\infty \frac{|x - n|}{\sqrt{n}}\, x^{n-1} e^{-x} dx,$$

which yields $I_n = \sqrt{2\pi n}\, n^n e^{-n}/\Gamma(n + 1)$. However, $\Gamma(n + 1) = n!$ and, as we have seen above, $I_n \to 1$ as $n \to \infty$. Finally we get $\sqrt{2\pi n}\, n^n e^{-n}/n! \to 1$ as $n \to \infty$, which is obviously equivalent to Stirling's formula.

24.40. It can be easily checked that $E\{|\xi|\}$ exists, but $V\{\xi\}$ does not exist. Therefore we shall use the following *theorem of Lévy-Feller-Khintchine* (see the book [32] cited at the end of this Manual): Let ξ_1, ξ_2, ... be i.i.d. r.v.'s whose common d.f. F is continuous and symmetric; i.e., $F(-x) = 1 - F(x)$, $x \geqslant 0$. Let $\Delta y = y^2(1 - F(y))/\int_0^y u^2 dF(u) \to 0$ as $y \to \infty$. Then there exists a sequence $\{b_n\}$, $0 < b_n$, such that

$$\lim_{n\to\infty} P\{b_n^{-1}(\xi_1 + \ldots + \xi_n) < x\} = \Phi(x), \quad x \in \mathbb{R}_1. \qquad \square \quad (24.15)$$

In our case for $y > 1$, we have

$$F(y) = 1 - (2y^2)^{-1}; \qquad \int_0^y u^2 dF(u) = \ln y; \qquad \Delta y = 1/(2 \ln y).$$

Obviously $\Delta y \to 0$ when $y \to \infty$. Put

$$\delta(y) = y^2(1 - F(y))/\int_0^y u^2 dF(u) \qquad \text{and} \qquad d_n^2 = n\int_{-c_n}^{c_n} x^2 dF(x),$$

where c_n is given by $n^2 = \delta(c_n)$. Then $\{b_n\}$ in (24.15) can be chosen arbitrarily, such that $\frac{d_n}{b_n} \to 1$, as $n \to \infty$ (one can simply take $b_n = d_n$). We have here $\delta(y) = 2y^2/\ln y$. Let c_n be taken as the only positive solution of the equation $n^2 = 2c_n^4/\ln c_n$. Obviously $c_n \uparrow \infty$ if n is sufficiently large. We find $d_n^2 = 2n \ln c_n$; i.e., $d_n = \sqrt{2n \ln c_n}$. The last relation, as well as the equation for c_n, shows that $b_n = (n \ln n)^{1/2}$, which implies that $d_n/b_n \to 1$ as $n \to \infty$. This completes the proof.

24.41. Since $E(\xi_k) = 0$ and $V(\xi_k) = 1$, we can easily find that $S_n = \eta_1 + \ldots + \eta_n$ has mean $A_n = E(S_n) = 0$ and variance $B_n^2 = V(S_n) = 15[\frac{1}{16} + (\frac{1}{16})^2 + (\frac{1}{16})^3 + \ldots + (\frac{1}{16})^n] = 1 - (\frac{1}{16})^n$. We are interested in the limit of the quantity $F_n(x) = P\left\{\frac{S_n - A_n}{B_n} < x\right\}$, $x \in \mathbb{R}_1$, when $n \to \infty$. It is easy check that $P\{|S_n| \leqslant 0.5\} = 0$ for every $n = 1, 2, \ldots$ However, $A_n = 0$, $B_n \approx 1$ for n sufficiently large; hence, it is not possible for $F_n(x)$ to converge, as $n \to \infty$, to the standard normal d.f. $\Phi(x)$ for every $x \in \mathbb{R}_1$. Thus $\{\xi_n\}$ does not obey the CLT.

24.42. For ξ_k we have $E(\xi_k) = 0$ and $V(\xi_k) = \frac{1}{2} + \frac{5}{3 \cdot 2^{2k+7}}$. Obviously for any k, $\frac{1}{2} < V(\xi_k) < 1$. Then Lindeberg's condition (24.2) holds, and hence $\{\xi_n\}$ obeys the CLT; i.e., for every $x \in \mathbb{R}_1$, $F_n(x) \to \Phi(x)$ as $n \to \infty$.

Next we shall study the limit behaviour of the density $p_n(x)$ when $n \to \infty$. Denote by $q_n(x)$, $x \in \mathbb{R}_1$, the density of $S_n = \xi_1 + \ldots + \xi_n$. Then for n = 2 (see the solution of Exercise 16.22),

$$q_2(x) = (f_1 * f_2)(x) = \int_{-\infty}^{\infty} f_1(z) f_2(x - z)dz.$$

Now we shall determine $q_2(x)$ for $x = \frac{1}{2}$. By the definition of ξ_1 and ξ_2, we have that:

$$f_1(z) \neq 0, \quad \text{if } -\frac{1}{8} \leqslant z \leqslant \frac{1}{8} \quad \text{or} \quad \frac{15}{16} < |z| < \frac{17}{16};$$

$$f_2(\tfrac{1}{2} - z) \neq 0, \quad \text{if } \frac{7}{16} \leqslant z \leqslant \frac{9}{16} \quad \text{or} \quad \frac{15}{32} < |z| < \frac{17}{32}.$$

The last relations yields $q_2(\frac{1}{2}) = 0$. Similarly we find that

$$q_3(\tfrac{1}{2}) = (q_2 * f_3)(x)\Big|_{x=1/2} = \int_{-\infty}^{\infty} q_2(z) f_3(x - z)dz\Big|_{x=1/2} = 0,$$

and more generally $q_n(1/2) = 0$ for all $n \geqslant 2$. However $x = \frac{1}{2}$ is not the only value for which $q_n(x) = 0$. By similar reasoning we obtain that $q_n(x) = 0$ for all x of the form $x = \frac{1}{2}(2m + 1)$, $m = 0, \pm 1, \pm 2, \ldots$ Even more, $q_n(x) = 0$ for all $x = \frac{1}{2}(2m + 1) + \delta$, where $m = 0, \pm 1, \pm 2, \ldots$, $|\delta| < \frac{1}{4}$. The density q_n of S_n and the density p_n of $\eta_n = S_n/B_n$ are related by $p_n(x) = B_n q_n(xB_n)$. We shall study the limiting behaviour of $B_n q_n(xB_n)$ as $n \to \infty$. We easily find

$$B_n^2 = V(S_n) = \frac{n}{2} + \frac{5}{1152}(1 - \frac{1}{2^{2n}}).$$

Now we put $x = \frac{1}{2}$ and obtain

$$B_n q_n(\frac{1}{2} B_n) = \sqrt{\frac{n}{2} + \frac{5}{1152}(1 - \frac{1}{2^{2n}})} \times q_n(\frac{1}{2}\sqrt{\frac{n}{2} + \frac{5}{1152}(1 - \frac{1}{2^{2n}})}).$$

Take $n = 2(2N + 1)^2$, and then the argument of q_n is

$$\frac{2N+1}{2}\sqrt{1 + \frac{5}{1152}\left(1 - \frac{1}{2^{2\cdot 2(2N+1)^2}}\right)\frac{1}{(2N+1)^2}} .$$

It is easily seen that for sufficiently large N the last quantity can be written as $\frac{1}{2}(2N + 1) + \delta$, where $|\delta| < \frac{1}{4}$; hence $B_n q_n(\frac{1}{2} B_n) = 0$ for all sufficiently large n. Obviously this implies that $p_n(\frac{1}{2}) = 0$ for large values of n.

Finally we see that the relation

$$\lim_{n\to\infty} p_n(x) = \phi_0(x) = \frac{1}{\sqrt{2\pi}} e^{-x^2/2}$$

does not hold for all $x \in \mathbb{R}_1$; hence $\{\xi_n\}$ does not obey the local CLT.

TABLE 1
Normal Distribution

For the normal distribution with mean 0 and variance 1, the values of the distribution function $\Phi(x) = \frac{1}{\sqrt{2\pi}} \int_{-\infty}^{-x} e^{-u^2/2}\, du$ and of the density $\varphi_0(x) = \frac{1}{\sqrt{2\pi}} e^{-x^2/2}$ for some x are given, $x \geqslant 0$. When $x < 0$ we use the relations $\varphi_0(-x) = \varphi(x)$, $\Phi(-x) = 1 - \Phi(x)$.

x	$\Phi(x)$	$\varphi_0(x)$	x	$\Phi(x)$	$\varphi_0(x)$
0.00	0.5000	0.3989	2.00	0.9772	0.0540
05	5199	3984	05	9798	0488
10	5398	3970	10	9821	0440
15	5596	3945	15	9842	0396
20	5793	3910	20	9861	0355
25	5987	3867	25	9878	0317
30	6179	3814	30	9898	0283
35	6368	3752	35	9906	0252
40	6554	3683	40	9918	0224
45	6736	3605	45	9929	0198
50	6915	3521	50	9938	0175
55	7088	3429	55	9946	0155
60	7257	3332	60	9953	0136
65	7422	3230	65	9960	0119
70	7580	3123	70	9965	0104
75	7734	3011	75	9970	0091
80	7881	2897	80	9974	0079
85	8023	2780	85	9978	0069
90	8159	2661	90	9981	0060
95	8289	2541	95	9984	0051
1.00	8413	2420	3.00	9987	0044
05	8531	2299	05	9989	0038
10	8643	2179	10	9990	0033
15	8749	2059	15	9992	0028
20	8849	1942	20	9993	0024
25	8944	1826	25	9994	0020
30	9032	1714	30	9995	0017
35	9115	1604	35	9996	0015
40	9192	1497	40	9997	0012
45	9265	1394	45	9997	0010
50	9332	1295	50	9998	0009
55	9394	1200	55	9998	0007
60	9452	1109	60	9998	0006
65	9505	1023	65	9999	0005
70	9554	0940	70	9999	0004
75	9599	0863	75	9999	0004
80	9641	0790	80	9999	0003
85	9678	0721	85	9999	0002
90	9713	0656	90	1.000	0002
95	9744	0596	95	1.000	0002
2.00	9772	0540	4.00	1.000	0001

TABLE 2
Poisson Distribution

The values of the probabilities $P_k = \frac{\lambda^k e^{-\lambda}}{k!}$ for some λ and k are given.

k \ λ	0.1	0.2	0.3	0.4	0.5	0.6
0	0.9048	8187	7408	6703	6065	5488
1	0905	1637	2222	2681	3033	3293
2	0045	0164	0333	0536	0758	0988
3	0002	0011	0033	0072	0126	0196
4	0000	0001	0003	0007	0076	0030
5		0000	0000	0001	0002	0004
6				0000	0000	0000

k \ λ	0.7	0.8	0.9	1.0	2.0	3.0
0	0.4966	4493	4066	3679	1353	0498
1	3476	3595	3659	3679	2707	1494
2	1217	1438	1647	1839	2707	2240
3	0284	0383	0494	0613	1804	2240
4	0050	0077	0111	0153	0902	1680
5	0007	0012	0020	0031	0361	1008
6	0001	0002	0003	0005	0120	0504
7	0000	0000	0000	0001	0034	0216
8				0000	0009	0081
9					0002	0027
10					0000	0008
11						0002
12						0001
13						0000

k \ λ	4.0	5.0	6.0	7.0	8.0	9.0
0	0.0183	0067	0025	0009	0003	0001
1	0733	0337	0149	0064	0027	0011
2	1465	0842	0446	0223	0107	0050
3	1954	1404	0892	0521	0286	0150
4	1954	1755	1339	0913	0573	0337
5	1563	1755	1606	1277	0916	0607
6	1042	1462	1606	1490	1221	0911
7	0595	1044	1377	1490	1396	1171
8	0298	0653	1033	1304	1396	1318
9	0132	0363	0688	1014	1241	1318
10	0053	0181	0413	0710	0993	1186
11	0019	0082	0225	0452	0722	0970
12	0006	0034	0113	0264	0481	0728
13	0002	0013	0052	0142	0296	0504
14	0001	0005	0022	0071	0169	0324
15	0000	0002	0009	0033	0090	0194
16		0000	0003	0014	0045	0109
17			0001	0006	0021	0058
18			0000	0002	0009	0029
19				0001	0004	0014
20				0000	0002	0006
21					0001	0003
22					0000	0001
23						0000

References

1. Billingsley, P. (1979): *Probability and measure*. John Wiley & Sons, New York.
2. Bolshev, L. N. and N. V. Smirnov (1983): *Tables in mathematical statistics*. Nauka, Moscow. (In Russian.)
3. Borovkov, A. A. (1986): *Course in probability theory*. 2nd ed. Nauka, Moscow. (In Russian.)
4. Breiman, L. (1968): *Probability*. Addison Wesley, Reading (MA).
5. Chow, Y. and H. Teicher (1978): *Probability theory*. Springer-Verlag, New York.
6. Chung, K. L. (1974): *A course in probability theory*. 2nd ed. Acad. Press, New York.
7. Ciucu, G., V. Craiu, and I. Sacuiu (1974): *Probleme de teoria probabilitatilor*. Tehnika, Bucharest.
8. Ciucu, G. and C. Tudor (1978-1979): *Probabilitati si procese stocastice*. Vols. 1 & 2. Academiei, Bucharest.
9. Dimitrov, B. and N. Yanev (1988): *Probability and statistics*. Sofia Univ. Press, Sofia. (In Bulgarian.)
10. Dorogovcev, A. Ya., D. S. Silvestrov, A. V. Skorohod, and M. I. Yadrenko (1980): *Theory of probability. Collection of problems*. Vishca shkola, Kiev. (In Russian.)
11. Emelyanov, G. V. and V. P. Skitovitch (1967): *Collection of problems in probability theory and mathematical statistics*. Leningrad Univ. Press, Leningrad. (In Russian.)
12. Feller, W. (1968 and 1971): *An introduction to probability theory and its applications*. Vol. 1, 3rd ed. and vol. 2, 2nd ed. John Wiley & Sons, New York.
13. Gnedenko, B. V. (1982): *The theory of probability*. 5th ed. Mir Publ., Moscow.
14. Hogg, R. and A. Craig (1978): *Introduction to mathematical statistics*. 4th ed. McMillan, New York.
15. Johnson, N. and S. Kotz (1977): *Urn models and their application*. John Wiley & Sons, New York.
16. Kingman, J. F. C. and S. J. Taylor (1966): *Introduction to measure and probability*. Cambr. Univ. Press, Cambridge.
17. Kolmogorov, A. N. (1956): *Foundations of the theory of probability*. Chelsea Publ. Co., New York. (German ed. 1933; Russian eds. 1936 and 1973).
18. Lamperti, J. (1966): *Probability*. Benjamin, New York.
19. Loève, M. (1977-1978): *Probability theory*. Vols. 1 & 2, 4th ed. Springer-Verlag, New York.
20. Lukacs, E. (1970): *Characteristic functions*. 2nd ed. Griffin, London.
21. Mališić, J. (1970): *Collection of exercises in probability theory with applications*. Gradjevinska Knjiga, Belgrade. (In Serbo-Chroatian.)
22. Meshalkin, L. D. (1973): *Collection of problems in probability theory*. Noordhoff Intern. Publ., Leiden. (Russian ed. 1960).
23. Mirazchiiski, I. and J. Stoyanov (1979): *Problems in probability*. Narodna Prosveta, Sofia. (In Bulgarian.)
24. Moran, P. A. P. (1968): *An introduction to probability theory*. Oxford Univ. Press, New York.
25. Mosteller, F. (1965): *Fifty challenging problems in probability with solutions*. Addison Wesley, Reading (MA).

26. Neuts, M. (1973): *Probability*. Allyn & Bacon, Boston.
27. Neveu, J. (1965): *Mathematical foundation of the calculus of probability*. Holden-Day, San Francisco. (French ed. 1963).
28. Obrechkoff, N. (1963): *Theory of probability*. Nauka i Izkustvo, Sofia. (In Bulgarian.)
29. Obretenov, A. (1974): *Theory of probability*. Nauka i Izkustvo, Sofia. (In Bulgarian.)
30. Penkov, B. (1964): *Combinatorics*. Narodna Prosveta, Sofia. (In Bulgarian.)
31. Prohorov, Yu. V and Yu. A. Rozanov (1969): *Probability theory*. Springer-Verlag, Berlin. (Russian ed. 1967).
32. Rényi, A. (1970): *Probability theory*. Akad. Kiadó, Budapest and North-Holland, Amsterdam.
33. Riordan, J. (1958): *An introduction to combinatorial analysis*. John Wiley & Sons, New York.
34. Roussas, G. (1973): *A first course in mathematical statistics*. Addison Wesley, Reading (MA).
35. Santaló, L. A. (1976): *Integral geometry and geometric probability*. Addison Wesley, Reading (MA).
36. Sevastyanov, B. A., V. P. Chistyakov, and A. M. Zubkov (1985): *Problems in the theory of probability*. Mir, Moscow. (Russian ed. 1980).
37. Shiryaev, A. N. (1984): *Probability*. Springer-Verlag, New York. (Russian ed. 1980).
38. Stojanow, J., I. Mirazczijski, C. Ignatov, and M. Tanuszev (1982): *Zbior zadan z rachunku prawdopodobienstwa*. PWN, Warsaw. (In Polish; transl. of the Bulgarian ed. 1976).
39. Stoyanov, J. (1987): *Counterexamples in probability*. John Wiley & Sons, Chichester.
40. Thomasian, A. (1969): *The structure of probability theory with applications*. McGraw Hill, New York-London.
41. Tsokos, C. (1972): *Probability distributions: An introduction to probability theory with applications*. Duxbury Press, Belmont (CA).
42. Tutubalin, V. N. (1972): *Theory of probability*. Moscow Univ. Press, Moscow. (In Russian.)
43. Vasić, P. (1974): *Problems in theory of probability*. Gradjevinska Knjiga, Belgrade. (In Serbo-Chroatian.)
44. Whittle, P. (1970): *Probability*. Penguin Books, Harmondsworth.
45. Yaglom, A. M. and Yaglom, I. M. (1964): *Challenging mathematical problems with elementary solutions*. Holden-Day, San Francisco.

SUBJECT INDEX

9 789027 726872